Manual of Musculoskeletal Ultrasound

Mark H. Greenberg • Alvin Lee Day
Suliman Alradawi

Manual of Musculoskeletal Ultrasound

A Self-Study, Protocol-Based Approach

Springer

Mark H. Greenberg
Division of Rheumatology
Department of Internal Medicine
University of South Carolina School of Medicine, Columbia, SC
USA

Alvin Lee Day
Division of Rheumatology
Columbia VA Health Care System
Department of Internal Medicine
University of South Carolina School of Medicine, Columbia, SC
USA

Suliman Alradawi
Department of Medicine, Division of Rheumatology
Advocate Christ Medical Center, Advocate Medical Group, Chicago, IL
USA

ISBN 978-3-031-37415-9 ISBN 978-3-031-37416-6 (eBook)
https://doi.org/10.1007/978-3-031-37416-6

This Springer imprint is published by the registered company Springer Nature Switzerland AG
The registered company address is: Gewerbestrasse 11, 6330 Cham, Switzerland

Paper in this product is recyclable.

To my wife, ultrasound model, partner and best friend, Brenda.
Thank you for your constant support, patience, humor, and sacrifice in creating this book.

—MHG

To my wife, Stani, who filters my ideas, champions my endeavors, and centers our home.

—ALD

To my father, Tayssir, who made me a curious lifelong learner. To my mother, Sameera, who always supported any project I set my mind to achieve without hesitation. To my siblings, Kenan, Shahlan, and Tala, who inspire me to be a better version of myself every day. To my best friend and colleague, Dr. Amisha Shah, whose encouragement and optimism helped inspire the creation of this work.

—SA

Foreword

Doctors Greenberg, Day, and Alradawi have created a first-of-its-kind book on learning how to perform musculoskeletal ultrasound. Despite musculoskeletal ultrasound holding great potential for many clinicians across a broad spectrum of specialties and subspecialties, it is presently underutilized. A primary reason for this underutilization is the difficulty level and time needed to learn a new clinical skill set that involves a wide variety of anatomical structures and clinical pathologies. This situation has been compounded by the limited availability of qualified instructors in musculoskeletal ultrasound.

The authors have expertly addressed these issues with a learning approach that is simple, straightforward, clinically-oriented, protocol-based, and self-directed. The development of this approach draws on their extensive experience in the clinical application of musculoskeletal ultrasound and the many lessons learned from teaching a wide variety of learners over the years. Once the basic terminology and principles of ultrasound scanning are presented in the first chapter, learners can readily make great use of the book by selecting anatomical regions, structures, and scanning protocols most relevant to their learning and clinical practice needs.

Chapters are based on standard approaches to specific musculoskeletal regions such as the anterior, posterior, medial, and lateral areas of the knee. Each chapter begins with a chapter abstract followed by sections covering the reasons to perform the particular ultrasound study, questions to be answered by the study, basic regional physical and sonographic anatomy, relevant clinical comments including discussion of pathology, as well as potential pitfalls. The chapter ends with easy-to-follow step-by-step protocols for scanning and checklists for simplicity and consistency then a list of references. The diagnostic strength of using the ultrasound protocols for identifying specific pathology is also indicated when relevant.

Emphasis is placed on learning to capture normal clinically relevant ultrasound images of the major musculoskeletal regions of the body and identifying important anatomical structures involved in common musculoskeletal pathology in that region. Learners are greatly aided in acquiring the critical knowledge base for scanning so many anatomical structures involved in musculoskeletal pathologies with the use of catchy mnemonics, simple yet memorable illustrations, cartoon images, and high-quality normal ultrasound images with relevant structures identified.

For those new to ultrasound a thorough appreciation of the normal is essential in learning to recognize and understand clinical pathology. The authors discuss essential aspects of common pathology and tips and tricks to recognize and remember them but do not present pathological ultrasound images. Instead, they provide extensive definitive references on pathology to supplement the content of each chapter.

This book will be of great value to those new to ultrasound, those with considerable experience performing ultrasound but new to musculoskeletal ultrasound, and those wanting to expand their musculoskeletal ultrasound knowledge and skill. The authors have not only delivered a powerful and exciting new way to learn musculoskeletal ultrasound but are also introducing a new approach to ultrasound education in general. Their contribution deserves special recognition for advancing ultrasound education and the role it can play in ultimately advancing patient care.

Professor of Medicine and Past Dean
University of South Carolina School of Medicine
Columbia, SC, USA

Richard A. Hoppmann, MD, FACP, FAIUM

Preface

While teaching musculoskeletal ultrasound, we observed that students learned best when imaging defined structures guided by commonsense instructions.

This experience led to our writing various protocols with logical instructions for each step, including probe movements.

After achieving the ability to capture normal images, the next step was to remind the reader why we were doing a particular examination, that is, what pathology to look for in an image.

Each chapter starts with the clinical questions we want an ultrasound examination to answer and basic anatomy. With so much anatomy to learn, we liberally used simple illustrations, whimsical cartoons, and often corny mnemonics to aid in recollection.

We discuss pathologic entities discernible on ultrasound, pitfalls to avoid, and imaging tricks of the trade.

Our goal is to guide the reader to proficiency in musculoskeletal ultrasound without an instructor being present.

Columbia, SC, USA — Mark H. Greenberg
Columbia, SC, USA — Alvin Lee Day
Chicago, IL, USA — Suliman Alradawi

Contents

Contributors

Suliman Alradawi (Illustrator) Department of Medicine, Division of Rheumatology, Advocate Christ Medical Center, Advocate Medical Group, Chicago, IL, USA

Alvin Lee Day Division of Rheumatology, Columbia VA Health Care System, Department of Internal Medicine, University of South Carolina School of Medicine, Columbia, SC, USA

Mark H. Greenberg Division of Rheumatology, Department of Internal Medicine, University of South Carolina School of Medicine, Columbia, SC, USA

Chapter 1
Introduction/Getting Started

Musculoskeletal ultrasound (MSUS) has grown by leaps and bounds over the past decade due to technological advances and the need for inexpensive point-of-care imaging. The modality is indispensable in physiatry, orthopedics, podiatry, and rheumatology. However, gaining operational proficiency can be challenging.

Why Learn Musculoskeletal Ultrasound?

Health care professionals who have achieved competence in MSUS will attest to the difference in their practices. Orthopedic surgeons use MSUS to understand the function and structure of injured joints and surrounding soft tissues to better plan, or even avoid, surgery. Podiatrists can now instantly confirm soft tissue diagnoses involving the heel and ankle. Physiatrists and sports medicine specialists use sonography for rapid and precise injury assessment, thus formulating effective rehabilitation and accurate return-to-activity timetables. Ultrasound is integral to the field of regenerative medicine. Rheumatologists employ MSUS to define structural, functional, and inflammatory conditions to inform better therapeutic decisions.

Why a Protocol-Driven Approach?

Protocol-driven examinations improve the efficiency of ultrasound (US) examinations [1]. We prefer protocols since we less frequently perform certain investigations, and consistently obtaining a set of standard views ensures that all necessary images are acquired.

The contents do not represent the views of the U.S. Department of Veterans Affairs or the United States Government.

M. H. Greenberg et al., *Manual of Musculoskeletal Ultrasound*,
https://doi.org/10.1007/978-3-031-37416-6_1

The Goals of This Manual

This Manual demonstrates MSUS protocols that evaluate conditions commonly found in clinical practice. We show image acquisition methods in the illustrations along with the goal image. Readers may freely use these protocols in their clinical practice. We designed the order of image acquisition and hints intended to prevent the examination from grinding to a halt.

We also attempt to demystify the US and smooth the way from "Point A" to "Point B," enabling the US student to examine a normal joint. It is challenging to recognize pathology if one cannot capture and appreciate the image of normal structures. While we demonstrate some pathological conditions, this book does not focus on pathology. Many excellent books will serve that purpose; we have listed but a few [2–5].

Before appreciating pathology, the practitioner of MSUS needs to understand the echogenic appearance of different tissue structures and the basic anatomy and function of the region-of-interest (ROI). The next step is to learn to acquire these normal images logically and consistently. We have employed the hints that our medical students, residents, and fellows have found useful when learning US.

The last step is to understand the context. What can an US exam tell us in a particular clinical circumstance? What questions can be answered by using US? In which situations is the US diagnostically strong, weak, or somewhere in between? What are the common pitfalls and normal variants?

Many excellent US books meticulously describe normal and abnormal sonoanatomy, but our focus is on clinical utility and image acquisition. The authors acknowledge that learning sonoanatomy is daunting. We employ various mnemonic devices, including cartoon images, acronyms, word associations, phrases, and connections to prior knowledge. The illustrations aspire to be simple yet memorable visualizations of key concepts.

Tools

The Ultrasound Machine

A basic understanding of the control buttons and knobs of your US machine is necessary to acquire optimal images. Thankfully, the controls are comparable on most devices, although some variations exist. Learning a vehicle's controls is the first step to driving. We suggest reviewing the reference below, in addition to your US machine's "quick-start" guide for further information [6]. Below is a brief description of a typical US machine controls:

Depth/frequency: This controls the depth of penetration of the US beam, determined by the individual transducer's frequency range. The image depth is noted in centimeters on the ultrasound viewscreen. Lower frequencies achieve higher depth scans but at the cost of less resolution, thus fewer details.

Gain: This controls the number of incoming echoes processed by the US machine. The higher the gain, the brighter the image, with the downside of increased noise, artifacts, and decreased contrast in fine detail. Some devices have time–gain controls (TGC) to adjust the returning ultrasound beam at specific depths. By changing the TGC, you can focus on enhancing the echotexture of a specific structure at a particular depth. However, a common mistake is adjusting the gain to focus on a particular depth when changing the frequency is required.

Color flow Doppler: This button activates color flow Doppler (CF), which detects tissue movement and direction.

Power Doppler: This button activates power Doppler (PD), which detects the strength of tissue movement, making it more sensitive to tissue (blood) flow.

Pulse repetition frequency: This knob works with Doppler imaging. The lower the pulse repetition frequency (PRF), the more Doppler signals are seen on the image, at the risk of increased noise and artifacts.

Freeze frame, image recording, cine: The freeze-frame button locks in the image so that you can save/record. Likewise, a video clip is helpful for dynamic US studies and documenting ultrasound-guided procedures. The sonographer can usually prolong the video clip duration to capture more extended studies. Video clip capture controls differ somewhat in each machine.

Distance measurements: Caliper buttons allow measurements between two points. Measuring the cross-sectional area (CSA) of the object requires measuring the height and width and letting the machine calculate the size. Alternatively, many devices permit the sonographer to directly outline the structure's circumference to calculate a CSA.

Presets: These are buttons with standard settings to make it simpler to achieve the settings needed to evaluate specific structures. For example, a preset for a hand will have a higher frequency than for a hip. Presets do not work on every person since you may have a thin individual in whom you can use a higher frequency to look at the hip.

Probe (Transducer)

The probe is commonly corded and interfaces the patient to the US machine. The probe emits sound waves to a target, receives the reflected waves, and transmits them to the US machine and viewscreen. Probes used for MSUS are often linear or curvilinear and have varied footprint sizes. Each probe has a predetermined frequency range.

Terms

Terms are jargon used to communicate in the US world.

B mode scanning is "brightness" mode, which demonstrates a grayscale image.

Attenuation is the weakening of the returning ultrasound beam with a loss of resolution.

Echotexture is the appearance of structures and tissues in US images [7].

Echogenicity (Fig. 1.1) describes a tissue's ability to reflect or transmit US waves compared to surrounding tissues and its subsequent appearance on the US screen [8]. Images can be **hyperechoic** (bright white), **hypoechoic** (gray), or **anechoic** (black). **Heterogeneous** is a mixture of light and dark. **Isoechoic** describes that the structure has a similar echogenic appearance to adjacent tissue, sometimes making it difficult to discern. We recognize body structures and tissues by their different sonographic features. Recognizable structures include bone, blood vessels, nerves, tendons, ligaments, muscles, cartilage, synovial tissue, and ligaments.

Anisotropy (Figs. 1.2a, b, and 1.3) is described as distortion and loss of clarity of an US image when the US wave does not fall upon a structure at precisely 90° [9]. The image will appear artifactually hypoechoic. Minute changes in the angle of the US wave (**insonation angle**) may cause a structure to "disappear." Structures that are affected include ligaments, nerves, and particularly tendons. Anisotropy and the degree of anisotropy can also help identify these structures.

The **frequency** of the US wave dictates image quality. The higher the frequency, the more detail is revealed, albeit at the expense of diminished depth penetration [8]. The goal is to visualize the ROI with the highest frequency probe possible to maximize sonographic detail. For MSUS, we typically employ higher-frequency probes (10–22 MHz).

Longitudinal axis (LAX) is a notation for a longitudinal probe direction in which images are parallel to or line up with the extremity. Many call this the long-axis view [8].

Transverse axis (TAX) is a notation for the transverse orientation of the image. This cross-sectional view is alternatively referred to as the **short-axis view** [8].

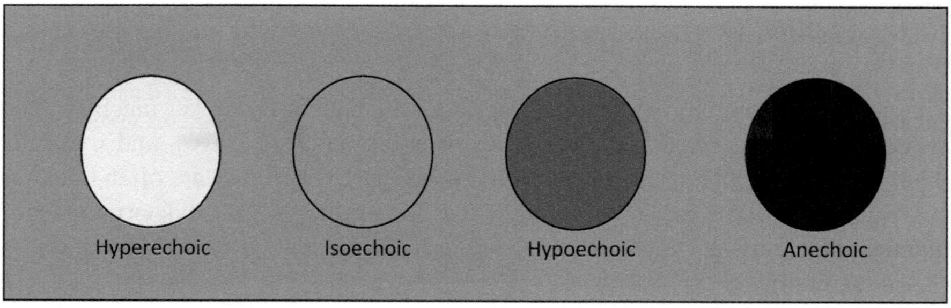

Fig. 1.1 Echogenicity

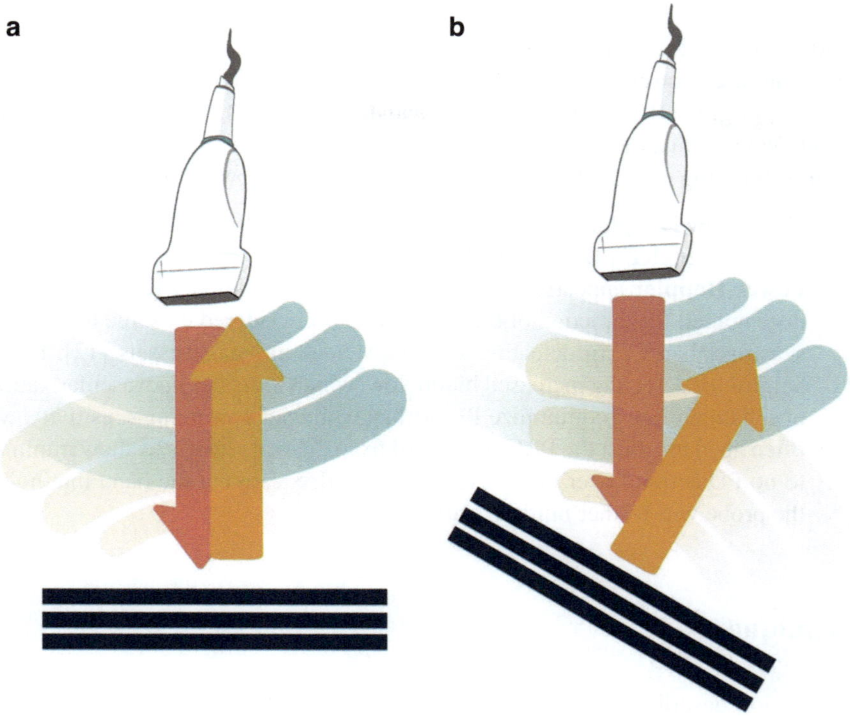

Fig. 1.2 Anisotropy is created when the angle of the probe is not perpendicular and some or all of the reflected ultrasound waves do not return to the probe. (**a**) No anisotropy since the returning ultrasound wave is fully acquired by the probe. (**b**) Anisotropy is produced since the returning ultrasound wave is not detected by the probe

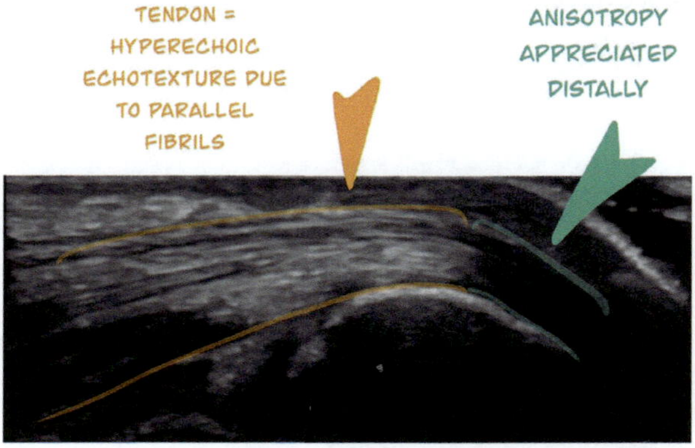

Fig. 1.3 Tendon in longitudinal view. Note the hyperechoic echotexture due to the parallel fibrils. Anisotropy is appreciated distally

Note that the classic definition of LAX or TAX probe orientation is with respect to an extremity's axis. Sometimes we may buck tradition by describing the probe's orientation with respect to a particular structure rather than the extremity in which the structure resides. When this does occur, we describe it in the text. An example would be when evaluating the subscapularis tendon, a structure oriented 90° to the proximal upper extremity.

Grayscale ultrasound is an US image of tissues comprised of black, white, and shades of gray [10]. Grayscale is the standard image in MSUS to which we may add Doppler imaging.

Doppler imaging uses the Doppler effect to detect blood flow through a vessel [11].

Color flow Doppler tells us the direction of blood flow [11]. A helpful mnemonic is "BART": Blue Away, Red Toward.

Power Doppler appears as a single color, most commonly red or orange. It does not tell the direction of blood flow and is considered by some to be more sensitive than CF in detecting slow flow rates in small vessels [12]. PD is helpful for detecting increased blood flow, which indicates neovascularization or inflammation. To minimize PD artifact while not sacrificing sensitivity, we often need to adjust the PRF downward from factory settings to show minimal to no PD activity over uninvolved bones [13]. Steadying the hand that holds the probe will further minimize artifacts.

Technique

This section describes how to use your US probe to obtain optimal images. The technique is critical. The examiner should hold the probe comfortably with at least one finger touching the patient for stability. Wrap the probe cord around the wrist or behind the neck to avoid a distracting tug (Fig. 1.4).

Both the patient and the examiner must be comfortable [14]. The view screen should be at eye level, directly in front of the examiner, to avoid head twisting. The probe direction or orientation is up to the examiner. The probe has a fin or dot, called

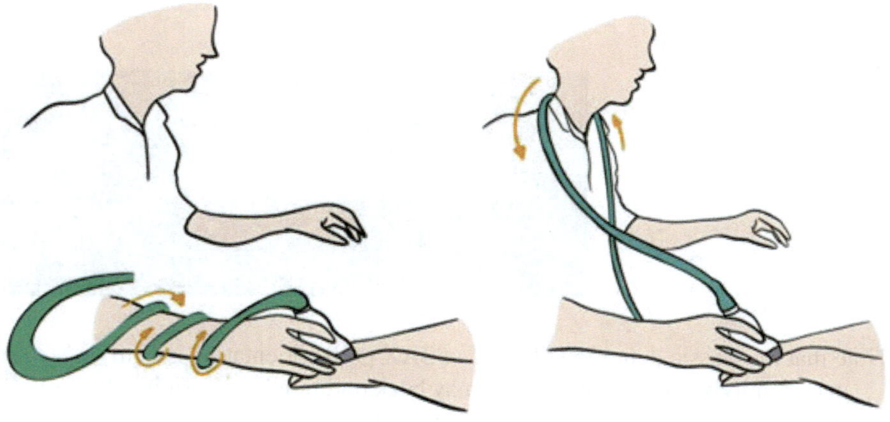

Fig. 1.4 Avoiding cord tugs

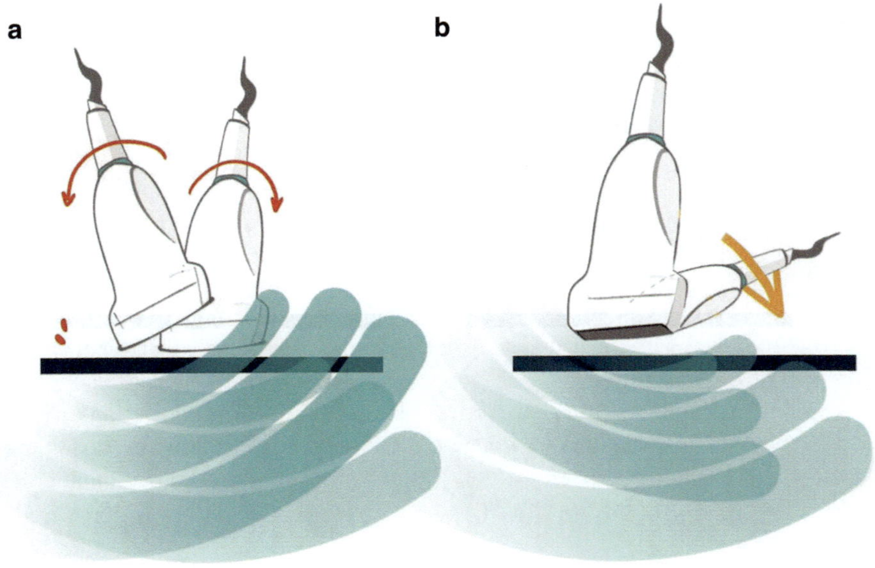

Fig. 1.5 Angling movements. (**a**) Heel-toe and (**b**) tilting

the probe marker, which corresponds to the dot on the viewscreen. Orient the probe in whichever direction facilitates easy recognition of the structures in the ROI. Routine **probe movements** are listed below. Of note, the terminology for these maneuvers may differ among sources.

Angling Movements

1. **Heel-toe** (Fig. 1.5a) means that the probe footprint remains stationary, but the transducer is angled about its long axis [10].
2. **Tilting** (Fig. 1.5b), alternatively called toggling, fanning, or rocking, means that the probe footprint remains stationary, but the transducer is angled about its short axis [15].

Dynamic Movements

1. **Longitudinal slide** (Fig. 1.6a) or **LAX slide** means to move the probe along the long axis of the probe.
2. **Transverse slide** (Fig. 1.6b) or **TAX slide** is like the LAX slide, but the direction of movement is along the short axis.
3. **Rotation** (Fig. 1.7) is when the probe is slowly spun to various degrees, with the axis of rotation being the center of the transducer [15]. A rotation variant keeps

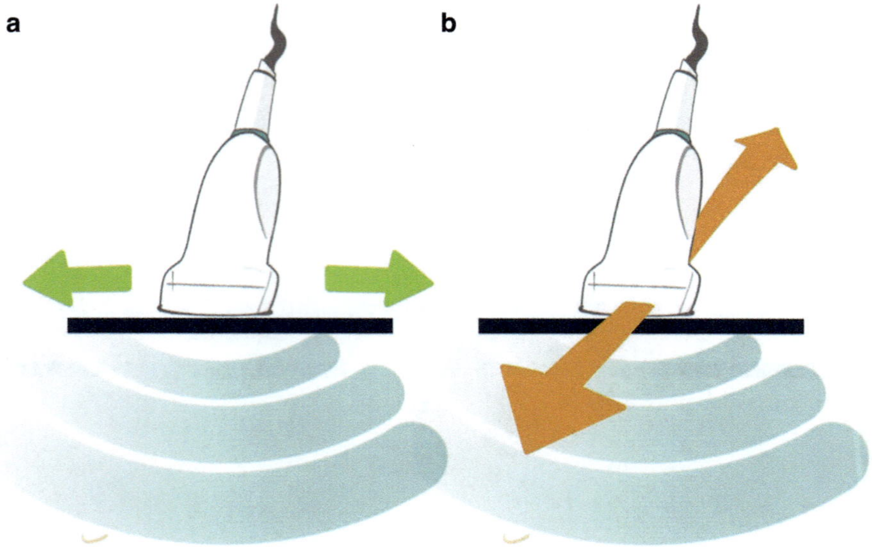

Fig. 1.6 Dynamic movements. (**a**) Longitudinal slide and (**b**) transverse slide

Fig. 1.7 Dynamic movement. Rotation

one end of the probe stationary and then rotates the other end in one direction and then the other. We find this maneuver invaluable to regain visualization of narrow longitudinal objects such as injection needles. Rotation is critical for obtaining orthogonal views of an ROI to confirm pathology [16]. Orthogonal views depict the shape of an object in two perpendicular dimensions to create a three-dimensional mental image of the object.

4. **Compression** occurs when the examiner firmly presses the transducer against the surface of the ROI. With this maneuver, veins collapse, but arteries do not. Also, compression will displace free fluid in a bursa or joint recess, but thickened synovial tissue will not shift much. Be mindful that compression may inadvertently reduce PD activity in an image.
5. **Floating the probe** maintains a thick layer of US gel between the skin surface and transducer to enhance the resolution of superficial structures.

Sonopalpation

Sonopalpation is a technique in which the examiner uses the transducer to delineate pathology. One type of sonopalpation is **compression**, described above. There are other types as well:

1. **Pain-localizing sonopalpation** is when the sonographer uses compression to detect the point of maximal tenderness by reproducing pain. Once localized, the examiner can sonographically interrogate the painful area for signs of pathology.
2. **Dynamic sonopalpation** is when the examiner holds the transducer in a fixed position. The body part is moved passively or actively to determine the abnormal movement of a ROI.
3. **Combined pain-localizing and dynamic sonopalpation** occurs when a maneuver will reproduce or exacerbate pain while the examiner directly observes the pathologic mechanism.

Tissues

Echogenic Features of Normal Tissues

Bone, being hyperechoic, is the most straightforward structure to visualize since it reflects sound waves in total, rendering only the bony surface visible [17] (Fig. 1.8).

Tendons (Fig. 1.9a, b) are hyperechoic due to their longitudinal, parallel fibrils [18]. In a longitudinal view, healthy tendons exhibit a "brush-like" pattern. However, the fibrils appear as numerous hyperechoic dots and dashes in a cross-sectional view.

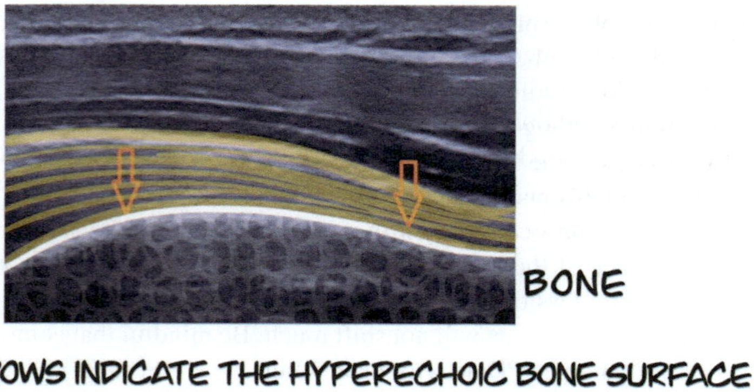

Fig. 1.8 Bone

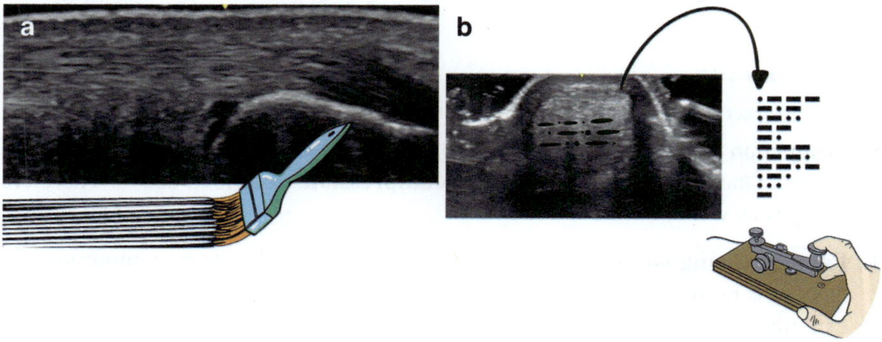

Fig. 1.9 (**a**) Tendon in longitudinal view illustrating hyperechoic echotexture due to the parallel fibrils. (**b**) Tendon in transverse view. Hyperechoic echotexture produces "dots and dashes"

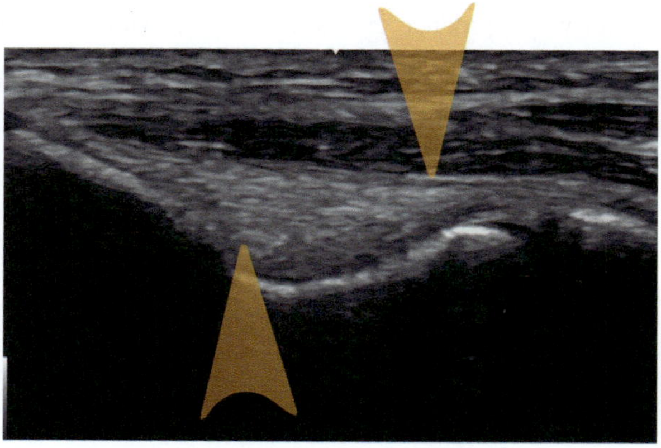

Fig. 1.10 Ligament (orange arrows) in longitudinal view showing similar, but less conspicuous echotexture as compared to tendon

Ligaments (Fig. 1.10) also have parallel longitudinal fibrils but are less hyperechoic than tendons in longitudinal and transverse views. Therefore, the echotexture of ligaments is less striking on US. Ligaments may be more difficult to view due to anisotropy from their varying attachment points.

Arteries and veins are anechoic to hypoechoic due to the blood inside (Fig. 1.11a) [19]. **Color flow Doppler** detects the direction of flow in arteries and veins and serves to delineate one from another (Fig. 1.11b). **Power Doppler** detects blood flow in arteries and veins (Fig. 1.11c). Probe compression is a practical means to differentiate veins from arteries, as veins will compress with probe compression (Fig. 1.11d).

Nerves (Fig. 1.12a, b) appear as a mixture of hyperechoic and hypoechoic areas, which convey a "honeycomb" appearance in TAX [20]. Alternatively, a normal nerve in TAX may demonstrate several wavy hyperechoic lines, not entirely parallel to each other. "Tram tracks" are often used to describe the sonographic appearance of normal nerves in LAX. The outer hyperechoic epineurium produces two parallel lines that resemble tram tracks [21]. Linear hypoechoic nerve fascicles alternate with hyperechoic connective tissue within the nerve itself, creating lines that parallel the epineurium. Nerves are more echogenic than muscles but less echogenic than tendons. Nerves also exhibit anisotropy, but not quite as much as tendons [22].

Muscle (Fig. 1.13a, b) is hypoechoic with hyperechoic fibroadipose perimysium interspersed within the muscle tissue [23]. In the longitudinal view, the perimy-

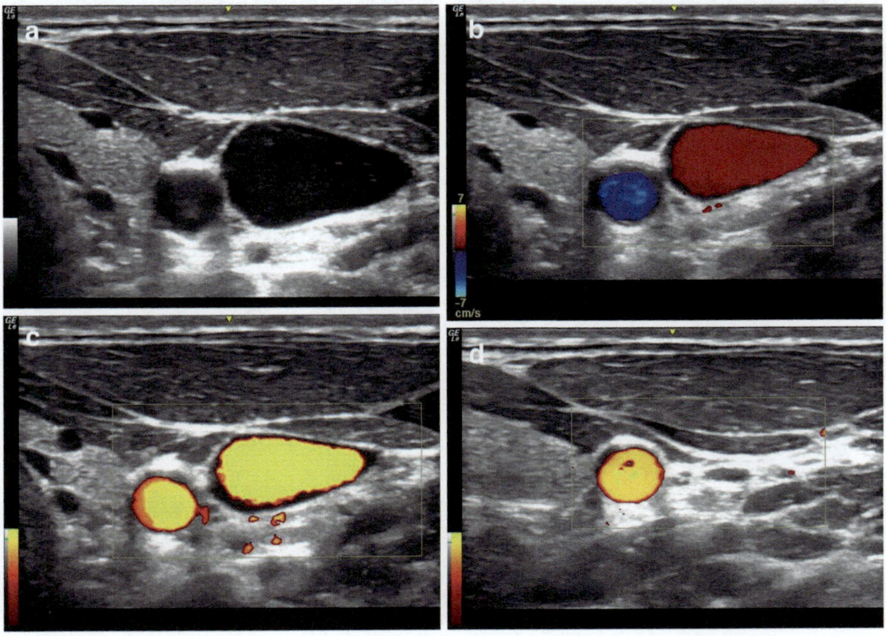

Fig. 1.11 (**a**) Transverse view of the common carotid artery and internal jugular vein, grayscale. (**b**) Transverse view of the common carotid artery and internal jugular vein, color flow Doppler. Vein is red, artery is blue. (**c**) Transverse view of the common carotid artery and internal jugular vein, power Doppler. (**d**) Transverse view of the common carotid artery and internal jugular vein, power Doppler, probe compression. The vein is compressed, whereas the lumen of the artery remains patent

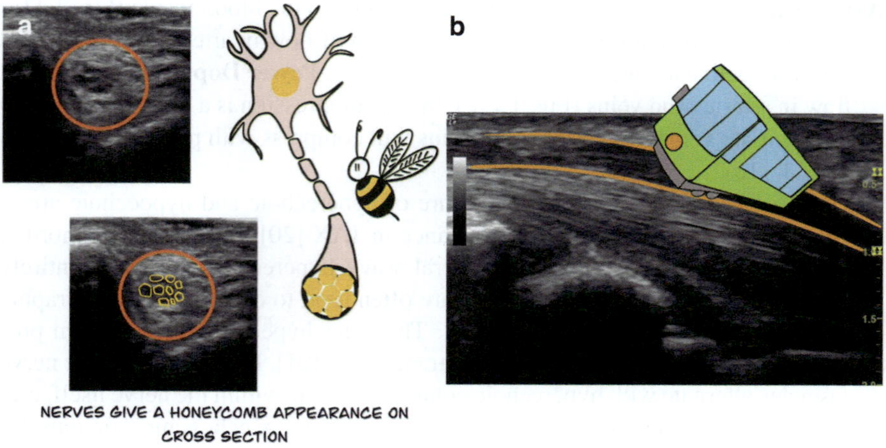

Fig. 1.12 (**a**) Nerve in transverse axis. Note the honeycomb appearance in cross section. (**b**) Nerve in long axis. Note the tram-tracks appearance

Fig. 1.13 (**a**) Muscle in long axis. Note the hypoechoic texture with interspersed hyperechoic perimysia. Pennate pattern. (**b**) Muscle in short axis. Note the hyperechoic perimysia. "Starry night" pattern. Alternatively, an "uncooked steak" appearance

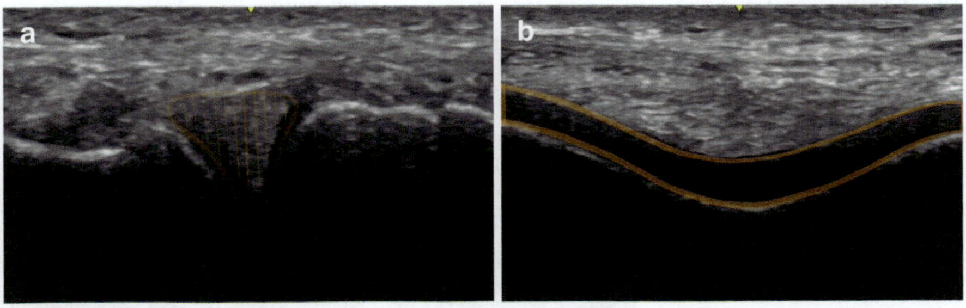

Fig. 1.14 (**a**) Fibrocartilage: hyperechoic triangular-shaped knee meniscus. (**b**) Hyaline cartilage: hypoechoic to anechoic, distal femoral cartilage

Fig. 1.15 Hyaline cartilage interface reflex

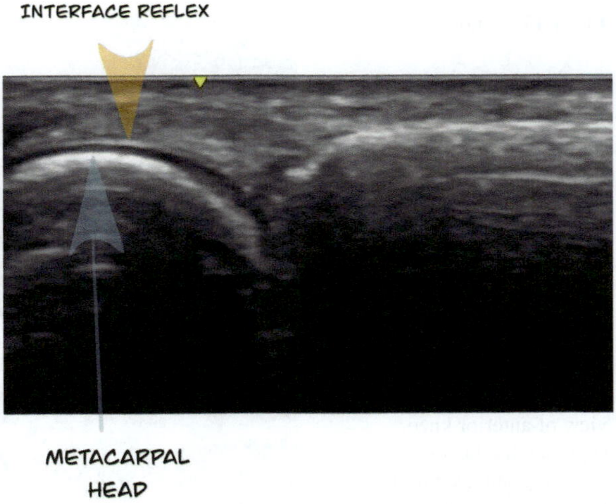

sia run lengthways, producing a pennate pattern, like a bird's feather. In cross section, longitudinal and speckled areas invoke a classic "starry night" appearance.

Cartilage (Fig. 1.14a, b) is hyperechoic or anechoic, depending on the type. Fibrocartilage is hyperechoic, whereas hyaline cartilage is hypoechoic to anechoic [24]. The surface of normal cartilage may have a thin, smooth, hyperechoic interface reflection signal when the US wave is perpendicular (Fig. 1.15).

Synovial tissue (Fig. 1.16) is hypoechoic intra-articular tissue that will not condense or shift with probe pressure [25].

Tenosynovium (Fig. 1.17) is a thin sheath encasing a tendon [26]. This sheath is composed of synovial cells and may not be apparent unless pathology is present [27].

The **paratenon**, considered a "false tendon synovial sheath," is separate from yet still surrounds the tendon [28, 29]. A paratenon is a loose collection of areolar connective tissue which supplies blood to a tendon. Tendons with paratenons include the Achilles tendon and the extensor tendons of the fingers [28, 30].

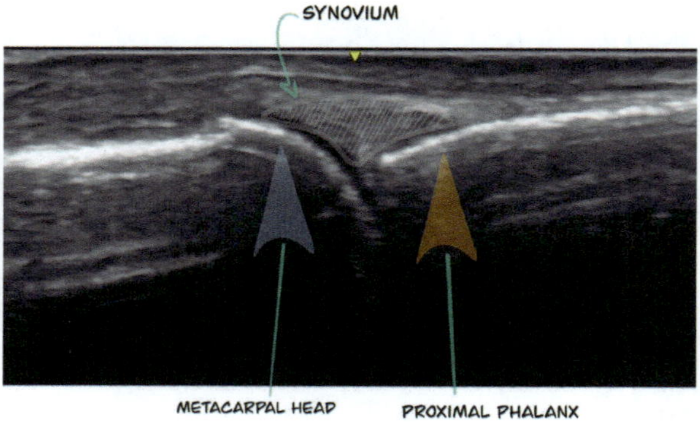

Fig. 1.16 Synovial tissue

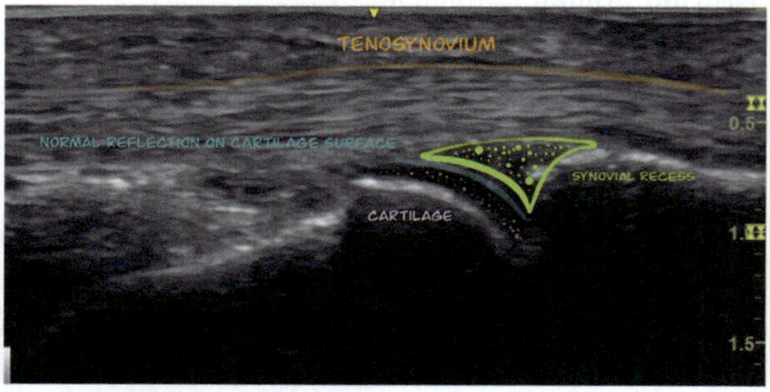

Fig. 1.17 Tenosynovium

Fig. 1.18 Longitudinal view of anterior knee, suprapatellar bursa. The bursa is connected to the knee joint and normally contains a small (physiologic) amount of synovial fluid

Synovial fluid (Fig. 1.18) is liquid located within a joint capsule, bursa, or tenosynovial sheath. It is anechoic and will shift (displace) with probe compression [31]. A **bursa** is a sac that contains synovial fluid [24]. There may be a slight hyperechoic coating surrounding anechoic synovial fluid. Some bursae typically have a small amount of physiologic synovial fluid, while others are not visualized unless distended by a pathologic amount of synovial fluid [32].

Sonographic Pathology by Structure

The examiner confirms pathology with images in two perpendicular, or orthogonal, planes. Images of pathology, with a few exceptions, are beyond the scope of this Manual. Please refer to other books which contain many pathologic US images [2–5]. The international network called OMERACT (Outcomes Measures in Rheumatology) has been vital in establishing uniform ultrasound definitions of pathology [16].

Bone

Erosion: An intra-articular discontinuity of the bone surface visible in two perpendicular planes [16].
Osteophyte: Appears as a step-up bony prominence [16].
New bone growth most often occurs in spondyloarthritis, in which extraneous bone called an enthesophyte might develop at the insertion of tendons or ligaments on bone.
Acute fracture may demonstrate discontinuity of the bone cortex with possible step-off deformity [33].

Arteries

Halo sign is a homogeneous, hypoechoic wall thickening of an artery, appearing as a dark halo surrounding a sharply defined arterial lumen, which may signify giant cell arteritis. The halo remains visible even with probe compression [16].

Veins

Venous thrombosis may be evident with vein resistance to compression by the transducer, best noted on CF. Echogenic material may be seen wholly or partially occluding the vessel [19].

Nerves

Entrapment: Nerves tend to enlarge proximal to compressions, as may occur when constricted within tight fibro-osseous canals such as the cubital tunnel and the carpal tunnel [34]. There may be focal or diffuse swelling of the nerve, with loss of the normal nerve echotexture.

A true **neuroma** following nerve injury results from a disorganized attempt to regenerate, culminating in a morass of nerve and scar tissue [34]. The unaffected portion of the nerve is continuous with the hypoechoic neuroma.

Cartilage

Fibrocartilage is usually hyperechoic. A hypoechoic cleft may represent a tear. Fibrocartilage may also exhibit hyperechoic deposits of varied sizes and shapes, raising suspicion of **calcium pyrophosphate crystal deposition disease (CPPD)** [16].

Hyaline cartilage is ordinarily anechoic or slightly hypoechoic. Like fibrocartilage, it may exhibit the same hyperechoic deposits of **CPPD** [16].

Osteoarthritic hyaline cartilage damage may be evident by thinning or loss of anechoic cartilage producing irregularity and loss of sharp margins [16]. Early signs of osteoarthritic cartilage may be a loss of clarity in the cartilage and a loss of sharpness at the outer border of the cartilage [35]. A subtle sign of hyaline cartilage surface damage may be the lack of a slight hyperechoic reflection.

Monosodium urate (MSU) crystals may form a thick layer superficial to the **hyaline cartilage** covering joint surfaces [36]. This hyperechoic, continuous, or discontinuous layer persists despite changing the angle of the probe (insonation angle) and is called the **double contour sign (DCS).** The MSU layer is often so thick that its width rivals the thickness of the bony cortex. Sonographic visualization of a DCS by itself is not unequivocally diagnostic for gout but is simply evidence for MSU deposition on the cartilage surface.

Tendons and Ligaments

Tendon damage often appears as a focal tendon defect (loss of fibers) [16]. The sonographer should position joints to enhance tendon tension to help visualize and grade tendon tears.

Complete tears may display two stumps, perhaps retracted from the ROI [37]. An absent tendon should prompt a thorough search for tendon stubs.

A **partial-thickness tear** presents as a focal anechoic region with loss of fibrillar pattern but the preservation of overall tendon continuity [38]. Be mindful of anisotropy mimicking a partial-thickness tear.

Full-thickness tears will exhibit features of partial-thickness tears, but an actual tendon gap is noted [38]. Understand that *a full-thickness tear does not mean that there is a complete tear of the tendon*, only that the tear has traversed the breadth of the tendon. Of course, a full-thickness tear may eventually progress to a complete tear.

Tendinosis, a degenerative process, is the preferred term over tendonitis or tendinopathy. In tendinosis, there may be a partial loss of fibrillar pattern, edema, hypoechoic areas, small anechoic clefts due to intrasubstance tearing, or perhaps calcium deposits [39]. Note that PD activity may be present with tendinosis, but realize that this may be neovascularization as a response to tendon injury rather than a sign of inflammation.

Tenosynovitis is tenosynovial hypertrophy signified by abnormal anechoic or hypoechoic tissue within the synovial sheath. Tendon sheaths develop tenosynovitis due to an external or internal source of irritation, which may also occur with tendinosis. There may be hyperemia on Doppler imaging and perhaps synovial fluid within the sheath [16].

Paratenonitis manifests as hypoechoic edematous changes adjacent to a tendon [29, 40]. The paratenon superficial to the extensor tendons of the fingers may also exhibit anechoic fluid collections [30]. There may be neovascularization on Doppler imaging [27].

Calcium pyrophosphate deposition crystals in tendons are hyperechoic, often linear formations that move along within the tendon during a dynamic assessment [16].

Monosodium urate crystals may also be present within tendons, represented as aggregates, hyperechoic areas, or tophi [41].

Ligament damage is analogous to that of tendons, as ligaments may also show tears, CPPD fragments, and degeneration. The ligament may enlarge and become hypoechoic in areas or may show loss of fibrillar echotexture with or without tears [39]. If the ligament appears to be intact, stress testing the ligament in comparison with the normal contralateral side may demonstrate the laxity of the damaged ligament. The US examiner may find it difficult to discern the retracted stumps of a complete ligament tear as opposed to those of complete tendon tears.

Enthesitis: The enthesis is where a tendon or ligament attaches to the bone. The sonographer delineates enthesitis as a hypoechoic or thickened insertion within 2 mm from the bony cortex [16]. Additionally, there may be bony erosion or calcifications/enthesophytes as a sign of structural damage. If active enthesitis is present, there may be a Doppler signal.

Muscle

Muscle strain or tear may be associated with swelling and loss of continuity of the perimysial hyperechoic fibers.

Muscle **hematomas,** initially hypoechoic, later become hyperechoic as the hematoma transitions to granulation tissue [42].

The sonographer detects **muscle atrophy** by noting a loss of muscle mass compared with the contralateral unaffected muscle [43]. The atrophic muscle may eventually demonstrate increased echogenicity.

Other Pathologic Findings

Excessive synovial fluid is indicated by detecting overabundant, commonly anechoic material within the joint capsule. This material is displaceable and compressible but does *not* exhibit a Doppler signal [31].

Hyperechoic deposits of CPPD may be present in the synovial fluid. The examiner may mobilize these deposits by joint movement and probe pressure [16]. **MSU aggregates and even tophi** may also be present within the synovial fluid.

Synovitis is synovial hypertrophy, defined as abnormal hypoechoic synovial tissue within the joint capsule that cannot be displaced, is poorly compressible, and may exhibit Doppler signal [16].

Loose bodies may appear as hyperechoic, rounded forms within the synovial fluid [44].

Cartilage interface sign: Hyaline cartilage may typically exhibit a slight hyperechoic superficial surface that may be exaggerated when an adjacent overlying hypoechoic torn tendon touches the cartilage surface. This hyperechogenicity is further augmented if the tear is associated with anechoic fluid, which rests upon the cartilage surface. This phenomenon may be seen with articular-sided supraspinatus tendon tears that extend to the cartilage [45].

Bursitis or bursal inflammation may result from an external source of irritation, such as excessive pressure on the bursa. Likewise, the inflammatory instigator may come from within, such as infection, crystal disease, or inflammatory arthropathy [46]. The excess synovial fluid may cause bursal distension.

Ganglion cysts may arise from a joint or tendon sheath and may be degenerative due to prior trauma or simply idiopathic [47]. They may appear as well-defined cysts or as multiloculated structures. Ganglia are noncompressible and contain hypoechoic to anechoic mucinous fluid.

Monosodium urate crystal deposition upon the surface of hyaline cartilage which produces DCS, has been detailed above. These deposits may also occur within synovial tissue or tendons in the form of aggregates, or tophi. MSU aggregates may appear as heterogeneous, hyperechoic foci exhibiting substantial reflectivity of sound waves even with changes in insonation angle [16]. Aggregates may form circumscribed tophi that are inhomogeneous, hyperechoic, or hypoechoic, perhaps surrounded by a slight anechoic halo. MSU aggregates and tophi may occasionally generate a posterior acoustic shadow and can be present in many locations.

Calcium deposits may occur within synovial fluid, tendons, ligaments, or cartilage and appear as hyperechoic structures varying from tiny specks to large blocks. Calcium can be abnormally deposited due to a metabolic issue, an autoimmune disease, dystrophic calcification due to tears, or a degenerative process [48]. The latter is called dystrophic calcification.

Artifacts: Friend or Foe?

Artifacts are the desert mirages of US. They are parts of images that do not genuinely exist but result from the technology used to create them. In this case, US waves strike a target and are reflected to the transducer for graphic interpretation by the machine. Remember that what you see on the US screen is the machine's virtual construct, not necessarily the truth. So, when we use US, we substitute virtual reality for actual reality. In MSUS, artifacts can be both confounding and convenient. There are many different artifacts. Below are the more common ones.

Anisotropy has been described in the **Terms** section above. See Figs. 1.2a, b, and 1.3.

Reverberation occurs when the US wave is perpendicular to two parallel, highly reflective surfaces [49]. The wave acts like a ping-pong ball, bouncing back and forth between the two surfaces; the transducer registers a portion of the wave with each ricochet. The US machine interprets the returning US waves as multiple parallel structures deep to the object(s). Wave reflection is dependent on the difference in acoustic impedance between objects, and the larger the difference, the more reflection occurs. The tissue's density and the speed of sound within the particular tissue determine the acoustic impedance [49].

Needle visualization exploits the reverberation artifact (Fig. 1.19). The two highly reflective surfaces on each side of the needle encourage the US beam to bounce back and forth within the entire needle, with some US waves returning to the transducer with each rebound. This scenario produces a reverberation artifact at integer multiples of the needle's width. This artifact occurs in solid and hollow needles and is not dependent on the presence of an inner lumen [50].

A subset of reverberation artifacts is the **comet-tail artifact** [51]. The objects producing this artifact are tiny in length and depth, and the two reflective surfaces are very close to each other. This type of reverberation trails off in a triangular pattern like a comet's tail. However, separate reverberation lines are difficult to resolve due to the narrow depth of the object. The B-line artifact in abnormal lungs is a typical example. Still, the comet-tail artifact can also occur with small foreign bodies, adenomyomatosis of the gallbladder, or thyroid colloid nodules [52, 53].

Mirror-image artifacts occur when a target object is near a deeper strong reflector [49, 54]. The US wave bounces off a robust reflective surface (such as the bony cortex); however, part of the US wave impacts the deepest side of the target object and is reflected to the strong reflector and then finally back to the transducer. The transducer detects the returning errant US wave, oblivious to its meandering path, and then constructs an image. The time differences in echoes returning to the transducer between the object and the strong reflector produce a mirror image of the target object, which deceptively appears deep to the reflecting surface. The deeper appearing mirror image is typically not as intense as the original in grayscale. Mirror images can also occur with CF or PD (Fig. 1.20).

Acoustic shadowing, also called **posterior acoustic shadowing,** occurs when the US wave is reflected, absorbed, or refracted by an object, resulting in an anechoic

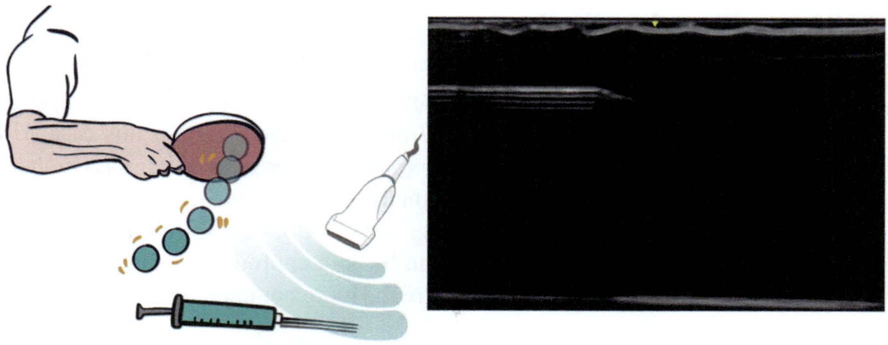

Fig. 1.19 Reverberation of a therapeutic needle in a practice gel phantom

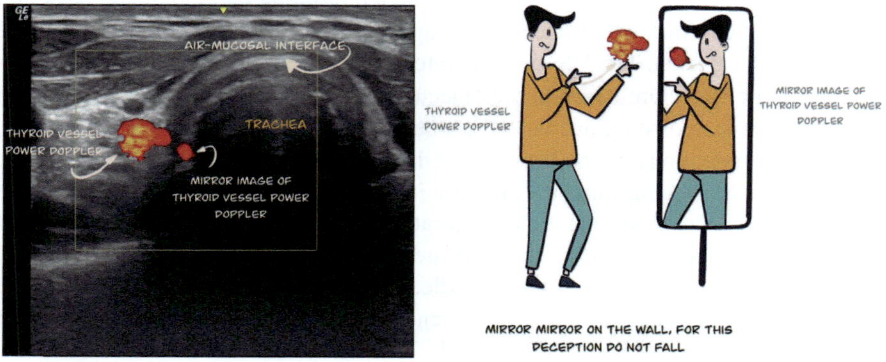

Fig. 1.20 Mirror-image artifact due to the strong reflector of the trachea (air). The color flow Doppler signal is mirrored on the other side of the trachea

or hypoechoic area deep to the object [51]. Acoustic shadowing most commonly occurs with bones. The object becomes a de facto soundwave umbrella, resulting in a hypoechoic or even anechoic shadow deep to the object.

Acoustic enhancement, also called **posterior acoustic enhancement**, occurs when a fluid-filled structure such as a cyst or bursa demonstrates increased echogenicity deep to the target structure, since US waves are less attenuated by the fluid than the surrounding soft tissues. The US machine misinterprets these relatively stronger soundwaves as an area beyond the target that has enhanced echogenicity [49]. Acoustic enhancement is helpful to verify that the structure in question is, in fact, fluid-filled. Figure 1.21 reveals acoustic enhancement deep to a normal fluid-filled bladder. Additionally, turning on CF can determine that the fluid-filled structure is neither an artery nor a vein.

Interface reflex (Fig. 1.15) occurs over hyaline cartilage. It is a thin, bright hyperechoic line along the edge of a smooth superficial cartilage surface, present only if the probe is precisely perpendicular to the surface [55]. The interface reflex (IR) should not be confused with gout's DCS. An IR may also be confused with a CIS but should be easily distinguishable since there would be no evidence of an

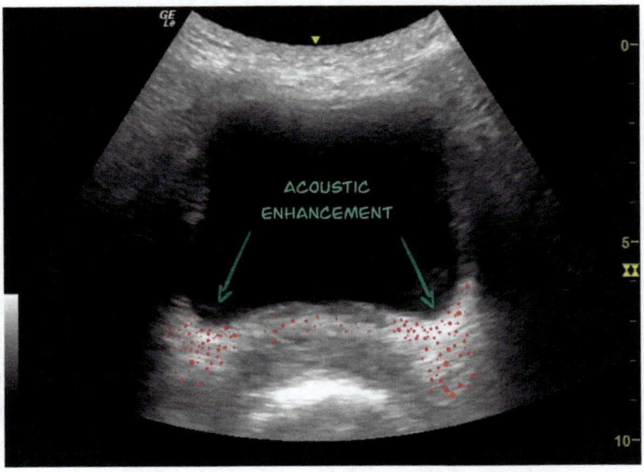

Fig. 1.21 Acoustic enhancement deep to a normal fluid-filled bladder

overlying tendon tear with an IR. The IR helps gauge the smoothness of the outer surface of cartilage, indicating which segments of cartilage are intact.

Organization of the Manual

We have divided his book into various examination sections designed to evaluate specific conditions. For instance, the volar wrist is often examined separately from the rest of the wrist when assessing for carpal tunnel syndrome. Most protocols are region-specific; however, protocols for enthesopathy and crystal arthropathy involve multiple body areas.

Limited Versus Complete Protocols

The MSUS protocols developed for this Manual explore ROIs of common clinical conditions. Limited protocols, typically 2–5 images, detect the presence of a sonographic finding. Complete protocols offer more detail regarding the joint space and additional structures in a region. Limited protocols can be pulled from the complete protocols as needed.

Finished Reports

Finished reports should contain standard information that often includes:

1. Report Title: Limited or Complete Exam, Area Examined (e.g., volar wrist) or Specialty Examination (e.g., enthesopathy)
2. Ultrasonographer's name

3. Impression
4. Examination side of the body
5. Indication for the examination
6. Study for diagnosis, prognosis, or treatment monitoring
7. Prior treatment failures
8. Technique and equipment used
9. Image Key or List of Images
10. Findings

The "**Impression**" appears toward the report's top since most readers want to know the results. The "**Image Key**" lists the images of the protocol in chronological order. If the examiner desires additional views, they can insert these views into the image list. Finally, "**Findings**" will contain the raw data noted on the exam. We find it helpful to create a template with a detailed routine examination of the ROI; the sonographer can later make changes to reflect pathology/variations found in the individual patient. This section also reminds us of the structures that need to be evaluated in ROI. We find this format helpful when first learning diagnostic US and doing examinations that we perform less frequently. A practice guideline for documentation of ultrasound reports is available [56].

Additional Suggestions

Regarding the US machine, you do not need to purchase the most expensive machine available. Manufacture-reconditioned/warranted machines are often acceptable if new machines are too expensive. For ergonomics, obtain a height-adjustable examination table with a removable armrest. While you are learning, set aside dedicated time for practice. You may not want to charge patients until you feel confident that the US report is meaningful. Make sure images are retrievable and backed up if needed.

When studying US, write down your questions as they occur. Seek the answers in a textbook or journal article, ask a mentor, or save your list to ask an instructor at a course. Always pick up the probe. It never hurts. Incorporating MSUS into your practice improves patient care and practitioner satisfaction.

The goal of this Manual is to acquire quality US images of *normal* structures. It is then up to the examiner to apply the same protocol steps when seeking pathology. We hope to make the notoriously steep learning curve of MSUS a bit gentler. Become adept at producing the images in this Manual. Try the protocols on yourself and different normal subjects to gain confidence and speed. You always have a willing test subject—yourself!

As you embark on learning this exciting modality, remember that MSUS is not done in a vacuum. Before a sonographic evaluation, a focused history and physical exam will help guide the US exam and suggest new diagnostic considerations. However, it is essential to keep in mind that abnormalities found by US may *not* be the cause of the complaint. The practitioner must place US findings into proper clinical context. Finally, and most importantly, MSUS requires direct interaction between clinician and patient. This laying on of hands is beneficial in many ways.

Acknowledgment While writing this Manual, we encountered several commonly used descriptive terms (e.g., tram-tracking, Popeye's sign, etc.) that have become part of the MSUS lexicon. Every effort has been made to track down the first usage of these terms to acknowledge their provenance and recognize their creator. Sometimes, however, we can not find the origin of these widely used terms. We do not assume credit for or imply authorship of these terms and would like to recognize the unheralded creators of these wonderfully evocative terms and expressions.

References

1. Brandli L. Benefits of protocol-driven ultrasound exams. Radiol Manag. 2007;29(4):56–9.
2. Jacobson JA. Fundamentals of musculoskeletal ultrasound. 3rd ed. Philadelphia, PA: Elsevier; 2018.
3. Bianchi S, Martinoli C. Ultrasound of the musculoskeletal system. New York: Springer; 2007.
4. Minagawa H, Wong KH. Musculoskeletal ultrasound: echo anatomy & scan technique. Amazon Digital Services; 2017.
5. Kohler MJ. Musculoskeletal ultrasound in rheumatology review. Cham: Springer International Publishing; 2016.
6. Enriquez JL, Wu TS. An introduction to ultrasound equipment and knobology. Crit Care Clin. 2014;30(1):25–45, v.
7. Kim SY, Cheon JH, Seo WJ, Yang GY, Choi YM, Kim KH. A pictorial review of signature patterns living in musculoskeletal ultrasonography. Korean J Pain. 2016;29(4):217–28.
8. Ihnatsenka B, Boezaart AP. Ultrasound: basic understanding and learning the language. Int J Shoulder Surg. 2010;4(3):55–62.
9. Connolly DJ, Berman L, McNally EG. The use of beam angulation to overcome anisotropy when viewing human tendon with high-frequency linear array ultrasound. Br J Radiol. 2001;74(878):183–5.
10. Jacobson JA. Introduction. In: Fundamentals of musculoskeletal ultrasound. 3rd ed. Philadephia, PA: Elsevier; 2018. p. 1–15.e1.
11. Boote EJ. AAPM/RSNA physics tutorial for residents: topics in US: Doppler US techniques: concepts of blood flow detection and flow dynamics. Radiographics. 2003;23(5):1315–27.
12. Torp-Pedersen ST, Terslev L. Settings and artefacts relevant in colour/power Doppler ultrasound in rheumatology. Ann Rheum Dis. 2008;67(2):143–9.
13. Torp-Pedersen S, Christensen R, Szkudlarek M, Ellegaard K, D'Agostino MA, Iagnocco A, et al. Power and color Doppler ultrasound settings for inflammatory flow: impact on scoring of disease activity in patients with rheumatoid arthritis. Arthritis Rheumatol. 2015;67(2):386–95.
14. Jaeger KA, Imfeld S. The damaging effect of ultrasound—to the examiner. Ultraschall Med. 2006;27(2):131–3.
15. Schmidt W. Introduction: musculoskeletal ultrasound indications and fundamentals. In: Kohler MJ, editor. Musculoskeletal ultrasound in rheumatology review. Cham: Springer International Publishing; 2016. p. 1–21.
16. Bruyn GA, Iagnocco A, Naredo E, Balint PV, Gutierrez M, Hammer HB, et al. OMERACT definitions for ultrasonographic pathologies and elementary lesions of rheumatic disorders 15 years on. J Rheumatol. 2019;46(10):1388.
17. Blankstein A. Ultrasound in the diagnosis of clinical orthopedics: the orthopedic stethoscope. World J Orthop. 2011;2(2):13–24.

18. Martinoli C, Derchi LE, Pastorino C, Bertolotto M, Silvestri E. Analysis of echotexture of tendons with US. Radiology. 1993;186(3):839–43.
19. Kory PD, Pellecchia CM, Shiloh AL, Mayo PH, DiBello C, Koenig S. Accuracy of ultrasonography performed by critical care physicians for the diagnosis of DVT. Chest. 2011;139(3):538–42.
20. Martinoli C, Serafini G, Bianchi S, Bertolotto M, Gandolfo N, Derchi LE. Ultrasonography of peripheral nerves. J Peripher Nerv Syst. 1996;1(3):169–78.
21. Ceponis A, Kissin E. Ultrasound of the hand and wrist. In: Kohler MJ, editor. Musculoskeletal ultrasound in rheumatology review. 2nd ed. Springer; 2021. p. 53–82.
22. Hamidi M, Kliot M, Gallagher T, Hopkins B, Younger J, Liu B, et al. The utility of multimodality perioperative imaging in peripheral nerve interventions. J Neurosurg Imaging Techn. 2017;2(1):106–17.
23. Franchi MV, Raiteri BJ, Longo S, Sinha S, Narici MV, Csapo R. Muscle architecture assessment: strengths, shortcomings and new frontiers of in vivo imaging techniques. Ultrasound Med Biol. 2018;44(12):2492–504.
24. Alves TI, Girish G, Kalume Brigido M, Jacobson JA. US of the knee: scanning techniques, pitfalls, and pathologic conditions. Radiographics. 2016;36(6):1759–75.
25. D'Agostino MA, Terslev L, Aegerter P, Backhaus M, Balint P, Bruyn GA, et al. Scoring ultrasound synovitis in rheumatoid arthritis: a EULAR-OMERACT ultrasound taskforce-part 1: definition and development of a standardised, consensus-based scoring system. RMD Open. 2017;3(1):e000428.
26. Ettema AM, Zhao C, Amadio PC, O'Byrne MM, An KN. Gliding characteristics of flexor tendon and tenosynovium in carpal tunnel syndrome: a pilot study. Clin Anat. 2007;20(3):292–9.
27. Mascarenhas S. A narrative review of the classification and use of diagnostic ultrasound for conditions of the achilles tendon. Diagnostics (Basel, Switzerland). 2020;10(11):944.
28. Benjamin M, Kaiser E, Milz S. Structure-function relationships in tendons: a review. J Anat. 2008;212(3):211–28.
29. Robinson P. Sonography of common tendon injuries. Am J Roentgenol. 2009;193(3):607–18.
30. Suzuki T, Shirai H. SAT0570 Clinical significance of finger extensor paratenonitis detected by musculoskeletal ultrasound. Ann Rheum Dis. 2020;79(Suppl 1):1243–4.
31. Wakefield RJ, Balint PV, Szkudlarek M, Filippucci E, Backhaus M, D'Agostino MA, et al. Musculoskeletal ultrasound including definitions for ultrasonographic pathology. J Rheumatol. 2005;32(12):2485–7.
32. Schmidt WA, Schmidt H, Schicke B, Gromnica-Ihle E. Standard reference values for musculoskeletal ultrasonography. Ann Rheum Dis. 2004;63(8):988–94.
33. Craig JG, Jacobson JA, Moed BR. Ultrasound of fracture and bone healing. Radiol Clin N Am. 1999;37(4):737–51, ix.
34. Kele H. Ultrasonography of the peripheral nervous system. Pers Med. 2012;1(1):417–21.
35. Grassi W, Lamanna G, Farina A, Cervini C. Sonographic imaging of normal and osteoarthritic cartilage. Semin Arthritis Rheum. 1999;28(6):398–403.
36. Gutierrez M, Schmidt WA, Thiele RG, Keen HI, Kaeley GS, Naredo E, et al. International consensus for ultrasound lesions in gout: results of Delphi process and web-reliability exercise. Rheumatology. 2015;54(10):1797–805.
37. Nazarian LN, Jacobson JA, Benson CB, Bancroft LW, Bedi A, McShane JM, et al. Imaging algorithms for evaluating suspected rotator cuff disease: Society of Radiologists in ultrasound consensus conference statement. Radiology. 2013;267(2):589–95.
38. Jacobson JA. Shoulder US: anatomy, technique, and scanning pitfalls. Radiology. 2011;260(1):6–16.
39. Smith J, Finnoff JT. Diagnostic and interventional musculoskeletal ultrasound: part 2. Clinical applications. PM R. 2009;1(2):162–77.
40. Dong Q, Fessell DP. Achilles tendon ultrasound technique. AJR Am J Roentgenol. 2009;193(3):W173.

41. Carroll M, Dalbeth N, Allen B, Stewart S, House T, Boocock M, et al. Ultrasound characteristics of the achilles tendon in tophaceous gout: a comparison with age- and sex-matched controls. J Rheumatol. 2017;44(10):1487–92.
42. Speer KP, Lohnes J, Garrett WE. Radiographic imaging of muscle strain injury. Am J Sports Med. 1993;21(1):89–96.
43. Pillen S, van Alfen N. Skeletal muscle ultrasound. Neurol Res. 2011;33(10):1016–24.
44. Bianchi S, Martinoli C. Detection of loose bodies in joints. Radiol Clin N Am. 1999;37(4):679–90.
45. Jacobson JA, Lancaster S, Prasad A, van Holsbeeck MT, Craig JG, Kolowich P. Full-thickness and partial-thickness supraspinatus tendon tears: value of US signs in diagnosis. Radiology. 2004;230(1):234–42.
46. Martinoli C, Bianchi S, Prato N, Pugliese F, Zamorani MP, Valle M, et al. US of the shoulder: non-rotator cuff disorders. Radiographics. 2003;23(2):381–401; quiz 534.
47. Zhang A, Falkowski AL, Jacobson JA, Kim SM, Koh SH, Gaetke-Udager K. Sonography of wrist ganglion cysts: which location is Most common? J Ultrasound Med. 2019;38(8):2155–60.
48. Abhishek A, Doherty M. Update on calcium pyrophosphate deposition. Clin Exp Rheumatol. 2016;34(4 Suppl 98):32–8.
49. Baad M, Lu ZF, Reiser I, Paushter D. Clinical Significance of US artifacts. Radiographics. 2017;37(5):1408–23.
50. Day AL, Greenberg MH. Reverberation artifacts. Unpublished Data; 2020.
51. Taljanovic MS, Melville DM, Scalcione LR, Gimber LH, Lorenz EJ, Witte RS. Artifacts in musculoskeletal ultrasonography. Semin Musculoskelet Radiol. 2014;18(1):3–11.
52. Feldman MK, Katyal S, Blackwood MS. US artifacts. Radiographics. 2009;29(4):1179–89.
53. Lichtenstein D. Lung ultrasound in the critically ill. Curr Opin Crit Care. 2014;20(3):315–22.
54. Bonhof JA. Ultrasound artifacts—part 1. Ultraschall Med. 2016;37(2):140–53; quiz 54–5.
55. Thiele RG. Rheumatologic findings. In: Daniels JM, Dexter WW, editors. Basics of musculoskeletal ultrasound. New York: Springer; 2013. p. 119–26.
56. AIUM practice guideline for documentation of an ultrasound examination. J Ultrasound Med. 2014;33(6):1098–102.

Chapter 2
Volar Wrist

Reasons to Do the Study
1. Carpal tunnel syndrome diagnosis and cause
2. Compression of the ulnar nerve in Guyon's canal
3. Wrist pain
4. Soft tissue swelling
5. Evaluation of the flexor tendons for tears, tendinosis, and tenosynovitis
6. Evaluation for possible tenosynovitis
7. Evaluation for inflammatory arthritis

Questions to Be Answered
1. Is there a structural or inflammatory cause for wrist pain?
2. Is there sonographic evidence of median or ulnar nerve compression? If so, what is the structure(s) causing this?
3. What is the true nature of an abnormal physical examination finding, and is this causing the discomfort?
4. What is the next step in alleviating the problem (e.g., splinting, injection, medication, surgery, or a combination)?

Basic Anatomy

1. Bones
2. Soft tissue structures (Fig. 2.1a, b)

The contents do not represent the views of the U.S. Department of Veterans Affairs or the United States Government.

M. H. Greenberg et al., *Manual of Musculoskeletal Ultrasound*,
https://doi.org/10.1007/978-3-031-37416-6_2

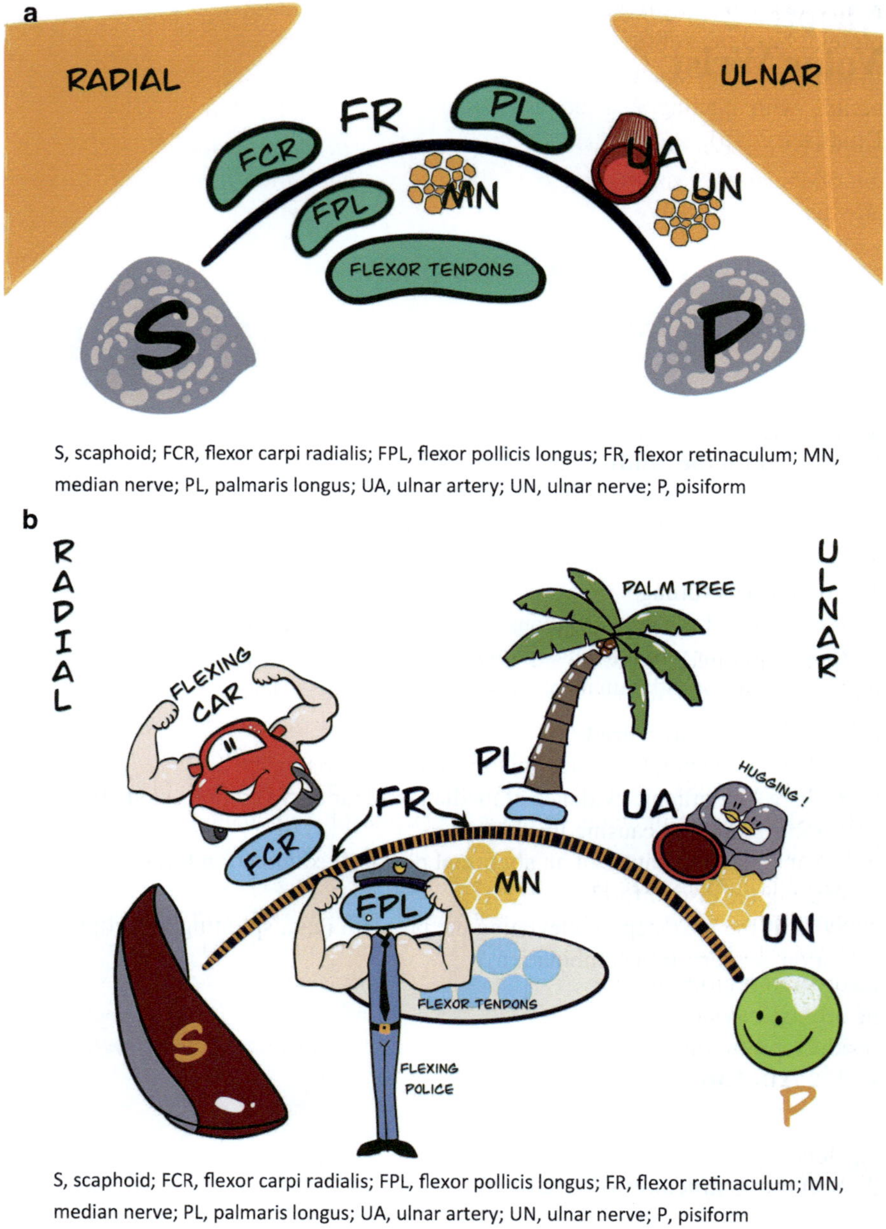

Fig. 2.1 (**a**) Carpal tunnel schematic. (**b**) Carpal tunnel schematic cartoon. *S* scaphoid, *FCR* flexor carpi radialis, *FPL* flexor pollicis longus, *FR* flexor retinaculum, *MN* median nerve, *PL* palmaris longus, *UA* ulnar artery, *UN* ulnar nerve, *P* pisiform

Clinical Comments

Ultrasound (US) can detect compression of the median nerve (MN) in the carpal tunnel (CT) and the ulnar nerve (UN) in the wrist and hand. It can also detect the presence of pertinent wrist variants, such as a persistent median artery (PMA), a bifid MN, and accessory muscles. Calcium and monosodium urate deposits, tenosynovitis, tendon pathology, soft tissue swelling, and hyperemia are readily apparent.

Carpal Tunnel Syndrome

Carpal tunnel syndrome (CTS) is a dysfunction of the MN within the CT due to injury, compression, or other causes. The CT is a passageway at the wrist through which multiple tendons and the MN traverse to reach the hand. The roof is the flexor retinaculum (FR), or transverse carpal ligament, while the carpal bones form the floor and edges. Patients may complain of numbness, tingling, or weakness of the first three digits of the hand [1].

US can detect a sonographic appearance of the MN consistent with CTS to help corroborate a clinical diagnosis. However, the US exam cannot "rule in" or "rule out" CTS. To make the diagnosis, the patient must have the clinical symptoms of the syndrome, irrespective of the sonographic appearance [2]. Ultrasound can detect MN compression as nerve swelling just proximal to the constriction point in the fibro-osseous canal; the swollen nerve is called a "pseudoneuroma" [3].

For the diagnosis of CTS, electrophysiologic studies (EPS) remain the tests of choice but may be normal in 10–25% of patients with clinical CTS [4, 5]. However, US may confirm CTS in approximately 30% of patients with a clinical diagnosis of carpal tunnel syndrome but *normal* EPS [6]. Bear in mind that US and EPS are complementary [2]. The two different studies approach the diagnosis of CTS from different viewpoints and thus are not always in agreement. Ultrasound demonstrates structural changes, while EPS describes nerve function. If compression of the MN is causing clinical CTS, US and EPS may frequently agree. However, EPS is better suited for evaluating noncompressive causes of mononeuropathy in either the MN or UN [2].

Ultrasound is a potent tool for evaluating CTS but does have limitations. A patient who does not meet sonographic criteria for CTS may still have the condition. Further, US cannot verify CT surgical release adequacy [7].

To evaluate for CTS using US, compare the amount of swelling of a constricted MN to a more proximal (presumably normal-sized) portion of the MN in the forearm (Fig. 2.2). The examiner identifies the MN by its deep location proximal to the CT in the protocol. Next, the sonographer compares the cross-sectional area (CSA)

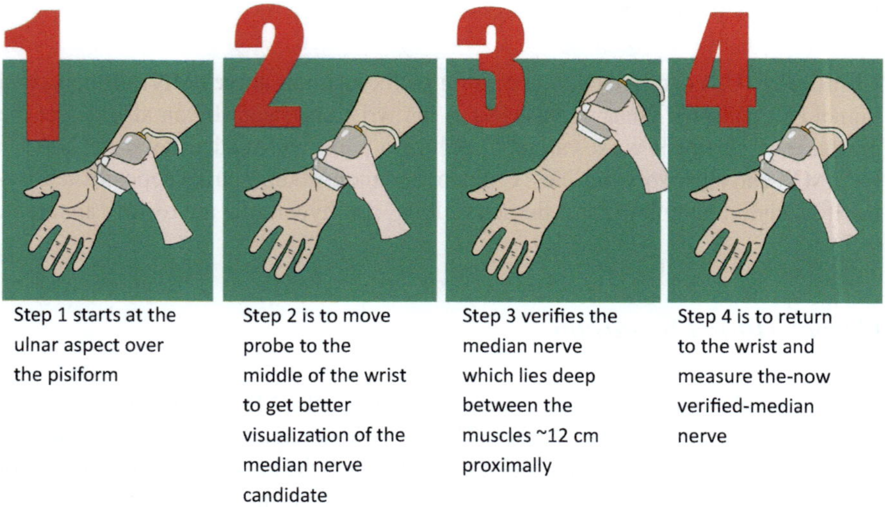

Fig. 2.2 Summary of the four steps to locate the median nerve and determine enlargement

of the MN at the wrist and the point ~12 cm proximal to the wrist by dividing the CSA of the MN at the wrist by the CSA ~12 cm proximal. This quotient is the wrist:forearm ratio (WFR). A WFR value greater than 1.4 strongly implies that the patient has MN compression within the CT [8]. Alternatively, if the CSA at the wrist is >2 mm^2 more than the MN at the point ~12 cm proximal to the wrist, this is 97% sensitive and 100% specific for CTS [9].

Other signs of CTS are:

1. Notch sign: a distinct area of compression of the MN, often seen in the longitudinal-axis (LAX) view [10].
2. Bulging or bowing outward of the FR [11].
3. Compressed or flattened MN [11].
4. Morphologic changes of the MN, including loss of distinct nerve echotexture, swelling of the MN, or thickening of the perineurium [12].

In some cases, MN enlargement proximal to the compression may *not* be the most salient feature. The examiner may observe MN compression from a thickened FR, flexor tenosynovitis, a ganglion cyst, or an anomalous muscle. The width of the midline FR, from proximal to distal, is about 14.4 ± 1.9 mm [13]. The FR thickness is usually 1–1.5 mm [11]. However, the FR may be more than 30% thicker in patients with CTS [14]. A thickened FR may point to repetitive action as a cause of CTS. Be on the alert for a thickened FR, flexor tenosynovial hypertrophy, a ganglion cyst, anomalous muscles, MN displacement, and two variants: a bifid MN and a PMA.

The examiner may begin with a limited protocol to determine if there is, in fact, sonographic evidence for MN compression. If needed, the examiner performs a complete volar wrist evaluation to include the UN since the deep motor branch

(DMB) of the UN innervates the adductor pollicis and the flexor pollicis brevis muscles [15]. Thus, compromise of the UN may cause thumb weakness, mimicking CTS.

After structural evaluation, look for a safe area to place an injection, should this be necessary. We perform CT injections using US to improve efficacy and safety [2, 16, 17]. When injecting the CT, be aware that the recurrent median nerve (RMN) is a branch of the MN that variably branches off the radial aspect of the nerve, sometimes within the CT. An 18 MHz high-frequency probe is crucial for detecting this small nerve branch [18].

This protocol focuses on the soft tissues of the volar wrist. If your examination does not elucidate the cause of volar wrist pain, proceed to evaluate the carpal bones, the distal ulna, and the radius. We can better evaluate ligamentous tears with dorsal wrist views. Other modalities such as radiographs, computerized tomography, and magnetic resonance imaging will better delineate bony abnormalities than the US examination.

Ulnar Nerve Damage

Ulnar neuropathy may cause paresthesia of the fifth and the ulnar aspect of the fourth fingers [19]. The DMB of the UN supplies the hypothenar muscles, interossei, adductor pollicis, and flexor pollicis brevis [15]. Therefore, compromise of the DMB may mimic median neuropathy by causing thumb weakness.

If the UN or one or both branches are compressed, there may be a loss of the normal echotexture. There are no specific sonographic criteria for UN dysfunction, but a helpful suggestion is to compare the maximal CSA of the affected nerve to a more proximal, normal portion of the UN [20].

If a nerve branch is compromised, determine what structure is responsible. The most common structure compressing the UN is a ganglion cyst [21]. Remember that Guyon's Canal is superficial to the FR and is separate from the CT. Note if a hypoechoic or anechoic structure, such as a ganglion cyst, is present, perhaps compressing the UN. If so, look for the root or stalk of the ganglion. Power Doppler (PD) or color flow Doppler (CF) may help delineate an ulnar artery aneurysm or pseudoaneurysm, which likewise might compress the UN.

Pitfalls

1. To avoid mistaking a flexor tendon for the MN, move the probe ~12 cm proximally up the forearm, and the true MN will arc in a radial direction and move deep to rest on the pronator quadratus muscle. Reversing the course of the probe will bring the image of the MN back into the CT, thus identifying the MN [7].

2. It is difficult for the US to determine if a CT release has been successful. The FR appearance will vary, and the MN size will not predict clinical outcomes [7, 22, 23].
3. An anechoic or hypoechoic rim surrounding a flexor tendon in the transverse axis (TAX) may be due to normal variant distal extension of a muscle, tenosynovial hypertrophy, or frank tenosynovitis. Power Doppler activity in the tendon sheath indicates more active tenosynovitis. Rotating the probe to LAX will determine whether or not this is simply normal tapering of the muscle as the tendon is formed [7].
4. When tracing the nerve to calculate the CSA, measure inside the hyperechoic epineurium [7].
5. Learn the sonographic criteria for the diagnosis of CTS based on measurement of the CSA to determine MN swelling:

   (a) CSA of MN at the CT >12 mm^2 [7, 24].
   (b) CSA >2 mm^2 in the CT compared with the CSA over the pronator quadratus (>4 mm^2 for bifid MN) [7, 9].
   (c) The WFR method of comparing the ratio of the CSAs of the swollen distal MN to the unaffected, more proximal MN [8]. Again, a WFR value greater than 1.4 strongly implies that the patient has MN compression within the CT.

Method

Begin the steps for identifying the MN and determining the presence of sonographic criteria for CTS (Fig. 2.2).

Protocol Image 1: Volar Wrist, Ulnar Aspect, Transverse (Fig. 2.3)

Begin with the patient's arm resting with the palm up and the wrist in a neutral position. The patient is facing the examiner. Place the probe on the volar wrist crease in TAX toward the ulnar aspect of the wrist to look for the rounded bony structure of the pisiform bone. See the pulsation of the UA. If this is not apparent, turn on PD to delineate the UA.

Deep to the UA, note the proximal edge of the FR, which is the hyperechoic, fibrillar band-like structure that stretches horizontally from the scaphoid bone and reaches the base of the pisiform bone. The FR may appear hypoechoic depending on the angle of insonation. The FR is located just deep to the pulsating UA. The FR is both the floor of Guyon's canal and the roof of the carpal tunnel.

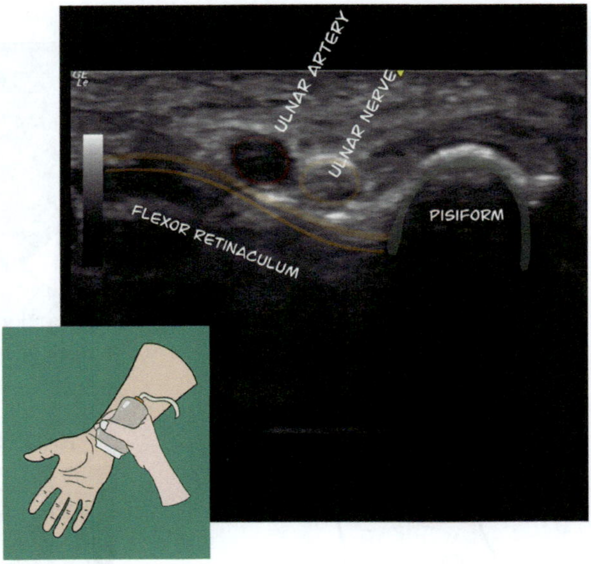

Fig. 2.3 Protocol Image 1: Volar wrist, ulnar aspect, transverse

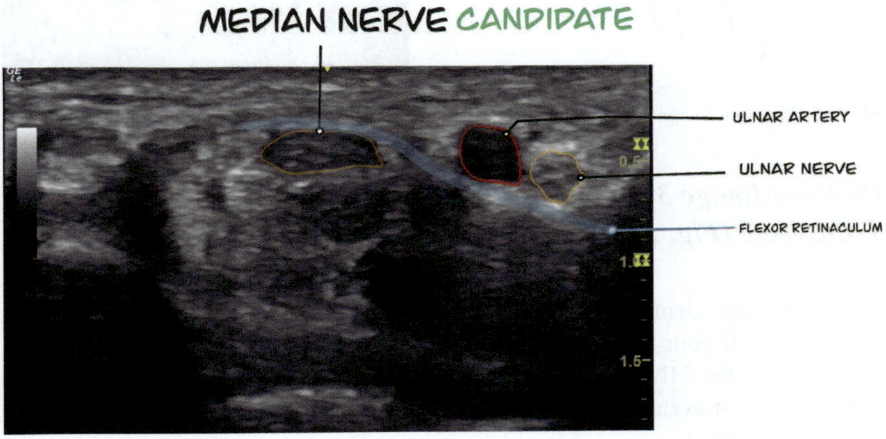

Fig. 2.4 Protocol Image 2: Volar wrist, mid-wrist, transverse

Protocol Image 2: Volar Wrist, Mid-Wrist, Transverse (Fig. 2.4)

Next, keep the probe in TAX orientation and perform an LAX slide in the radial direction. You will see several ovoid structures as you seek the MN. Tilt the transducer in either direction to better delineate these structures. Select one deep to the FR and consider it the "median nerve candidate," since we want to verify that this is the MN. The structure should have an elliptical and honeycomb appearance and exhibit moderate anisotropy. It may look more like parallel wavy lines than a honeycomb. A surrounding hyperechoic epineurium is a piece of additional evidence that the structure is a nerve.

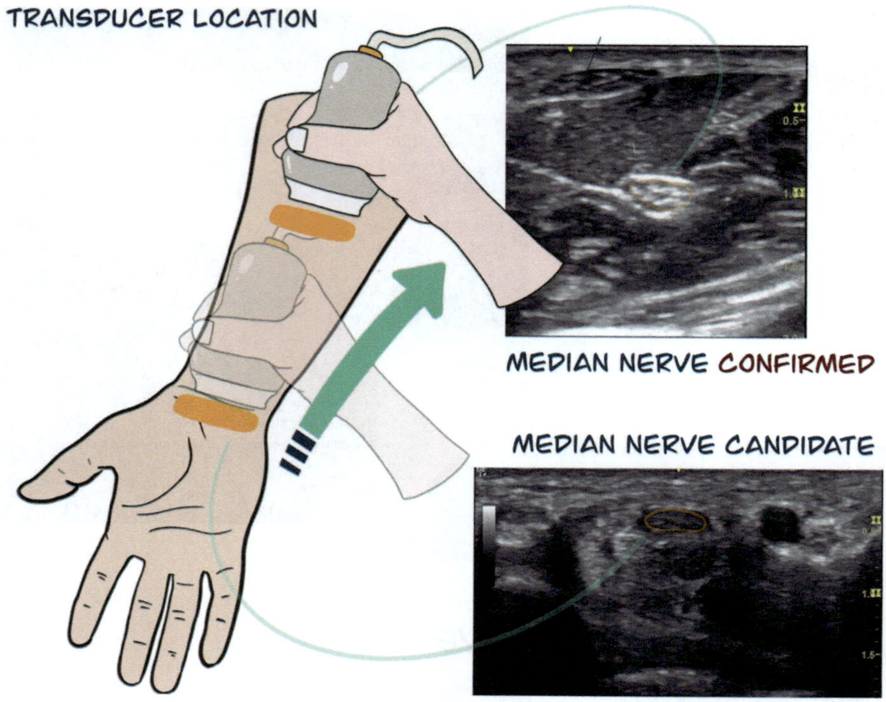

Fig. 2.5 Protocol Image 3: Forearm, 12 cm proximal to wrist crease, transverse

Protocol Image 3: Forearm, 12 cm Proximal to Wrist Crease, Transverse (Fig. 2.5)

Verify the MN identity by performing a proximal TAX slide of approximately 12 cm to see if your prime MN candidate dives deeper as you move the probe. In this location, the MN lies superficial to the pronator quadratus muscle and has a more circular or even triangular appearance. The MN is hyperechoic relative to the surrounding muscle.

Protocol Image 4: Median Nerve, 12 cm Proximal to Wrist Crease, Transverse, Cross-Sectional Area Measurement (Fig. 2.6)

Center the MN and freeze the image. Select the measure function and trace the inner circumference within the epineurium [7]. The MN, roughly 12 cm proximal to the wrist, will not be constricted in this location [9]. The typical CSA is 4–9 mm^2. An alternative to free measurement is the preset elliptical measurement function. The

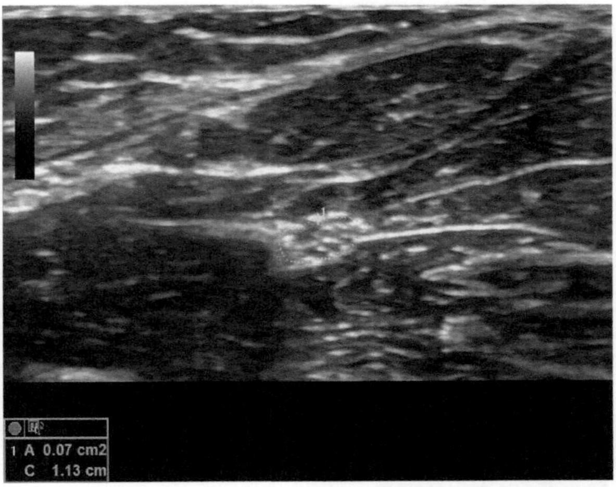

Fig. 2.6 Protocol Image 4: Median nerve, 12 cm proximal to wrist crease, transverse, cross-sectional area measurement

maximal height and width of the MN are measured, enabling the US machine to calculate the CSA.

Protocol Image 5: Proximal Carpal Tunnel, Transverse (Fig. 2.7)

Move the probe distally in a TAX slide to follow the MN back to the wrist crease. It will become more superficial and ovoid as it eventually nestles beneath the FR. Look at the MN at the proximal edge of the FR. Capture the image of the verified MN that has the largest CSA and well-defined edges. Look for loss of normal MN echotexture, FR thickening, MN compression or flattening under the FR, and hyperechoic swelling or thickening of the epineurium surrounding the MN. Note if there is a variant split (bifid) MN (Fig. 2.8). Look at the surrounding flexor tendons to see if there is swelling of the tendon or its surrounding sheath (the tenosynovium). Inspect for other structures within the CT that might compress or displace the MN, such as a ganglion cyst.

You can also check the flexor pollicis longus (FPL) in the CT radial to the MN. When you think that a particular tendon might be the FPL, manually flex and extend the thumb to verify its identity. Likewise, look for the flexor carpi radialis (FCR), which is superficial to the CT. Manually flex and extend the wrist to verify that this is the correct tendon. The region near the FCR and the radial artery is a common area for ganglion cysts, which often originate from the radiocarpal joint capsule. The FCR itself may develop tenosynovitis with inflammatory arthritis. The FCR tendon may exhibit tendinosis or tendon tears associated with osteoarthritis of the scaphoid-trapezium-trapezoid (triscaphe) joint [7].

Look for variants within the CT, including an extension of the flexor digitorum superficialis muscle, an encroaching lumbrical muscle, or a palmaris longus (PL)

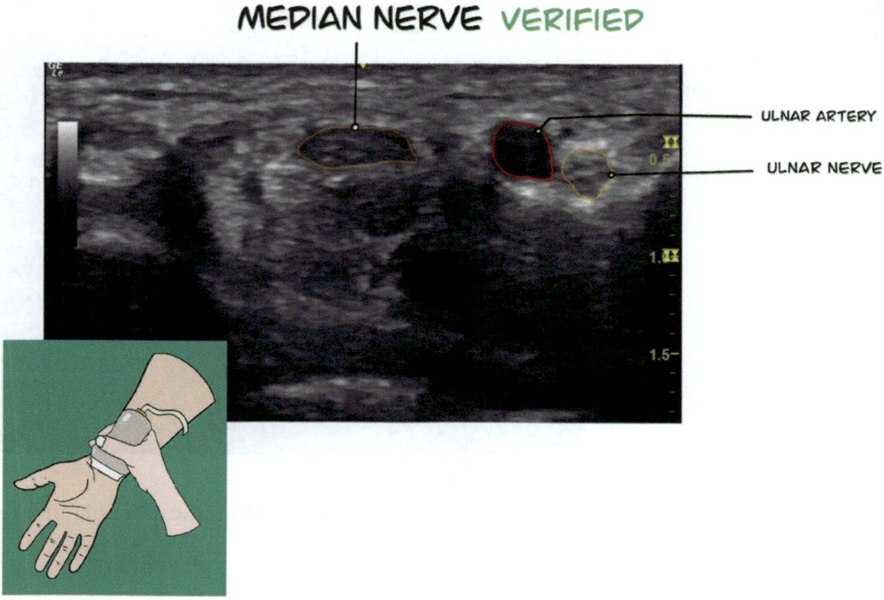

Fig. 2.7 Protocol Image 5: Proximal carpal tunnel, transverse

Fig. 2.8 Bifid median nerve

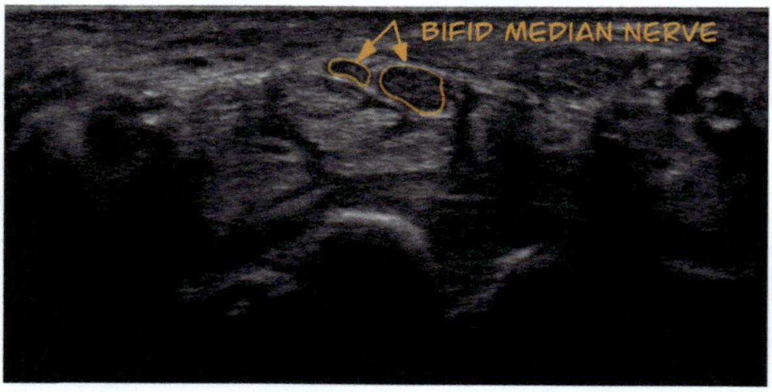

tendon within the CT. Oddly, the PL can sometimes be "reversed" where the muscle is distal to the tendon [25].

Protocol Image 6: Median Nerve, Proximal Carpal Tunnel, Transverse, Cross-Sectional Area Measurement (Fig. 2.9)

Measure the largest CSA at the wrist. Be sure to measure within the hyperechoic epineurium surrounding the MN. Compare the MN CSA at the wrist to the MN ~12 cm proximal to the wrist by dividing the CSA of the MN at the wrist by the

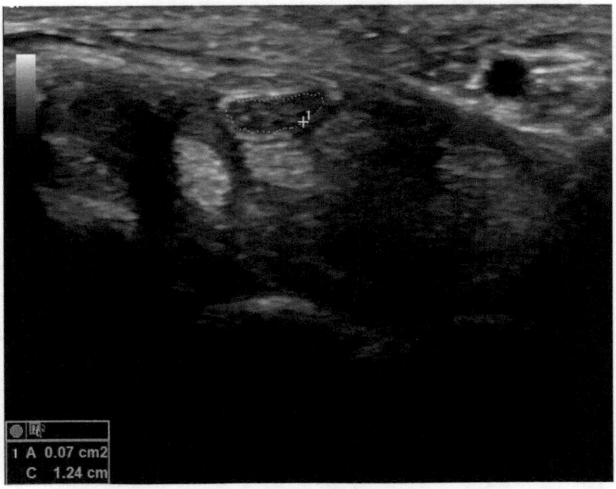

Fig. 2.9 Protocol Image 6: Median nerve, proximal carpal tunnel, transverse, cross-sectional area measurement

CSA ~12 cm proximal. This quotient is the WFR. A WFR value greater than 1.4 strongly implies that the patient has MN compression within the CT [8]. Alternatively, if the CSA at the wrist is >2 mm^2 more than the MN at the point ~12 cm proximal to the wrist, this is 97% sensitive and 100% specific for CTS [9].

Protocol Image 7: Proximal Carpal Tunnel, Transverse + Power Doppler (Fig. 2.10)

Turn on PD to determine if there is a PMA or evidence of active tenosynovitis. Scan at the edge of the CT and throughout the entire CT. Do not forget to lower the pulse repetition frequency (PRF) to enhance the detection of active tenosynovitis. However, avoid dropping the PRF so much that extraneous PD artifacts emerge. When using PD, remember to use a light touch and plenty of transmission gel. Only the UA or the radial artery will exhibit PD activity in most cases.

If you are considering a carpal tunnel injection, an option is to check PD or CF to look for blood vessels that might complicate the procedure. From this information, determine if and from what approach you would like to inject the carpal tunnel, bearing in mind that there is a radial branch of the MN (recurrent or thenar branch of the MN), which may not be apparent on US [26]. This smaller nerve sometimes resides within the CT and argues for injecting via the ulnar aspect.

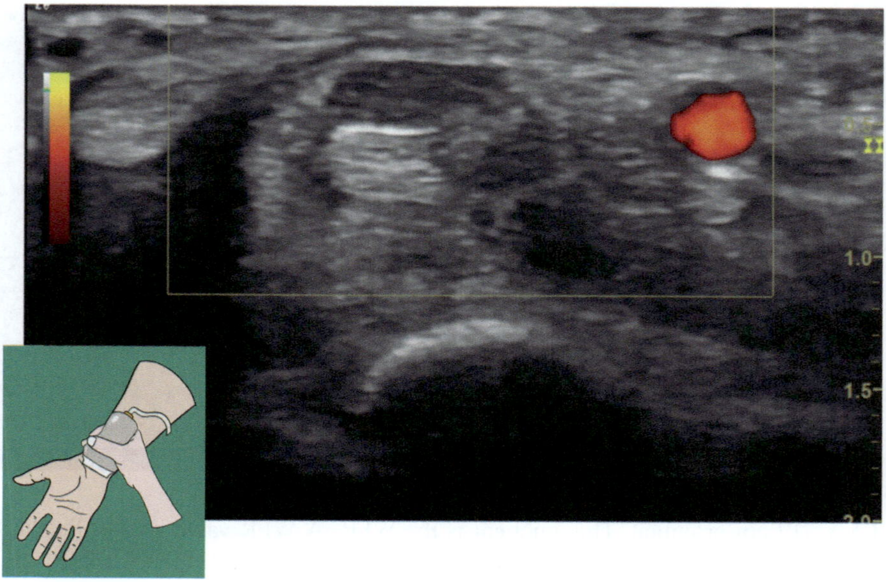

Fig. 2.10 Protocol Image 7: Proximal carpal tunnel, transverse + power Doppler

Protocol Image 8: Median Nerve, Longitudinal ± Power Doppler (Fig. 2.11)

Next, center the probe over the MN and rotate the probe 90° to look at the MN in LAX. Observe any compression and FR thickening. A trick is to center the probe on the MN in TAX so that when you rotate the probe, you see an ovoid MN expanding out and becoming a longitudinal structure. The MN has the typical "train or tram tracks" of a nerve, best observed proximally, where the MN assumes a more normal appearance.

Follow the nerve down into the CT, performing a heel-to-toe maneuver if needed to keep the probe parallel to the MN. Constriction of the MN may produce a "dumb-bell sign" in this view. Placing a rolled-up towel under the metacarpals will prevent wrist hyperextension [20]. Try to follow the MN distally.

Deep to the flexor tendons, note the hyperechoic volar aspects of the distal radius, lunate, and capitate bones. An additional image might be the MN in LAX with PD to look for tenosynovitis of the flexor tendons or aneurysms present in the CT. The above steps comprise the limited evaluation of the CT, which rapidly determines if there is evidence for MN compression.

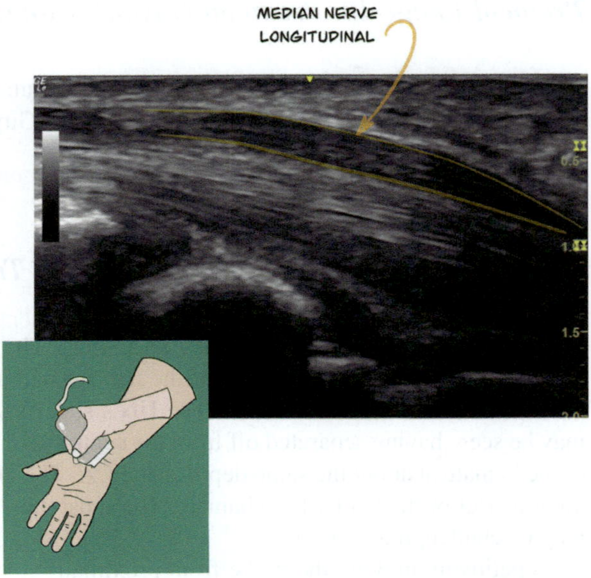

Fig. 2.11 Protocol Image 8: Median nerve, longitudinal ± power Doppler

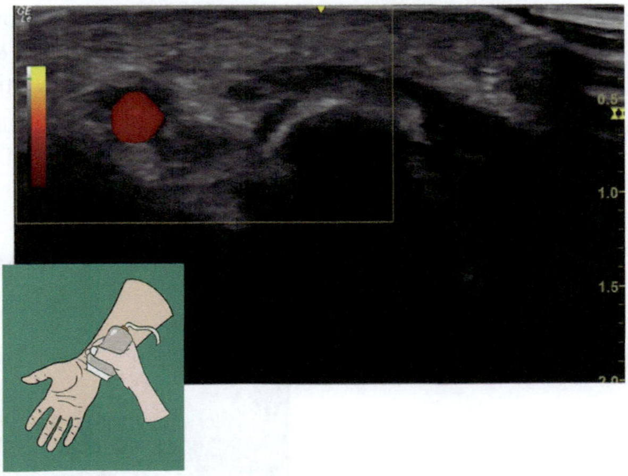

Fig. 2.12 Protocol Image 9: Proximal Guyon's canal, transverse + power Doppler

Protocol Image 9: Proximal Guyon's Canal, Transverse + Power Doppler (Fig. 2.12)

To complete the full volar wrist soft tissue examination, identify the UN in Guyon's canal by again placing the probe in TAX at the level of the pisiform, as was done in **Protocol Image 1**. Identify the UA with the help of the PD and the rounded hyperechoic pisiform bone that delineates the proximal portion of Guyon's canal.

Protocol Image 10: Proximal Guyon's Canal, Transverse

Turn off the PD to improve resolution. The UN abuts the UA on its ulnar aspect. Again, note that the FR is both the floor of Guyon's canal, and the roof of the CT.

Protocol Image 11: Distal Guyon's Canal, Transverse (Fig. 2.13)

Move the probe in TAX more distally (TAX slide). The pisiform will disappear, and then the deeper hyperechoic hook of the hamate will appear. The two branches of the UN are evaluated at this level. The UN's superficial sensory (SSB) and DMB may be seen, having separated off from the central UN. The SSB may be superficial to the hamate at about the same depth as the UA. The DMB may be seen just on the ulnar aspect of the hook of the hamate. The separation of the two branches distally may be challenging to discern.

Repetitively moving the probe from proximally to distally may enhance appreciation of the split. One or both tiny nerve branches may be hypoechoic as you move the probe due to anisotropy. Once each component is identified, correct for anisotropy by tilting the probe. The DMB may be located slightly more ulnar than the SSB; alternatively, the two branches may stack up vertically [27].

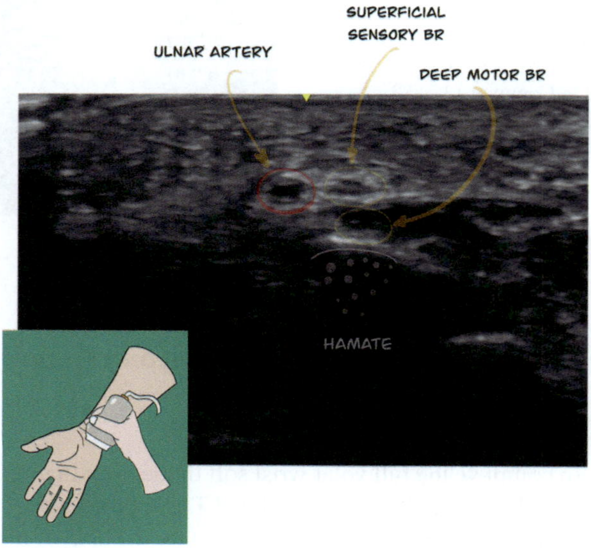

Fig. 2.13 Protocol Image 11: Distal Guyon's canal, transverse

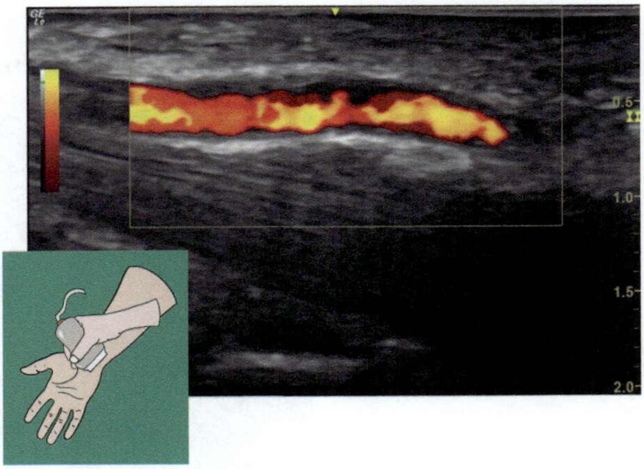

Fig. 2.14 Protocol Image 12: Ulnar artery at Guyon's canal, longitudinal + power Doppler

Protocol Image 12: Ulnar Artery at Guyon's Canal, Longitudinal + Power Doppler (Fig. 2.14)

Center the probe on the UA in TAX. Turn the transducer 90° to visualize the UA in LAX. Alternatively, visualize what you believe is the UN in LAX, turn on the PD, and then perform a very slight TAX slide in the radial direction to visualize part of the pulsating UA.

Protocol Image 13: Ulnar Nerve at Guyon's Canal, Longitudinal (Fig. 2.15)

Once you visualize the UA, turn off the PD and move the probe slightly in an ulnar direction to capture and verify the identity of the UN in LAX. If normal, the nerve should exhibit "train or tram tracks" echotexture. You may be able to discern the more distal splitting of the UN into its superficial and deep branches.

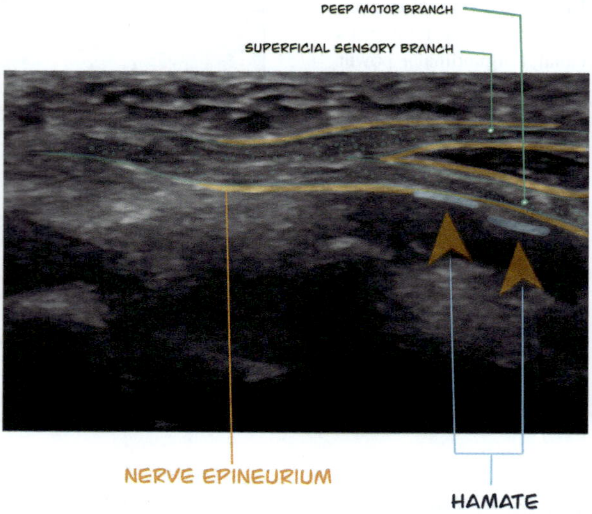

Fig. 2.15 Protocol Image 13: Ulnar nerve at Guyon's canal, longitudinal

Complete Volar Wrist Ultrasonic Examination Checklist

☐ Protocol Image 1: Volar wrist, ulnar aspect, transverse
☐ Protocol Image 2: Volar wrist, mid-wrist, transverse
☐ Protocol Image 3: Forearm, 12 cm proximal to wrist crease, transverse
☐ Protocol Image 4: Median nerve, 12 cm proximal to wrist crease, transverse, cross-sectional area measurement
☐ Protocol Image 5: Proximal carpal tunnel, transverse
☐ Protocol Image 6: Median nerve, proximal carpal tunnel, transverse, cross-sectional area measurement
☐ Protocol Image 7: Proximal carpal tunnel, transverse + power Doppler
☐ Protocol Image 8: Median nerve, longitudinal ± power Doppler
☐ Protocol Image 9: Proximal Guyon's canal, transverse + power Doppler
☐ Protocol Image 10: Proximal Guyon's canal, transverse
☐ Protocol Image 11: Distal Guyon's canal, transverse
☐ Protocol Image 12: Ulnar artery at Guyon's canal, longitudinal + power Doppler
☐ Protocol Image 13: Ulnar nerve at Guyon's canal, longitudinal

References

1. Padua L, Coraci D, Erra C, Pazzaglia C, Paolasso I, Loreti C, et al. Carpal tunnel syndrome: clinical features, diagnosis, and management. Lancet Neurol. 2016;15(12):1273–84.
2. Greenberg MH, Greer J, Fant JW. Using ultrasound to diagnose carpal tunnel syndrome. The Rheumatologist. 2018.
3. Kele H. Ultrasonography of the peripheral nervous system. Pers Med. 2012;1(1):417–21.
4. Werner RA, Andary M. Electrodiagnostic evaluation of carpal tunnel syndrome. Muscle Nerve. 2011;44(4):597–607.
5. Atroshi I, Gummesson C, Johnsson R, Ornstein E. Diagnostic properties of nerve conduction tests in population-based carpal tunnel syndrome. BMC Musculoskelet Disord. 2003;4:9.

6. Koyuncuoglu HR, Kutluhan S, Yesildag A, Oyar O, Guler K, Ozden A. The value of ultrasonographic measurement in carpal tunnel syndrome in patients with negative electrodiagnostic tests. Eur J Radiol. 2005;56(3):365–9.
7. Chiavaras MM, Jacobson JA, Yablon CM, Brigido MK, Girish G. Pitfalls in wrist and hand ultrasound. AJR Am J Roentgenol. 2014;203(3):531–40.
8. Hobson-Webb LD, Massey JM, Juel VC, Sanders DB. The ultrasonographic wrist-to-forearm median nerve area ratio in carpal tunnel syndrome. Clin Neurophysiol. 2008;119(6):1353–7.
9. Klauser AS, Halpern EJ, De Zordo T, Feuchtner GM, Arora R, Gruber J, et al. Carpal tunnel syndrome assessment with US: value of additional cross-sectional area measurements of the median nerve in patients versus healthy volunteers. Radiology. 2009;250(1):171–7.
10. Lee D, van Holsbeeck MT, Janevski PK, Ganos DL, Ditmars DM, Darian VB. Diagnosis of carpal tunnel syndrome. Ultrasound versus electromyography. Radiol Clin N Am. 1999;37(4):859–72, x.
11. Bianchi S, Martinoli C. Wrist. In: Ultrasound of the musculoskeletal system. Springer; 2007. p. 425–92.
12. Martinoli C, Serafini G, Bianchi S, Bertolotto M, Gandolfo N, Derchi LE. Ultrasonography of peripheral nerves. J Peripher Nerv Syst. 1996;1(3):169–78.
13. Pacek CA, Chakan M, Goitz RJ, Kaufmann RA, Li ZM. Morphological analysis of the transverse carpal ligament. Hand (N Y). 2010;5(2):135–40.
14. Marquardt T, Gabra J, Evans P, Seitz W, Li Z-M. Thickness and stiffness adaptations of the transverse carpal ligament associated with carpal tunnel syndrome. J Musculoskelet Res. 2017;19:1650019.
15. Bini N, Leclercq C. Anatomical study of the deep branch of the ulnar nerve and application to selective neurectomy in the treatment of spasticity of the first web space. Surg Radiol Anat. 2019;42:253.
16. Evers S, Bryan AJ, Sanders TL, Selles RW, Gelfman R, Amadio PC. Effectiveness of ultrasound-guided compared to blind steroid injections in the treatment of carpal tunnel syndrome. Arthritis Care Res (Hoboken). 2017;69(7):1060–5.
17. Greenberg MH, Fant JW Jr. Comparison of the effectiveness of ultrasound-guided steroid injections and blind steroid injections in the treatment of carpal tunnel syndrome: comment on the article by Evers et al. Arthritis Care Res (Hoboken). 2018;70(11):1717–8.
18. Riegler G, Pivec C, Platzgummer H, Lieba-Samal D, Brugger P, Jengojan S, et al. High-resolution ultrasound visualization of the recurrent motor branch of the median nerve: normal and first pathological findings. Eur Radiol. 2017;27(7):2941–9.
19. Coraci D, Loreti C, Piccinini G, Doneddu PE, Biscotti S, Padua L. Ulnar neuropathy at wrist: entrapment at a very "congested" site. Neurol Sci. 2018;39(8):1325–31.
20. Ceponis A, Kissin E. Ultrasound of the hand and wrist. In: Kohler MJ, editor. Musculoskeletal ultrasound in rheumatology review. 2nd ed. Springer; 2021. p. 53–82.
21. Kwak K-W, Kim M-S, Chang C-H, Kim S-H. Ulnar nerve compression in Guyon's canal by ganglion cyst. J Korean Neurosurg Soc. 2011;49(2):139–41.
22. Lee CH, Kim TK, Yoon ES, Dhong ES. Postoperative morphologic analysis of carpal tunnel syndrome using high-resolution ultrasonography. Ann Plast Surg. 2005;54(2):143–6.
23. Naranjo A, Ojeda S, Rua-Figueroa I, Garcia-Duque O, Fernandez-Palacios J, Carmona L. Limited value of ultrasound assessment in patients with poor outcome after carpal tunnel release surgery. Scand J Rheumatol. 2010;39(5):409–12.
24. Peetrons PA, Derbali W. Carpal tunnel syndrome. Semin Musculoskelet Radiol. 2013;17(1):28–33.
25. Murabit A, Gnarra M, Mohamed A. Reversed palmaris longus muscle: anatomical variant—case report and literature review. Can J Plast Surg. 2013;21(1):55–6.
26. Smith J, Barnes DE, Barnes KJ, Strakowski JA, Lachman N, Kakar S, et al. Sonographic visualization of thenar motor branch of the median nerve: a cadaveric validation study. PM R. 2017;9(2):159–69.
27. Jacobson JA. Wrist and hand ultrasound. In: Fundamentals of musculoskeletal ultrasound. 3rd ed. Elsevier; 2018.

Chapter 3
Dorsal Wrist (Radial, Dorsal, Ulnar)

Reasons to Do the Study

1. Wrist pain or stiffness
2. Possible de Quervain's tenosynovitis
3. First extensor compartment tenosynovitis, tendinosis, tears
4. Proximal compartment syndrome
5. Evaluation of soft tissue swelling or masses
6. Pain or weakness after trauma
7. Loss of finger extension capability
8. Paresthesia on the hand dorsum

Questions We Want Answered

1. What is the cause of wrist pain?
2. Is tenosynovitis present, and if so, what is the cause?
3. Is there an underlying inflammatory condition affecting the hand, and is it a systemic process such as spondyloarthropathy, rheumatoid arthritis, or crystal disease?
4. If there is an inflammatory condition, is it eroding cartilage or bone?
5. Is there a structural, functional, or repetitive injury problem causing the hand pain?
6. What is causing soft tissue swelling?
7. What is this lump on the hand dorsum?
8. Are the tendons, ligaments, and cartilage surfaces normal in appearance?
9. Is there evidence for tendon compartment intersection syndromes?
10. Is there evidence for a triangular fibrocartilage complex tear?
11. Is there any evidence for calcium deposition in the triangular fibrocartilage that might indicate calcium pyrophosphate deposition disease?

The contents do not represent the views of the U.S. Department of Veterans Affairs or the United States Government.

M. H. Greenberg et al., *Manual of Musculoskeletal Ultrasound*,
https://doi.org/10.1007/978-3-031-37416-6_3

12. Is there evidence for a ganglion? If so, is the origin a joint or tendon sheath?
13. Is there evidence for radial nerve compression?

Necessary Basic Anatomy

Carpal Bones (Fig. 3.1a, b)

1. **Proximal carpal bones (radial to ulnar)**
    (a) Scaphoid: Boat-like
    (b) Lunate: Crescent-shaped, like the moon
    (c) Triquetrum: Triangular
    (d) Pisiform: Pea-shaped
2. **Distal carpal bones (ulnar to radial)**
    (a) Hamate: Bone with a hook (palmar directed)
    (b) Capitate: Head
    (c) Trapezoid: Like the trapezium but smaller in size
    (d) Trapezium: Like a table

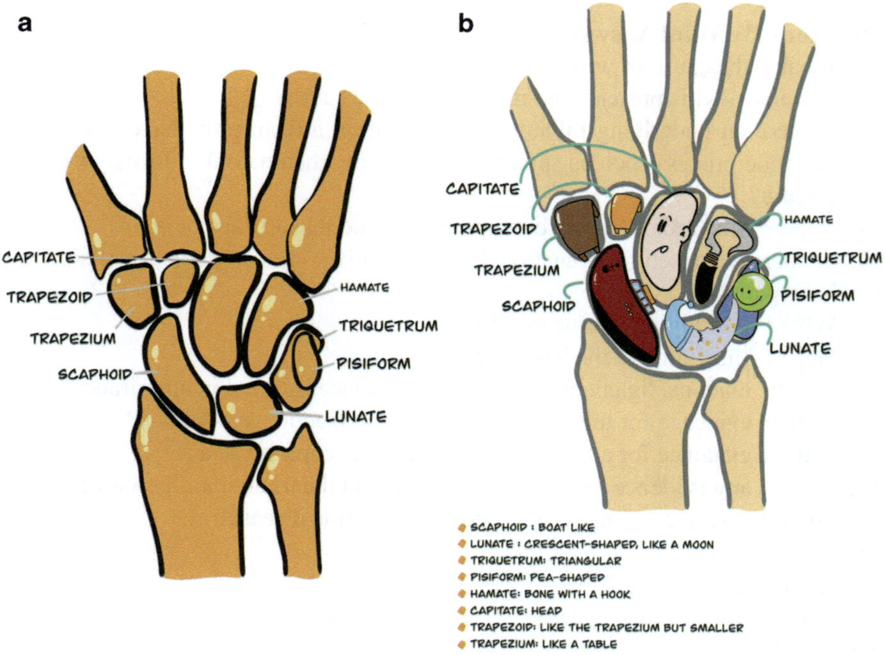

Fig. 3.1 (**a**) Basic anatomy. (**b**) Basic anatomy cartoons

A mnemonic for the carpal bones, starting at the proximal row on the radial side and moving in a circular direction:

Straight **L**ine **T**o **P**inky, **H**alf **C**ircle **T**o **T**humb

Joints of the Wrist (Fig. 3.2)

1. **Radiocarpal joint (RCJ)**: This is the junction of the distal radius and two of the carpal bones, the scaphoid and lunate. The RCJ is best seen in longitudinal (LAX) on ultrasound (US).
2. **Intercarpal (midcarpal) joint**: Formed between the proximal and distal rows of carpal bones, it is also best seen in LAX on US.
3. **Distal radioulnar joint (DRUJ)**: Located between the two distal bones of the forearm, its best view on US is in transverse (TAX).

Lister's Tubercle (Fig. 3.2)

Lister's tubercle (LT) is a palpable bony prominence on the dorsum of the distal radius [1]. It is well-delineated on US and an invaluable sonographic landmark between the second and third extensor compartments. Along the ulnar aspect of LT runs the third extensor compartment (EC), which contains the extensor pollicis

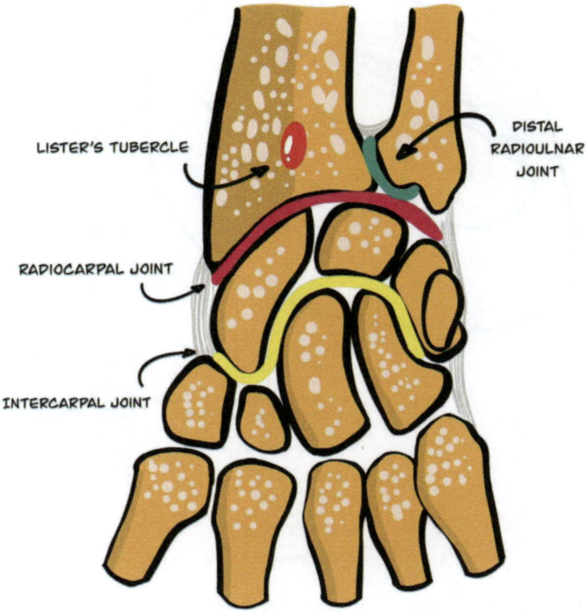

Fig. 3.2 Wrist joints and Lister's tubercle

longus (EPL). Lister's tubercle acts as a pulley for the EPL since this tendon angles back toward the first digit [2].

Extensor Compartments (Fig. 3.3a)

The primary soft tissue structures of the dorsal wrist are the various extensor and abductor tendons of the six wrist compartments. The dorsal (extensor) wrist tendons are separated into six fibro-osseous compartments (radial to ulnar) [3]:

1. Abductor pollicis longus (APL) and extensor pollicis brevis (EPB)
2. Extensor carpi radialis longus and brevis (ECRL and ECRB)
3. Extensor pollicis longus (EPL)
4. Extensor digitorum (ED) and extensor indicis (EI)
5. Extensor digiti minimi (EDM)
6. Extensor carpi ulnaris (ECU)

A story about an apple helps us remember the hand extensor compartments and tendons (Fig. 3.3b).

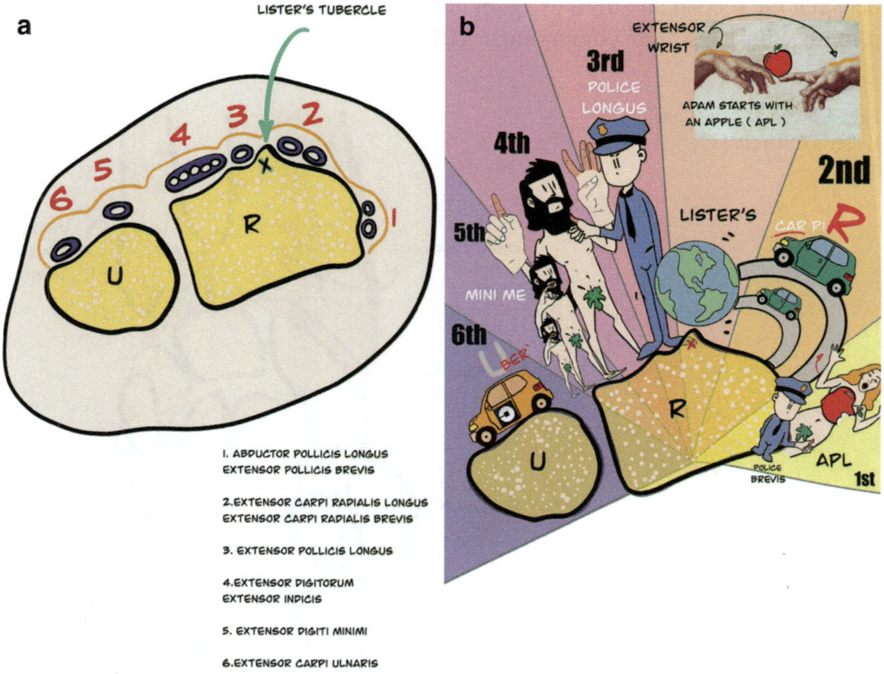

Fig. 3.3 (a) Extensor compartments (b) Extensor compartment cartoons

1. **Extensor compartment #1**: Eve, having taken a bite of the APL (abductor pollicis longus), is confronted by a **short policeman** (extensor pollicis brevis). He informs her that she and Adam must leave the Garden of Eden.
2. **Extensor compartment #2**: They drive their car to earth along a **long** and then a **brief** route (extensor carpi radialis longus and brevis).
3. **Extensor compartment #3**: Wearing no clothes, Adam and Eve are immediately arrested by a **tall policeman** (extensor pollicis longus).
4. **Extensor compartment #4**: They are given an **extens**ive sentence of **community** service for **indec**ent exposure (extensor digitorum communis and indicis).
5. **Extensor compartment #5**: Nine months later, Eve bears a son whom they like to call Adam's mini-me (extensor digiti minimi).
6. **Extensor compartment #6**: Adam becomes an **U**ber driver to support his growing family (extensor carpi ulnaris).

The first EC is the most radial. The APL is most ventral, that is, closer to the radial aspect of the palm. Logically, the **A**PL abducts the thumb, and **E**PB extends the thumb. You can manually move the thumb (or another finger) to verify the tendon's identity at any point in the US exam.

In the first EC tunnel, the floor is the bony groove of the radial styloid. Variations include a vertical septum splitting the first EC into separate sub-compartments, one containing the APL, the other the EPB. Another variation is that the APL may be a single tendon or consist of multiple tendon slips. We also find the mnemonic "The second EC is double-crossed" to be helpful. The second EC is crossed twice: initially by the first EC proximal to the wrist and again by the third EC as it moves obliquely to the thumb distal to the wrist. The ECU tendon is found in the ulnar groove and is technically not located over the wrist dorsum.

Soft Tissue Structures

1. **Extensor retinaculum (ER)**: Located superficially to the extensor tendons, it forms the roof of the extensor tendons.
2. **Scapholunate ligament (SLL)**: The SLL is located between the scaphoid and lunate carpal bones and is a triangular structure that is mechanically important as a wrist stabilizer [4]. Sonographically detectable tears may occur here. The SLL is best evaluated from the dorsal approach in TAX but can also be assessed from the volar aspect.
3. **Triangular fibrocartilage complex (TFCC)**: The triangular fibrocartilage runs from the ulnar side of the distal radius to the base of the ulnar styloid [5]. It separates the RCJ from the DRUJ. There are multiple components to the TFCC, but the chief sonographic components are the articular disc (AD) and the meniscal homologue (MH) (Fig. 3.4). The AD appears triangular on US, with the deep portion tapering in part due to attenuation from structures that lie superficially. The MH seems to be almost ovoid-shaped. It may be difficult to discern the normal AD and MH from the surrounding ligaments on US.

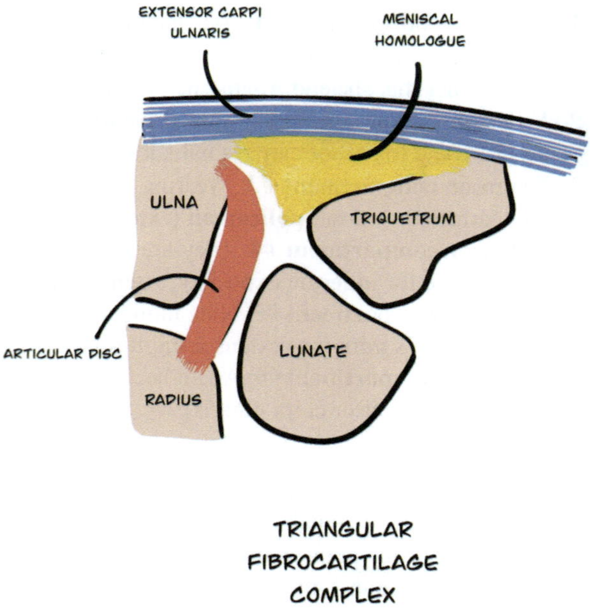

Fig. 3.4 Triangular fibrocartilage complex simplified schematic

Superficial Radial Nerve

Ultrasound enables the identification of the superficial radial nerve (SRN) at the level of the distal radius [6]. More proximally, at the level of the distal forearm, the SRN is close to the cephalic vein [6, 7]. A high-frequency probe of at least 17 MHz may be needed to follow the small branches of the SRN.

Clinical Comments

Always start with hand or wrist radiographs. Include clenched fist views if you suspect an SLL tear. As always, radiographs give the lay of the land before the US exam and will provide essential clues such as joint space narrowing, fracture, osteonecrosis, TFCC calcium deposits, and bone alignment abnormalities. Joint space narrowing implies cartilage erosion or thinning that the examiner should detect on US.

de Quervain's Tenosynovitis

de Quervain's stenosing tenosynovitis (dQT), in which the first EC tendon sheath and the covering ER become irritated and inflamed, often has a mechanical origin due to overuse [8]. Repetitively lifting a baby may produce dQT, termed baby wrist

[9]. Symptoms include pain at the distal radius radiating to the forearm or thumb, local swelling, sensitivity to pressure, and crepitus [10, 11]. Finkelstein's test is when the examiner places the thumb in the fist and flexes the wrist in an ulnar direction. This maneuver reproduces the symptoms of dQT and is the clinical examination of choice [12].

Ultrasound may reveal thickening of the tenosynovial sheath and overlying ER, tendinopathy, or fluid within the tendon sheath. The normal ER is barely visible, so thickening is a significant sign. Other sonographic findings include a hypoechoic peritendinous halo sign (due to edema) and, if in the acute phase, power Doppler (PD) activity in the affected soft tissues [13].

Due to tendinopathy, the affected tendon(s) may have changed shape from oval to more circular in the TAX view [8]. Dynamic US in LAX view may reveal tendon obstruction under the thickened ER impeding smooth tendon gliding. If a variant vertical septum exists between the two tendons, then only one tendon may be involved with dQT, more commonly the EPB.

The first EC may also develop tenosynovitis associated with inflammatory arthropathies such as rheumatoid arthritis, psoriatic arthritis, other spondyloarthropathies, sarcoidosis, or chronic crystal arthropathy. Tenosynovitis may present as an effusion within the synovial sheath or a thickened tendon sheath. Probe pressure may cause fluid displacement, helping to differentiate an effusion from tenosynovial thickening. If a septum exists within the compartment, then corticosteroids may need to be injected into each separate tendon sheath to treat the condition effectively. Ultrasound-guided needle injection helps target the affected tendon sheath accurately [14].

Intersection Syndromes

Remember that the first EC crosses the second EC close to the wrist. This can cause the proximal intersection syndrome (PIS), which is different from the distal intersection syndrome (DIS), which happens when the third EC crosses the second EC farther from the wrist (Fig. 3.5). In each intersection syndrome, the compartments develop inflammation as they cross paths.

Proximal intersection syndrome: Proximal intersection syndrome results from overuse, producing pain and swelling at the distal forearm. Proximal intersection syndrome is also known as crossover syndrome, peritendinitis crepitans, or oarsman's wrist [8]. Pain occurs approximately 4–8 cm proximal to the wrist on the dorsal forearm, where the first and second extensor compartments cross [8, 11, 15]. Ultrasound findings at the site of PIS can include edema, loss of the hyperechoic plane dividing the two different compartments, tenosynovial effusion, tendinosis, hypervascularity on Doppler, and sometimes ganglion cysts [8, 11]. The cause and pathologic mechanisms of PIS are not well understood.

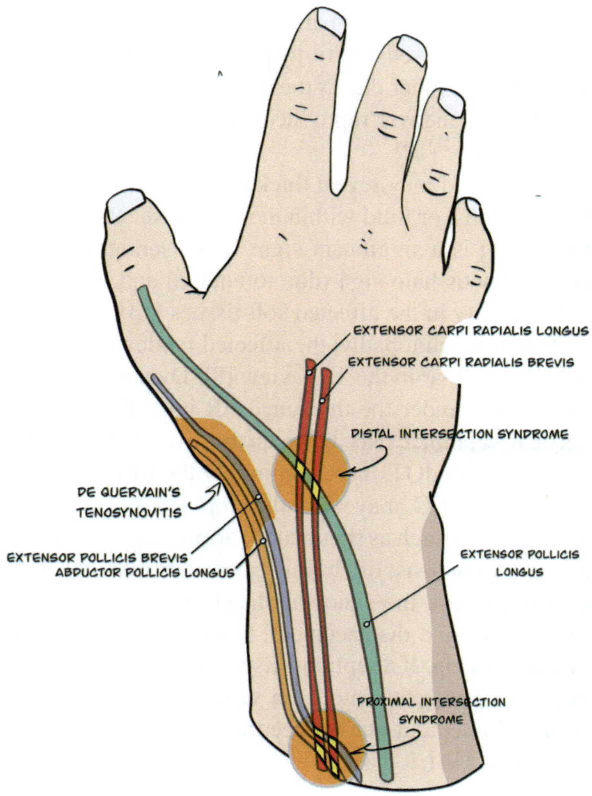

Fig. 3.5 Intersection syndrome locations

Distal intersection syndrome: This occurs just distal to LT where the third EC crosses over the second EC [8]. Repeated flexion and extension of the wrist produce irritation between the two tendons. Ultrasound reveals tenosynovial effusion of the EPL proximal and distal to the intersection. With EPL tenosynovitis, consider other causes of impingement, such as underlying osteophytes, scaphoid or distal radius fractures, or surgical hardware [8].

Fourth Compartment

Fourth compartment tenosynovitis may occur with inflammatory arthropathy and infectious disease [16]. Ultrasound reveals tenosynovial and retinacular thickening with PD activity [8]. Tenosynovitis may also arise with hardware impingement, such as after radial fracture repair due to an impinging screw tip. The hyperechoic screw tip with reverberation is often clearly seen. Such impingement may cause a complete tear of the tendon in the fourth EC; retraction may result in the nonvisualization of a tendon within its sheath. Another cause of tenosynovitis in the fourth EC occurs when the EI muscle has not yet formed into a tendon [17]. The anomalous muscle may cause compression within the fourth EC.

Fifth Compartment

Anatomic variation of the EDM may be associated with tenosynovitis. Further, DRUJ arthritis may also cause EDM tenosynovitis due to the contiguity of the two structures [8, 18].

Sixth Compartment

Although the ER covers the ECU tendon, an additional, deeper subsheath secures the ECU tendon within the groove of the ulnar head [16, 19]. Repetitive stress on the tendon may lead to stenosing tenosynovitis of the subsheath. In such cases, US reveals thickening of the ER, fibrosis, and reactive tenosynovial effusion. Attenuation or tearing of the subsheath may result in ECU tendon instability or even dislocation out of the ulnar groove [19]. Such ECU tendon dislocation differs from the normal partial subluxation of the ECU tendon with forearm supination. Tears of the subsheath occur with a recurrent injury in racquet sports, abrupt twisting, and severe DRUJ arthropathy. Tendonitis of the ECU may mimic TFCC injury. Tenosynovitis of the ECU may also be an early sign of rheumatoid arthritis and may predict erosion progression [20].

Scapholunate Ligament Tear

The SLL is an important wrist stabilizer [21]. Tears of the SLL may occur with sports-related injuries and, if unrecognized, may result in enduring wrist instability and arthritis termed scapholunate advanced collapse [22, 23]. An MRI or CT arthrogram is the gold standard for evaluating SLL tears [22]. The SLL is best seen on the wrist dorsum on US and is identifiable as a hyperechoic fibrillar structure [24]. However, the normal dorsal SLL may be invisible to US. In one US study, the normal dorsal SLL was completely visible in 48%, partly visible in 30%, and barely visible or invisible in 23% [25]. Sprains may reveal a hypoechoic and thickened ligament. With a complete tear, the SLL may not be visible, although, again, the lack of visualization of the SLL may be usual for some individuals.

Ulnar deviation of the wrist creates tension on the SLL and may demonstrate widening (dissociation) of the scapholunate interval in the presence of a complete SLL tear [26]. Some advocate a clenched fist dynamic maneuver as more reliable [27]. The normal distance between the scaphoid and the lunate bones is 2.9–4.5 mm [28]. Sonography may be a reasonable screen for SLL tears [29]. The sonographic finding of a normal, intact SLL essentially excludes scapholunate dissociation. If an SLL tear is clinically suspected, dynamic sonography may reveal the diagnosis and preclude the need for an MR arthrogram [29]. If you suspect a SLL tear, use US to compare with the contralateral wrist. Stress US or radiographs (patient grips an object firmly or makes a fist with maximal ulnar deviation) compared with neutral views may show a gap (>4 mm on radiographs) between the two bones with an SLL tear. This finding is known as the Terry Thomas sign, named after the comedian with the large gap between his teeth [30].

Again, sonographic evaluation of the SLL is still considered inferior to MR or CT evaluation [31]. Scapholunate ligament degeneration may occur with inflammatory arthropathy, particularly calcium pyrophosphate deposition disease (CPPD) and rheumatoid arthritis and is associated with intercarpal synovitis. Sonographic detection of hyperechoic calcium crystals in the SLL is supportive evidence for CPPD [32].

Triangular Fibrocartilage Complex Injury/Calcium Deposition

The TFCC, composed of cartilage and ligaments, stabilizes the ulnar aspect of the wrist during rotational movements. This wide variety of wrist movements separates humans from lower primates [33]. Causes of TFCC injuries include falling onto a hyperextended wrist, using a bat or racquet, or a power drill injury torquing the wrist. Damage to the TFCC presents with ulnar-sided wrist pain after an injury. While a high-resolution US may be a suitable screen for TFCC tears, magnetic resonance arthrogram imaging is the most accurate diagnostic tool [29]. Nonetheless, TFCC lesions may still be challenging to diagnose, even with MR imaging [34].

On US, the AD typically has an elongated-to-triangular shape and lies just distal to the ulnar head. It has a homogenous hypoechoic or slightly hyperechoic echotexture. A tear manifests as a hypoechoic to anechoic fissure in the structure. Patients with active rheumatoid arthritis in the wrist who demonstrate high PD scores are prone to TFCC tears [35]. The TFCC is vital to evaluate in rheumatology since there may be conspicuous calcium deposition in the AD and perhaps the MH. Such chondrocalcinosis is a solid clue to CPPD presence. In one study of patients with CPPD, there was substantial agreement in the finding of chondrocalcinosis of the TFCC comparing US and CT [32].

Ganglia

Ganglia are the most common masses of the wrist and hand [36]. These are cysts filled with thick fluid and lined with a fibrous capsule [37]. They may be round, soft, firm, painless, or tender on clinical examination. A ganglion has no synovial lining and hence differs from a synovial cyst, in which the synovial membrane herniates through a joint capsule. Many clinicians use the terms ganglion, ganglion cyst, and synovial cyst interchangeably. However, US cannot reliably differentiate between the two [38]. Ganglia may be caused by trauma, arthritis of any cause, tendon injury, or tenosynovitis.

Ganglia may connect to joints, tendon sheaths, ligaments, joint capsules, or bursae. Clinically undetectable (occult) ganglia are frequently detected by US [37]. Most ganglia occur at the wrist from either the dorsal capsule near the SLL or the radial aspect of the volar wrist between the radial artery and the flexor carpi radialis;

it is debatable which of these two locations is most common [39]. Other ganglia locations include the ulnar aspect of the wrist associated with the triangular fibrocartilage complex tear. Ganglia sonographically appear hypoechoic to anechoic, are well-defined, and may show septa [37]. Increased posterior through-transmission may be noted [40]. There may be Doppler activity, which may or may not correlate with symptoms [37]. The differential diagnosis for ganglia includes tenosynovial giant cell tumor, epidermoid cyst, lipoma, tenosynovitis, rheumatoid nodule, gouty tophus, tendon xanthoma, and synovial sarcoma [41]. If you detect a mass that may be a ganglion, then look for a stalk and attempt to follow it down to its point of origin. A stalk's presence helps confirm the diagnosis while determining the underlying structural issue responsible for the ganglia.

Tenosynovitis

Tenosynovitis may clinically appear as a mass if there is significant hypertrophy [37]. Ultrasound reveals a tendon surrounded by anechoic fluid or hypoechoic and hypertrophied synovium (see Chap. 5).

Superficial Radial Nerve

The superficial branch of the radial nerve (SRN), a purely sensory nerve, becomes superficial about 9 cm proximal to the styloid process of the radius [11, 42]. The nerve bifurcates to supply the first and second webspaces and the dorsolateral thumb. The practitioner should suspect SRN injury with a radius fracture or hardware, fracture fixation pins, penetrating trauma, handcuffs, compression from tight jewelry or a cast, and cephalic vein cannulation [11, 43, 44]. Additionally, the SRN may be compressed between the brachioradialis muscle and the ECRL tendons, known as Wartenberg's Syndrome [7, 45]. Damage to the SRN produces lateral wrist and thumb pain, numbness, and paresthesia similar to dQT symptoms [11]. Further, Wartenberg's Syndrome may also be associated with dQT. A positive Tinel's test over the affected nerve may help to clinically discriminate between these pathologic entities [44]. Ultrasound reveals a local thickening of the nerve, particularly when compared with the contralateral side [11]. Ultrasound will distinguish Wartenberg's syndrome from dQT or trapeziometacarpal joint arthritis [7].

Pitfalls

1. Before performing an US exam for dorsal wrist pain, always mark the location of the pain since this is useful in making the diagnosis.
2. When looking at the TFCC, do not mistake the MH for the AD, the latter being thinner and abutting the ulnar head.

3. The examiner should delineate a dorsal ganglion cyst from a distended RCJ recess [46].
4. If you detect a mass that may be a ganglion, look for a stalk and attempt to follow it down to its point of origin to confirm the diagnosis. You may also be able to determine the underlying structural issue responsible for the ganglia.
5. When examining the dorsal wrist in LAX, do not mistake the anisotropy of the more superficial ER for true tenosynovitis or tenosynovial fluid. Such misidentification of tissue for synovitis is called pseudotenosynovitis [46].
6. Another cause of pseudotenosynovitis in the LAX view of a tendon is the normal hypoechoic muscle tapering at the musculotendinous junction, mimicking a thickened tenosynovium [46]. Avoid this pitfall by verifying what appears to be tenosynovitis in two orthogonal planes.
7. Do not neglect to evaluate the tendon sheath of the ECU for inflammatory tenosynovitis since this is a high-yield area when considering rheumatoid arthritis [46–48].
8. Scapholunate ligament tears can be challenging to diagnose. With a complete tear, the SLL may not be visible, but remember that nonvisualization of the SLL may be expected in some individuals. Stress-view radiographs and US may be helpful.
9. Do not mistake the hypoechoic synovium superficial to the scaphoid and lunate bones for an intact SLL in the TAX view [49].
10. A hypoechoic septum may exist between the APL and the EPB in the first EC. Remember this if you contemplate injecting the first compartment, since both sub-compartments may need to be separately injected.
11. In the first EC, accessory APL or EPB tendons may confuse the situation and are essential to recognize since they may have separate tendon sheaths. If present, accessory tendons may predispose to dQT and complicate potential surgical decompression. One trick is to look at the surrounding area and manually abduct and extend the thumb to see if there are extra tendons present.
12. Locating the SRN may require a higher-frequency probe than 12 MHz, typically at least 17 MHz [6, 7].
13. Note that the SRN is ventral to the APL, so if you plan to inject the first EC or the fascial plane between the first EC and the second EC, avoid a ventral approach, which might injure the radial nerve and artery.
14. Tendonitis of the ECU may clinically mimic TFCC injury.

Method

The patient is seated opposite you, the wrist is neutral, and the forearm is halfway between supination and pronation. The sonographer lines the thumb dorsum up with the forearm. A rolled towel under the wrist may give added support. Ask the patient to momentarily abduct the thumb to identify the anatomic snuff box, defined by two extensor compartments at the base of the thumb on the radial aspect. In pen, draw one line on each of the two tendons of the snuff box—the one closest to the palm is the first EC, and the other one is the third EC.

Next, palpate the dorsal distal radius and mark off the slight bony ridge, LT. Use at least a 12 MHz linear probe for the examination, but be mindful that locating the SRN may require a higher-frequency probe, perhaps 17 MHz or higher [6, 7]. Use PD to find arteries and evaluate hyperemia.

Protocol Image 1: First Extensor Compartment at Scaphoid, Transverse + Power Doppler (Fig. 3.6)

Center the probe over the first EC in TAX and find the radial styloid. It is bony, superficial, and almost pointed. Perform a slight TAX slide distally until you see a drop-off of bone (the radioscaphoid joint) and, continuing distally, see the deeper bony scaphoid. Turn on the PD to see the large radial artery (RA) that should be central and deep to the APL and EPB. Center the probe over the RA. The first EC is superficial to the RA in this location. The APL is closer to the palm and may be divided into three slips. The cephalic vein, detected on Color Flow, is at the same level as the RA but is more dorsal.

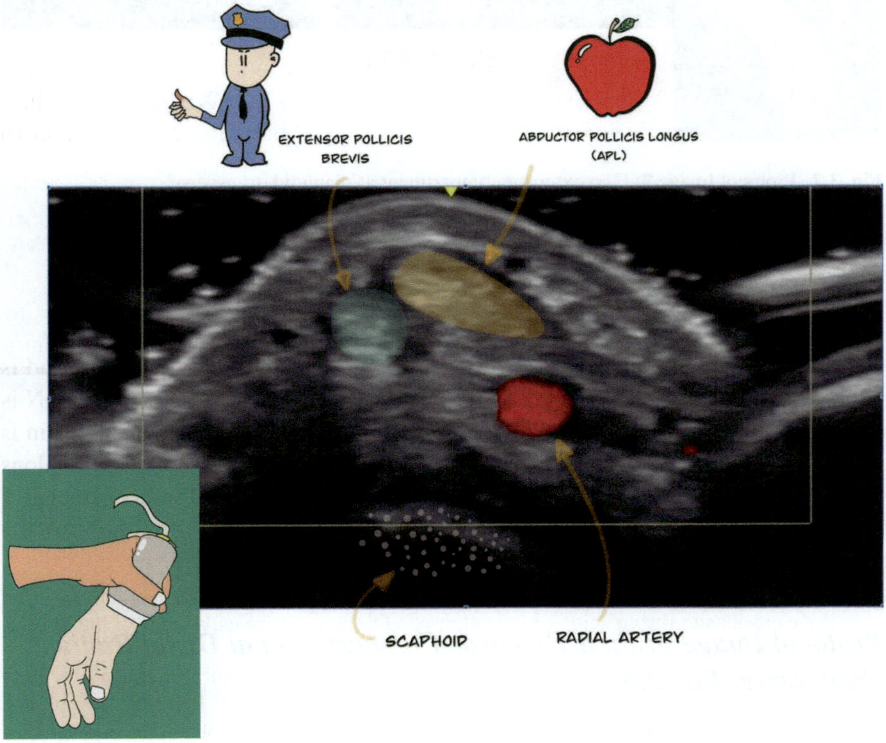

Fig. 3.6 Protocol Image 1: First extensor compartment at scaphoid, transverse + power Doppler

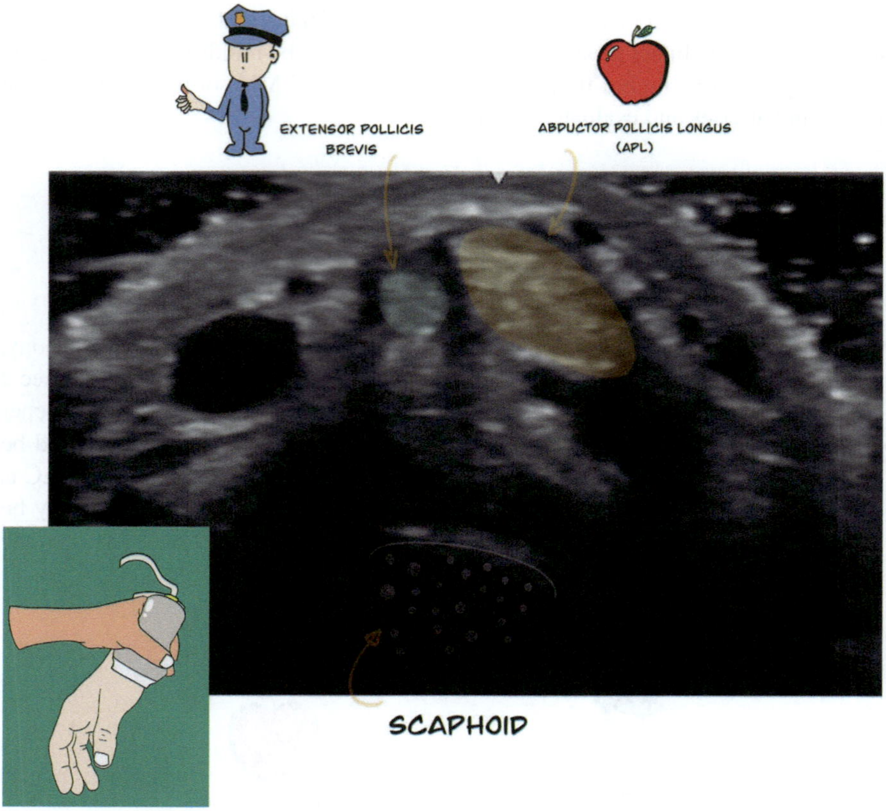

Fig. 3.7 Protocol Image 2: First extensor compartment at scaphoid, transverse

Protocol Image 2: First Extensor Compartment at Scaphoid, Transverse (Fig. 3.7)

Turn PD off to focus on the APL. The APL is often divided into three slips and is ventral to the EPB. The RA is central to and deep to the APL, whereas the SRN is superficial to the APL. The nerve may not be well-visualized. The cephalic vein is more dorsal and closer to the EPB. The APL and EPB may have accessory tendons that predispose them to tenosynovitis of the first EC. Note that Protocol Images 1 and 2 are distal to the ER.

Protocol Image 3: First Extensor Compartment at Distal Radius, Transverse (Fig. 3.8)

Do a TAX slide proximally back to the bony distal radius to see the scaphoid bone disappear and the distal radius reappear. Stop here. The thin hypoechoic line superficial to the tendons of the first EC is the ER. The slender ER may be

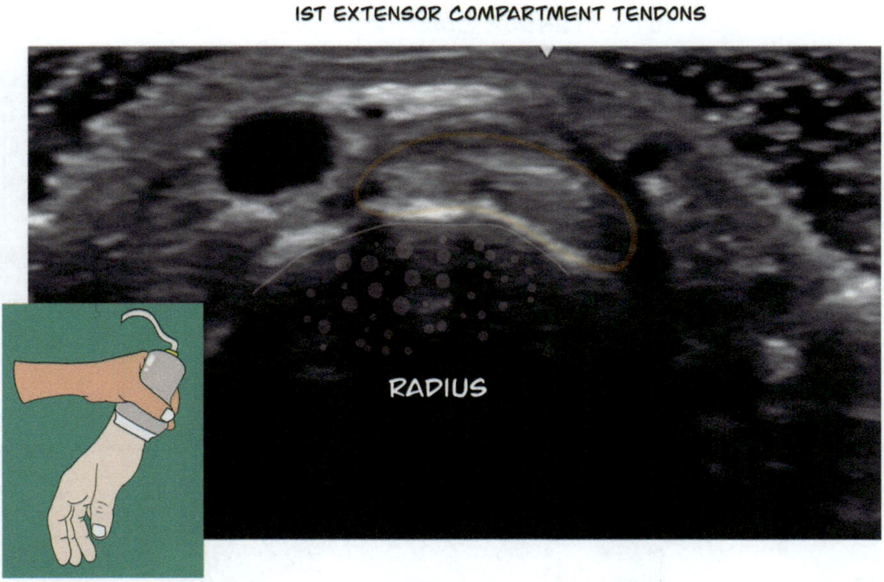

Fig. 3.8 Protocol Image 3: First extensor compartment at distal radius, transverse

challenging to visualize unless pathologically thickened, such as occurs with dQT.

The three slips of the APL have merged into a single tendon. Many people will have a hypoechoic septum between the APL and EPB. The septum may have a slight associated bone elevation on the radius, a clue to its presence. The APL is closer to the palm and is larger than the EPB. The two tendons are usually ovoid-shaped, but they may be more rounded if tendinosis is present.

de Quervain's tenosynovitis affects the tendon sheath and overlying ER. There may be anechoic fluid in the tendon sheath, tendinosis, and ER thickening. Remember, dQT is due to entrapment deep to the ER, often associated with overactivity of thumb movement. Make sure that you are aware of the location of both the RA and SRN before injecting. The SRN is next to the APL, closer to the palm (more ventral), and outside the tendon sheath. Damage to the SRN may cause pain that can mimic dQT.

Once you visualize the SRN in TAX, follow the nerve proximally or distally to look for an area of damage or compression. The RA is now in a ventral/ulnar position deep to the first EC. Accessory APL or EPB tendons may confuse the situation and are very important to recognize since they may have separate tendon sheaths. One trick is to look at the surrounding area and manually abduct and extend the thumb to see if there are extra tendons present.

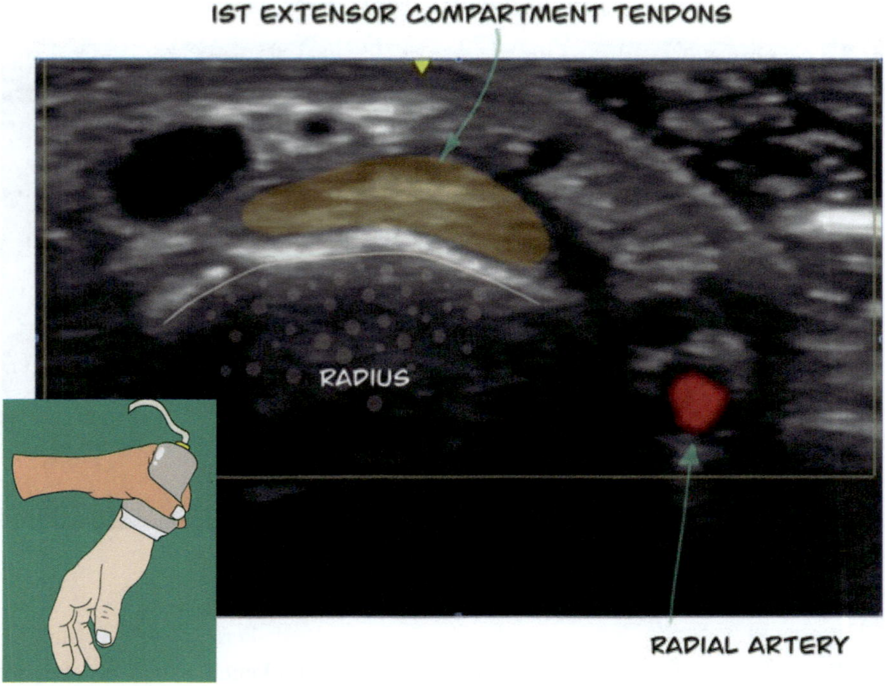

Fig. 3.9 Protocol Image 4: First extensor compartment at distal radius, transverse + power Doppler

Protocol Image 4: First Extensor Compartment at Distal Radius, Transverse + Power Doppler (Fig. 3.9)

Next, turn on the PD to look at the RA and assess for PD activity that indicates possible acute tenosynovitis.

Protocol Image 5: First and Second Extensor Compartments at Proximal Intersection, Transverse (Fig. 3.10)

Turn the PD off, then move the probe in TAX proximally, observing the APL as it moves in an ulnar direction over the deeper second EC tendons. The crossing of the first and second ECs, about 3–5 cm proximal to the wrist crease, is the potential location for PIS. This intersection is more proximal to the site of dQT. The contents of the first EC may look more muscular than tendon-like at this location since it is

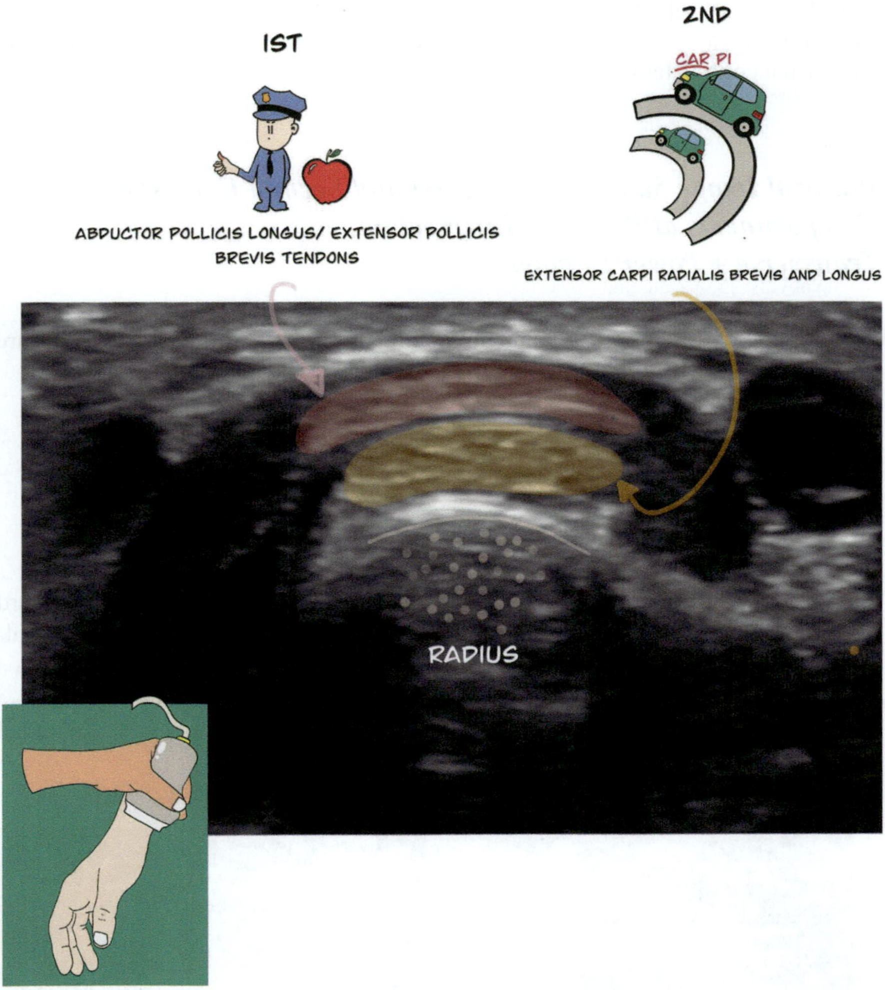

Fig. 3.10 Protocol Image 5: First and second extensor compartments at proximal intersection, transverse

near the myotendinous junction. The second EC contents are typically formed tendons at this location. If you have difficulty discriminating the first from the second EC, tilt the probe to use anisotropy to your advantage.

If there is pain over the intersection of the first and second ECs, look for a thickened fascia between the two compartments and possible tendinosis. Sonopalpatory pain may be precipitated or aggravated by actively resisting the patient's wrist extension. Always compare what you think might be an abnormal US finding with the normal contralateral limb.

Protocol Image 5a (Optional): First and Second Extensor Compartments at Proximal Intersection, Transverse + Power Doppler

Turn the PD on to look for inflammation and RA variants or branches. Make sure that you are aware of the location of the RA and the SRN before injecting.

Protocol Image 6: Second Extensor Compartment at Distal Radius, Transverse (Fig. 3.11)

Next, have the patient place the palm down and slightly flex the wrist downward over the edge of the exam table. Do a TAX slide distally to the wrist crease to visualize the first EC once again at the distal radius (**Protocol Image 3**). Then perform a slight LAX slide in the ulnar direction to visualize LT, which should appear as a sharp bony peak at the distal radius. The second EC is nestled next to the radial aspect of LT with the ECRB tendon abutting LT.

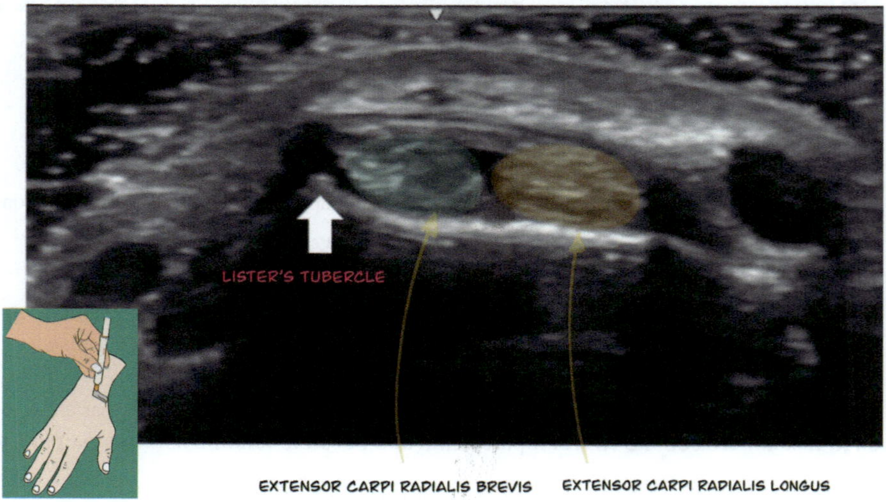

Fig. 3.11 Protocol Image 6: Second extensor compartment at distal radius, transverse

Protocol Image 6a (Optional): Second Extensor Compartment at Distal Radius, Transverse + Power Doppler

Turn on the PD to look for hyperemia of the tenosynovium of the second EC.

Protocol Image 7: Third Extensor Compartment at Distal Radius, Transverse (Fig. 3.12)

With the PD off, move the probe slightly in an ulnar direction and center it on the first oval structure on the ulnar side of LT. A single tendon, the EPL, inhabits the third EC. Verify that you are looking at the EPL by extending the first digit.

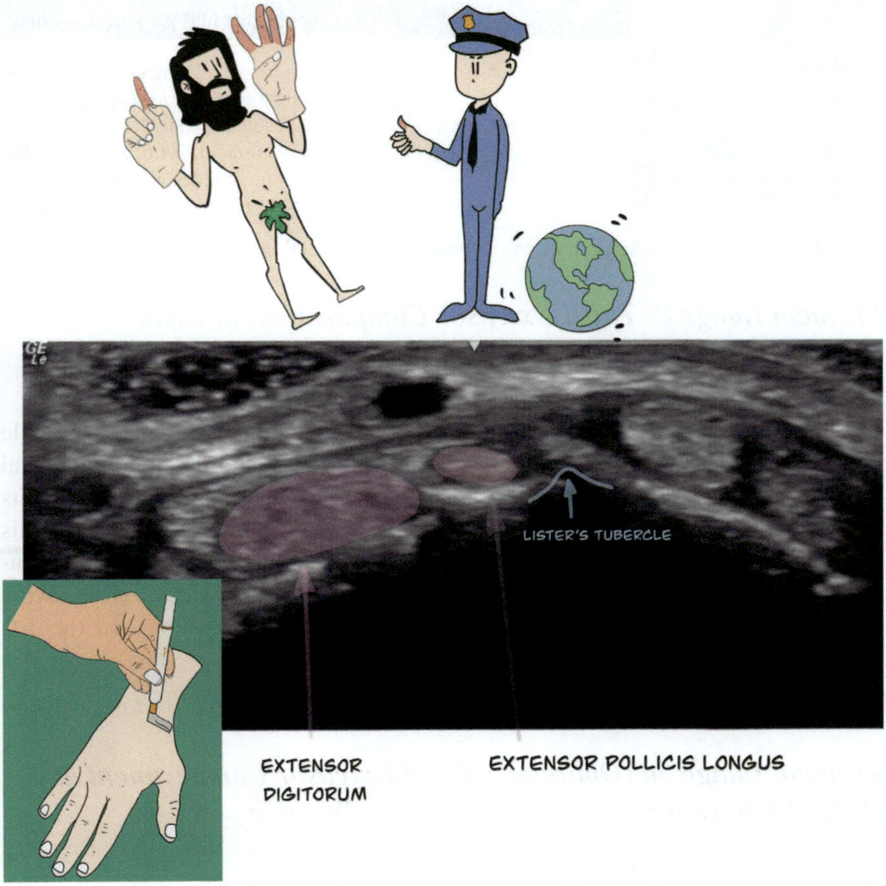

Fig. 3.12 Protocol Image 7: Third extensor compartment at distal radius, transverse

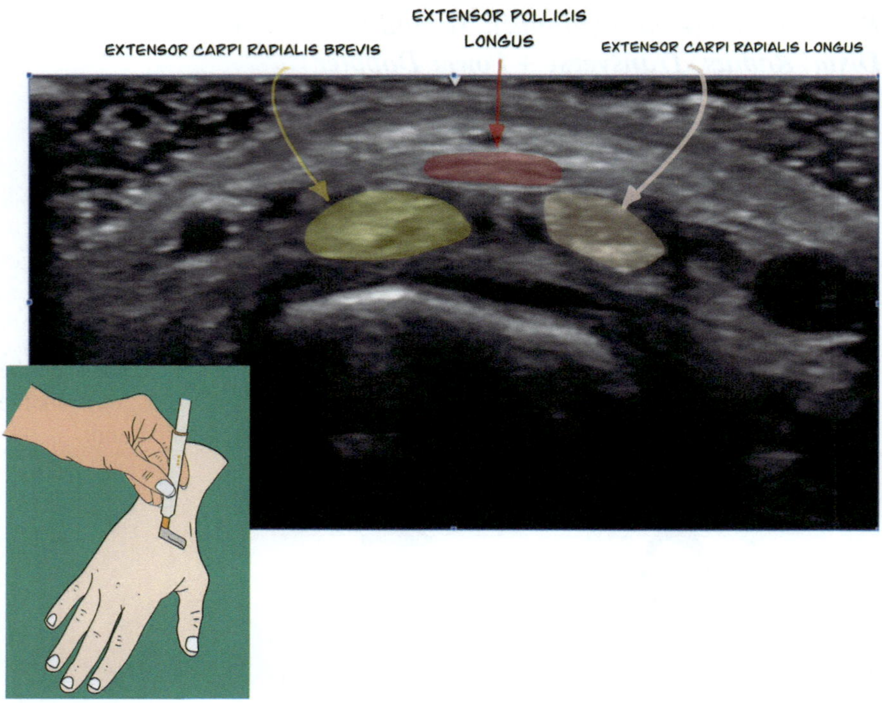

Fig. 3.13 Protocol Image 8: Third extensor compartment at distal intersection, transverse

Protocol Image 8: Third Extensor Compartment at Distal Intersection, Transverse (Fig. 3.13)

The thin EPL runs diagonally in a radial direction to the first digit; you may be able to follow this in TAX or LAX to its insertion on the dorsum of the first digit, distal phalanx. The third EC comprises the ulnar aspect of the anatomic snuff box. On its way to its insertion, the EPL is superficial to the second EC as it crosses over. This crossing may cause DIS. The EPL and ECRB share a joint opening with their tendon sheaths, and in DIS, both the second and third ECs will show tenosynovitis [8]. You may find it challenging to visualize the EPL in LAX beyond the distal radius.

Protocol Image 8a (Optional): Third Extensor Compartment at Distal Intersection, Transverse + Power Doppler

Turn on the PD to look for tenosynovitis if DIS is suspected.

Protocol Image 9: Scapholunate Joint, Transverse ± Stress View (Fig. 3.14)

Go back to the distal radius in TAX and center the probe over LT. Move the transducer distally, and the radius will drop out of view. Continue moving the TAX slide distally until you see two bones with an intervening space. The scaphoid is radial, and the lunate is on the ulnar aspect. You are now looking for the SLL that should bridge these two carpal bones. You can see the ligament in about 50% of healthy people. Therefore, you may normally see some or none of the SLL. If you visualize it completely, that is strong evidence that no significant tears are present. If you suspect an SLL tear, measure the distance between these two bones with and without stress (the patient makes a fist and flexes the wrist in an ulnar direction). Compare this distance with the contralateral side. Do not misidentify the hypoechoic synovium superficial to the scapholunate joint as the SLL.

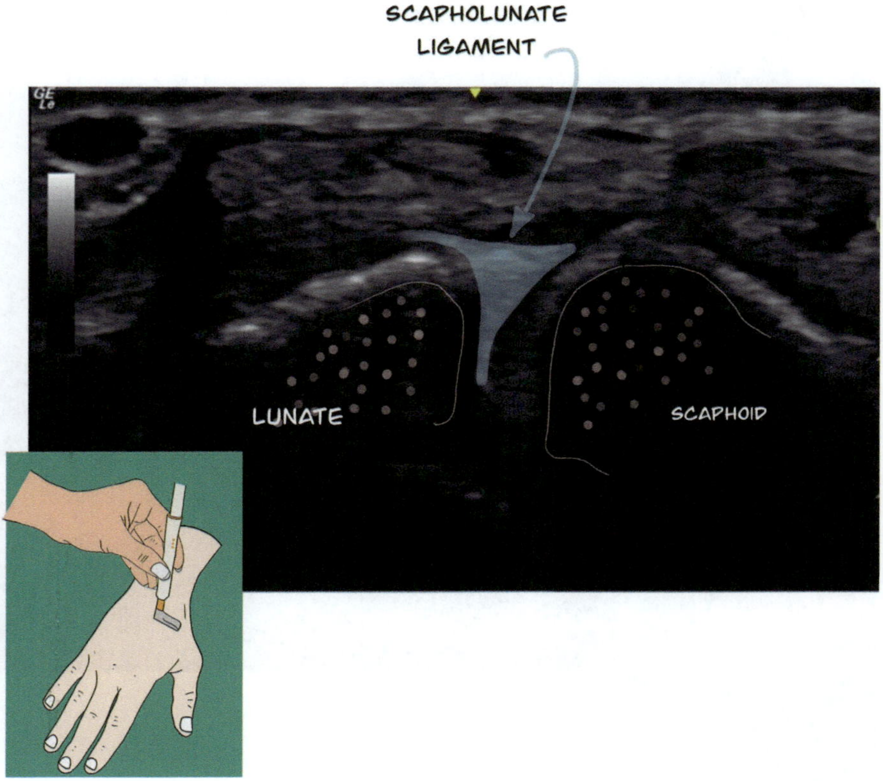

Fig. 3.14 Protocol Image 9: Scapholunate joint, transverse ± stress view

Protocol Image 10: Fourth and Fifth Extensor Compartments at Distal Radius, Transverse (Fig. 3.15)

Next, with the wrist back in a neutral position, perform a slight TAX slide proximally back to LT and then an LAX slide in the ulnar direction. You will see the third EC and the adjacent fourth and fifth ECs. You can verify the fifth EC identity since it is superficial to the DRUJ. The fourth EC contains ED and EI, and the fifth EC contains EDM. The second and fifth fingers each have a second extensor tendon, the EI and EDM, respectively. Move the digits manually to verify which tendons are in the fourth and fifth compartments. If necessary, add additional images for any tendon in LAX and TAX, with and without PD.

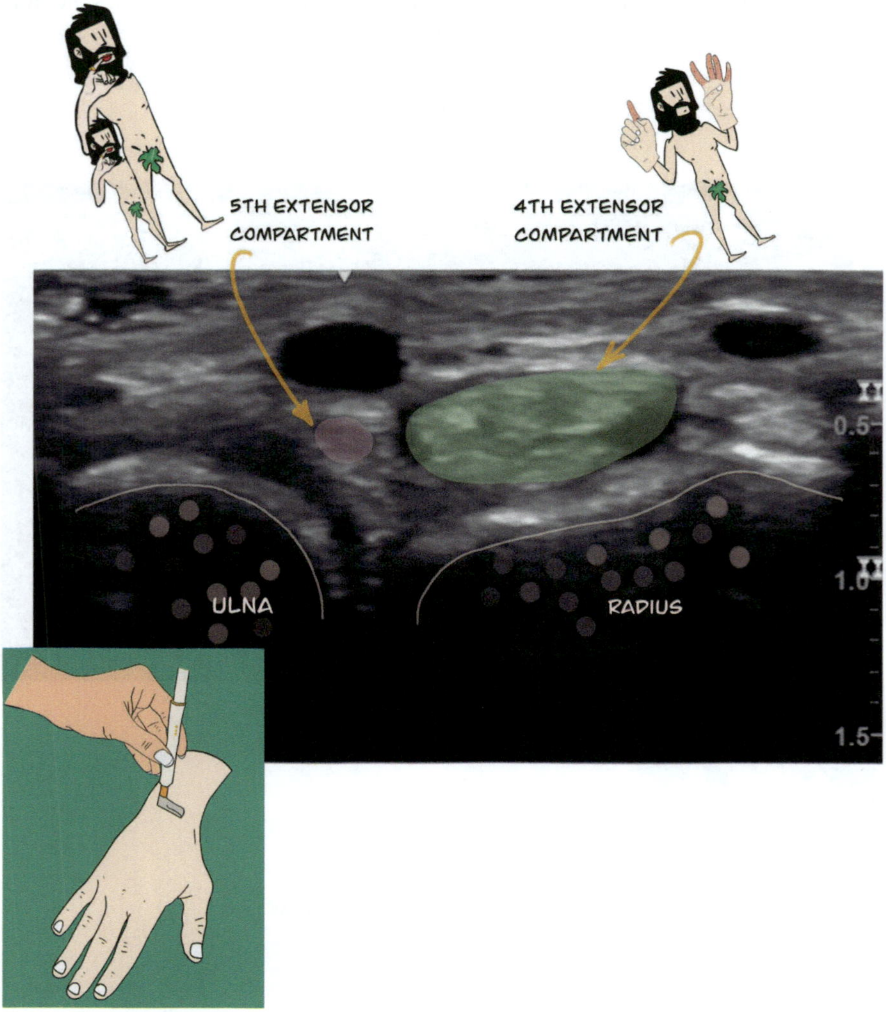

Fig. 3.15 Protocol Image 10: Fourth and fifth extensor compartments at distal radius, transverse

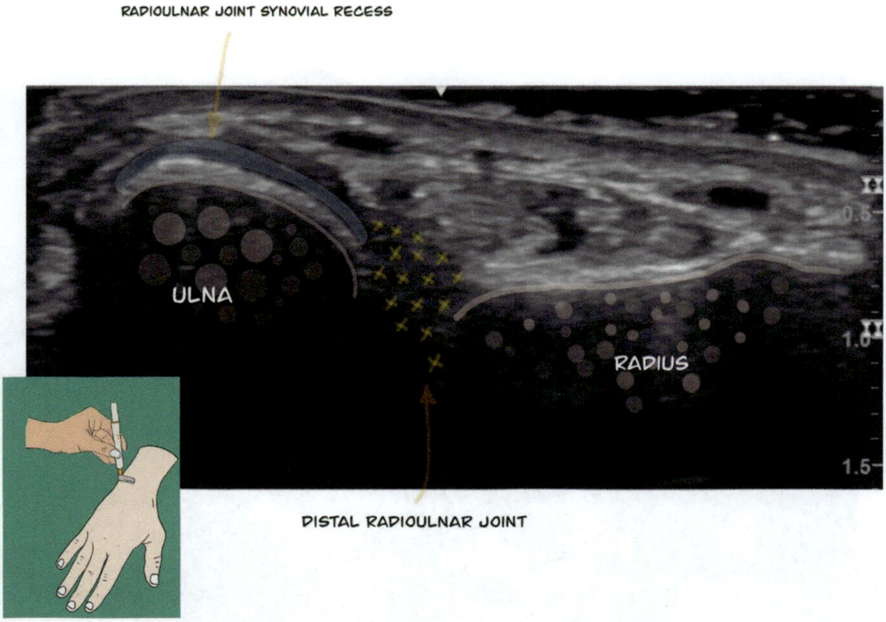

Fig. 3.16 Protocol Image 11: Distal radioulnar joint at distal radius, transverse ± power Doppler

Protocol Image 11: Distal Radioulnar Joint at Distal Radius, Transverse ± Power Doppler (Fig. 3.16)

Next, focus on the flat distal radius and a curved, semi-circular ulnar head. The fifth EC sits at this junction, superficial to the joint space of the DRUJ. Perform a TAX slide proximally and then distally to fully evaluate the DRUJ. Use PD if you are looking for synovitis.

Protocol Image 12: Sixth Extensor Compartment at Distal Ulna, Transverse + Dynamic Exam (Fig. 3.17)

Place a rolled towel beneath the wrist to elevate it, and place the wrist in a slight radial deviation. Perform a LAX slide in an ulnar direction to center the probe over the ulnar bone, then perform a TAX slide distally. You should see a curved semi-circular bony structure, the ulnar groove for the ovoid ECU. Observe the ER, which is superficial to the tendon. You may want to perform slight distal and proximal TAX slides as you rotate the transducer around the distal ulna bone to inspect for rheumatoid bone erosion [50]. Make sure to look at the tenosynovium surrounding the ECU for thickening since tenosynovitis of this tendon sheath may be an early sign of rheumatoid arthritis and may predict erosion progression [20].

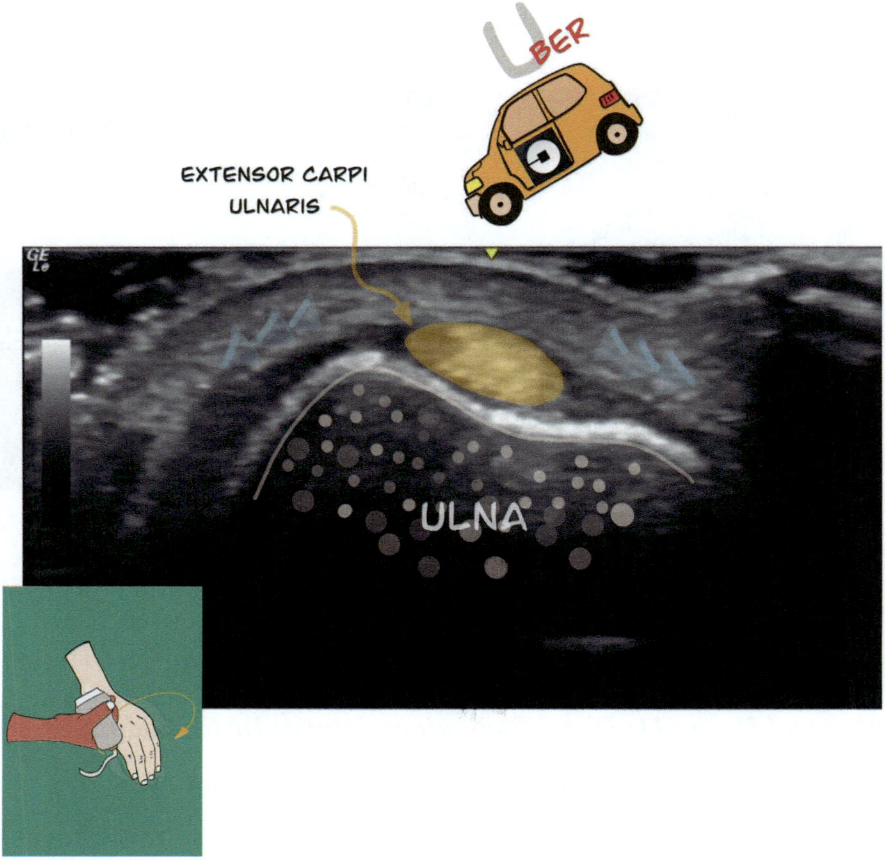

Fig. 3.17 Protocol Image 12: Sixth extensor compartment at distal ulna, transverse + dynamic exam

The ECU location typically changes with pronation and supination. To dynamically evaluate for ECU subluxation, the patient plants the elbow on the exam table, forearm in a vertical position, palm facing away. Place the probe in TAX across the distal ulna to visualize the ECU. Have the patient slowly rotate the palm inward to a neutral position and then toward the patient. Slowly move the transducer along with the rotating forearm by keeping the ECU in focus and observing its relation to the underlying distal ulna. The ECU should move to a dorsal location but may dislocate out of the groove if the subsheath deep to the ER is damaged. When doing dynamic testing, apply the probe lightly in TAX to avoid tethering the ECU. Remember that many ordinary people will have some degree of asymptomatic ECU subluxation.

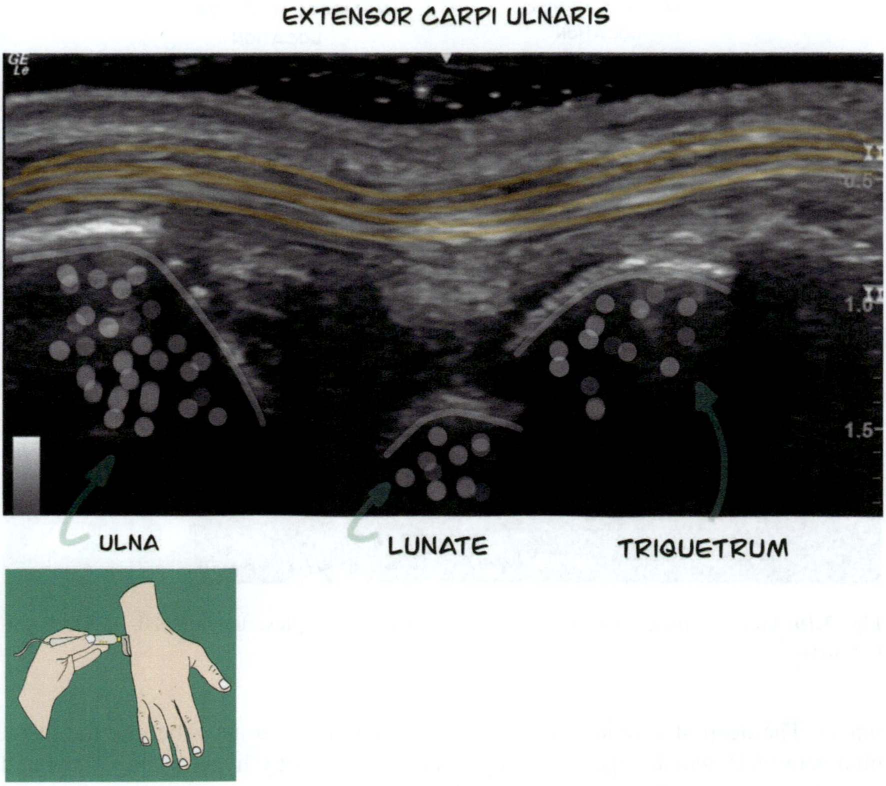

Fig. 3.18 Protocol Image 13: Sixth extensor compartment at distal ulna, longitudinal

Protocol Image 13: Sixth Extensor Compartment at Distal Ulna, Longitudinal (Fig. 3.18)

Go back to the initial position in Protocol Image 12, using a rolled towel beneath the wrist. Pronate the forearm, center the ECU in the image in TAX, and then rotate the probe 90° to look at the ECU in LAX. Then perform a TAX slide around the ulnar styloid (toward the palm) to look at the entire tendon. Visualize the hyperechoic lunate in a moderately deep location. This slide around the distal ulna may reveal an ulnar styloid fracture or bone erosion not apparent on radiographs.

Protocol Image 14: Triangular Fibrocartilage Complex, Longitudinal + Radial Deviation Wrist (Fig. 3.19)

With the probe still in LAX, have the patient radially flex the wrist to facilitate the evaluation of the soft tissues of the TFCC. The proximal bone is the distal ulna, and the triquetrum is the distal bone. The moderately deep, slightly curved bone is the

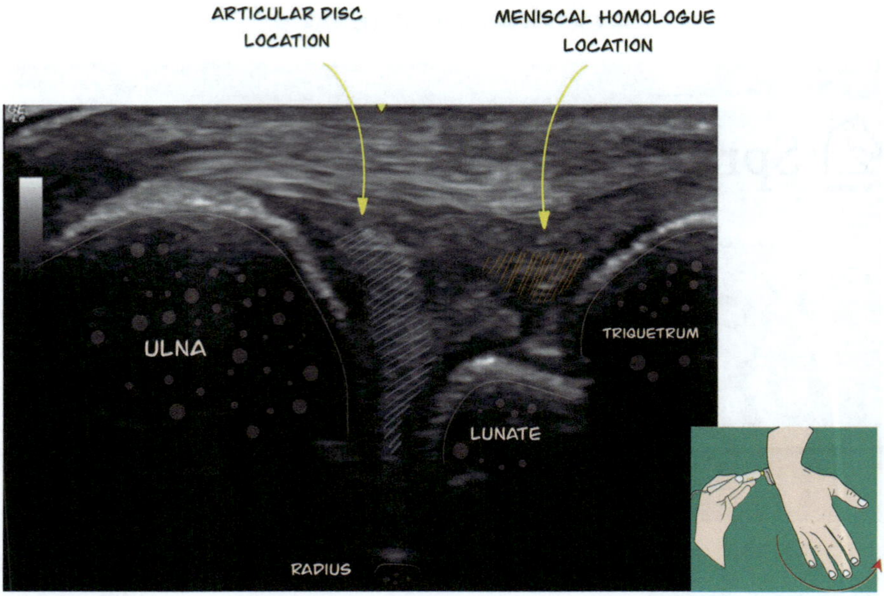

Fig. 3.19 Protocol Image 14: Triangular fibrocartilage complex, longitudinal, radial deviation wrist

lunate. The deepest visible bone, however, is the distal radius. Abutting the distal ulna is the AD, which appears as a hypoechoic (or slightly hyperechoic) elongated triangle in a vertical orientation toward the radius. Deep to the ECU tendon and between the AD and triquetrum is the location of the MH. A slight tilting of the probe helps to delineate these structures. Look for calcium deposits in the AD, a finding that may indicate CPPD. The AD and MH may be difficult to delineate in the absence of calcification.

Protocol Image 15: Radiocarpal Joint at Scaphoid and Trapezoid, Longitudinal ± Power Doppler (Fig. 3.20)

Next, remove the towel and place the wrist in a pronated, flat, and neutral position directly on the examination table. Go back to the TAX view of the dorsal scaphoid and lunate (**Protocol Image 9**). Move the probe in an LAX slide in a radial direction to center it over the scaphoid. Rotate the probe 90° to LAX to see the scaphoid and the more distal trapezoid in LAX. Turn on PD to look for joint recess synovitis if needed.

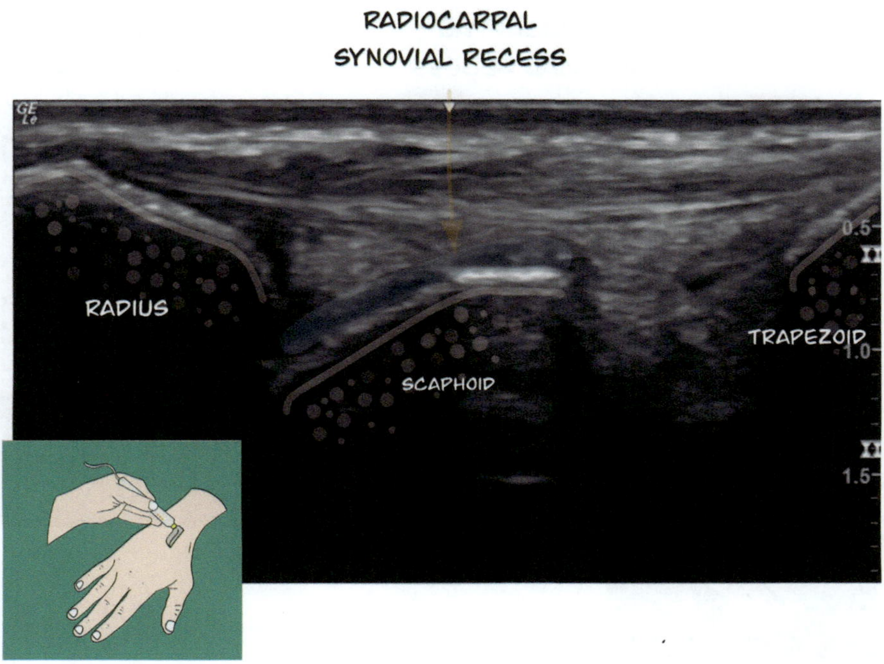

Fig. 3.20 Protocol Image 15: Radiocarpal joint at scaphoid and trapezoid, longitudinal ± power Doppler

Protocol Image 16: Radiocarpal and Intercarpal Joints, Longitudinal ± Power Doppler (Fig. 3.21)

Perform a slow TAX slide in the ulnar direction. The scaphoid disappears, and you then see the lunate and, more distally, the capitate bone. Examine the dorsal recesses and the extensor tendons, looking for crystal deposition and effusion in the recesses. Turn on the PD if necessary to delineate synovitis. Be wary of feeder blood vessels (located superficial to the joint recesses) mimicking PD activity (see Chap. 5 for further information).

Complete Dorsal Wrist Ultrasonic Examination Checklist

☐ Protocol Image 1: First extensor compartment at scaphoid, transverse + power Doppler.
☐ Protocol Image 2: First extensor compartment at scaphoid, transverse.
☐ Protocol Image 3: First extensor compartment at distal radius, transverse.
☐ Protocol Image 4: First extensor compartment at distal radius, transverse + power Doppler.
☐ Protocol Image 5: First and second extensor compartments at proximal intersection, transverse.

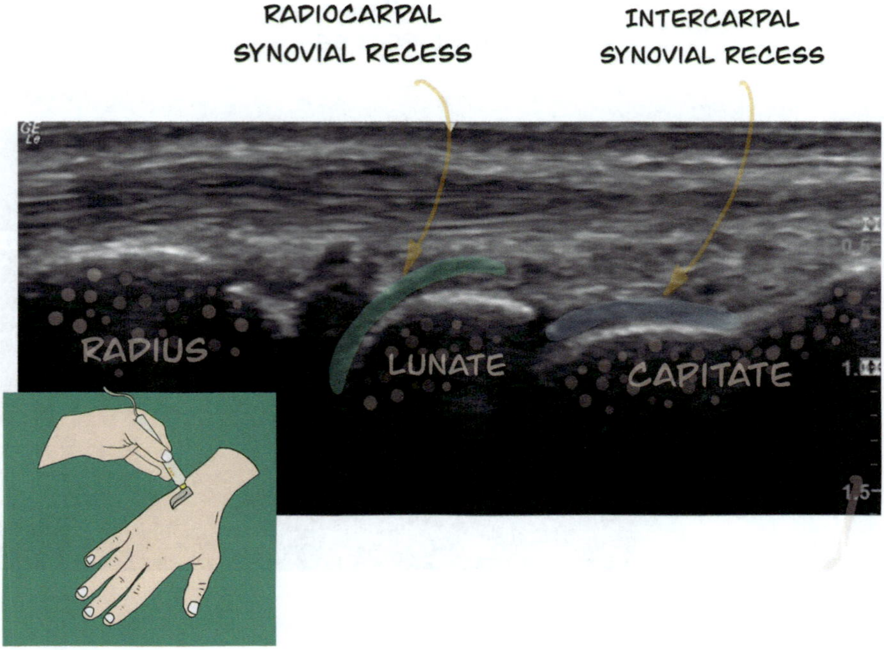

Fig. 3.21 Protocol Image 16: Radiocarpal and intercarpal joints, longitudinal ± power Doppler

- ☐ Protocol Image 5a (optional): First and second extensor compartments at proximal intersection, transverse + power Doppler.
- ☐ Protocol Image 6: Second extensor compartment at distal radius, transverse.
- ☐ Protocol Image 6a (optional): Second extensor compartment at distal radius, transverse + power Doppler.
- ☐ Protocol Image 7: Third extensor compartment at distal radius, transverse.
- ☐ Protocol Image 8: Third extensor compartment at distal intersection, transverse.
- ☐ Protocol Image 8a (optional): Third extensor compartment at distal intersection, transverse + power Doppler.
- ☐ Protocol Image 9: Scapholunate joint, transverse ± stress view.
- ☐ Protocol Image 10: Fourth and fifth extensor compartments at distal radius, transverse.
- ☐ Protocol Image 11: Distal radioulnar joint at distal radius, transverse ± power Doppler.
- ☐ Protocol Image 12: Sixth extensor compartment at distal ulna, transverse + dynamic exam.
- ☐ Protocol Image 13: Sixth extensor compartment at distal ulna, longitudinal.
- ☐ Protocol Image 14: Triangular fibrocartilage complex, longitudinal, radial deviation wrist.
- ☐ Protocol Image 15: Radiocarpal joint at scaphoid and trapezoid, longitudinal ± power Doppler.
- ☐ Protocol Image 16: Radiocarpal and intercarpal joints, longitudinal ± power Doppler.

References

1. Ağır I, Aytekin MN, Küçükdurmaz F, Gökhan S, Cavuş UY. Anatomical localization of lister's tubercle and its clinical and surgical importance. Open Orthop J. 2014;8:74–7.
2. Chan WY, Chong LR. Anatomical variants of lister's tubercle: a new morphological classification based on magnetic resonance imaging. Korean J Radiol. 2017;18(6):957–63.
3. Ramage JL, Varacallo M. Anatomy, shoulder and upper limb, wrist extensor muscles. In: StatPearls. Treasure Island, FL: StatPearls Publishing; 2022.
4. Sokolow C, Saffar P. Anatomy and histology of the scapholunate ligament. Hand Clin. 2001;17(1):77–81.
5. Skalski MR, White EA, Patel DB, Schein AJ, RiveraMelo H, Matcuk GR Jr. The traumatized TFCC: an illustrated review of the anatomy and injury patterns of the triangular fibrocartilage complex. Curr Probl Diagn Radiol. 2016;45(1):39–50.
6. Choi SJ, Ahn JH, Ryu DS, Kang CH, Jung SM, Park MS, et al. Ultrasonography for nerve compression syndromes of the upper extremity. Ultrasonography. 2015;34(4):275–91.
7. Tagliafico A, Cadoni A, Fisci E, Gennaro S, Molfetta L, Perez MM, et al. Nerves of the hand beyond the carpal tunnel. Semin Musculoskelet Radiol. 2012;16(2):129–36.
8. Rosskopf AB, Martinoli C, Sconfienza LM, Gitto S, Taljanovic MS, Picasso R, et al. Sonography of tendon pathology in the hand and wrist. J Ultrason. 2021;21(87):e306–17.
9. Goel R, Abzug JM. de Quervain's tenosynovitis: a review of the rehabilitative options. Hand (N Y). 2015;10(1):1–5.
10. de Quervain F. On a form of chronic tendovaginitis by Dr. Fritz de Quervain in la Chaux-de-Fonds. 1895. Am J Orthop (Belle Mead NJ). 1997;26(9):641–4.
11. De Maeseneer M, Marcelis S, Jager T, Girard C, Gest T, Jamadar D. Spectrum of normal and pathologic findings in the region of the first extensor compartment of the wrist: sonographic findings and correlations with dissections. J Ultrasound Med. 2009;28(6):779–86.
12. Wu F, Rajpura A, Sandher D. Finkelstein's test is superior to Eichhoff's test in the investigation of de Quervain's disease. J Hand Microsurg. 2018;10(2):116–8.
13. Diop AN, Ba-Diop S, Sane JC, Tomolet Alfidja A, Sy MH, Boyer L, et al. [Role of US in the management of de Quervain's tenosynovitis: review of 22 cases]. J Radiol. 2008;89(9 Pt 1):1081–4.
14. Kamel M, Moghazy K, Eid H, Mansour R. Ultrasonographic diagnosis of de Quervain's tenosynovitis. Ann Rheum Dis. 2002;61(11):1034–5.
15. Draghi F, Bortolotto C. Intersection syndrome: ultrasound imaging. Skelet Radiol. 2014;43(3):283–7.
16. Plotkin B, Sampath SC, Sampath SC, Motamedi K. MR imaging and US of the wrist tendons. Radiographics. 2016;36(6):1688–700.
17. Kim CH. Anomalous extensor indicis proprius muscle. Arch Plast Surg. 2013;40(1):79–81.
18. Yoo MJ, Chung KT, Kim JP, Kim MJ, Lee KJ. Tendon impingement of the extensor digiti minimi: clinical cases series and cadaveric study. Clin Anat. 2012;25(6):755–61.
19. Rosskopf AB, Taljanovic MS, Sconfienza LM, Gitto S, Martinoli C, Picasso R, et al. Pulley, flexor, and extensor tendon injuries of the hand. Semin Musculoskelet Radiol. 2021;25(2):203–15.
20. Lillegraven S, Bøyesen P, Hammer HB, Østergaard M, Uhlig T, Sesseng S, et al. Tenosynovitis of the extensor carpi ulnaris tendon predicts erosive progression in early rheumatoid arthritis. Ann Rheum Dis. 2011;70(11):2049–50.
21. Mohr A, Guermazi A, Genant HK. Value of sonography of the scapholunate ligament. Am J Roentgenol. 2003;181(1):275–7.
22. Bianchi S, Almusa E, Chick G, Bianchi E. Ultrasound in wrist and hand sport injuries. Aspetar Sports Med J. 2013;3:274.
23. Shah CM, Stern PJ. Scapholunate advanced collapse (SLAC) and scaphoid nonunion advanced collapse (SNAC) wrist arthritis. Curr Rev Musculoskelet Med. 2013;6(1):9–17.

24. Jacobson JA, Oh E, Propeck T, Jebson PJL, Jamadar DA, Hayes CW. Sonography of the scapholunate ligament in four cadaveric wrists: correlation with MR arthrography and anatomy. Am J Roentgenol. 2002;179(2):523–7.
25. Griffith JF, Chan DP, Ho PC, Zhao L, Hung LK, Metreweli C. Sonography of the normal scapholunate ligament and scapholunate joint space. J Clin Ultrasound. 2001;29(4):223–9.
26. Rodriguez RM, Ernat JJ. Ultrasonography for dorsal-sided wrist pain in a combat environment: technique, pearls, and a case report of dynamic evaluation of the scapholunate ligament. Mil Med. 2020;185(1–2):e306–11.
27. Reckelhoff KE, Clark TB, Kettner NW. The sonographic squeeze test: assessing the reliability of the dorsal scapholunate ligament. J Med Ultrasound. 2013;21(3):138–42.
28. Kendi AT, Güdemez E. Sonographic evaluation of scapholunate ligament: value of tissue harmonic imaging. J Clin Ultrasound. 2006;34(3):109–12.
29. Taljanovic MS, Sheppard JE, Jones MD, Switlick DN, Hunter TB, Rogers LF. Sonography and sonoarthrography of the scapholunate and lunotriquetral ligaments and triangular fibrocartilage disk: initial experience and correlation with arthrography and magnetic resonance arthrography. J Ultrasound Med. 2008;27(2):179–91.
30. Frankel VH. The Terry-Thomas sign. Clin Orthop Relat Res. 1978;135:311–2.
31. Falsetti P, Conticini E, Baldi C, Bardelli M, Al Khayyat SG, D'Alessandro R, et al. Ultrasound evaluation of the scapholunate ligament and scapholunate joint space in patients with wrist complaints in a rheumatologic setting. J Ultrason. 2021;21(85):e105–e11.
32. Cipolletta E, Smerilli G, Mashadi Mirza R, Di Matteo A, Carotti M, Salaffi F, et al. Sonographic assessment of calcium pyrophosphate deposition disease at wrist. A focus on the dorsal scapholunate ligament. Joint Bone Spine. 2020;87(6):611–7.
33. Palmer AK, Werner FW. The triangular fibrocartilage complex of the wrist—anatomy and function. J Hand Surg Am. 1981;6(2):153–62.
34. Wu WT, Chang KV, Mezian K, Naňka O, Yang YC, Hsu YC, et al. Ulnar wrist pain revisited: ultrasound diagnosis and guided injection for triangular fibrocartilage complex injuries. J Clin Med. 2019;8(10):1540.
35. Chen JH Sr, Huang KC, Huang CC, Lai HM, Chou WY, Chen YC Sr. High power Doppler ultrasound score is associated with the risk of triangular fibrocartilage complex (TFCC) tears in severe rheumatoid arthritis. J Investig Med. 2019;67(2):327–30.
36. Bianchi S, Abdelwahab IF, Zwass A, Giacomello P. Ultrasonographic evaluation of wrist ganglia. Skelet Radiol. 1994;23(3):201–3.
37. Bianchi S, Della Santa D, Glauser T, Beaulieu JY, van Aaken J. Sonography of masses of the wrist and hand. AJR Am J Roentgenol. 2008;191(6):1767–75.
38. Teh J. Ultrasound of soft tissue masses of the hand. J Ultrason. 2012;12(51):381–401.
39. Zhang A, Falkowski AL, Jacobson JA, Kim SM, Koh SH, Gaetke-Udager K. Sonography of wrist ganglion cysts: which location is most common? J Ultrasound Med. 2019;38(8):2155–60.
40. Teefey SA, Dahiya N, Middleton WD, Gelberman RH, Boyer MI. Ganglia of the hand and wrist: a sonographic analysis. AJR Am J Roentgenol. 2008;191(3):716–20.
41. De Keyser F, Isaac Z, Curtis MR (2021) Ganglion cysts of the wrist and hand. UpToDate.
42. Ikiz ZA, Uçerler H. Anatomic characteristics and clinical importance of the superficial branch of the radial nerve. Surg Radiol Anat. 2004;26(6):453–8.
43. Braidwood AS. Superficial radial neuropathy. J Bone Joint Surg Br. 1975;57(3):380–3.
44. Saba EKA. Superficial radial neuropathy: an unobserved etiology of chronic dorsoradial wrist pain. Egypt Rheumatol Rehabil. 2021;48(1):29.
45. Lanzetta M, Foucher G. Entrapment of the superficial branch of the radial nerve (Wartenberg's syndrome). A report of 52 cases. Int Orthop. 1993;17(6):342–5.
46. Chiavaras MM, Jacobson JA, Yablon CM, Brigido MK, Girish G. Pitfalls in wrist and hand ultrasound. AJR Am J Roentgenol. 2014;203(3):531–40.
47. Ohrndorf S, Halbauer B, Martus P, Reiche B, Backhaus TM, Burmester GR, et al. Detailed joint region analysis of the 7-joint ultrasound score: evaluation of an arthritis patient cohort over one year. Int J Rheumatol. 2013;2013:493848.

48. Navalho M, Resende C, Rodrigues AM, Ramos F, Gaspar A, Pereira da Silva JA, et al. Bilateral MR imaging of the hand and wrist in early and very early inflammatory arthritis: tenosynovitis is associated with progression to rheumatoid arthritis. Radiology. 2012;264(3):823–33.
49. Ceponis A, Kissin E. Ultrasound of the hand and wrist. In: Kohler MJ, editor. Musculoskeletal ultrasound in rheumatology review. 2nd ed. Springer; 2021. p. 53–82.
50. Zayat AS, Ellegaard K, Conaghan PG, Terslev L, Hensor EM, Freeston JE, et al. The specificity of ultrasound-detected bone erosions for rheumatoid arthritis. Ann Rheum Dis. 2015;74(5):897–903.

Chapter 4
Fingers

Reasons to Do the Study

1. Finger pain, stiffness, or dysfunction.
2. Evaluation of soft tissue swelling or masses.
3. Pain or weakness after trauma.
4. Loss of finger extension or extension capability.

Questions We Want Answered

1. What is the cause of finger pain? Is it injury, a functional issue, an inflammatory process, or a tumor?
2. Is synovitis, tenosynovitis, or enthesitis present, and if so, what is the cause?
3. Is there an underlying inflammatory condition affecting the hand, and if so, is it a systemic process such as spondyloarthropathy, rheumatoid arthritis, or crystal disease?
4. If there is an inflammatory condition, is it eroding cartilage or bone?
5. Is there a structural, functional, or repetitive injury problem?
6. What is causing soft tissue swelling?
7. What is this lump on the finger?
8. Are tendons/ligaments/cartilage normal in appearance?
9. Is there evidence for ganglion cysts, and if so, is the origin the joint or the tendon sheath?
10. What is causing a triggering finger?
11. What is the first carpometacarpal joint's status?

The contents do not represent the views of the U.S. Department of Veterans Affairs or the United States Government.

M. H. Greenberg et al., *Manual of Musculoskeletal Ultrasound*,
https://doi.org/10.1007/978-3-031-37416-6_4

Fig. 4.1 Finger joints

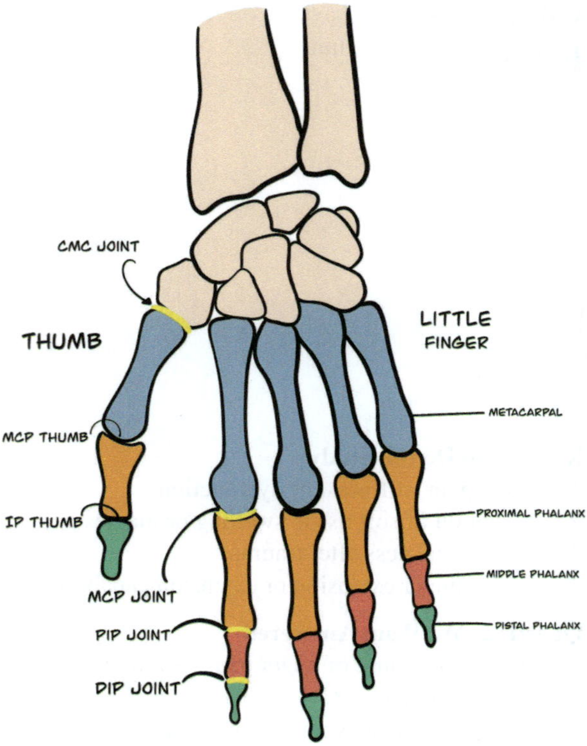

Necessary Basic Anatomy (Fig. 4.1)

The remarkable dexterity of the human finger is directly related to its complex anatomy.

Digits 2 Through 5

For digits 3 and 4, think of the number 3 [1]:

Three joints: metacarpophalangeal, proximal interphalangeal, and distal interphalangeal

Three phalanges: proximal, middle, and distal

Three of the five annular pulleys (A1, A3, and A5) are over the flexor surfaces of the three finger joints in each digit.

Three cruciate pulleys (C1–3) are poorly seen on US due to anisotropy

Three tendons: flexor digitorum superficialis (FDS), flexor digitorum profundus (FDP), and the extensor tendon (ET)

Digits 2 and 5 each have an additional extensor tendon (the extensor indicis (EI) and extensor digiti minimi (EDM), respectively).

Extensor Tendons (Fig. 4.2)

The extensor tendons run along the dorsal aspect of the finger. The ETs are incompletely ensheathed in a tenosynovium, and a paratenon covers the portion of the ET over the metacarpophalangeal joint (MCPJ) [2]. Distal to the MCPJ, the ET splits into a central and two lateral slips (bands) [3]. The central slip inserts into the dorsal

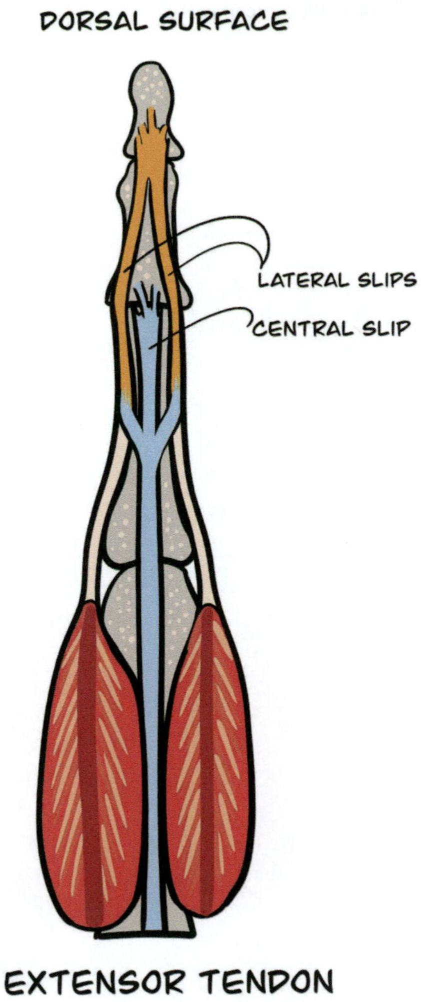

Fig. 4.2 Extensor tendon simplified

aspect of the base of the middle phalanx, while the lateral slips insert into the dorsal part of the distal phalangeal base.

Flexor Tendons (Fig. 4.3)

The two flexor tendons (FTs), one deep (FDS) and one shallow (FDP), are stacked upon each other on the volar aspect of the MCPJ [1]. A tenosynovium surrounds the FTs. The FDS splits in two at the proximal phalanx. Each segment of the FDS moves along the side of the FDP and then inserts into the middle phalanx, deep into the FDP. The FDP goes on to insert on the distal phalanx base.

Fig. 4.3 Flexor tendons simplified

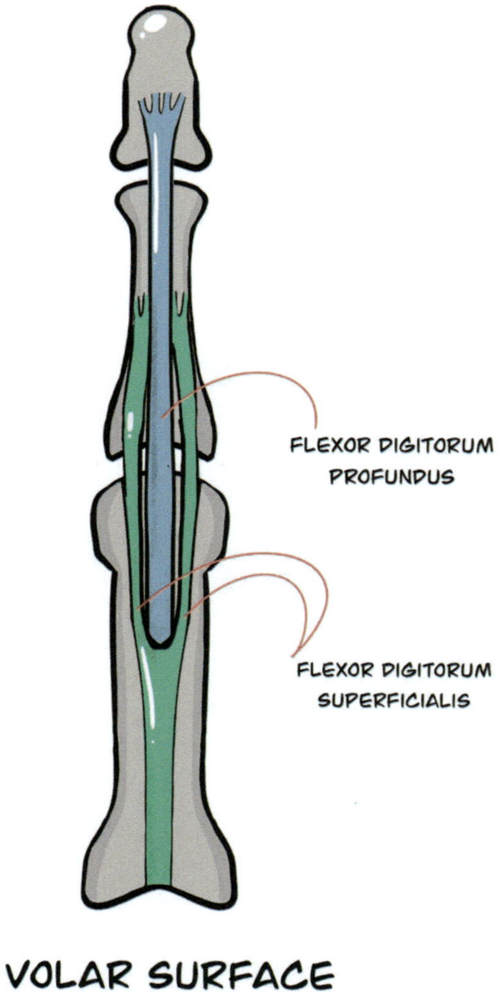

Finger Pulleys (Fig. 4.4)

The flexor tendon sheath is strapped down to the bone via the annular and cruciate pulleys [3]. The pulley system prevents bowstringing of the FTs during finger flexion. The Cl pulley is classically designated between the A2 and A3 annular pulleys. Some authors delineate the cruciate pulley between the A1 and A2 pulleys as C1 and renumber the cruciate ligament system as C1–C4 [4]. For consistency, we have chosen the original designation of three cruciate ligaments as C1–C3, with the C1 located between the A2 and A3 pulleys.

Annular pulleys A1–A4 are usually visible with high-frequency transducers [5]. These annular pulleys appear in longitudinal (LAX) view as elongated hyperechoic structures; however, they may be hypoechoic depending on anisotropy. In the transverse (TAX) view, each annular pulley is slightly hyperechoic at the most superficial portion of the flexor tendon but becomes hypoechoic to anechoic as the pulley drapes down along the edges of the flexor tendons [5]. The sonographer can visualize the C1 cruciate pulley in LAX view nearly 50% of the time, appearing as a thickened longitudinal hyperechoic structure. Annular pulley A5 and cruciate pulleys C2 and C3 are usually not visible sonographically [5].

Fig. 4.4 Volar finger annular and cruciate pulleys

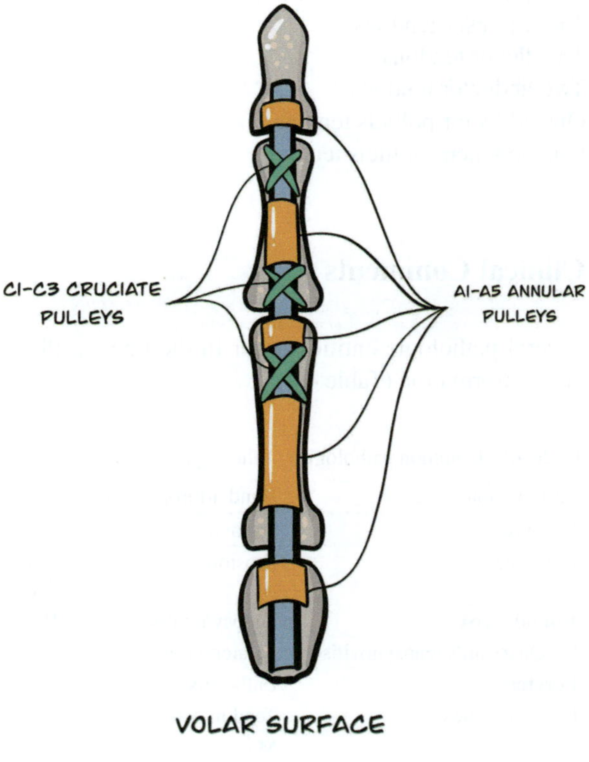

Volar Plates

All finger joints have thick fibrocartilage volar, or palmar, plates to limit finger hyperextension [3]. In LAX view, these are seen as a hyperechoic triangular structure superficial to the articular cartilage. The volar plate over the metacarpal head is the easiest to visualize.

Extensor Hood

The extensor hood (EH), a fibrous tissue that covers the ET, keeps the ET in place at the MCPJ [1]. Sagittal bands from the EH travel along both sides of the MCPJ to the volar plate, thus stabilizing the ET.

First Digit

Three pulleys: two annular and one oblique
Two phalanges: proximal and distal
Two joints: the first metacarpophalangeal and the interphalangeal
Two extensor tendons
Two flexor tendons
Two abductor tendons
One adductor pollicis tendon
One opponens pollicis tendon

Clinical Comments

Several pathologic entities occur in the fingers. Please refer to other chapters for more information (Table 4.1).

Table 4.1 Common pathologies of the fingers by chapter

Introduction	Hand arthropathies	Crystalline disease
Fracture	Synovitis	Gout
Neuroma	Erosions	Calcium pyrophosphate deposition disease
Dorsal wrist	Tenosynovitis	Hydroxyapatite deposition disease
De Quervain's tenosynovitis	Paratenonitis	
Forefoot	Enthesitis	
Foreign bodies	Tendon damage	
	Masses	

Pathology by Location

It is clinically helpful to classify finger pathology according to dorsal, ventral, and lateral locations.

Dorsal Aspect

Extensor Tendon Tears

Extensor tendon tears may result from open or closed injury [6]. The examiner should describe the tear location and whether it is complete or partial thickness. For partial-thickness tears, define the percentage of tendon involved; for complete tendon tears, describe the degree of tendon retraction and any associated avulsed bony fragment [6]. Full-thickness tears present sonographically as tendon discontinuity, and the stumps may show loss of normal fibrillar echotexture with evidence of tendinopathy [6]. If retracted, there may be a complete absence of the tendon. Use dynamic views to evaluate the gap between tendon stumps.

Partial-thickness tears sonographically appear as incomplete discontinuities or focal losses of fibular echotexture [6]. If acute, there may be hypoechoic to anechoic tendon abnormalities, perhaps with tendon swelling or a hematoma. Chronic partial-thickness tears reveal hypoechogenicity with surrounding fibrosis [6]. Bony fragments within the tendon are hyperechoic with posterior shadowing.

Mallet Finger

Also called baseball finger, this occurs when sudden, forceful extreme flexion of the extended distal interphalangeal joint (DIPJ) injures the ET at its insertion on the base of the distal phalanx [3]. For example, this injury may occur when a baseball hits a fingertip. Patients present with pain, swelling, and a loss of ability to extend the DIPJ [3]. Untreated, a swan neck deformity may ensue [6]. Ultrasound (US) reveals the disrupted insertion of the ET on the distal phalanx and tendon swelling [3, 6]. There may also be a retracted tendon in the form of a soft tissue mass, with a detached avulsed bone fragment.

Central Slip Injury

The central slip of the ET at the base of the middle phalanx may be injured due to sudden flexion of the proximal interphalangeal joint (PIPJ), direct trauma to the dorsal surface of the middle phalanx, or volar dislocation of the PIPJ [6]. This injury acutely presents with pain and swelling, along with a diminished ability to extend the PIPJ. Some extension capability is initially retained at the PIPJ if the lateral slips remain intact [3].

The unaffected lateral slips eventually migrate in a volar direction, producing PIPJ flexion and DIPJ extension and, ultimately, the boutonniere deformity [3]. Therefore, overlooking a central slip lesion may result in a permanent boutonniere deformity [6]. To confirm the pathology, place the probe in LAX over the dorsal PIPJ to look for a central ET defect and abnormal gliding of the central tendon slip with finger flexion and extension.

Boxer's Knuckle

Sagittal bands are part of the dorsal EH located at the MCPJ. The bands connect the volar plate to the ET and prevent ET subluxation during movement [6]. Sagittal band disruption commonly occurs with direct trauma to the proximal phalanx of the dorsal MCPJ, such as in boxing [3]. The resultant radial or ulnar subluxation or dislocation of the ET adversely affects MCPJ extension [3, 6]. Ultrasound reveals a hypoechogenic abnormal sagittal band. The ET may be normal or swell and lose fibrillar pattern [3]. Dynamic US in the TAX view shows subluxation or dislocation of the ET between the metacarpal heads with MCPJ flexion [6].

Paratenonitis

Extensor tendons of the fingers do not have an actual tenosynovial sheath; however, a paratenon surrounds the portion of the ET dorsal to the MCPJ. Inflammatory conditions, overuse, and trauma may cause paratenonitis [6].

Volar Aspect

Flexor Tendon Injury

Tendon sprain can result in tendinopathy and increased FT thickness, most reliably evaluated near the A1 pulley [6, 7]. Comparison to the contralateral FT is useful. Since the FDP inserts on the distal phalanx base and the FDS inserts on the middle phalanx, it is relatively easy to differentiate the two tendons on both TAX and LAX views [3]. Passively flexing and extending the PIPJ will activate both FTs, whereas moving the DIPJ in isolation will actuate only the FDP. Dynamic US is standardly done with active, passive, and resisted FT evaluations to delineate partial and complete tears [3]. Ultrasound shows partial tears as hypoechoic fusiform swelling of the tendon with focal loss of fibrillar echotexture. You may not find the tendon in a complete tear, only stumps [3].

Jersey Finger

The distal FDP tendon insertion may be injured when the DIPJ is bent and then forcefully hyperextended, such as in sports when a player grabs a jersey, and the opponent accelerates away [6]. The fourth finger is most often affected. A complete tear may cause the tendon to retract, leaving an empty tendon sheath. There may also be a bone avulsion [6]. This injury results in an inability to flex the DIPJ.

Trigger Finger

Trigger finger is a form of stenosing tenosynovitis. It most commonly arises at the A1 pulley, superficial to the palmar aspect of the MCPJ. As the patient bends the finger, there is a brief blockage of the FT and subsequent painful snapping when it is extended [8]. A thickened A1 pulley compresses the FT, provoking tendon impingement, swelling, and tenosynovitis. The first, third, and fifth fingers are most involved [8]. In most cases, trigger fingers are idiopathic or due to repetitive microtrauma. However, athletes who use rackets may be prone to direct pressure on the A1 pulley. Other conditions that may predispose to trigger finger include trauma, diabetes, rheumatoid arthritis (see below), amyloid, hypothyroidism, and acromegaly [8]. Less typically, trigger finger may result from scarring, tumors, or peritendinous ganglion cysts [8].

Hypoechoic thickening of the A1 pulley is best seen on the TAX view. An A1 pulley thickness greater than 0.62 mm is a cutoff value in healthy adults [9]. Along with intermittent locking during extension, the trigger finger may present with a palmar tender nodular area due to A1 pulley thickening [8]. Potential sonographic findings with trigger fingers [3] are as follows:

(a) Diffuse hypoechoic thickening of the A1 pulley.
(b) Swollen, rounded appearance of the FT in TAX view.
(c) Real-time visualization of the locking/snapping of the flexor tendon at the MCPJ.

Sometimes, it is just the FDS, contiguous to the A1 pulley, that will catch while the FDP movement remains unhindered. A ganglion cyst uncommonly arises from the FT sheath near the A1 pulley but may cause a trigger finger; ganglion removal may be necessary to prevent the recurrence of the trigger finger [10]. Ultrasound can evaluate trigger fingers and guide treatment decisions [11].

(See Chap. 5 for more information.)

Tenosynovitis

Flexor tenosynovitis may be due to trauma, inflammatory arthropathy, overuse, or infection [6]. Ultrasound reveals acute changes such as tenosynovial effusion, distended tenosynovium, and hyperemia on power Doppler (PD). The FT may lose its normal fibrillar echotexture and become thickened [6]. In more chronic flexor tenosynovitis, the synovial sheath may thicken, producing a "blurred image" of the tendon [6, 12]. Foreign bodies from penetrating trauma may also cause tenosynovitis [6]. Chronic inflammatory arthropathy may be associated with small echogenic particles called rice bodies. Chronic overuse may produce a thickened pulley, leading to stenosing tenosynovitis and trigger finger. Ultrasound provides needle injection guidance and may assess the efficacy of treatment for the inflammatory condition [6].

Rock Climber Finger

Ultrasound reveals the normal annular pulley to be 0.3–0.5 mm in thickness [6]. In LAX view, the pulley looks like a hyperechoic line on top of an oblong center that is anechoic to hyperechoic and rests on the outer edge of the FDS tendon. On the TAX view, the pulley appears hyperechoic at the volar aspect and hypoechoic at the lateral aspect due to anisotropy [6].

In rock climbers, the combination of PIPJ flexion coupled with DIPJ hyperextension generates excessive stress on the pulley. This injury most commonly affects the A2 pulley of the third and fourth digits [6]. Baseball players, bowlers, and patients receiving multiple corticosteroid injections near the pulley may also develop an A2 pulley injury [4]. Rock climbers routinely have thickened annular pulleys [6]. The pulley may be subject to strains or partial or complete tears, with the latter causing FT bowstringing.

Measuring the tendon-to-bone distance (TBD) in LAX view between the FT and bone is a sensitive method to detect milder degrees of bowstringing [6]. Compare the volar tendon position at rest and dynamically in resisted flexion. The normal TBD at the A2 pulley ligament is less than 2 mm [6]. There are various opinions on what absolute TBD value is consistent with a complete A2 tear, but values greater than 2 mm should heighten suspicion for an A2 injury [4]. Partial pulley tears produce minimal to no FT displacement. Still, the pulley may appear hypoechoic and swollen on US, perhaps exhibiting hyperemia on PD with a more acute injury [6].

In addition to bowstringing, sonographic findings associated with A2 pulley injuries include the frank absence of the pulley in LAX view, an FT sheath cyst, tenosynovial fluid, the presence of fibrous tissue, PIPJ or DIPJ effusion, or thickening of the A2 pulley [4]. One literature review describes that US and MRI have similar accuracy for diagnosing A2 pulley injuries; however, US offers a dynamic assessment of the FT pulley system and easy side-to-side comparison [4].

Volar Plate Injuries (Fig. 4.5)

The volar plate is a fibrous structure located just superficially to the cartilage of the volar PIPJ and MCPJ, abutting the FT. It sonographically appears as a hyperechoic, curvilinear form covering the joint cartilage. A high-frequency transducer is necessary to evaluate correctly. If suddenly hyperextended, the volar plate, particularly

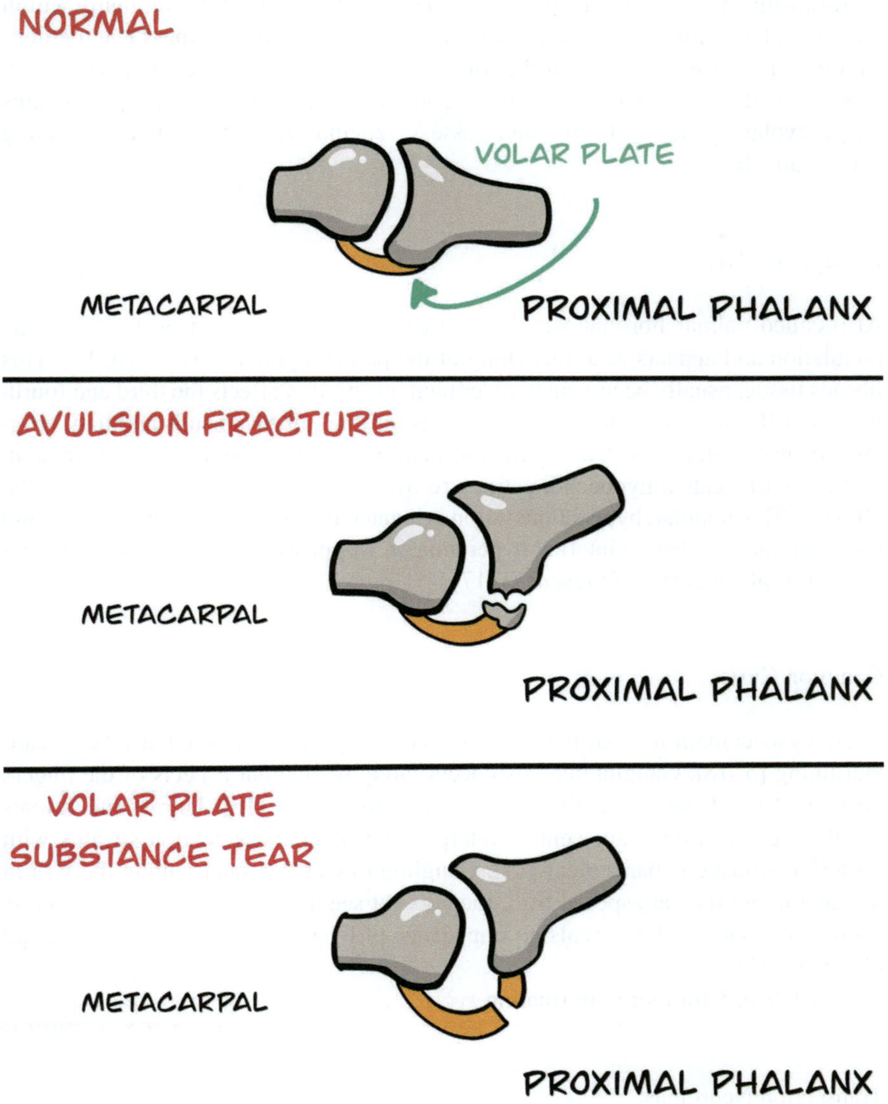

Fig. 4.5 Volar plate injuries

the PIPJ plate, may sustain a substance tear or, more commonly, a bony avulsion of the distal insertion on the middle phalanx with an intact volar plate [3]. Lateral radiographs may suggest a bony avulsion; US confirms a hyperechoic cortical bony fragment. A hypoechoic cleft within a swollen volar plate signifies a volar plate substance tear [3]. A full-thickness tear through the volar plate is less common than a bone avulsion at the distal insertion. If the avulsion fracture at the distal insertion of the PIPJ volar plate is substantial and involves the collateral ligaments, the middle phalanx may become unstable and chronically dislocated [13].

A full-thickness volar plate tear will appear as a hypoechoic discontinuity within the volar plate on US [14]. The plate may retract with a full-thickness tear, demonstrating an increased gap when the joint is stressed in hyperextension. Treatment depends on the size of the bony avulsion and joint instability [15]. Ultrasound helps gauge volar plate stability and assess edema resolution when treating conservatively.

Dupuytren's Disease

Also called palmar fibromatosis, Dupuytren's disease occurs in up to 2% of the population and appears as a thickening of the palmar aponeurosis [3, 16, 17]. This fibrous tissue, usually occurring in older men, most often affects the third and fourth fingers [17]. It can present as nodules, cords, or thick fibrous bands and may progress to involve the FTs causing flexion contractures that limit finger movement. Ultrasound reveals a hypoechoic structure overlying and possibly adhering to the FT. [17]. This nodular, hypoechoic lesion is located between the FT and the skin and has sharp margins but no internal hyperemia on Doppler [3]. Dynamic finger movements reveal adherence if present [3, 17].

Ganglion Cysts

These cysts contain mucoid material surrounded by fibrous tissue but lack a synovial lining [3, 18]. Ganglia may also occur along the palmar aspects of the fingers near the A1 pulleys, presenting as painful, firm masses [16]. Ultrasound reveals small anechoic lesions adjacent to interphalangeal joints or tendon sheaths with posterior acoustic enhancement [19]. Ganglion cysts often occur along the FTs in the lateral and medial aspects. Still, they do not seem to directly affect the tendon itself since dynamic US reveals no impedance of FT movement or position change of the cyst [3].

(See Chap. 5 for more information.)

Transection Neuromas

Small caliber palmar digital nerves are prone to injury during finger laceration [20]. High-frequency US helps diagnose traumatic digital nerve injuries. The TAX view may reveal a loss of nerve continuity that can be confirmed in the LAX view [20]. An injured nerve may eventually form a posttraumatic neuroma. These may occur as a stump neuroma or a neuroma-in-continuity and appear as a hypoechoic, well-defined mass lacking internal vascularity [20].

Glomus Tumor

Glomus tumors are rare benign vascular tumors that occur most commonly on the fingertips and under the fingernail [21]. These are extremely painful nodules and are often temperature-sensitive [3]. The US appearance is that of a small solid mass beneath the fingernail with homogeneous hypoechogenicity and perhaps erosion of underlying phalangeal bone. Doppler activity may reveal a mass with hyperemia, but a lack of Doppler activity does not rule out the presence of this lesion [16, 21, 22]. While there are no specific sonographic findings for glomus tumors, excruciating pain with a hypervascular mass beneath the fingernail is diagnostically useful [3, 23]. MRI remains the best diagnostic test to differentiate the mass from other soft tissue tumors [3, 24].

Tenosynovial Giant Cell Tumors

Giant cell tumor (GCT) is a localized form of pigmented villonodular synovitis, which affects the tendon sheath and is often nodular [16]. These slow-growing, firm masses are common soft tissue tumors of the hand and wrist, primarily affecting the distal fingers' volar aspect. On US, a tenosynovial GCT is homogeneous, hypoechoic, well-defined, in close contact with a tendon, and has a variable Doppler appearance [16]. These tumors have internal echoes and lack posterior acoustic enhancement [3]. In addition, these masses may be lobulated, do not move with tendon movement, and produce pressure erosions of the bony cortex. Pathologic verification is required [16].

Lateral Aspects

The volar plate and collateral ligaments stabilize the MCPJs and PIPJs [3]. With high-frequency transducers, we can evaluate the collateral ligaments of the interphalangeal joints. The MCPJ collateral ligaments are sonographically inaccessible in the LAX view except for the first MCPJ, the radial aspect of the second MCPJ, and the ulnar part of the fifth MCPJ [3]. The anisotropic collateral ligaments appear on US as hyperechoic to hypoechoic bands traversing the joint.

Ulnar Collateral Ligament Injuries (Fig. 4.6)

The first digit ulnar collateral ligament (UCL) at the MCPJ, integral for joint function and stability, may be injured with sudden hyperabduction and hyperextension [25, 26]. This injury typically occurs during a fall while gripping a ski pole, but years ago it resulted from a repetitive injury from strangling wounded rabbits ("gamekeeper's thumb"). Immediate imaging is necessary to determine if surgery is needed since delay may result in chronic pain, osteoarthritis, and decreased hand

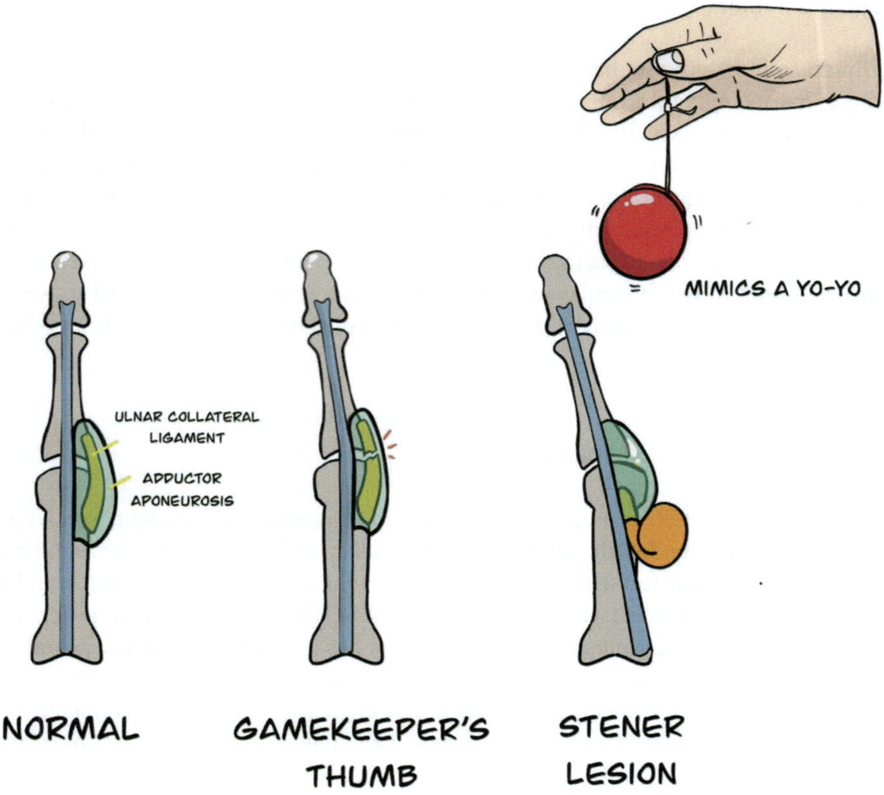

Fig. 4.6 Ulnar collateral ligament tears

function [26]. MRI is the gold standard for characterizing UCL tears, but US has also proven highly accurate [27]. A UCL tear may range from a simple sprain to partial and full-thickness tears [26].

A partially torn UCL appears on US as hypoechoic and thickened [3]. Nondisplaced complete UCL tears have a stump near the proximal or distal insertion of the ligament, which remains in anatomic alignment. This nondisplaced UCL is a gamekeeper's or skier's thumb [3]. However, when the proximal stump of a completely torn UCL displaces out of anatomic alignment, it ceases to be called a gamekeeper's thumb. It is now characterized as a Stener lesion that will not heal with conservative management [3, 26, 28].

The defining feature of a Stener lesion is that the rolled-up stump is either at the proximal edge or superficial to the adductor aponeurosis. The US appearance of a Stener lesion demonstrates the proximal UCL stump to be a hypoechoic mass adjacent to the metacarpal neck and head; the hyperechoic adductor aponeurosis appears to be leading into the mass [26, 27]. This appearance resembles a "yo-yo on a string." The yo-yo is the retracted proximal UCL stump, folded up upon itself, and the string is the adductor aponeurosis [26]. It is noteworthy that the rolled-up UCL remnant location is not always superficial to the adductor aponeurosis, yet is still characterized as a Stener lesion [28].

Ultrasound may also detect bony avulsions since radiographs are not sensitive in this regard [26]. A pitfall is that an injured adductor aponeurosis may mimic intact UCL fibers. Gently passively flexing and extending the interphalangeal joint will identify the aponeurosis by showing the movement of the adductor aponeurosis gliding smoothly over the MCPJ [28].

Ulnar Digital Neuroma

Traumatic irritation of the ulnar digital nerve of the first digit may occur in bowlers and less commonly in baseball players, massage therapists, and jewelers [29]. This repetitive trauma produces a painful soft tissue nodule along the ulnar/volar aspect of the thumb in the web space between the first two digits. In addition, paresthesia may extend along the medial aspect of the thumb, and Tinel's sign might be positive [29]. Ultrasound may successfully detect a neuroma when MRI fails, although, historically, MRI has been the imaging modality of choice.

First Carpometacarpal Osteoarthritis

First carpometacarpal joint osteoarthritis commonly causes pain and disability, particularly in older populations. In a study of 93 participants, plain radiographs and US comparably detected osteophytes, with US being more sensitive [30]. The detection of erosions was also similar, although US may not depict some erosions due to occluding osteophytes. Power Doppler activity in the first carpometacarpal joint (CMCJ) correlated with increased severity of thumb base pain [30]. However, pain and function are poorly associated with radiographic and US structural findings.

Rheumatoid Arthritis

Rheumatoid arthritis (RA) may affect fingers in myriad ways. Inflammation may produce paratenonitis over the MCPJ, tenosynovitis of the flexor tendons, and synovitis of the MCPJs and PIPJs [2]. Rheumatoid nodules may develop along tendon sheaths and in subcutaneous tissue [31, 32]. Flexor and extensor tendon rupture may occur due to chronic tenosynovitis causing secondary tendinopathy or from the abrasive effect of eroded bone or osteophytes. Synovitis of the MCPJs may loosen the joint capsule and surrounding ligaments, resulting in ulnar drift, exacerbated by tendon subluxation or dislocation [31]. Likewise, the instability of PIPJs and DIPJs may culminate in swan neck or boutonniere deformity. Rheumatoid arthritis increases the predisposition to trigger fingers under the A1 pulley due to localized synovitis, tenosynovitis, and rheumatoid nodules. Do not mistake the smooth groove-like curve proximal to the end of the dorsal metacarpal head as a bony erosion [28]. This "pseudoerosion" is the normal attachment point for the synovial membrane.

Pitfalls

1. When dynamically evaluating tendons with US, test passive, active, and resisted movement.
2. Full-thickness tears may result in a loss of continuity or the absence of the tendon. The retracted tendon may form a soft tissue mass with a detached avulsed bone fragment. Do not neglect to search for stumps.
3. Acute finger injuries need rapid evaluation to avoid permanent dysfunction or deformity. Do not hesitate to obtain an MRI to confirm a sonographic diagnosis. Timely referral to hand surgery is of paramount importance.
4. An extensor tendon central slip injury may present with a retained ability to extend the PIPJ due to intact lateral slips of the extensor tendon. Do not incorrectly perceive this as a lack of damage to the central slip. Overlooking a central slip injury may result in a permanent boutonniere deformity.
5. Direct injury to the MCPJs may disrupt the extensor hood sagittal band, causing lateral subluxation of the extensor tendon. Do not neglect to perform dynamic US in TAX view to diagnose tendon subluxation or dislocation.
6. With a volar finger laceration, suspect a traumatic digital nerve injury. Perform careful US interrogation in TAX with confirmation in LAX view. Transected nerves need to be promptly diagnosed with an immediate referral to a hand surgeon.
7. Do not neglect to assess the FDS and FDP in isolation. Passively flexing and extending the middle and distal phalanges at the PIPJ activates both flexor tendons, while bending the isolated DIPJ will move only the FDP.
8. A ganglion cyst arising from the flexor tendon sheath near the A1 pulley in the presence of a trigger finger often requires the removal of the ganglion cyst to prevent the recurrence of the trigger finger.
9. In any sudden hyperextension injury, examine the volar plate to look for evidence of a substance tear or bony avulsion.
10. First MCPJ UCL injury diagnostic pitfalls:

    (a) Do not mistake an injured abductor aponeurosis for intact UCL fibers. Gentle passive extension and flexion of the interphalangeal joint will demonstrate the gliding of an intact adductor aponeurosis. Overzealous joint manipulation may convert a partial UCL tear into a complete one.
    (b) The UCL remnant may be superficial to the adductor aponeurosis or located at the proximal edge.
    (c) Small bony avulsions noted on US but undetected radiographically may be diagnostically helpful.

11. Beware of the dorsal metacarpal head "pseudoerosion," which is a normal anatomic indentation.

Method

The patient faces the examiner with the pronated hand flat on a table or arm of the chair. Start with the dorsal surface of the fingers. A rolled-up towel placed under the fingers may help slightly bend the fingers to improve joint visualization. For the volar aspect of the fingers, the hand is supinated, palm up, with extended fingers. The examination for digits 2 through 5 is identical for each finger, although the second and fifth digits offer additional radial and ulnar views. For the thumb, views of the radial first CMCJ and the ulnar first MCPJ are most productive.

Use a linear probe with a frequency of at least 12 MHz. Evaluation of entheses often requires a higher frequency, typically 16–22 MHz. Evaluate each joint for fluid and synovitis at the dorsal recess. Pathologic conditions involving tendons are confirmed with dynamic active, passive, and resisted maneuvers under US. Although the protocol below only shows an individual digit, it is recommended that any digits of interest be included.

Protocol Image 1: Dorsal Metacarpophalangeal Joint, Extension + Flexion, Longitudinal ± Power Doppler (Fig. 4.7)

Place the transducer over the extended MCPJ in LAX. Perform TAX slides in radial and ulnar directions to fully evaluate the cartilage and the synovial recess. Evaluate the three margins of the joint recess: proximal, distal, and superficial. The superficial margin can be defined by passively flexing and extending the finger to identify the moving extensor tendon, which borders the superior recess margin. Synovial hypertrophy will expand these margins and is graded [1–3] (see Chap. 5). Look for synovial fluid in the joint recess.

Examine the smoothness of the superficial surface of the cartilage covering the metacarpal head (MCH). Look at the MCH itself and determine (a) if cartilage surrounds the distal and dorsal portions of the MCH, (b) the constitution of the cartilage (focal or diffuse thinning, loss of sharpness of the superficial border), and (c) if there is a normal hyperechoic interface reflex. A long-standing erosive process may result in cartilage damage ranging from minimal to severe.

Passively flexing and extending the MCPJ will also enable a thorough inspection of the cartilage covering the MCH. With a slightly flexed MCPJ, perform radial and ulnar TAX slides to evaluate the cartilage more fully. Rock to probe as needed to counter anisotropy, keeping the cartilage in focus. Remember, hyperechoic synovium may have migrated into the area where cartilage previously existed.

If the radiographs indicate possible bone erosion, do a focused US examination on that area to verify. Look at the bony surfaces of the MCH and the proximal phalanx for irregularities. Do not mistake the normal smooth groove-like concavity proximal to the MCH as a bony erosion since this is the proximal attachment site for the synovial membrane [28]. If you suspect that there truly is erosion in this area, look for contiguous synovial hypertrophy.

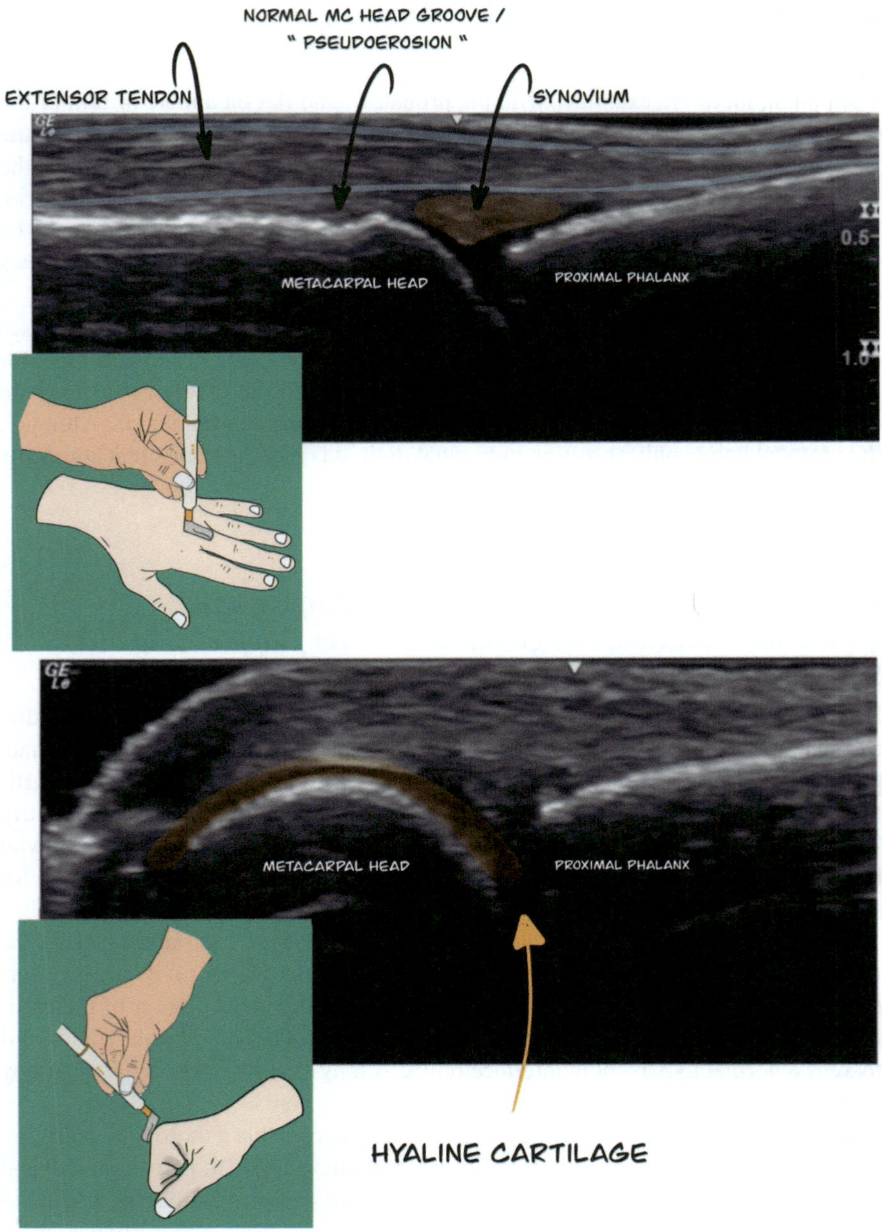

Fig. 4.7 Protocol Image 1: Dorsal metacarpophalangeal joint, extension + flexion, longitudinal ± power Doppler

Remember to verify synovitis, cartilage erosion, and bone erosion in TAX view (Protocol Image 2, below). The effusion may be present along with synovitis. The synovitis may have receded in long-standing cartilage-erosive arthropathy, leaving only residual cartilage or bone erosion. Turn on the PD for another image. Power Doppler activity may be seen within the synovial tissue.

Also, examine the more superficial finger extensor tendon for absence, tendon damage, or paratenonitis. Paratenonitis sonographically presents as hyperemia or anechoic fluid along the extensor tendons [2]. It is most often associated with an underlying inflammatory polyarthropathy. With paratenonitis, the underlying MCPJ frequently has evidence of synovitis; however, paratenonitis may occur as an isolated finding, perhaps due to joint deformity or overuse [2].

Protocol Image 2: Dorsal Metacarpophalangeal Joint, Transverse, ± Power Doppler (Fig. 4.8)

Rotate the probe 90° to evaluate the slightly flexed MCPJ in TAX. Perform small proximal and distal TAX slides to verify any pathology visualized in LAX. The PD will delineate the dorsal digital arteries and may verify hyperemia noted on the LAX view. The extensor tendon is ovoid and exhibits anisotropy. The TAX orientation offers another view of the cartilage covering the MCH but remembers to slightly rock the probe to sharpen the view of the cartilage.

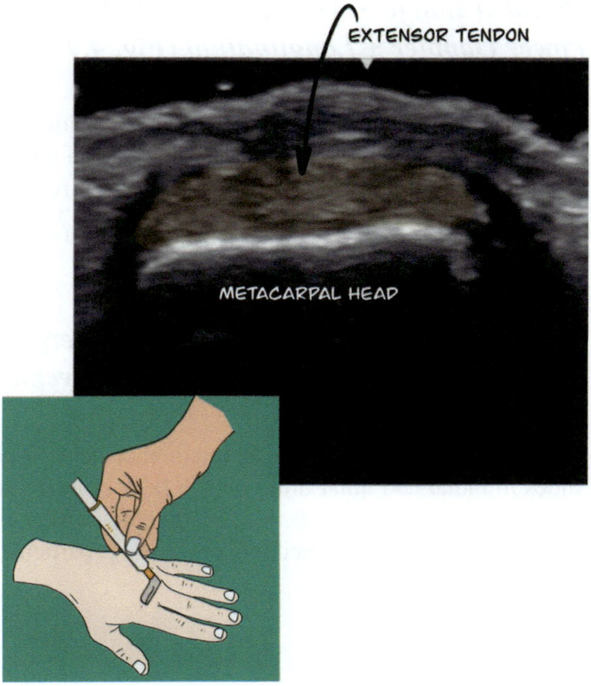

Fig. 4.8 Protocol Image 2: Dorsal metacarpophalangeal joint, transverse, ± power Doppler

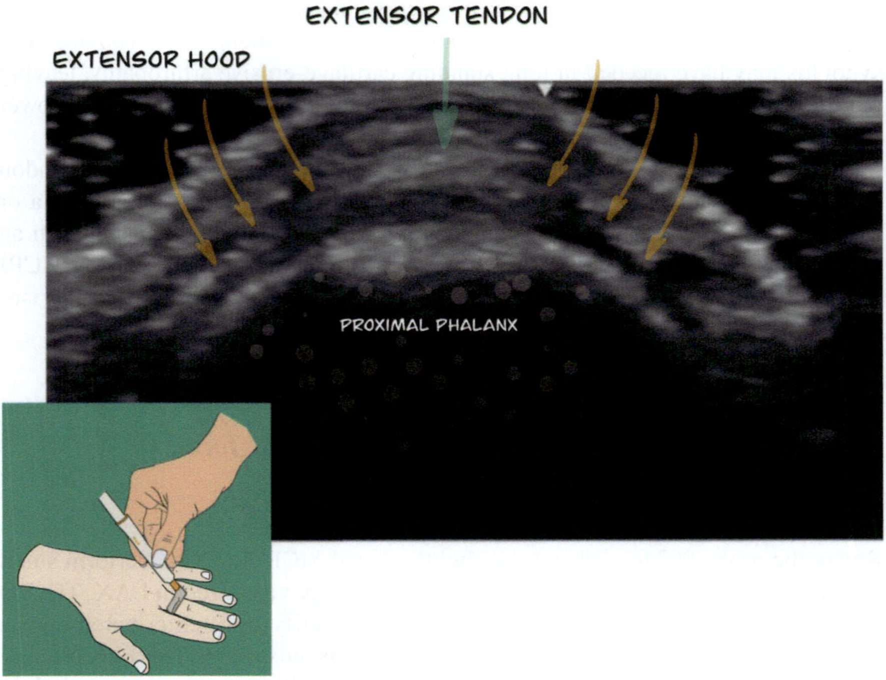

Fig. 4.9 Protocol Image 3: Dorsal proximal phalanx, transverse ± power Doppler ± longitudinal

Protocol Image 3: Dorsal Proximal Phalanx, Transverse ± Power Doppler ± Longitudinal (Fig. 4.9)

Still in TAX orientation, perform a distal TAX slide to see the hyperechoic cortex of the proximal phalanx. Turn on the PD to see the dorsal digital arteries. Superficial to the ET is the hypoechoic EH. If an abnormality is detected, rotate the probe 90° to examine the ET and cortex in LAX.

Protocol Image 4: Dorsal Proximal Interphalangeal Joint, Longitudinal ± Power Doppler ± Transverse (Fig. 4.10)

With the probe again in LAX orientation, center it over the PIPJ and perform TAX slides in radial and ulnar directions to fully evaluate the cartilage and the synovial recess. If you passively flex the joint, the cartilage may have better visibility. Osteophytes may be the predominant finding, particularly in the older age group.

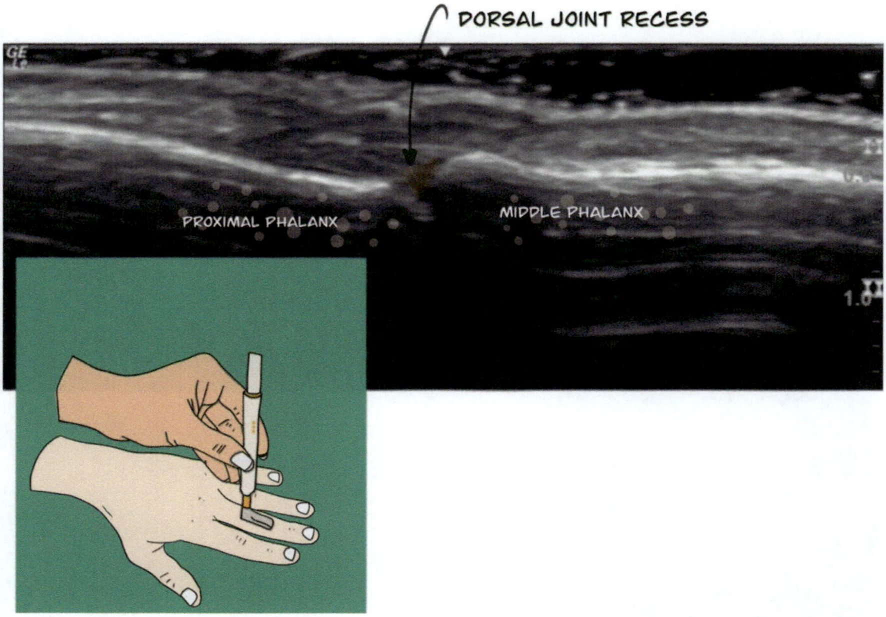

Fig. 4.10 Protocol Image 4: Dorsal proximal interphalangeal joint, longitudinal ± power Doppler ± transverse

The dorsal recess may be enlarged and extend proximally in the presence of synovitis. Such synovial hypertrophy may or may not have PD activity. Erosion of bone may be evident with rheumatoid arthritis, psoriatic arthritis, and perhaps in advanced gouty arthropathy. Verify any suspected bone erosion in the TAX view. While in LAX orientation, turn on the PD for another image. Power Doppler activity may be positive within the synovial tissue. Verify abnormal findings in TAX by rotating the probe.

Protocol Image 4a (Optional): Dorsal Middle Phalanx, Transverse ± Longitudinal

Rotate the probe to TAX orientation and perform a TAX slide to see the hyperechoic cortex of the middle phalanx. The central slip of the ET may be discernible in the LAX view.

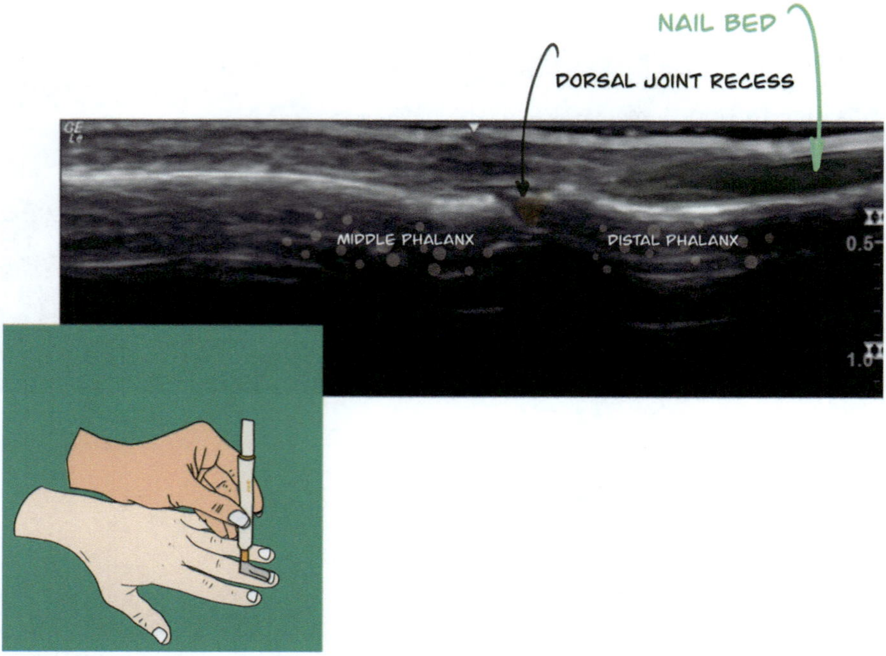

Fig. 4.11 Protocol Image 5: Dorsal distal interphalangeal joint, longitudinal ± transverse

Protocol Image 5: Dorsal Distal Interphalangeal Joint, Longitudinal ± Transverse (Fig. 4.11)

With the probe in LAX orientation, perform an LAX slide distally and center the transducer over the DIPJ. Passive flexion of the joint may better expose the cartilage. Bony osteophytes may be evident. Abnormal findings should be verified in TAX by rotating the probe.

Protocol Image 6: Dorsal Distal Phalanx Extensor Tendon Insertion Longitudinal + Power Doppler (Fig. 4.12)

Examine for enthesopathy and enthesitis by looking for PD activity along the insertion of the ET on the distal phalanx. This hyperechoic bone lies deep into the reverberating fingernail. Use copious transmission gel and the highest frequency probe available.

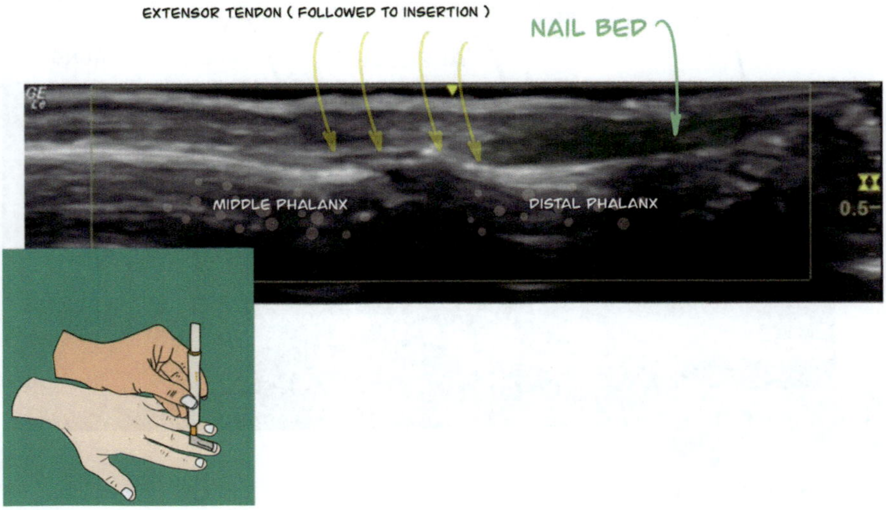

Fig. 4.12 Protocol Image 6: Dorsal distal phalanx extensor tendon insertion longitudinal + power Doppler

Protocol Image 7: Volar Metacarpophalangeal Joint, Dynamic Flexion, Longitudinal, ± Power Doppler ± Transverse (Fig. 4.13)

The patient places the hand palm side up with fingers extended on the examining surface. Place the probe in LAX over the palmar surface of the MCPJ. For each joint, perform a TAX slide in an ulnar direction and then a radial direction to fully evaluate. Look for synovitis, tenosynovitis, cartilage erosion, and bone erosion. Manually flex the finger to visualize the flexor tendons and surrounding tenosynovium. Isolated DIPJ passive flexion will activate only the FDP. The volar plate is superficial to the joint cartilage, and the anisotropic A1 pulley is superficial to the FDS. With a trigger finger, the superficial layer may be the only layer that becomes snagged beneath a thickened A1 pulley. Turn on the PD to look for a signal.

When examining the volar MCPJ for synovitis, note that the synovial recess extends proximally and should also warrant Doppler analysis (35). If there is suspicion of a volar plate tear, look for a hyperechoic bony avulsion, a hypoechoic cleft, or swelling of the volar plate. The gap of a complete volar plate tear may widen with gentle passive hyperextension of the joint during a dynamic US examination. Turn the probe 90° to look at the MCPJ in TAX to verify the pathology depicted on the LAX view and gain another perspective on the A1 pulley. When looking for A1 pulley thickening, compare the A1 pulley thickness to that of the same contralateral, non-triggering finger.

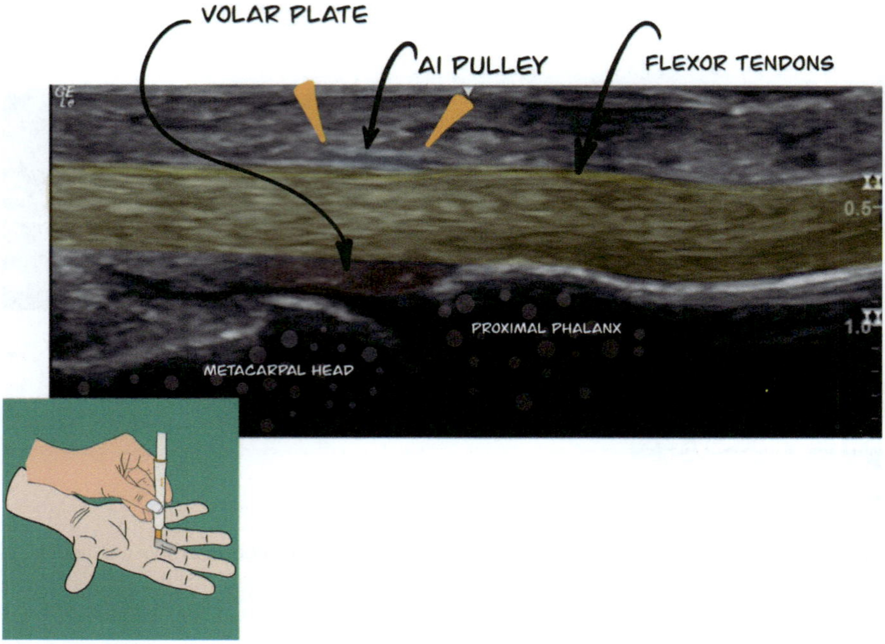

Fig. 4.13 Protocol Image 7: Volar metacarpophalangeal joint, dynamic flexion, longitudinal, ± power Doppler ± transverse

Protocol Image 8: Volar Proximal Phalanx, Longitudinal ± Transverse (Fig. 4.14)

With the probe back in LAX, perform an LAX slide distally to visualize the hyperechoic cortex of the proximal phalanx (PP). Perform slight TAX slides in radial and ulnar directions to encompass a full volar view of this bone. Look for a step-off deformity that may indicate a fracture. In LAX view, the A2 pulley appears as an oblong anechoic to hypoechoic structure resting on the superficial hyperechoic edge of the FDS tendon. The pulley is normally 0.3–0.5 mm thick and covered by a thin hyperechoic line on US [6].

To evaluate for bowstringing indicative of an A2 pulley tear, use a small-footprint transducer. Place the PIPJ at least 40° of flexion and the DIPJ at 10° of flexion [2]. The examiner resists active flexion of the finger by applying fingertip-to-fingertip resistance. Measure the TBD. Values greater than 2 mm may indicate an A2 pulley tear [2]. Compare the volar tendon position at rest and dynamically in resisted flexion.

Other signs of an A2 pulley injury are the absence of the pulley in LAX view, tenosynovial fluid in the flexor tendon sheath, irregular thickening of the A2 pulley,

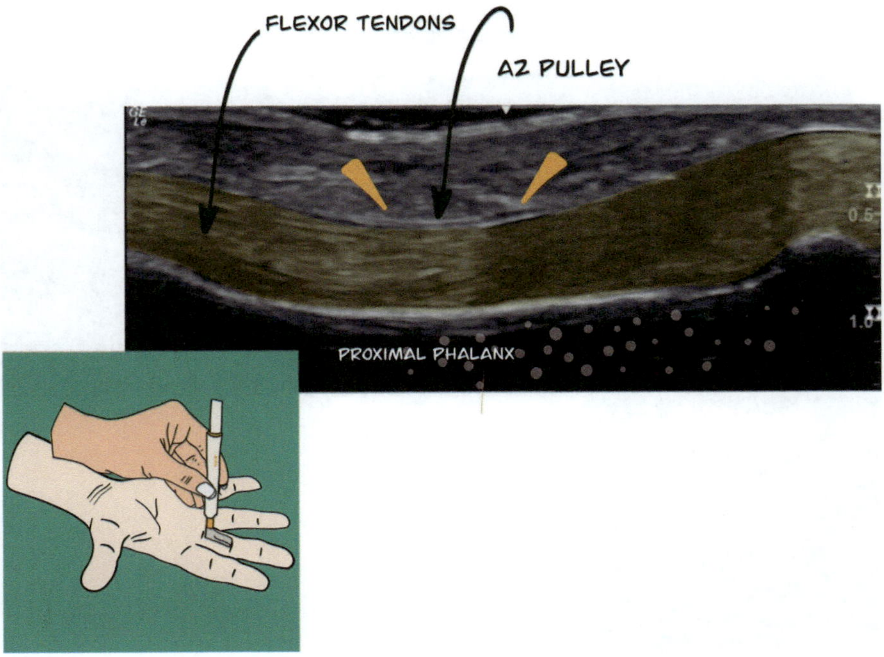

Fig. 4.14 Protocol Image 8: Volar proximal phalanx, longitudinal ± transverse

a cyst in the flexor tendon sheath, scar tissue, and effusion of the PIPJ [2]. Partial pulley tears produce minimal to no tendon displacement. Still, the pulley may appear hypoechoic and swollen on US, perhaps exhibiting hyperemia on PD with a more acute injury.

Rotate the probe 90° for the TAX view. In this view, the pulley appears to be slightly hyperechoic at the volar aspect and hypoechoic at the lateral aspect. The TAX view of the proximal phalanx demonstrates the FDS splitting, with each slip moving away from the deeper, more central FDP.

Protocol Image 9: Volar Proximal Interphalangeal Joint, Longitudinal ± Transverse (Fig. 4.15)

Next, move the probe, in LAX, to the PIPJ. Perform TAX slides in the ulnar and radial directions to fully evaluate the joint, synovial recess, and flexor tendon. The A3 pulley is superficial to the PIPJ. Turn the probe 90° to TAX to verify pathology.

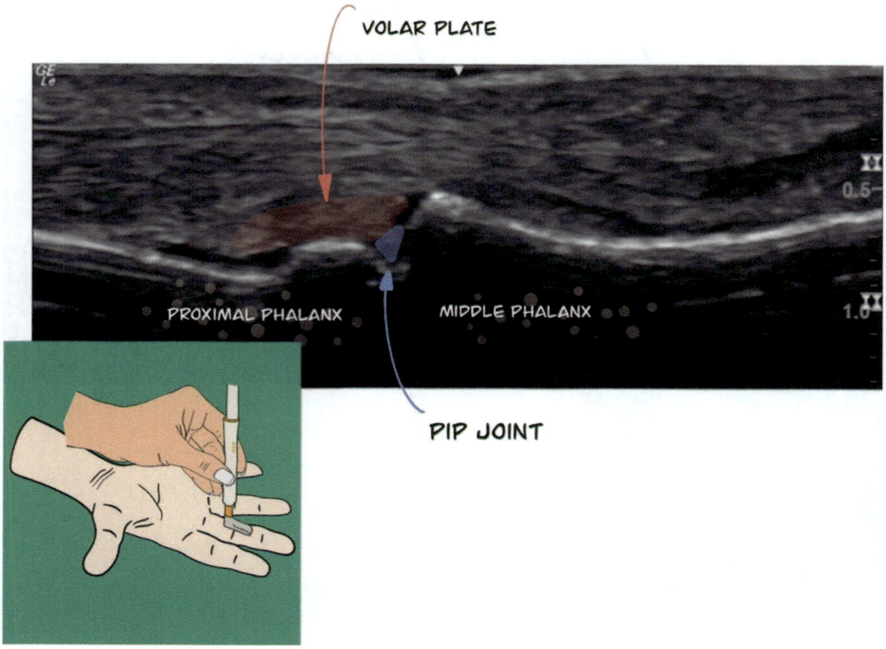

Fig. 4.15 Protocol Image 9: Volar proximal interphalangeal joint, longitudinal ± transverse

Protocol Image 9a (Optional): Volar Middle Phalanx, Longitudinal ± Transverse

With the probe back in LAX, perform an LAX slide distally to visualize the hyperechoic cortex of the proximal phalanx. Perform slight TAX slides in radial and ulnar directions to encompass a full volar view of this bone. Look for a step-off deformity that may indicate a fracture. Depending upon the angle of the probe, the anisotropic A4 pulley may be appreciated superficially to the FDS. Rotating the probe 90° affords a TAX view of the flexor tendons, which have reversed their relative positions. The FDS, now deep into the FDP, inserts into the middle phalanx.

Protocol Image 10: Volar Distal Interphalangeal Joint, Longitudinal ± Transverse (Fig. 4.16)

Next, move the probe in LAX to the DIPJ. Perform small TAX slides in the ulnar and radial directions to fully evaluate the joint, synovial recess, and flexor tendon. The A5 pulley is superficial to the DIPJ, although it may not be visible.

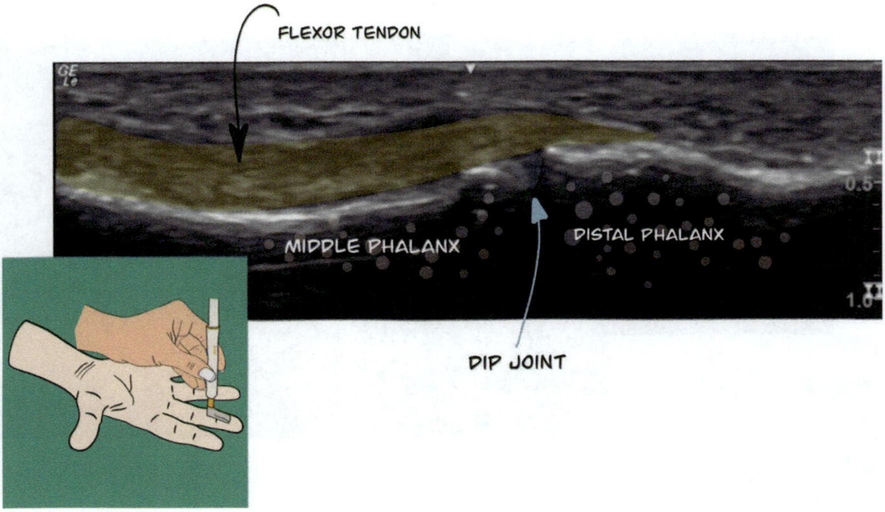

Fig. 4.16 Protocol Image 10: Volar distal interphalangeal joint, longitudinal ± transverse

Protocol Image 10a (Optional): Flexor Tendon Enthesis, Longitudinal + Power Doppler

For this view, it is best to use a small footprint probe with a frequency of at least 18 MHz. Extend the finger against a flat surface and place the probe in LAX orientation. Move the transducer distally to the FDP insertion on the distal phalanx. Small TAX slides in the ulnar and radial directions enhance the search for PD activity indicative of active enthesitis. Remember, a light touch and abundant transmission gel are mandatory.

Protocol Image 11: Ulnar Aspect of the First Metacarpophalangeal Joint, Longitudinal ± Transverse (Fig. 4.17)

The pronated hand is placed upon a curved bolster with the first and second fingers draped on opposite sides, thus abducting the thumb. Place a high-frequency probe in LAX along the ulnar aspect of the thumb base, focusing on the bony discontinuity, which is the first MCPJ. To identify the UCL, first identify the hyperechoic bony surfaces of the distal MC and proximal phalanx [27]. The normal UCL is a homogeneous fibrillar structure spanning the joint in this coronal view. The UCL may

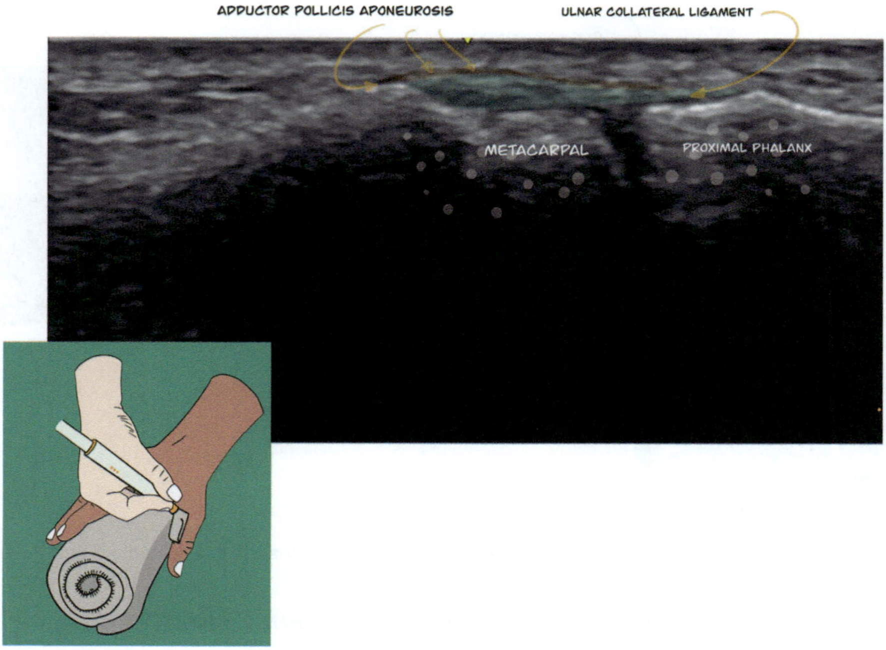

Fig. 4.17 Protocol Image 11: Ulnar aspect of the 1st metacarpophalangeal joint, longitudinal ± transverse

appear hypoechoic, particularly the proximal portion. Rock the probe to look for fibrillar echotexture. The adductor aponeurosis is superficial to the UCL; it is an oblong, thin longitudinal structure that is hypoechoic to hyperechoic depending on the degree of anisotropy.

Gently and passively flex the first interphalangeal joint to dynamically confirm the identity and integrity of the adductor aponeurosis, which normally glides over the MCPJ [27]. If the joint moves excessively compared with the contralateral joint or the UCL shows partial or complete disruption, this may be a gamekeeper's thumb. With a complete UCL tear, the ligament remnant may displace off the joint in a rolled-up ball proximal to the MCPJ, now termed a Stener lesion. With a Stener lesion, the UCL remnant may be located superficial to the adductor aponeurosis. Verify any sonographic findings in the TAX view.

Protocol Image 12: Radial Aspect of the First Carpometacarpal Joint, Longitudinal ± Power Doppler (Fig. 4.18)

The hand is relaxed and prone, with the first digit placed along the edge of the examination table. The transducer is placed in LAX along the thumbs; base dorsum, with the long axis of the transducer aligned with the thumb. Identify the

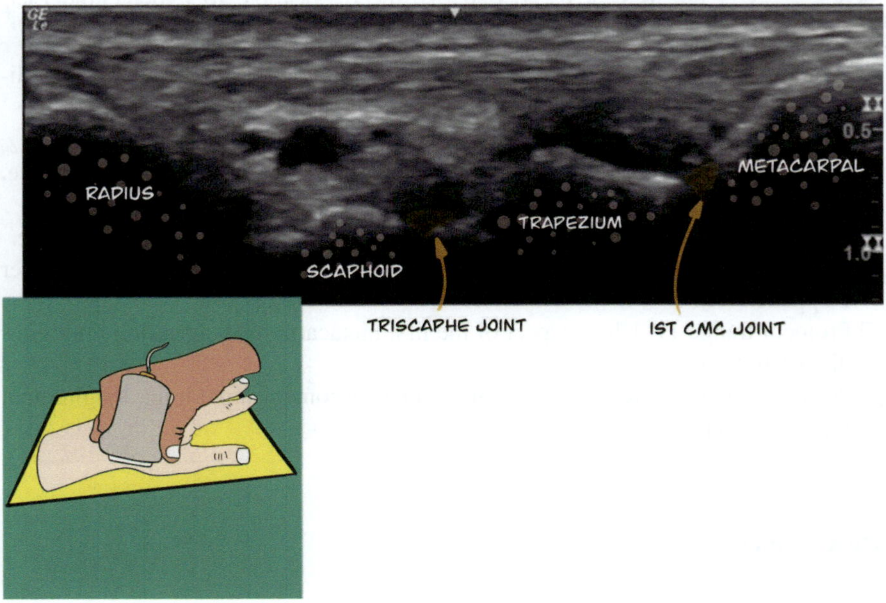

Fig. 4.18 Protocol Image 12: Radial aspect of the first carpometacarpal joint, longitudinal ± power Doppler

hyperechoic longitudinal bony surface of the first metacarpal and look for the gap between the first metacarpal base and the hyperechoic trapezium; this is the first CMCJ. The joint space proximal to the first CMCJ is the scaphotrapeziotrapezoid, or triscaphe, joint.

Gently grasp the thumb with your free hand and bend the first CMCJ. The first CMCJ moves, but the intercarpal joint does not. Since the first CMCJ is frequently affected by osteoarthritis, the joint may be narrowed, and osteophytes may hamper clear-cut identification of the joint. Move the probe in a TAX slide in each direction to obtain a more expansive view of the joint. The entire radial aspect of this joint is sonographically visible. Turn on the PD to look for hyperemia.

Complete Fingers Ultrasonic Examination Checklist

☐ Protocol Image 1: Dorsal metacarpophalangeal joint, extension + flexion, longitudinal ± power Doppler.

☐ Protocol Image 2: Dorsal metacarpophalangeal joint, transverse, ± power Doppler.

☐ Protocol Image 3: Dorsal proximal phalanx, transverse ± power Doppler ± longitudinal.

☐ Protocol Image 4: Dorsal proximal interphalangeal joint, longitudinal ± power Doppler ± transverse.

☐ Protocol Image 4a (optional): Dorsal middle phalanx, transverse ± longitudinal.

☐ Protocol Image 5: Dorsal distal interphalangeal joint, longitudinal ± transverse.

☐ Protocol Image 6: Dorsal distal phalanx extensor tendon insertion longitudinal + power Doppler.
☐ Protocol Image 7: Volar metacarpophalangeal joint, dynamic flexion, longitudinal, ± power Doppler ± transverse.
☐ Protocol Image 8: Volar proximal phalanx, longitudinal ± transverse.
☐ Protocol Image 9: Volar proximal interphalangeal joint, longitudinal ± transverse.
☐ Protocol Image 9a (optional): Volar middle phalanx, longitudinal ± transverse.
☐ Protocol Image 10: Volar distal interphalangeal joint, longitudinal ± transverse.
☐ Protocol Image 10a (optional): Flexor tendon enthesis, longitudinal + power Doppler.
☐ Protocol Image 11: Ulnar aspect of the first metacarpophalangeal joint, longitudinal ± transverse.
☐ Protocol Image 12: Radial aspect of the first carpometacarpal joint, longitudinal ± power Doppler.

References

1. Gupta P, Lenchik L, Wuertzer SD, Pacholke DA. High-resolution 3-T MRI of the fingers: review of anatomy and common tendon and ligament injuries. AJR Am J Roentgenol. 2015;204(3):W314–23.
2. Suzuki T, Shirai H. SAT0570 clinical significance of finger extensor paratenonitis detected by musculoskeletal ultrasound. Ann Rheum Dis. 2020;79(Suppl 1):1243–4.
3. Lee SA, Kim BH, Kim SJ, Kim JN, Park SY, Choi K. Current status of ultrasonography of the finger. Ultrasonography. 2016;35(2):110–23.
4. Berrigan W, White W, Cipriano K, Wickstrom J, Smith J, Hager N. Diagnostic imaging of A2 pulley injuries: a review of the literature. J Ultrasound Med. 2022;41:1047.
5. Boutry N, Titécat M, Demondion X, Glaude E, Fontaine C, Cotten A. High-frequency ultrasonographic examination of the finger pulley system. J Ultrasound Med. 2005;24(10):1333–9.
6. Rosskopf AB, Martinoli C, Sconfienza LM, Gitto S, Taljanovic MS, Picasso R, et al. Sonography of tendon pathology in the hand and wrist. J Ultrason. 2021;21(87):e306–17.
7. Klauser A, Stadlbauer KH, Frauscher F, Herold M, Klima G, Schirmer M, et al. Value of transducer positions in the measurement of finger flexor tendon thickness by sonography. J Ultrasound Med. 2004;23(3):331–7.
8. Bianchi S, Gitto S, Draghi F. Ultrasound features of trigger finger: review of the literature. J Ultrasound Med. 2019;38(12):3141–54.
9. Spirig A, Juon B, Banz Y, Rieben R, Vögelin E. Correlation between sonographic and in vivo measurement of A1 pulleys in trigger fingers. Ultrasound Med Biol. 2016;42(7):1482–90.
10. Altaf W, Attarde D, Sancheti P, Shyam A. Triggering of thumb by a ganglion cyst of the flexor tendon sheath at A1 pulley: a case report. J Orthop Case Rep. 2021;11(3):10–2.
11. Kara M, Ekiz T, Sumer HG. Hand pain and trigger finger due to ganglion cyst: an ultrasound-guided diagnosis and injection. Pain Physician. 2014;17(6):E786.
12. Plotkin B, Sampath SC, Sampath SC, Motamedi K. MR imaging and US of the wrist tendons. Radiographics. 2016;36(6):1688–700.
13. Yoong P, Johnson CA, Yoong E, Chojnowski A. Four hand injuries not to miss: avoiding pitfalls in the emergency department. Eur J Emerg Med. 2011;18(4):186–91.

14. Almusa E, Peterson WM, Bianchi S, Jacob D, Hoffman D. Radiological investigations. In: Chick G, editor. Acute and chronic finger injuries in ball sports. Paris: Springer; 2013.
15. Leclère FM, Mathys L, Juon B, Vögelin E. The role of dynamic ultrasound in the immediate conservative treatment of volar plate injuries of the PIP joint: a series of 78 patients. Plast Surg (Oakv). 2017;25(3):151–6.
16. Bianchi S, Della Santa D, Glauser T, Beaulieu JY, van Aaken J. Sonography of masses of the wrist and hand. AJR Am J Roentgenol. 2008;191(6):1767–75.
17. Robbin MR, Murphey MD, Temple HT, Kransdorf MJ, Choi JJ. Imaging of musculoskeletal fibromatosis. Radiographics. 2001;21(3):585–600.
18. Garcia J, Bianchi S. Diagnostic imaging of tumors of the hand and wrist. Eur Radiol. 2001;11(8):1470–82.
19. Bianchi S, Martinoli C. Ultrasound of the musculoskeletal system. New York: Springer; 2007.
20. Endo Y, Sivakumaran T, Lee SC, Lin B, Fufa D. Ultrasound features of traumatic digital nerve injuries of the hand with surgical confirmation. Skelet Radiol. 2021;50(9):1791–800.
21. Fan Z, Wu G, Ji B, Wang C, Luo S, Liu Y, et al. Color Doppler ultrasound morphology of glomus tumors of the extremities. Springerplus. 2016;5(1):1319.
22. Reynolds DL Jr, Jacobson JA, Inampudi P, Jamadar DA, Ebrahim FS, Hayes CW. Sonographic characteristics of peripheral nerve sheath tumors. AJR Am J Roentgenol. 2004;182(3):741–4.
23. Fornage BD. Glomus tumors in the fingers: diagnosis with US. Radiology. 1988;167(1):183–5.
24. Ham KW, Yun IS, Tark KC. Glomus tumors: symptom variations and magnetic resonance imaging for diagnosis. Arch Plast Surg. 2013;40(4):392–6.
25. Hergan K, Mittler C, Oser W. Pitfalls in sonography of the Gamekeeper's thumb. Eur Radiol. 1997;7(1):65–9.
26. Ebrahim FS, De Maeseneer M, Jager T, Marcelis S, Jamadar DA, Jacobson JA. US diagnosis of UCL tears of the thumb and Stener lesions: technique, pattern-based approach, and differential diagnosis. Radiographics. 2006;26(4):1007–20.
27. Moore BJ, Iafrate JL, Kakar S, Wisniewski SJ, Murthy NS, Smith J. Accuracy of ultrasound compared to magnetic resonance imaging in the diagnosis of thumb ulnar collateral ligament injuries: a prospective case series. J Ultrasound Med. 2021;40(6):1251–7.
28. Chiavaras MM, Jacobson JA, Yablon CM, Brigido MK, Girish G. Pitfalls in wrist and hand ultrasound. AJR Am J Roentgenol. 2014;203(3):531–40.
29. Wajid H, LeBlanc J, Shapiro DB, Delzell PB. Bowler's thumb: ultrasound diagnosis of a neuroma of the ulnar digital nerve of the thumb. Skelet Radiol. 2016;45(11):1589–92.
30. Oo WM, Deveza LA, Duong V, Fu K, Linklater JM, Riordan EA, et al. Musculoskeletal ultrasound in symptomatic thumb-base osteoarthritis: clinical, functional, radiological and muscle strength associations. BMC Musculoskelet Disord. 2019;20(1):220.
31. Papp SR, Athwal GS, Pichora DR. The rheumatoid wrist. J Am Acad Orthop Surg. 2006;14(2):65–77.
32. Kotob H, Kamel M. Identification and prevalence of rheumatoid nodules in the finger tendons using high frequency ultrasonography. J Rheumatol. 1999;26(6):1264–8.

Chapter 5
Hand Arthropathies

Reasons to Do the Study
1. Hand pain
2. Hand stiffness
3. Hand swelling, diffuse, or localized

Questions We Want Answered
1. What is the cause of hand pain?
2. Is there an underlying inflammatory condition affecting the hand, and if so, is it a systemic process such as spondyloarthropathy, rheumatoid arthritis, or crystal disease?
3. If there is an inflammatory condition, is it eroding cartilage or bone?
4. Is there a structural, functional, or repetitive injury causing the hand pain?
5. Is tenosynovitis present, and if so, what is the cause?
6. What is causing soft tissue swelling?
7. What is this lump in the hand?

Necessary Anatomy

Joints

The joint capsule is the outer covering of the synovial recess; the capsule contains a superficial fibrous layer and an inner synovial lining [1]. The joint capsule, ordinarily thin and slightly hyperechoic on ultrasound (US), may or may not be easily discernable. The synovial recess is typically hypoechoic, and we focus on the synovial recess for synovitis grading.

The contents do not represent the views of the U.S. Department of Veterans Affairs or the United States Government.

M. H. Greenberg et al., *Manual of Musculoskeletal Ultrasound*,
https://doi.org/10.1007/978-3-031-37416-6_5

Clinical Comments

The eye may be "the window to the soul," but the hand is the window to the diagnosis. Ultrasound is superb for detecting subclinical synovitis, tenosynovitis, enthesitis, cartilage and bone erosion, crystal deposition, and bony deformity [2–9]. Additionally, US elucidates the presence of ganglion cysts, soft tissue masses, and the tendons' functional abnormalities. The pattern of involvement may lead to a diagnosis [10]. Diagnostically, US combined with radiographs, clinical evaluation, and laboratory data enables clinicians to address the most significant challenges in evaluating polyarthropathy:

1. Differentiating inflammatory from noninflammatory diseases.
2. Delineating the precise inflammatory disease.

The task is daunting when obesity obscures synovitis, in mild or early-onset disease, and when serologies are negative [10]. Diagnosing early seronegative rheumatoid arthritis (RA) and psoriatic arthritis (PsA) has been facilitated by US [11, 12]. Other patients have a history compatible with an inflammatory condition but an unimpressive physical examination. For all these patients, ultrasonic visualization of pathology is invaluable.

So, where does US fit into the algorithm of polyarthropathy evaluation? The answer is beyond the scope of this text but is superbly discussed elsewhere [10]. At our institution, we evaluate the dorsal wrist, volar wrist, hand, and, if necessary, finger entheses for diagnostic purposes. We follow a standard protocol but also focus on areas of pain and physical examination abnormalities. Our protocol is constantly evolving to reflect changes in technology and concepts in this developing area. At this point, US is *not* recommended to guide the treatment intensity in RA [13, 14]. However, for the individual patient with polyarthritis, US can be beneficial for:

Diagnosis: Ultrasound detects synovitis, cartilage, and bone erosion, tenosynovitis, power Doppler (PD) activity, enthesitis, and crystal disease.
Treatment: Ultrasound can detect persistent subclinical active synovitis in patients who are not improving clinically.
Prognosis: Ultrasound can assess the presence of bone and cartilage erosion.
Disease progression: Ultrasound can gauge the rate of progression of cartilage and bone erosion.

Synovitis

Synovitis and Ultrasound

OMERACT definitions [15]:

Grayscale (GS): Hypoechoic abnormal hypertrophied synovial tissue within the capsule that is not displaceable and poorly compressible; effusion does not need to be present.

Power Doppler: Power Doppler activity may or may not be present; PD does not need to be present to diagnose synovitis.

Ultrasound is superior to clinical examination for the detection of subclinical synovitis [4, 16, 17]. In RA, US is superior to plain radiographs and comparable to MRI for the detection of synovitis [16, 18]. Disease activity response to treatment in RA is also reflected in changes in synovitis on US [18–23]. Power Doppler findings of the hand and wrist have shown excellent correlation with other clinical measurements of disease activity in RA [24, 25]. Notably, most RA patients in clinical remission may continue to have joint synovitis that is undetected by all clinical parameters [26]. The presence and activity of synovitis correlate directly with the progression of joint erosion.

Since US informs about the presence and persistence of synovitis, and such activity seems to correlate with response to treatment and potential for structural damage, it would seem logical that US might be a reasonable means to guide the treatment intensity of RA patients. However, two studies did not demonstrate improved patient outcomes when using PD activity in joints to guide treatment intensity [13, 14]. Another study using US findings to guide early RA treatment intensity did not reduce MRI inflammation or result in less structural damage [27]. Perhaps future studies of longer duration may better clarify whether US has a role in treatment guidance. We find US invaluable for detecting synovitis as a diagnostic tool. For an individual patient, we perform US when we suspect disease activity that is not clinically apparent to help decide whether to alter treatment [14]. We do not use it routinely to guide treatment intensity, however.

Ultrasonic Grading of Synovitis

There is no standardized scoring system or even agreement about which joints to evaluate. We describe several scoring systems for synovitis, erosion of cartilage and bone, tenosynovitis, and enthesitis. Over the years, several synovitis scoring systems have been described [28]. This discussion pertains to individual joint synovitis scoring. Composite multiple-joint scoring to derive a total score is a separate issue beyond the scope of this Manual [29].

The EULAR-OMERACT individual joint scoring system has shown moderate to good reliability in the metacarpophalangeal (MCPJ) [30]. We use this scoring system in our practice. Although most methods have been validated for RA, some rheumatologists apply them for practical purposes to PsA. The US findings of synovitis, effusion, and erosion detection are similar in these two diseases [31]. Scoring systems propose rules to encourage reliable readings among sonographers, characterize disease severity, and monitor treatment efficacy in multi-institutional studies. However, in our institution, we simply note the presence and degree of synovitis.

We grade (score) each synovial joint recess for synovial thickening on GS (grade 0–3) and PD (grade 0–3) [11, 32, 33]. Recognizing the lack of synovitis (grade 0) or severe synovitis (grade 3) is relatively easy. If we are uncertain whether the joint has a grade 1 or grade 2 score, we sometimes score it as grade 1–2. To avoid observer interpretation drift, we compare previous and current images. In our clinic, a hand synovitis examination consists of the dorsal radiocarpal joint (RCJ) and the intercarpal (or midcarpal) joint (ICJ), the dorsal MCPJs of digits 2, 3, and 5, the dorsal proximal interphalangeal joints (PIPJs) of digits 2 and 3, and the dorsal distal ulna (DU). We add an evaluation of the triangular fibrocartilage complex (TFCC) when considering calcium pyrophosphate deposition disease (CPPD). For finger enthesitis evaluation, we examine the distal dorsal entheses of the extensor tendons in digits 2 and 3. Other symptomatic or grossly anomalous sites are similarly evaluated.

If considering tenosynovitis, we examine the extensor tendons at the wrist, the extensor carpi ulnaris (ECU), and the flexor tendons of the volar wrist and MCPJs. We also grade any PD activity near the DU, which may indicate active synovitis [11, 34]. This protocol is subject to change based on validity testing and the clinical utility of future novel scoring systems.

Grayscale Synovitis Grading (Table 5.1)

Synovitis of the dorsal MCPJ (independent of an effusion) is scored by GS as [30, 33] follows:

0: Absence of hypoechoic areas within the joint
1: Minimal (hypoechoic synovial hypertrophy [SH] up to the level of the imaginary horizontal line connecting the (superficial) bone surfaces between the metacarpal head and the proximal phalanx)
2: Moderate (hypoechoic SH extending above the horizontal joint line but with the upper surface of the SH being flat or concave)
3: Severe or marked (hypoechoic SH extending beyond the joint line with the upper surface of the SH being convex)

The EULAR-OMERACT ultrasound task force excluded synovial fluid when scoring synovitis, recognizing that effusion, albeit a proxy measure of inflammation, did not add additional information regarding synovitis definition and severity [33]. In actual practice, you will observe structures that may appear as hypoechoic or even anechoic fluid but may be synovitis. Probe compression within a closed space, such as a joint capsule, may not differentiate synovial fluid from synovitis. Thus, we include the presence and extent of such structure in our report if there is no other clear-cut evidence of synovitis.

Table 5.1 Summary of grayscale synovitis scoring [30, 33]

Summary of grayscale synovitis scoring
0 = Absence (no enlargement of the recess)
1 = Mild (minimal effusion or synovial hypertrophy)
2 = Moderate (moderate effusion/hypertrophy; the joint capsule bulges beyond the normal confines)
3 = Severe or marked (extensive effusion/hypertrophy)

Table 5.2 Summary of power Doppler synovitis scoring [30, 33]

Summary of power Doppler synovitis scoring
0 = Absence (no intra-articular color signal)
1 = Mild (up to three color signals, two single and one confluent signal, or two confluent signals in the intra-articular area)
2 = Moderate (greater than grade 1 to ≤50% of the intra-articular area filled with color signals)
3 = Marked (>50% of the intra-articular area filled with color signals)

Power Doppler Grading (Table 5.2)

Synovitis is scored by PD activity within the GS-defined synovium as [11, 33] follows:

0: Absence (no intra-articular color signal)
1: Mild (up to three color signals, two single and one confluent signal, or two confluent signals in the intra-articular area)
2: Moderate (greater than grade 1 but ≤50% of the intra-articular area filled with color signals)
3: Marked (>50% of the intra-articular area filled with color signals)

Combined Scoring System

Combining the GS and the PD scoring has been proposed as another approach to quantifying synovitis sonographically [33].

Summary of combined synovitis scoring scale:

0: Normal (no SH on GS + no PD signal)
1: Minimal (grade 1 SH on GS + ≤ grade 1 PD signal)
2: Moderate (grade 2 SH on GS + ≤ grade 2 PD—or—grade 1 SH on GS + grade 2 PD signal)
3: Severe (grade 3 SH on GS + ≤ grade 3 PD—or—grade 1 or 2 SH on GS + grade 3 PD signal)

Practical Synovitis Grading of Specific Joints in the Wrist and Hand

The joint capsule is graded for synovial proliferation by GS and neovascularization by PD [11, 33]. We recommend the atlas-based approach for synovitis grading; the sonographer directly compares the US image with various examples of GS and PD synovitis [34].

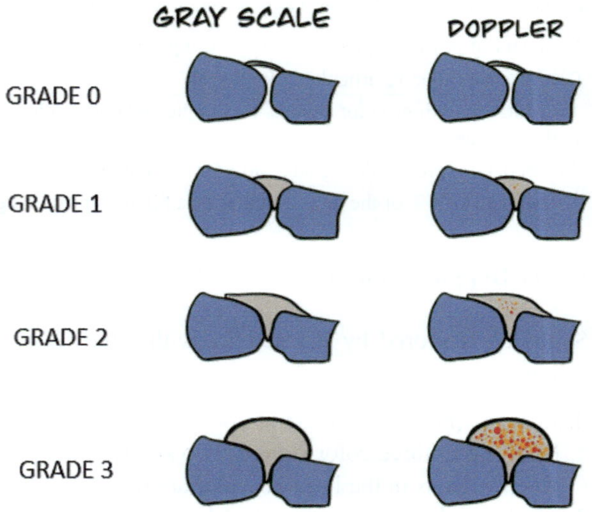

Fig. 5.1 Dorsal metacarpophalangeal joint synovitis in grayscale and power Doppler

Dorsal Metacarpophalangeal Joint Synovitis (Fig. 5.1)

Synovial hypertrophy in the dorsal recess will expand superficially to or beyond an imaginary line between the superficial metacarpal (MC) head and the proximal phalanx; further hypertrophy occurs in a proximal and distal direction [33]. Other joints follow a similar pattern. Please refer to the atlas for specific examples [34].

Dorsal Radiocarpal and Intercarpal Joint Synovitis (Fig. 5.2)

We evaluate these two joints in longitudinal (LAX) to define GS synovitis based on joint capsule distension and synovial swelling superficial to an imaginary line (beta-line) between the superficial distal radius and the proximal superficial capitate (Fig. 5.3) [35]. Furthermore, we score the RCJ and the ICJ separately for both the GS activity and PD activity. However, be aware that many experts score these two joints together. In the wrist, GS reflects SH as hypoechoic; we are interested in PD activity only within the area of SH as defined by GS [11]. Since we are seeking PD activity *within* the joint recess, PD activity superficial to the joint recess may simply represent normal vasculature rather than actual neovascular change within the joint recess [35].

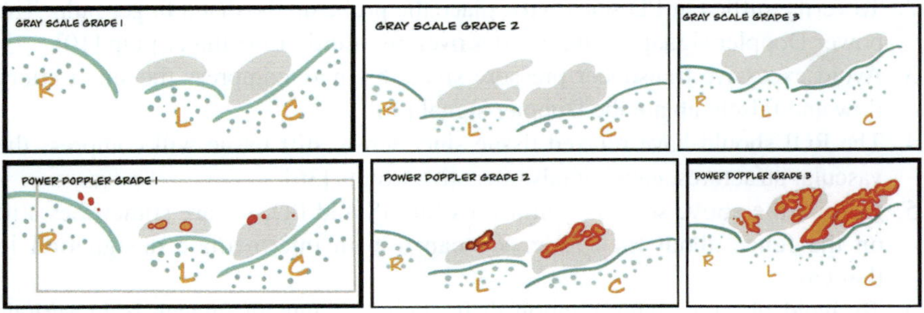

Fig. 5.2 Dorsal radiocarpal and intercarpal joint synovitis in grayscale and power Doppler

Fig. 5.3 Beta-line

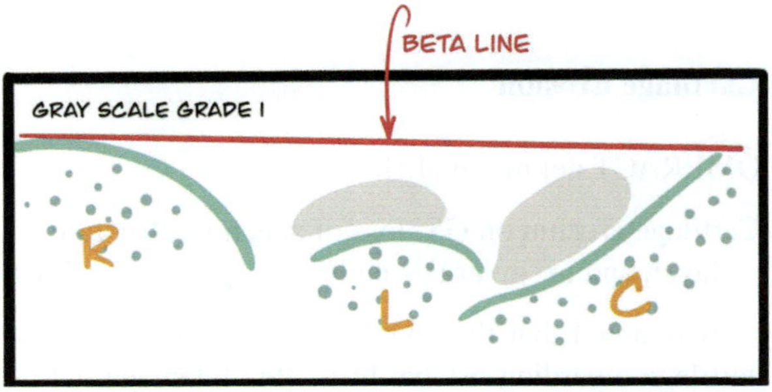

Synovitis Grading Hints

1. Optimize the PD settings.

   The goal is to increase Doppler signal detection sensitivity while minimizing random electrical noise and vessel wall motion artifacts [36, 37].

   (a) Turn up the gain until random noise is encountered, and then slowly lower it until the noise mostly or entirely disappears [37, 38].

   (b) Use the lowest pulse repetition frequency (PRF) that does not produce excessive noise. The lower the PRF, the lower the linked wall filter, and thus the higher the sensitivity to blood flow. Since our goal is to detect minimal blood flow, we want to maximize sensitivity but not so much as to create excessive random noise [37].

   (c) Use the smallest Doppler box possible to encompass the region of interest (ROI).

   (d) Whenever possible, recalibrate Doppler settings for each patient. However, use identical settings on the same US device when following the Doppler activity of an individual patient longitudinally [37].

2. To verify sufficient PD sensitivity, place the probe on the distal finger pulp. The power Doppler signal should register over one-third of the finger pulp [10].
3. Avoid excessive transducer pressure since this may compress the small artery flow and falsely ablate the Doppler signal [36].
4. The ROI should have relaxed tissue since tense, stiff tissue will compress the vascular structures, causing false-negative results [36].
5. Despite the above scoring definitions for GS and PD, we are sometimes still on the fence about the grade. We can use an intermediate designation in that case.
6. To avoid observer interpretation drift, directly compare images from various timeframes to gauge any change.
7. In our experience, the joint with the highest diagnostic yield is the MCPJ.

Cartilage Erosion

OMERACT definition [15]:

Cartilage erosion on GS: loss of anechoic structure and/or thinning of the cartilage layer, and irregularities and/or sharpness of at least one cartilage margin.

It is noted that the above definition, as defined by OMERACT, is for hyaline cartilage regarding osteoarthritis (OA) damage. Ultrasound reliably detects MCPJ cartilage pathology [39]. Evaluation of erosions can be done concurrently with synovitis grading. A recently proposed grading system for cartilage erosion is a bit different from other synovitis grading systems in that there are only three categories, namely, 0, 1, and 2 (Fig. 5.4) [40].

Grade 0 (normal) shows robust anechoic/hypoechoic cartilage of uniform thickness and a sharp outer edge in two orthogonal (90°) views.

Grade 1 (minimal change) shows cartilage with some focal thinning in areas or incomplete loss of cartilage.

Grade 2 shows cartilage thinning or complete loss of cartilage, either focally or diffusely.

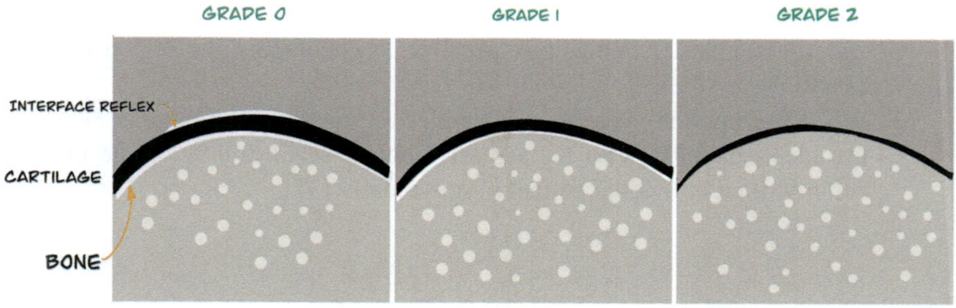

Fig. 5.4 Scoring cartilage erosion

This novel system is a pragmatic grading system since it is relatively easy to recognize normal, unadulterated cartilage (grade 0) and severely damaged cartilage (grade 2). Grade 1 lies somewhere in between those two extremes. Examining the MCPJ using this grading system has shown good inter- and intra-reader reliability [40]. Whichever system is used, it is most important to perform orthogonal views of as many cartilage segments as possible.

A technique we use to evaluate the MC head cartilage more thoroughly is for the examiner to gently grasp the patient's prone hand from palm side to palm side. The examiner places the fingertips under the volar aspect of the middle phalanges and, with the opposite hand, gently holds the probe over the dorsal MCPJ. The MCPJ is flexed and extended by the examiner to demonstrate a more considerable extent of the cartilage dynamically. We look at the cartilage interface reflex to delineate the superficial aspect and smoothness of each segment of cartilage perpendicular to the transducer. This technique has not been validated to our knowledge.

Normative US-derived values of MC head cartilage thickness are available [41]. Males have thicker cartilage than females. Interestingly, about 18% of normal people had low-grade cartilage changes. Be aware that some athletes (such as weightlifters) may have increased MC head thickness [42]. It is important to note that cartilage damage at the MCPJs may occur at least as often in OA as it does in RA; however, in OA, the cartilage damage is more often distributed equally at MCPJs 2 through 5, whereas RA-induced cartilage damage predominantly affects MCPJs 2 and 3 [43].

Bone Erosion (Fig. 5.5)

OMERACT definition [15]:

Bone erosion on GS: Intra- and/or extra-articular discontinuity of bone surface visible in two perpendicular planes.

Look for step-offs, divots, or gaps in areas of bone that are usually smooth and uninterrupted [44]. In RA, bone erosion has diagnostic utility and is indicative of a more severe disease [45]. Standard radiographs lack sensitivity compared to US for detecting bone erosion [46, 47]. One study, using CT as the gold standard, determined the sensitivity of US and radiographs for detecting RA bone erosion was 43% and 19%, respectively [48].

The highest yielding areas for RA erosions are the lateral quadrants of the second and fifth MC heads and the fifth metatarsal (MT) head; the US examination of these areas found bone erosion in 60% of patients with early RA. A cutoff of ≥ 2.5 mm for erosion size maximizes sensitivity and specificity for RA erosions [49].

Another study described US-detected erosions with a false positivity rate of up to 29% due to the high sensitivity to cortical irregularity that can occur when osteophytes produce false erosions (pseudoerosions) [50]. A second type of pseudoerosion on US may result from normal bone contours being misinterpreted as erosion, such as in the dorsal MC heads, the lunate, triquetrum, and ulna [51].

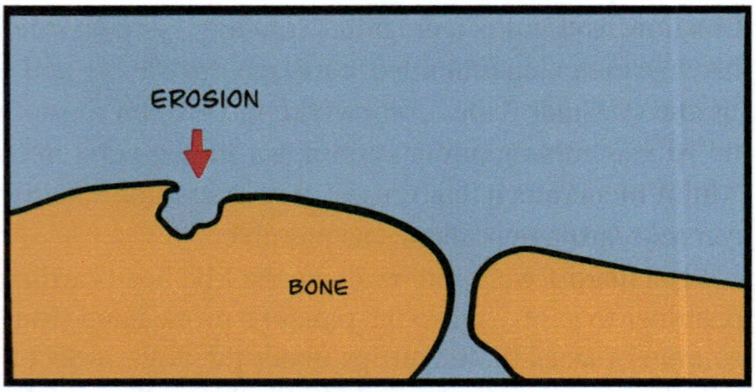

Fig. 5.5 Bone erosion

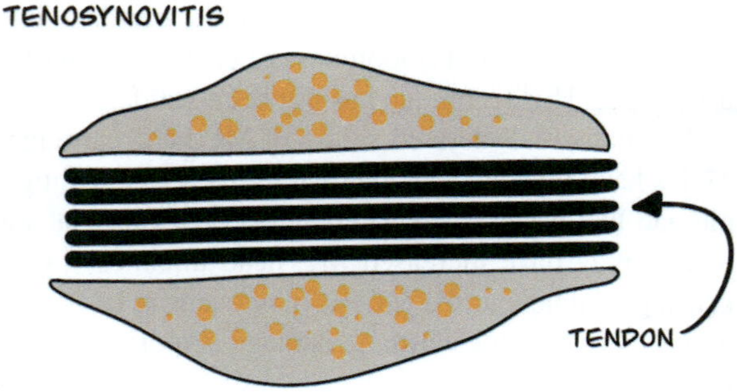

Fig. 5.6 Tenosynovitis

Tenosynovitis (Fig. 5.6)

The tenosynovium is the synovial-lined sheath encasing a tendon that may become inflamed due to repetitive trauma or systemic inflammation.

OMERACT definition [15]:

Tenosynovitis on GS: Tenosynovial hypertrophy is present when abnormal anechoic and/or hypoechoic (relative to tendon fibers) tissue within the synovial sheath is non-displaceable and poorly compressible, and seen in two perpendicular planes. There is abnormal tendon sheath widening due to abnormal tenosynovial fluid and/or hypertrophy.

Tenosynovitis on PD: The tissue within the tenosynovium may or may not exhibit Doppler signals. However, consider the Doppler signal to be significant only in the presence of peritendinous synovial sheath widening on GS and only if:

1. It is noted in two perpendicular planes.
2. It is present *within* the peritendinous synovial sheath.
3. It is not due to normal feeding vessels entering the synovial sheath from surrounding tissues.

Tenosynovitis may also occur due to mechanical compression of the tendon and tenosynovial sheath, known as stenosing tenosynovitis. The classic example of tenosynovitis is de Quervain's tenosynovitis (also called stenosing tenosynovitis) of the first extensor compartment (EC) at the wrist. Ultrasound is beneficial for diagnosing and treating this condition [52]. Another example of stenosing tenosynovitis is the trigger finger, involving the flexor tendons. Tenosynovitis may occur with inflammatory arthropathies such as rheumatoid arthritis, spondyloarthropathies, and sarcoidosis. Tenosynovitis may cause compression of the median nerve within the carpal tunnel.

Ultrasound is more accurate than clinical examination when detecting flexor tenosynovitis [53]. Ultrasound of the flexor tendon sheath has helped predict which patients will develop persistent RA with polyarthritis [54]. In addition, extensor tenosynovitis in RA correlates with increased disease activity [55]. OMERACT has developed a reliable US scoring system for tenosynovitis in RA [56]. Both GS and PD have four grades (0–3: normal, minimal, moderate, and severe). This scoring system has been shown to reliably detect changes over time [54]. In a small multicenter study, tenosynovitis score changes did show substantial responsiveness to RA treatment, with a good correlation with the Disease Activity Score but not with the Health Assessment Questionnaire—Disability Index [57]. The most involved tendon sheath in this RA study was that of the ECU. It remains to be seen if *tenosynovitis scores* should guide treatment intensity in RA.

Paratenonitis

A paratenon, tissue lacking a proper synovial membrane, surrounds the finger extensor tendons superficial to the dorsal MCP joints. Paratenonitis sonographically presents as hyperemia or anechoic fluid along the extensor tendons [58]. It is most often associated with an underlying inflammatory polyarthropathy such as RA. The underlying joint frequently has evidence of synovitis on US; however, paratenonitis may occur as an isolated finding, perhaps due to joint deformity or overuse.

Enthesitis

The enthesis (Fig. 5.7) is the insertional point for attachment of not only tendons but also ligaments, fascia, muscles, or joint capsules [59].

OMERACT definition [15]:

Enthesitis on GS: Hypoechoic and/or thickened tendon insertion <2 mm from the bony insertion; it may be accompanied by enthesophytes, calcifications, or bony erosions.

Enthesitis on PD: Doppler signal <2 mm from the bony insertion.

For the hand, the prime location to examine the enthesis is at the tendon insertion onto the bone in the distal phalanges. Sonographically accessible areas for enthesis evaluation include the insertions of the flexor and extensor tendons on the phalanges [60–62]. Enthesitis is predominantly the domain of spondyloarthropathy (SpA), in particular PsA. Given that 15% of patients with PsA have no evidence of psoriasis before the arthritis presentation [63], confirmation of enthesitis may be supportive

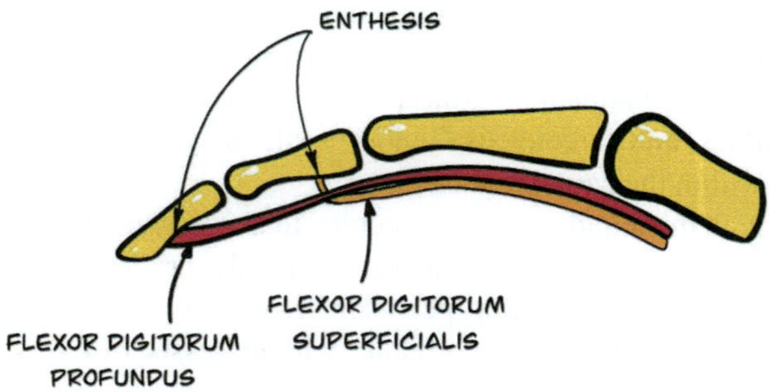

Fig. 5.7 Enthesis

evidence for the disease. In PsA, enthesitis reportedly occurs in 35–50% of patients [64]. However, enthesitis may also occur in RA, gout, crystalline diseases, sarcoidosis, juvenile idiopathic arthritis, diffuse idiopathic skeletal hyperostosis, and OA, albeit less frequently than with SpA.

In the latest classification system for SpA, enthesitis is a significant criterion for diagnosing both peripheral and axial SpA. However, for this system, the presence or absence of enthesitis was derived clinically, not sonographically [59]. Enthesitis is clinically underdiagnosed by physical examination [59]. Ultrasound has emerged as the preferred method for enthesitis detection due to its sensitivity, structural delineation, and functional evaluation [64, 65]. Ultrasound may reveal subclinical enthesitis.

A scoring system for enthesitis has been developed, using 0–3 grading [66]:

Grayscale grades:

0: Normal
1: Hypoechogenicity
2: Hypoechogenicity + thickening + calcifications or enthesophytes
3: Grade 2 features + erosions

Power Doppler grades:

0: No PD signal
1: <2 punctate signals
2: 2–4 punctate signals or 1 confluent signal
3: >4 punctate signals or >1 confluent signal

A Doppler signal within 2 mm of the bony cortex is arbitrarily accessed for scoring, although enthesitis may generate PD even 5 mm from cortical bone [67]. When evaluating for enthesitis, focus on the tendon's insertional points, particularly the extensor tendons of fingers. Look for loss of tendon echotexture, the presence of bony enthesophytes, and power Doppler activity. If present, score the enthesitis as described above.

For psoriatic arthritis, US gives important diagnostic clues [68]:

1. Enthesitis may occur at the insertion of the extensor tendon on the bony distal phalanx. Nail abnormalities in PsA are related to enthesopathy beneath the nail bed [69, 70].
2. Enthesitis may present at the insertion of the central slip of the extensor tendon on the PIPJ.

3. Paratenonitis, that is, inflammation (hypoechoic swelling with or without PD), especially surrounding both the extensor tendon and paratenon over the MCPJ [71].
4. Dactylitis is a sausage digit that is hypoechoic with diffuse soft tissue swelling, especially surrounding the flexor tendons [72, 73].

In our experience, sonographic evaluation of fingers for enthesopathy requires a high-definition probe set at a minimum of 16 MHz to look at the extensor and flexor entheses of the fingers. Grade enthesitis according to the proposed scoring system and scrutinize for PD activity. A gentle examiner's touch is needed to preclude a reduction in the Doppler signal. Likewise, the finger should be in a relaxed, flat, neutral position since putting tension on the tendon at the enthesis might diminish PD activity [74]. If sonography documents enthesitis with positive PD, place this information into the appropriate clinical context. In summation, US may confirm a clinical impression of enthesitis and detect subclinical enthesitis. Nonetheless, sonographic demonstration of enthesitis supports a SpA diagnosis but is not pathognomonic.

(See Chap. 25 for more information.)

Crystal Deposition

In the hand, we look for crystal deposition within the TFCC and the MC head cartilage. Perform US on any suspicious area noted on radiographs or any site described by the patient as painful.

Gout

More commonly, gout starts in the lower extremity, such as the first metatarsophalangeal joint (MTPJ), and later involves the upper extremities. In the hand, gout may lead to acute, subacute, or chronic inflammation of the dorsal MCPJ and PIPJs, mimicking RA [74]. Alternatively, structural deformity from chronic tophi may be the prominent clinical feature. In patients with gout, monosodium urate (MSU) crystals may deposit on cartilage surfaces forming the sonographic double contour sign (DCS). This thick hyperechoic line on the cartilage surface parallels the bony cortex, giving the appearance of an inverse sandwich cookie. The DCS is the most sensitive US sign of gout in the hand and wrist [75].

OMERACT definition of DCS [15]:

Grayscale: An abnormal hyperechoic band over the superficial margin of the articular hyaline cartilage, independent of the angle of insonation, which may be either irregular or regular, continuous, or intermittent, and can be distinguished from the cartilage interface sign.

Gout may also form tophi, which are MSU crystals surrounded by inflammatory cells and connective tissue that resemble granulomas [76]. Tophi are appreciated by US.

OMERACT definition of tophus [15]:

Grayscale: A circumscribed, inhomogeneous, hyperechoic, and/or hypoechoic aggregation (which may have a posterior acoustic shadow) and may be surrounded by a small anechoic rim (halo).

Gout tophi may be seen in any location, including synovial recesses, tendon sheaths, and even the finger pads [77]. Tophi on radiographs may imitate erosions from RA but appear as an asymmetric bony punched-out region with a narrow cortex and overhanging bony edge. Severe joint and bone destruction may occur with chronic tophaceous gout. Although gouty tophi may affect tendons, less commonly will gout cause tenosynovitis [77].

Gout aggregates may also be appreciated by US.

OMERACT definition of gout aggregates [15] the following:

Grayscale: Heterogeneous hyperechoic foci that maintain their high degree of reflectivity, even when the gain setting is minimized, or the insonation angle is changed, and which occasionally may generate a posterior acoustic shadow.

Calcium Pyrophosphate Deposition Disease.

Acute CPPD may present in the hands as localized inflammation and pain, often confused with a gouty flare, hence the name "pseudogout" [78].

OMERACT definitions [15]:

CPPD in fibrocartilage on GS: Hyperechoic deposits of variable shape, localized within the fibrocartilage structure, that remain fixed or move along with the fibrocartilage during a dynamic assessment.

CPPD in hyaline cartilage on GS: Hyperechoic deposits of variable size and shape, without posterior shadowing, localized within the hyaline cartilage that remain fixed and move along with the hyaline cartilage during a dynamic assessment.

CPPD in tendons on GS: Hyperechoic, linear structures generally without posterior shadowing, localized within the tendon, that remain fixed and move along with the tendon during a dynamic assessment.

In the hand, CPPD tends to involve the RCJ, the ICJ, and all the MCPJs, joints not typically involved with OA. There may be concomitant osteophytes and wrist tenosynovitis, but without the marginal erosions seen in RA [79]. Involvement of the MCPJs may show clinical swelling, localized synovitis, and cartilage destruction, thus imitating RA ("pseudorheumatoid") [80]. Suspect CPPD if the soft tissue distal to the ulnar styloid is calcified on a radiograph [79]. However, most patients with this radiographic finding have no symptoms [81]. Bony hooks on the radial side of the MC head may occur with CPPD [82].

If findings in the hand and wrist suggest CPPD, US imaging of other areas, such as the knee, may help confirm evidence of more diffuse calcium deposition. In one study of CPPD patients, US demonstrated calcium deposition in femoral cartilage at a rate of three to four times that of the wrist [83]. Although US is becoming more recognized as a useful imaging modality for CPPD [84], distinguishing the DCS of gout from chondrocalcinosis may be challenging [85].

(See Chap. 24 for more information.)

Masses

A suggested approach to the evaluation of masses in the hand:

1. Assess the clinical setting, such as a history of polyarthropathy, lipomas in other areas, penetrating trauma, repetitive movement history, precipitating/aggravating activities, and the presence of pain.
2. Capture sonographic orthogonal images of the mass, noting echotexture, shape, compressibility, border definition, the presence of a stalk, Doppler activity, relation to adjacent structures, and the effect of nearby dynamic movement.
3. Based on the degree of confidence in the sonographic and clinical diagnosis, choose to follow clinically, pursue further imaging, or refer to a hand surgeon.
4. Sonographic measurement in two dimensions may help determine size change over time [86].
5. Note that a palpable mass may not be easily detectable on US if it is isoechoic with the surrounding tissue.

Ganglion Cysts

Ganglion cysts are the most common masses of the wrist and hand [87]. These cysts are filled with a thick fluid, lined with a fibrous capsule, and may be round, soft or firm, painless, or tender on clinical examination [86]. Ganglia have no synovial lining and are hence different from synovial cysts. With a synovial cyst, the synovial membrane herniates through a joint capsule. However, US cannot reliably differentiate between the two [88]. Many clinicians use the terms ganglion, ganglion cyst, and synovial cyst interchangeably.

Ganglion cysts may occur due to trauma, arthritis of any cause, tendon injury, or tenosynovitis. They result from connective tissue herniation from joints, tendon sheaths, ligaments, joint capsules, or bursae. We frequently find clinically undetectable (occult) ganglion cysts [86]. Most ganglion cysts occur at the wrist from either the dorsal capsule near the scapholunate ligament or the radial aspect of the volar wrist between the radial artery and the flexor carpi radialis. It is debatable which of these two locations is the most common [89].

Sonographically, ganglion cysts appear hypoechoic or anechoic, are well-defined, and may show septa [86]. Increased posterior enhancement may be noted [90]. There may also be Doppler activity, which may or may not correlate with symptoms [86]. The volar ganglion may occur within the carpal tunnel and compress the median nerve. Flexor tenosynovitis may mimic a volar ganglion cyst [86]. The differential diagnosis for any ganglion includes tenosynovial giant cell tumor, epidermoid cyst, lipoma, tenosynovitis, rheumatoid nodule, gouty tophus, tendon xanthoma, and synovial sarcoma [91]. If you detect a mass that may be a ganglion, search for a stalk and follow it down to its point of origin to confirm the diagnosis and the underlying structural issue responsible for the ganglia.

Accessory Muscles

Accessory muscles may mimic a tumor and compress nerves. The most common accessory muscle of the wrist is the accessory abductor digiti minimi, present in approximately 4% of healthy people [92]. This particular accessory muscle may compress the ulnar nerve during contraction within Guyon's canal [93]. The ulnar nerve, usually round, may appear ovoid, especially with abduction of the fifth finger, if it is impinged by this accessory muscle [86].

Extensor digitorum brevis manus is present in about 2% of healthy adults and inserts on the index and middle finger dorsal aspect; it appears to be a fusiform lump alongside the extensor tendon mimicking a ganglion, extensor tenosynovitis, or a soft tissue tumor [86]. A third pseudotumor due to a muscle variant is the inverted palmaris longus, where the muscle belly is distal and the tendon is proximal. The muscle belly is present in the volar wrist and may present as a mass [94, 95].

Traumatic Disorders

Tendon Tear

With a complete tendon tear, the ends of the retracted tendons may appear as localized masses [86]. Ultrasound delineates these tendon stumps' locations and helps determine surgical repair procedures.

Foreign Bodies

Foreign bodies may appear as localized masses surrounded by either an abscess or chronic, granulomatous tissue [86]. An abscess manifests as a heterogeneous, poorly defined fluid collection with hypervascular changes noted on color Doppler. A granuloma appears as a hypoechoic halo around a foreign body.

(See Chap. 15 for more information.)

Hematomas

These appear as fluid collections within subcutaneous tissues and may demonstrate a fluid-fluid level [86].

Tumors, Pseudotumors, Other Lumps, and Bumps

Tenosynovial Giant Cell Tumors

(See Chap. 4 for more information.)

Schwannomas and Neurofibromas

Schwannomas and neurofibromas may affect nerves in the hand and wrist. These are also well-defined hypoechoic masses contiguous with a nerve and may have posterior acoustic enhancement [96]. You may see the nerve entering the tumor and exiting distally. Differentiation between schwannomas and neurofibromas is challenging if based exclusively on US findings [86].

Hemangiomas

These may be intramuscular and appear as anechoic to hypoechoic lesions with mixed internal echoes and vascular channels associated with hyperechoic fat [86]. Doppler typically shows low flow without evidence of arteriovenous shunting.

Lipomas

Lipomas are typically soft and painless masses. Echogenicity is variable and relates to adjacent soft tissue [97]. Lipomas are well-demarcated, homogeneous, and most commonly hyperechoic. Differentiating a lipoma from a liposarcoma can be difficult, and pathologic verification is required [86].

Pseudoaneurysms

Pseudoaneurysms are false aneurysms that develop when an artery is damaged by blunt or penetrating trauma [98]. An example of chronic blunt trauma causing a pseudoaneurysm includes hypothenar or thenar hammer syndrome. A sac forms outside the artery wall as blood dissects from the vessel. Doppler reveals high flow in the artery adjacent to the cystic mass. Pseudoaneurysms may involve the palmar arch but rarely directly affect the fingers, although emboli can cause digital ischemia and gangrene [98, 99]. The best diagnostic procedure is an angiogram, but US can be considered an initial modality for evaluation [99–101].

Rheumatoid Nodules

Rheumatoid nodules from RA may occur in any location. Most, but not all, patients with RA nodules are seropositive [102]. The nodules, often adjacent to bone, sonographically demonstrate slightly blurred edges with mixed hypoechoic echotexture [103]. There may be a central hypoechoic area, perhaps due to central necrosis. The nodules are compressible with no posterior shadowing and no Doppler activity.

Osteophytes

Osteoarthritis typically involves the interphalangeal finger joints and the first carpometacarpal joint (CMCJ). Osteoarthritis less commonly affects the MCPJs and the wrist. First CMCJ OA commonly causes pain and disability, particularly in older populations. In a study of 93 participants, plain radiographs and US comparably detected osteophytes, with US being more sensitive [104]. Erosion detection was also similar, although US may not detect some erosions due to occluding osteophytes. Power Doppler activity in the first CMCJ correlated with increased severity of thumb base pain [104]. However, pain and function are poorly associated with radiographic and US structural findings. An osteoarthritic osteophyte appears as a step-up bony prominence on US [15]. Enthesophytes, new bone growth associated with tendon insertions, most often occur in spondyloarthritis and may mimic osteophytes on US. However, plain radiographs will often differentiate between an enthesophyte and an osteophyte.

Glomus Tumor

(See Chap. 4 for more information.)

Neuromas

Small-caliber palmar digital nerves are prone to injury during finger laceration [105]. The injured nerve responds by forming a neuroma. High-frequency US helps diagnose traumatic digital nerve injuries. The TAX view may reveal a loss of nerve continuity confirmed in the LAX view [105]. The neuroma itself may appear as a hypoechoic, well-defined mass, lacking internal vascularity at the end of a normal-appearing nerve. Sonopalpation over the neuroma may reproduce pain.

Using Ultrasound to Aid in the Diagnosis of Arthropathy

Certain arthropathies tend to involve distinct areas of the hands, and sonographic results should be placed in the proper clinical context.

Specific Conditions

Rheumatoid Arthritis

Rheumatoid arthritis often involves the ECU tenosynovium and may cause ulnar styloid bone erosion. Synovitis and cartilage erosion may also occur with RA at the wrist, MCPJs, and PIPJs. Look for bone erosion along the radial aspect of the second MC head and the ulnar aspect of the fifth MC head since these areas are amenable to fuller US interrogation [10, 47, 106].

Spondyloarthropathy

Spondyloarthropathy (SpA), in particular PsA, may involve the finger and wrist joints and cause enthesitis of the extensor tendon, the collateral ligaments, and the palmar plate, paratenonitis, tenosynovitis, bone/cartilage erosion, and bone proliferation [10].

Gout

Look for a double contour sign over cartilage, urate aggregates at tendons/ligaments, and tophi in any location. Tophi may cause bone erosion with their classic overhanging edges. The wrist joint may be involved with gout. Again, gout typically starts in the foot/ankle, so there is often evidence of lower extremity involvement by the time it affects the hand [10].

Calcium Pyrophosphate Deposition Disease (CPPD)

Evaluate for calcification within cartilage, including the TFCC, the MCPJs, and the scapholunate ligament. The wrist is the most commonly affected by CPPD in the hand [10].

Lupus

Lupus may cause synovitis of the wrist, the MCPJs, and the PIPJs. The concept that lupus arthritis is non-erosive has recently been questioned [17, 107, 108].

Osteoarthritis

Osteoarthritis may damage the MC head cartilage in all five MCPJs compared to RA, which tends to affect the hyaline cartilage covering MC heads 2 and 3. Osteoarthritis-associated synovitis may be seen on GS and color Doppler at the first

CMCJ and interphalangeal joints [109]. Osteophytes may easily be seen at the first CMCJ, the PIPJs, and the DIPJs.

Specific Ultrasound Findings

1. Flexor tenosynovitis within the carpal tunnel may produce carpal tunnel syndrome.
2. Osteoarthritis may have a degree of synovitis and sometimes even PD activity, which can be confused with a true systemic inflammatory arthropathy.
3. Differentiating seronegative RA from PsA can be challenging, particularly if there is no history of psoriasis.
4. Hyperuricemia may be present in a patient with hand polyarthritis, but gout may not be causing inflammation. Demonstrating a double contour sign in the hand or elsewhere does not necessarily mean hand arthropathy symptoms are due to gout.
5. If you detect synovitis of the wrist or MCPJs, look for specific high-yield US findings, which will trigger a search for other sonographic diagnostic clues of inflammatory arthropathy.
   (a) Uncovering ECU tenosynovitis should heighten the suspicion of RA.
   (b) The finding of calcium deposits at the TFCC, the scapholunate ligament, or the hyaline cartilage of the MCPJs brings to mind the possibility of CPPD. Immediate sonography of the distal femoral cartilage may confirm chondrocalcinosis.
   (c) If you come across a double contour sign or tophus in the hands, look for evidence of uric acid deposition at the distal femoral cartilage, the tibiotalar joint, and the first MTPJs.
   (d) Discovering enthesitis at the distal extensor tendons of the fingers, although not pathognomonic for PsA, should prompt a thorough evaluation of SpA.

Pitfalls

1. Do not neglect to recalibrate Doppler settings for each patient; use the lowest gain and lowest PRF that do not produce excessive artifacts or noise.
2. Use identical settings on the same US device when following the Doppler activity of an individual patient longitudinally [37].
3. Verify sufficient PD sensitivity by placing the probe on the distal finger pulp. The power Doppler signal should register over one-third of the finger pulp [10].
4. When evaluating PD activity, use minimal probe compression to avoid dampening blood flow [95].
5. Minimize the ROI box when using PD since this increases the frame rate and helps detect hyperemia. In other words, use the smallest PD activity box possible [95].

6. Do not be fooled by a feeder vessel entering the lunate mimicking PD activity and thus synovitis [110].
7. Intra- and interobserver grading drift and disparities are overcome by comparing current and prior images.
8. When examining the dorsal wrist in LAX, do not mistake the anisotropy of the more superficial extensor retinaculum for tenosynovitis or tenosynovial fluid [95].
9. Do not neglect to evaluate the tendon sheath of the ECU for inflammatory tenosynovitis since this is a high-yield area when considering RA [95, 111, 112].
10. Confirm putative bone erosions with orthogonal views on US. If the presence of an erosion affects treatment, then additional imaging modalities may be necessary.
11. Be careful of the classic "pseudoerosion" of the metacarpal head, which is the normal smooth concavity on the dorsal aspect of the distal metacarpal where the bone shaft joins the metacarpal head [95]. If erosion is suspected, then the presence of contiguous synovial hypertrophy would argue for actual erosion.
12. Interpreting the meaning of isolated SH on GS requires careful consideration of the distribution of the synovitis and the patient history, physical examination, laboratory tests, and radiographs [95]. Such synovitis may or may not indicate underlying inflammatory arthritis.
13. Do not misinterpret a pulsating volar ganglion cyst adjacent to the radial artery as a pseudoaneurysm [95].
14. Do not mistake a pseudoaneurysm for a soft tissue tumor.
15. The dorsal arch of the hand may have feeder vessels superficial to the synovium. These vessels may mimic synovitis but are above the synovium [110]. Make sure you look for PD activity within the boundaries of the synovium as defined on GS.
16. Remember that the synovial recess reflection direction is distal for the radiocarpal and intercarpal joints and proximal for the radioulnar joint, the volar wrist, the MCPJs, and the PIPJs.
17. Synovitis of the MCPJ is often due to RA but can also be due to PsA or crystal disease. Occasionally, erosive OA, also known as inflammatory osteoarthritis, may cause some degree of synovitis in the MCPJs.
18. Do not use US, including Doppler, to guide treatment intensity in RA; this may result in overtreatment.
19. When examining the volar MCPJ for synovitis, note that the usual synovial recess extends proximally and should be included with Doppler analysis.
20. Another cause of pseudotenosynovitis is the normal hypoechoic muscle tapering at the musculotendinous junction. Avoid this pitfall by verifying what appears to be tenosynovitis in two orthogonal planes.
21. Bone erosions ≥ 2.5 mm have maximal sensitivity and specificity for RA bone erosions [49].
22. Always delineate a dorsal ganglion cyst from a distended radiocarpal joint recess. The ganglion has minimal compressibility, whereas the distended recess is more compressible.

Method

The patient faces you with the forearm pronated to expose the dorsal wrist. The forearm rests on a flat surface; the wrist is in a neutral position. The fingers may be placed on a slightly convex, curved surface to flex the finger joints and improve visualization minimally. Use a linear probe with a frequency of at least 12 MHz. Optimal evaluation of entheses often requires a higher frequency, at least 16 MHz. The protocol below has the sonographer turning on and off the PD for each joint. Some examiners prefer an alternative sequence of performing the entire examination using the GS and then repeating the whole examination with the PD.

Protocol Image 1: Dorsal Radiocarpal Joint, Intercarpal Joint, Longitudinal ± Power Doppler (Fig. 5.8)

Start with the probe in LAX along the dorsal radial aspect of the wrist to look for the radiocarpal joint. Next, do a TAX slide in a slight ulnar direction and observe the scaphoid's disappearance and then the bony lunate's emergence. You have located the RCJ (radiolunate) and ICJ (lunocapitate). Move the probe in TAX slides in

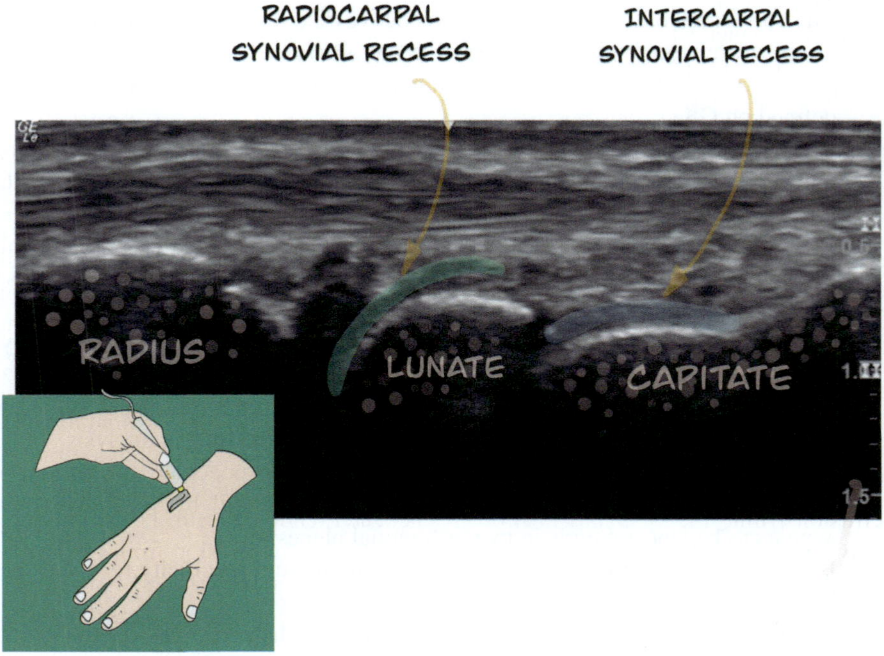

Fig. 5.8 Protocol Image 1: Dorsal radiocarpal joint, intercarpal joint, longitudinal ± power Doppler

either direction to sharply focus the bones and the dorsal synovial recesses. Record the image.

The GS evaluation looks at the hypoechoic dorsal synovium, the curved, mild-moderately hyperechoic joint capsule, and the hyperechoic bone. Observe for thickening of the joint capsule, particularly of the RCJ, and evidence of GS synovial hypertrophy. In addition, observe for effusion and ganglion cysts. Synovial hypertrophy, if present, should be graded as described in the **Clinical Comments** section. Note the extensor tendon that is superficial to the joints. Do not mistake the anisotropy of the extensor retinaculum for tenosynovitis or tenosynovial fluid.

Turn on the PD. Note that the PD signal that we focus on is *within* the joint recess, deep into the joint capsule. See Clinical Comments for the grading (0–3) system. Power Doppler activity superficial to the joint capsule may be due to normal feeder vessels and does not reflect PD activity directly in the joints. However, note if PD activity exists in the tenosynovium of the overlying extensor tendon.

Protocol Image 2: Dorsal Second Metacarpophalangeal Joint, Longitudinal ± Power Doppler (Fig. 5.9)

Lift the probe and place it over the second MCPJ in LAX. Perform slow TAX slides in radial and ulnar directions to fully evaluate the cartilage and the synovial recess. The radial aspect of this MCPJ is freely accessible and should be included in the examination. Evaluate the three margins of the recess—proximal, distal, and superficial. The superficial margin can be defined by passively flexing and extending the finger at the MCPJ to identify the moving extensor tendon, which borders the superior recess margin. Synovial hypertrophy will expand these margins and can be graded.

Examine the smoothness of the superficial surface of the cartilage covering the MC head. Look at the MC head itself and determine (a) if cartilage surrounds the distal and dorsal portions of the head; (b) the constitution of the cartilage (overall thickness degree, uniformity, sharpness of the external border); (c) if there is a slight hyperechoic reflective surface (interface reflex) indicating a smooth cartilage surface; and (d) if there is focal or diffuse cartilage thinning, which would indicate cartilage erosion. Passively flexing and extending the MCPJ will enable a thorough inspection of the MC head's cartilage. Remember that exuberant synovium may have migrated into the area where cartilage previously existed.

Look at the bony surfaces of the MC head and the proximal phalanx for irregularities, which may indicate bone erosion. If the radiographs suggest possible bone erosion, do a focused US examination on that area to verify. An effusion may be present along with synovitis. Prior synovitis may have receded in long-standing cartilage-erosive arthropathy, leaving only residual cartilage or bone erosion. PD activity may be noted within the synovial tissue. Remember to verify synovitis, cartilage erosion, and bone erosion in TAX views; these can be saved as needed.

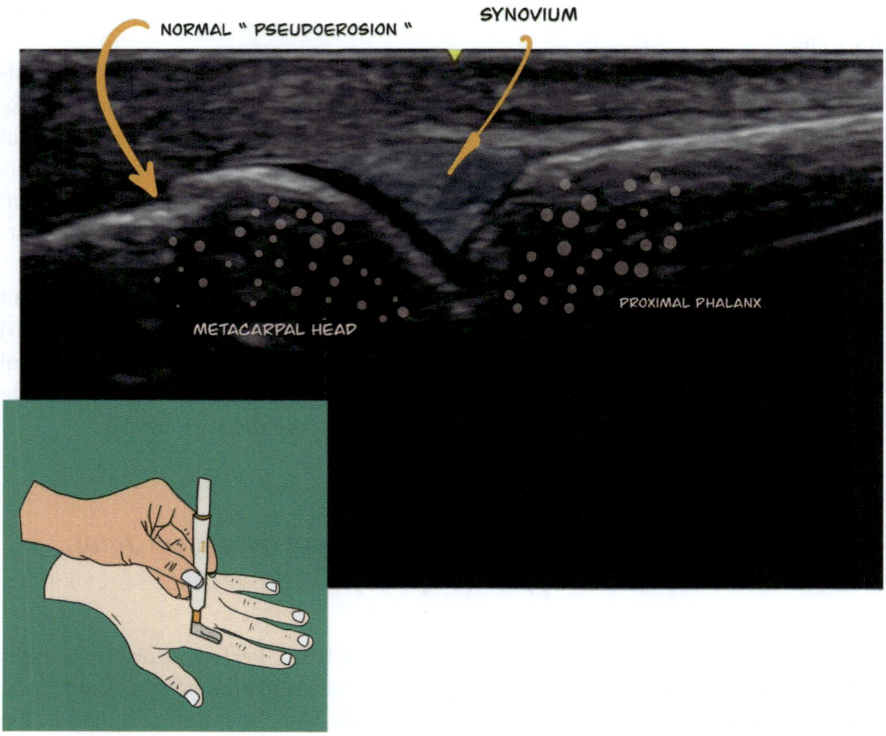

Fig. 5.9 Protocol Image 2: Dorsal 2nd metacarpophalangeal joint, longitudinal ± power Doppler

Protocol Image 3: Dorsal Second Proximal Interphalangeal Joint, Longitudinal ± Power Doppler (Fig. 5.10)

Place the probe over the second PIPJ in LAX. Perform slow TAX slides in radial and ulnar directions to fully evaluate the cartilage and the synovial recess. Osteophytes may be seen primarily in the older age group; the dorsal recess may be enlarged and extended proximally. Synovial hypertrophy can be graded with or without PD activity. Erosion of bone may be seen in aggressive polyarthritis such as RA, PsA, and perhaps in advanced gouty arthropathy. Again, verify suspected pathologies in the TAX view. Power Doppler activity may be positive within the synovial tissue.

Protocol Image 3a (Optional): Dorsal Distal Second Phalanx Extensor Tendon Insertion, Longitudinal ± Power Doppler

Examine for enthesopathy and enthesitis. Use the highest frequency probe available (at least 16 MHz) and recall that while the extensor tendon inserts on the bony distal phalanx, some fibers extend superficially to form the fingernail bed matrix.

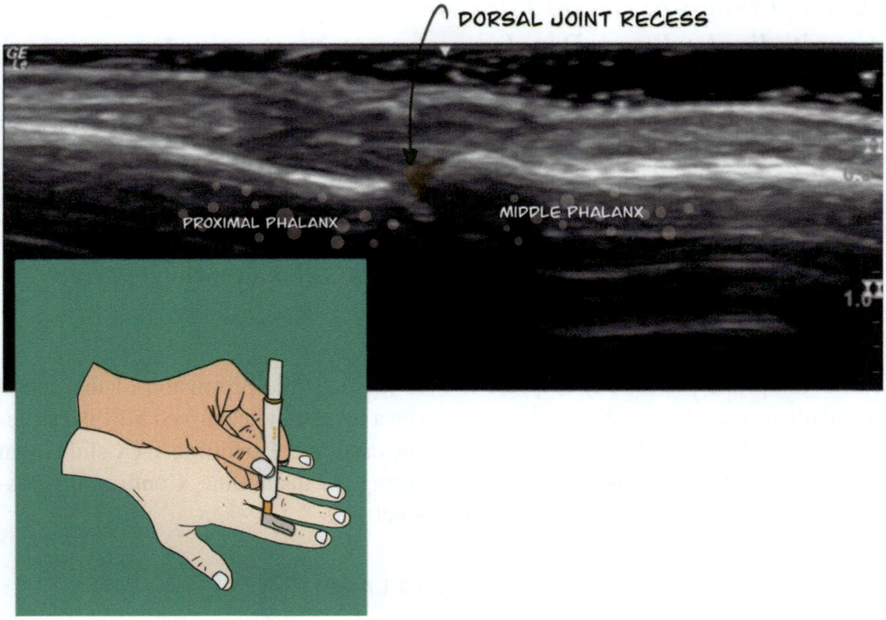

Fig. 5.10 Protocol Image 3: Dorsal 2nd proximal interphalangeal joint, longitudinal ± power Doppler

Protocol Image 4: Dorsal Third Metacarpophalangeal Joint, Longitudinal ± Power Doppler

Perform the same examination as per Protocol Image 2 above.

Protocol Image 5: Dorsal Third Proximal Interphalangeal Joint, Longitudinal ± Power Doppler

Perform the same examination as per Protocol Image 3 above.

Protocol Image 5a (Optional): Dorsal Distal Third Phalanx Extensor Tendon Insertion, Longitudinal ± Power Doppler

Perform the same examination as per Protocol Image 3a above.

Protocol Image 6: Dorsal Fifth Metacarpophalangeal Joint, Longitudinal ± Power Doppler

Perform the same examination as per Protocol Image 2 or 4 above. The ulnar aspect of the fifth MCPJ is easily accessible and should be included.

Protocol Image 7: Distal Ulna, Longitudinal ± Power Doppler (Fig. 5.11)

Move the probe proximally to place it over the palpable protrusion of the dorsal distal ulnar styloid in LAX. It will appear to be a hyperechoic, curved hill-like structure, generally covered with a rim of anechoic cartilage. Perform a TAX slide in an ulnar direction to examine the cartilage covering the distal ulna. Confirm any possible erosions in TAX. Turn on PD to assess activity.

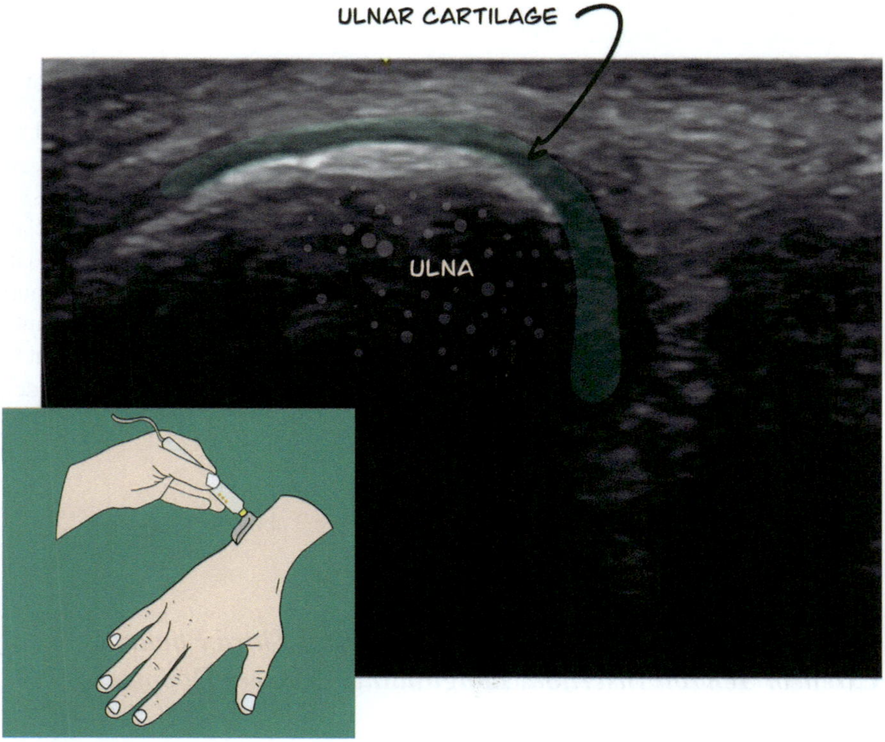

Fig. 5.11 Protocol Image 7: Distal ulna, longitudinal ± power Doppler

Protocol Image 8: Ulnocarpal Joint, Longitudinal ± Power Doppler (Fig. 5.12)

The patient turns the fingers radially to expose the ulnocarpal joint and extend the ECU tendon. Place the probe in LAX along the ulnar aspect of the wrist, with one end on the distal ulna and the other on the proximal carpal bone, specifically the triquetrum. The lunate bone is the hyperechoic curved bone deep to and adjacent to the triquetrum. You may see a flat, even deeper bone closer to the distal ulna; this is the radius. Look at the TFCC articular disk and the more distal meniscal homolog. Determine if calcium is embedded in the TFCC. Usually, the TFCC is barely visible unless it contains some hyperechoic calcium. Also, look at the superficially located ECU tendon and its synovial sheath. Confirm any possible pathology in TAX. Turn on PD to assess activity.

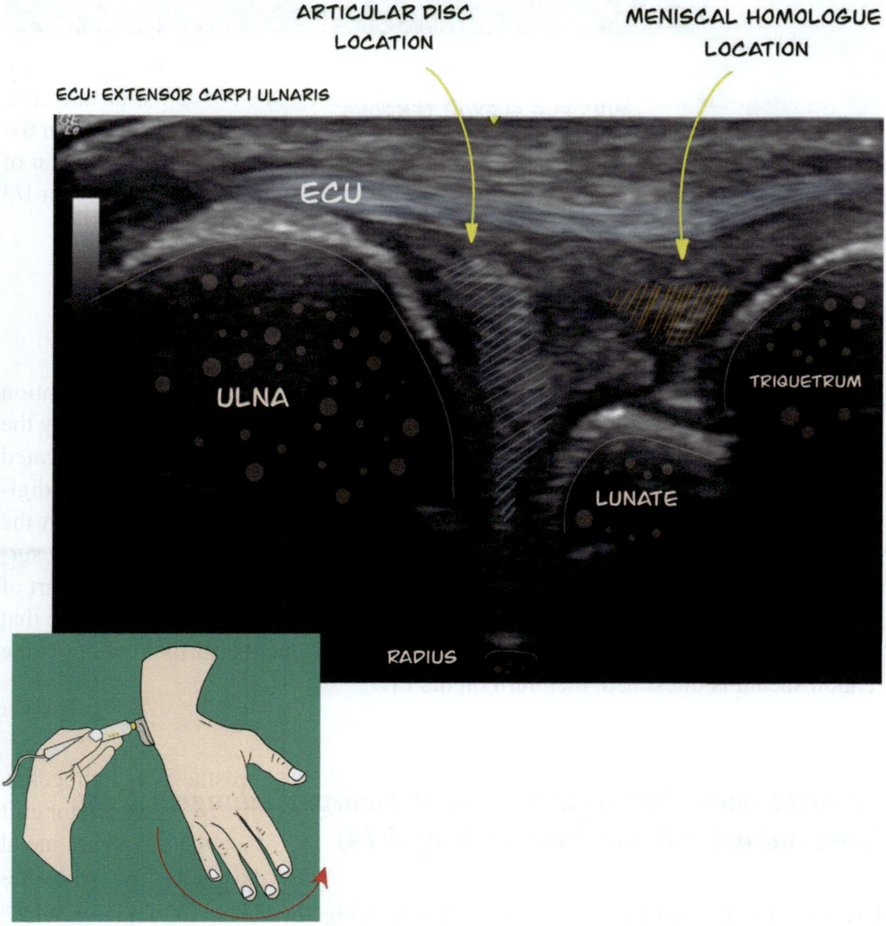

Fig. 5.12 Protocol Image 8: Ulnocarpal joint, longitudinal ± power Doppler

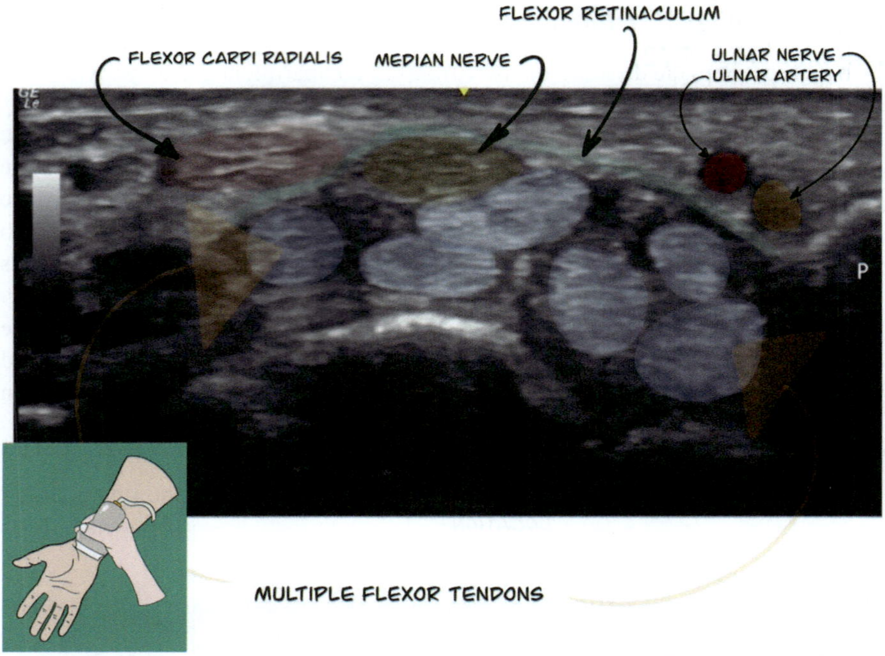

Fig. 5.13 Protocol Image 9: Volar wrist, transverse ± power Doppler

Protocol Image 9: Volar Wrist, Transverse ± Power Doppler (Fig. 5.13)

The patient supinates the forearm and rests the hand dorsum on the examination surface, extending the fingers. Place the probe in TAX across the wrist. Identify the median nerve and the flexor retinaculum. The flexor carpi radialis tendon, located superficially to the carpal tunnel, has a surrounding tendon sheath. The flexor digitorum superficialis and deeper flexor digitorum profundus tendons are within the CT. Also within the CT is the flexor pollicis longus tendon. A tenosynovium surrounds these tendons. If there is an anechoic or hypoechoic rim surrounding part of a tendon, it might be tenosynovitis. To verify, rotate the probe to LAX to ensure that you are not simply seeing the normal myotendinous junction of the tendon. If the tendon sheath is thickened, then turn on the PD.

Protocol Image 10: Volar Second Metacarpophalangeal Joint, Longitudinal ± Power Doppler (Fig. 5.14)

Examine the second MCPJ in LAX when looking for synovitis, tenosynovitis, cartilage, and bone erosions. For each joint, perform a TAX slide in an ulnar and then radial direction to fully evaluate the joint. Manually flex the finger to

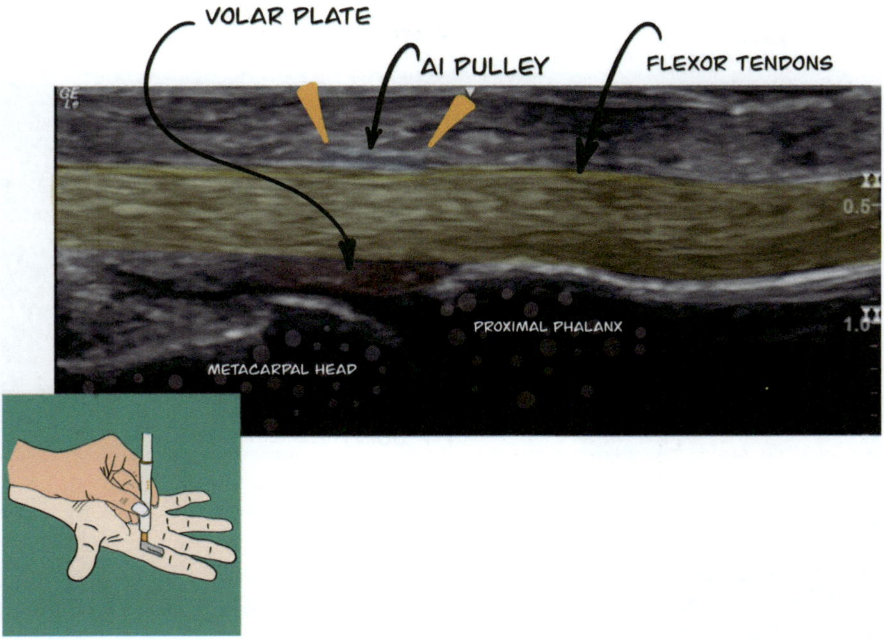

Fig. 5.14 Protocol Image 10: Volar second metacarpophalangeal joint, longitudinal ± power Doppler

visualize the flexor tendon and surrounding tenosynovium. The A1 pulley and volar plate may be visible superficially to the MCPJ. The volar plate is immediately superficial to the joint cartilage, and the A1 pulley is just superficial to the flexor tendon. Remember that the flexor tendon has a superficial (flexor digitorum superficialis) and deep (flexor digitorum profundus) layer. The superficial layer may be the only layer that becomes snagged beneath the A1 pulley with a trigger finger.

Turn on the PD to look for activity. When examining the volar MCPJ for synovitis, keep in mind that the normal synovial recess extends proximally and should also be included with Doppler analysis [113].

Protocol Image 11: Volar Second Proximal Interphalangeal Joint, Longitudinal ± Power Doppler (Fig. 5.15)

Next, move the probe, still in LAX, to the second PIPJ. Perform a TAX slide in ulnar and radial directions to fully evaluate the joint, synovial recess, and flexor tendon.

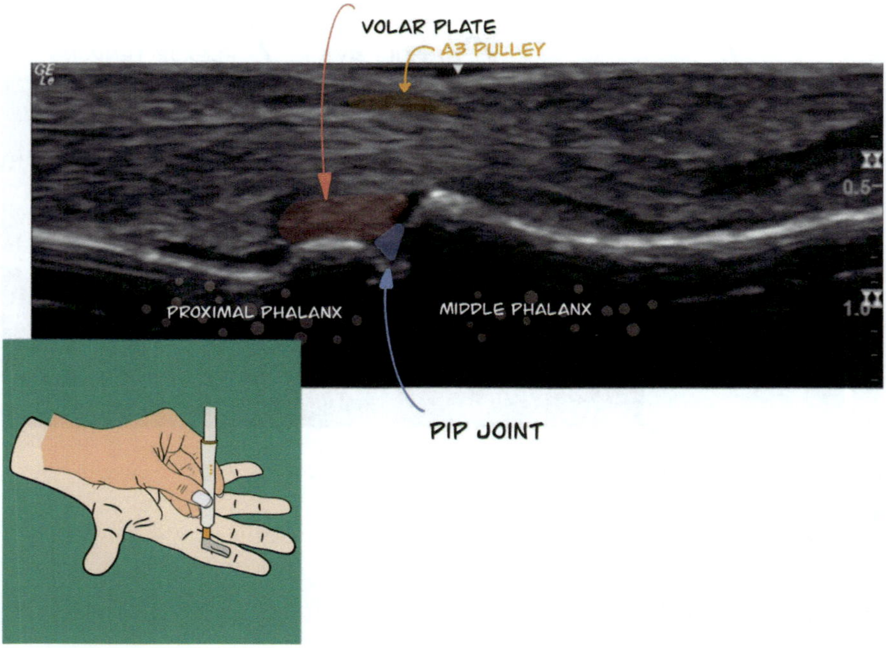

Fig. 5.15 Protocol Image 11: Volar second proximal interphalangeal joint, longitudinal ± power Doppler

Protocol Image 12 (Optional): Volar Third Metacarpophalangeal Joint, Longitudinal ± Power Doppler

Next, move the probe to the third MCPJ. Repeat the examination as per Protocol Image 10.

Protocol Image 13 (Optional): Volar Third Proximal Interphalangeal Joint, Longitudinal ± Power Doppler

Next, move the probe, still in LAX, to the third PIPJ. Repeat the evaluation as in Protocol Image 11.

Protocol Image 14 (Optional): Flexor Tendon Enthesis, Longitudinal + Power Doppler

For this view, it is best to use a high-frequency probe (at least 16 MHz). Remain in LAX. For each of the fingers, move the transducer distally to examine the deep flexor tendon insertion on the distal phalanx. Turn on PD.

Complete Hand Arthropathies Ultrasonic Examination Checklist

☐ Protocol Image 1: Dorsal radiocarpal joint, intercarpal joint, longitudinal ± power Doppler.
☐ Protocol Image 2: Dorsal second metacarpophalangeal joint, longitudinal ± power Doppler.
☐ Protocol Image 3: Dorsal second proximal interphalangeal joint, longitudinal ± power Doppler.
☐ Protocol Image 3a (optional): Dorsal distal second phalanx extensor tendon insertion, longitudinal ± power Doppler.
☐ Protocol Image 4: Dorsal third metacarpophalangeal joint, longitudinal ± power Doppler.
☐ Protocol Image 5: Dorsal third proximal interphalangeal joint, longitudinal ± power Doppler.
☐ Protocol Image 5a (optional): Dorsal distal third phalanx extensor tendon insertion, longitudinal ± power Doppler.
☐ Protocol Image 6: Dorsal fifth metacarpophalangeal joint, longitudinal ± power Doppler.
☐ Protocol Image 7: Distal ulna, longitudinal ± power Doppler.
☐ Protocol Image 8: Ulnocarpal joint, longitudinal ± power Doppler.
☐ Protocol Image 9: Volar wrist, transverse ± power Doppler.
☐ Protocol Image 10: Volar second metacarpophalangeal joint, longitudinal ± power Doppler.
☐ Protocol Image 11: Volar second proximal interphalangeal joint, longitudinal ± power Doppler.
☐ Protocol Image 12 (optional): Volar third metacarpophalangeal joint, longitudinal ± power Doppler.
☐ Protocol Image 13 (optional): Volar third proximal interphalangeal joint, longitudinal ± power Doppler.
☐ Protocol Image 14 (optional): Flexor tendon enthesis, longitudinal + power Doppler.

References

1. Ralphs JR, Benjamin M. The joint capsule: structure, composition, ageing and disease. J Anat. 1994;184(Pt 3):503–9.
2. Epis O, Paoletti F, d'Errico T, Favalli E, Garau P, Mancarella L, et al. Ultrasonography in the diagnosis and management of patients with inflammatory arthritides. Eur J Intern Med. 2014;25(2):103–11.
3. Freeston JE, Coates LC, Nam JL, Moverley AR, Hensor EM, Wakefield RJ, et al. Is there subclinical synovitis in early psoriatic arthritis? A clinical comparison with gray-scale and power Doppler ultrasound. Arthritis Care Res (Hoboken). 2014;66(3):432–9.
4. Wakefield RJ, Green MJ, Marzo-Ortega H, Conaghan PG, Gibbon WW, McGonagle D, et al. Should oligoarthritis be reclassified? Ultrasound reveals a high prevalence of subclinical disease. Ann Rheum Dis. 2004;63(4):382–5.
5. Eder L, Barzilai M, Peled N, Gladman DD, Zisman D. The use of ultrasound for the assessment of enthesitis in patients with spondyloarthritis. Clin Radiol. 2013;68(3):219–23.

6. Anandarajah A. Imaging in psoriatic arthritis. Clin Rev Allergy Immunol. 2013;44(2):157–65.
7. Wakefield RJ, Balint PV, Szkudlarek M, Filippucci E, Backhaus M, D'Agostino MA, et al. Musculoskeletal ultrasound including definitions for ultrasonographic pathology. J Rheumatol. 2005;32(12):2485–7.
8. Fodor D, Nestorova R, Vlad V, Micu M. The place of musculoskeletal ultrasonography in gout diagnosis. Med Ultrason. 2014;16(4):336–44.
9. Patil P, Dasgupta B. Role of diagnostic ultrasound in the assessment of musculoskeletal diseases. Ther Adv Musculoskelet Dis. 2012;4(5):341–55.
10. Kaeley GS, Bakewell C, Deodhar A. The importance of ultrasound in identifying and differentiating patients with early inflammatory arthritis: a narrative review. Arthritis Res Ther. 2020;22(1):1.
11. Mendonça JA, Yazbek MA, Laurindo IM, Bertolo MB. Wrist ultrasound analysis of patients with early rheumatoid arthritis. Braz J Med Biol Res. 2011;44(1):11–5.
12. Hassan R, Hussain S, Bacha R, Gillani SA, Malik SS. Reliability of ultrasound for the detection of rheumatoid arthritis. J Med Ultrasound. 2019;27(1):3–12.
13. Haavardsholm EA, Aga A-B, Olsen IC, Lillegraven S, Hammer HB, Uhlig T, et al. Ultrasound in management of rheumatoid arthritis: ARCTIC randomised controlled strategy trial. BMJ. 2016;354:i4205.
14. Dale J, Stirling A, Zhang R, Purves D, Foley J, Sambrook M, et al. Targeting ultrasound remission in early rheumatoid arthritis: the results of the TaSER study, a randomised clinical trial. Ann Rheum Dis. 2016;75(6):1043–50.
15. Bruyn GA, Iagnocco A, Naredo E, Balint PV, Gutierrez M, Hammer HB, et al. OMERACT definitions for ultrasonographic pathologies and elementary lesions of rheumatic disorders 15 years on. J Rheumatol. 2019;46(10):1388.
16. Szkudlarek M, Narvestad E, Klarlund M, Court-Payen M, Thomsen HS, Østergaard M. Ultrasonography of the metatarsophalangeal joints in rheumatoid arthritis: comparison with magnetic resonance imaging, conventional radiography, and clinical examination. Arthritis Rheum. 2004;50(7):2103–12.
17. Iagnocco A, Ossandon A, Coari G, Conti F, Priori R, Alessandri C, et al. Wrist joint involvement in systemic lupus erythematosus. An ultrasonographic study. Clin Exp Rheumatol. 2004;22(5):621–4.
18. Fiocco U, Ferro F, Vezzù M, Cozzi L, Checchetto C, Sfriso P, et al. Rheumatoid and psoriatic knee synovitis: clinical, grey scale, and power Doppler ultrasound assessment of the response to etanercept. Ann Rheum Dis. 2005;64(6):899–905.
19. Ribbens C, André B, Marcelis S, Kaye O, Mathy L, Bonnet V, et al. Rheumatoid hand joint synovitis: gray-scale and power Doppler US quantifications following anti-tumor necrosis factor-alpha treatment: pilot study. Radiology. 2003;229(2):562–9.
20. Terslev L, Torp-Pedersen S, Qvistgaard E, Kristoffersen H, Røgind H, Danneskiold-Samsøe B, et al. Effects of treatment with etanercept (Enbrel, TNRF:Fc) on rheumatoid arthritis evaluated by Doppler ultrasonography. Ann Rheum Dis. 2003;62(2):178–81.
21. Filippucci E, Iagnocco A, Salaffi F, Cerioni A, Valesini G, Grassi W. Power Doppler sonography monitoring of synovial perfusion at the wrist joints in patients with rheumatoid arthritis treated with adalimumab. Ann Rheum Dis. 2006;65(11):1433–7.
22. Hau M, Schultz H, Tony HP, Keberle M, Jahns R, Haerten R, et al. Evaluation of pannus and vascularization of the metacarpophalangeal and proximal interphalangeal joints in rheumatoid arthritis by high-resolution ultrasound (multidimensional linear array). Arthritis Rheum. 1999;42(11):2303–8.
23. Backhaus TM, Ohrndorf S, Kellner H, Strunk J, Hartung W, Sattler H, et al. The US7 score is sensitive to change in a large cohort of patients with rheumatoid arthritis over 12 months of therapy. Ann Rheum Dis. 2013;72(7):1163–9.
24. Vlad V, Micu M, Porta F, Radunovic G, Nestorova R, Petranova T, et al. Ultrasound of the hand and wrist in rheumatology. Med Ultrason. 2012;14(1):42–8.
25. Naredo E, Möller I, Cruz A, Carmona L, Garrido J. Power Doppler ultrasonographic monitoring of response to anti-tumor necrosis factor therapy in patients with rheumatoid arthritis. Arthritis Rheum. 2008;58(8):2248–56.

26. Brown AK, Quinn MA, Karim Z, Conaghan PG, Peterfy CG, Hensor E, et al. Presence of significant synovitis in rheumatoid arthritis patients with disease-modifying antirheumatic drug-induced clinical remission: evidence from an imaging study may explain structural progression. Arthritis Rheum. 2006;54(12):3761–73.
27. Sundin U, Aga AB, Skare Ø, Nordberg LB, Uhlig T, Hammer HB, et al. Conventional versus ultrasound treat to target: no difference in magnetic resonance imaging inflammation or joint damage over 2 years in early rheumatoid arthritis. Rheumatology (Oxford). 2020;59(9):2550–5.
28. Ohrndorf S, Glimm A-M, Burmester G-R, Backhaus M. Musculoskeletal ultrasound scoring systems: Assessing disease activity and therapeutic response in rheumatoid arthritis. Int J Clin Rheumatol. 2011;6:57–65.
29. Backhaus M, Ohrndorf S, Kellner H, Strunk J, Backhaus TM, Hartung W, et al. Evaluation of a novel 7-joint ultrasound score in daily rheumatologic practice: a pilot project. Arthritis Rheum. 2009;61(9):1194–201.
30. Terslev L, Naredo E, Aegerter P, Wakefield RJ, Backhaus M, Balint P, et al. Scoring ultrasound synovitis in rheumatoid arthritis: a EULAR-OMERACT ultrasound taskforce-part 2: reliability and application to multiple joints of a standardised consensus-based scoring system. RMD Open. 2017;3(1):e000427.
31. Dubash SR, De Marco G, Wakefield RJ, Tan AL, McGonagle D, Marzo-Ortega H. Ultrasound imaging in psoriatic arthritis: what have we learnt in the last five years? Front Med (Lausanne). 2020;7:487.
32. Scheel AK, Hermann KG, Kahler E, Pasewaldt D, Fritz J, Hamm B, et al. A novel ultrasonographic synovitis scoring system suitable for analyzing finger joint inflammation in rheumatoid arthritis. Arthritis Rheum. 2005;52(3):733–43.
33. D'Agostino MA, Terslev L, Aegerter P, Backhaus M, Balint P, Bruyn GA, et al. Scoring ultrasound synovitis in rheumatoid arthritis: a EULAR-OMERACT ultrasound taskforce-part 1: definition and development of a standardised, consensus-based scoring system. RMD Open. 2017;3(1):e000428.
34. Hammer HB, Bolton-King P, Bakkeheim V, Berg TH, Sundt E, Kongtorp AK, et al. Examination of intra and interrater reliability with a new ultrasonographic reference atlas for scoring of synovitis in patients with rheumatoid arthritis. Ann Rheum Dis. 2011;70(11):1995–8.
35. Ben-Artzi A, Kaeley GS, Ranganath VK. Ultrasound scoring of joint synovitis. Springer; 2021.
36. Smith E, Azzopardi C, Thaker S, Botchu R, Gupta H. Power Doppler in musculoskeletal ultrasound: uses, pitfalls and principles to overcome its shortcomings. J Ultrasound. 2021;24(2):151–6.
37. Torp-Pedersen ST, Terslev L. Settings and artefacts relevant in colour/power Doppler ultrasound in rheumatology. Ann Rheum Dis. 2008;67(2):143–9.
38. Martinoli C, Derchi LE. Gain setting in power Doppler US. Radiology. 1997;202(1):284–5.
39. Iagnocco A, Conaghan PG, Aegerter P, Möller I, Bruyn GA, Chary-Valckenaere I, et al. The reliability of musculoskeletal ultrasound in the detection of cartilage abnormalities at the metacarpo-phalangeal joints. Osteoarthr Cartil. 2012;20(10):1142–6.
40. Mandl P, Studenic P, Filippucci E, Bachta A, Backhaus M, Bong D, et al. Development of semiquantitative ultrasound scoring system to assess cartilage in rheumatoid arthritis. Rheumatology (Oxford). 2019;58(10):1802–11.
41. Cipolletta E, Filippucci E, Matteo AD, Hurnakova J, Carlo MD, Pavelka K, et al. FRI0634 standard reference values of metacarpal head cartilage thickness measurement by ultrasound in healthy subjects. Ann Rheum Dis. 2019;78(Suppl 2):1014–5.
42. İskender Ö, Kaymak B, Kara M, Akıncı A, Ülkar B, Özçakar L. AB0085 ultrasonographic evaluation of the metacarpal cartilage thickness in weight lifters and volleyball players. Ann Rheum Dis. 2019;78(Suppl 2):1505–6.
43. Hurnakova J, Filippucci E, Cipolletta E, Di Matteo A, Salaffi F, Carotti M, et al. Prevalence and distribution of cartilage damage at the metacarpal head level in rheumatoid arthritis and osteoarthritis: an ultrasound study. Rheumatology (Oxford). 2019;58(7):1206–13.

44. Tămaş MM, Filippucci E, Becciolini A, Gutierrez M, Di Geso L, Bonfiglioli K, et al. Bone erosions in rheumatoid arthritis: ultrasound findings in the early stage of the disease. Rheumatology (Oxford). 2014;53(6):1100–7.
45. Möttönen TT. Prediction of erosiveness and rate of development of new erosions in early rheumatoid arthritis. Ann Rheum Dis. 1988;47(8):648–53.
46. Grassi W, Filippucci E, Farina A, Salaffi F, Cervini C. Ultrasonography in the evaluation of bone erosions. Ann Rheum Dis. 2001;60(2):98–103.
47. Hammer HB, Haavardsholm EA, Bøyesen P, Kvien TK. Bone erosions at the distal ulna detected by ultrasonography are associated with structural damage assessed by conventional radiography and MRI: a study of patients with recent onset rheumatoid arthritis. Rheumatology (Oxford). 2009;48(12):1530–2.
48. Døhn UM, Ejbjerg BJ, Court-Payen M, Hasselquist M, Narvestad E, Szkudlarek M, et al. Are bone erosions detected by magnetic resonance imaging and ultrasonography true erosions? A comparison with computed tomography in rheumatoid arthritis metacarpophalangeal joints. Arthritis Res Ther. 2006;8(4):R110.
49. Zayat AS, Ellegaard K, Conaghan PG, Terslev L, Hensor EM, Freeston JE, et al. The specificity of ultrasound-detected bone erosions for rheumatoid arthritis. Ann Rheum Dis. 2015;74(5):897–903.
50. Finzel S, Ohrndorf S, Englbrecht M, Stach C, Messerschmidt J, Schett G, et al. A detailed comparative study of high-resolution ultrasound and micro-computed tomography for detection of arthritic bone erosions. Arthritis Rheum. 2011;63(5):1231–6.
51. Falkowski AL, Jacobson JA, Kalia V, Atinga A, Gandikota G, Thiele RG. Ultrasound characterization of pseudoerosions and dorsal joint recess morphology of the hand and wrist in 100 asymptomatic subjects. Eur J Radiol. 2020;124:108842.
52. Kamel M, Moghazy K, Eid H, Mansour R. Ultrasonographic diagnosis of de Quervain's tenosynovitis. Ann Rheum Dis. 2002;61(11):1034–5.
53. Hmamouchi I, Bahiri R, Srifi N, Aktaou S, Abouqal R, Hajjaj-Hassouni N. A comparison of ultrasound and clinical examination in the detection of flexor tenosynovitis in early arthritis. BMC Musculoskelet Disord. 2011;12:91.
54. Sahbudin I, Pickup L, Nightingale P, Allen G, Cader Z, Singh R, et al. The role of ultrasound-defined tenosynovitis and synovitis in the prediction of rheumatoid arthritis development. Rheumatology (Oxford). 2018;57(7):1243–52.
55. Ramrattan LA, Kaeley GS. Sonographic characteristics of extensor tendon abnormalities and relationship with joint disease activity in rheumatoid arthritis: a pilot study. J Ultrasound Med. 2017;36(5):985–92.
56. Naredo E, D'Agostino MA, Wakefield RJ, Möller I, Balint PV, Filippucci E, et al. Reliability of a consensus-based ultrasound score for tenosynovitis in rheumatoid arthritis. Ann Rheum Dis. 2013;72(8):1328–34.
57. Ammitzbøll-Danielsen M, Østergaard M, Naredo E, Iagnocco A, Möller I, D'Agostino MA, et al. The use of the OMERACT ultrasound tenosynovitis scoring system in multicenter clinical trials. J Rheumatol. 2018;45(2):165–9.
58. Suzuki T, Shirai H. SAT0570 clinical significance of finger extensor paratenonitis detected by musculoskeletal ultrasound. Ann Rheum Dis. 2020;79(Suppl 1):1243–4.
59. Kehl AS, Corr M, Weisman MH. Review: Enthesitis: new insights into pathogenesis, diagnostic modalities, and treatment. Arthritis Rheumatol. 2016;68(2):312–22.
60. D'Agostino MA, Said-Nahal R, Hacquard-Bouder C, Brasseur JL, Dougados M, Breban M. Assessment of peripheral enthesitis in the spondylarthropathies by ultrasonography combined with power Doppler: a cross-sectional study. Arthritis Rheum. 2003;48(2):523–33.
61. D'Agostino MA, Olivieri I. Enthesitis. Best Pract Res Clin Rheumatol. 2006;20(3):473–86.
62. D'Agostino MA, Aegerter P, Bechara K, Salliot C, Judet O, Chimenti MS, et al. How to diagnose spondyloarthritis early? Accuracy of peripheral enthesitis detection by power Doppler ultrasonography. Ann Rheum Dis. 2011;70(8):1433–40.

63. Catanoso M, Pipitone N, Salvarani C. Epidemiology of psoriatic arthritis. Reumatismo. 2012;64(2):66–70.
64. Kaeley GS, Eder L, Aydin SZ, Gutierrez M, Bakewell C. Enthesitis: a hallmark of psoriatic arthritis. Semin Arthritis Rheum. 2018;48(1):35–43.
65. Bandinelli F, Prignano F, Bonciani D, Bartoli F, Collaku L, Candelieri A, et al. Ultrasound detects occult entheseal involvement in early psoriatic arthritis independently of clinical features and psoriasis severity. Clin Exp Rheumatol. 2013;31(2):219–24.
66. Terslev L, Naredo E, Iagnocco A, Balint PV, Wakefield RJ, Aegerter P, et al. Defining enthesitis in spondyloarthritis by ultrasound: results of a Delphi process and of a reliability reading exercise. Arthritis Care Res (Hoboken). 2014;66(5):741–8.
67. Tom S, Zhong Y, Cook R, Aydin SZ, Kaeley G, Eder L. Development of a preliminary ultrasonographic enthesitis score in psoriatic arthritis—GRAPPA Ultrasound Working Group. J Rheumatol. 2019;46(4):384–90.
68. Zabotti A, Idolazzi L, Batticciotto A, De Lucia O, Scirè CA, Tinazzi I, et al. Enthesitis of the hands in psoriatic arthritis: an ultrasonographic perspective. Med Ultrason. 2017;19(4):438–43.
69. Tan AL, Benjamin M, Toumi H, Grainger AJ, Tanner SF, Emery P, et al. The relationship between the extensor tendon enthesis and the nail in distal interphalangeal joint disease in psoriatic arthritis—a high-resolution MRI and histological study. Rheumatology (Oxford). 2007;46(2):253–6.
70. Aydin SZ, Castillo-Gallego C, Ash ZR, Marzo-Ortega H, Emery P, Wakefield RJ, et al. Ultrasonographic assessment of nail in psoriatic disease shows a link between onychopathy and distal interphalangeal joint extensor tendon enthesopathy. Dermatology. 2012;225(3):231–5.
71. Gutierrez M, Filippucci E, De Angelis R, Filosa G, Kane D, Grassi W. A sonographic spectrum of psoriatic arthritis: "the five targets". Clin Rheumatol. 2010;29(2):133–42.
72. Bandinelli F, Denaro V, Prignano F, Collaku L, Ciancio G, Matucci-Cerinic M. Ultrasonographic wrist and hand abnormalities in early psoriatic arthritis patients: correlation with clinical, dermatological, serological and genetic indices. Clin Exp Rheumatol. 2015;33(3):330–5.
73. Fournié B, Margarit-Coll N, Champetier de Ribes TL, Zabraniecki L, Jouan A, Vincent V, et al. Extrasynovial ultrasound abnormalities in the psoriatic finger. Prospective comparative power-doppler study versus rheumatoid arthritis. Joint Bone Spine. 2006;73(5):527–31.
74. Gutierrez M, Filippucci E, Grassi W, Rosemffet M. Intratendinous power Doppler changes related to patient position in seronegative spondyloarthritis. J Rheumatol. 2010;37(5):1057–9.
75. Klauser AS, Halpern EJ, Strobl S, Abd Ellah MMH, Gruber J, Bellmann-Weiler R, et al. Gout of hand and wrist: the value of US as compared with DECT. Eur Radiol. 2018;28(10):4174–81.
76. Chhana A, Dalbeth N. The gouty tophus: a review. Curr Rheumatol Rep. 2015;17(3):19.
77. Fitzgerald BT, Setty A, Mudgal CS. Gout affecting the hand and wrist. J Am Acad Orthop Surg. 2007;15(10):625–35.
78. Rosenthal AK, Ryan LM. Calcium pyrophosphate deposition disease. N Engl J Med. 2016;374(26):2575–84.
79. Rosales-Alexander JL, Balsalobre Aznar J, Magro-Checa C. Calcium pyrophosphate crystal deposition disease: diagnosis and treatment. Open Access Rheumatol. 2014;6:39–47.
80. Paalanen K, Rannio K, Rannio T, Asikainen J, Hannonen P, Sokka T. Prevalence of calcium pyrophosphate deposition disease in a cohort of patients diagnosed with seronegative rheumatoid arthritis. Clin Exp Rheumatol. 2020;38(1):99–106.
81. Hirose J. Clinical presentation and diagnosis of calcium deposition diseases. Int J Clin Rheumatol. 2010;5(1):117–28.
82. Adamson TC 3rd, Resnik CS, Guerra J Jr, Vint VC, Weisman MH, Resnick D. Hand and wrist arthropathies of hemochromatosis and calcium pyrophosphate deposition disease: distinct radiographic features. Radiology. 1983;147(2):377–81.
83. Vele P, Simon SP, Damian L, Felea I, Muntean L, Filipescu I, et al. Clinical and ultrasound findings in patients with calcium pyrophosphate dihydrate deposition disease. Med Ultrason. 2018;20(2):159–63.

84. Miksanek J, Rosenthal AK. Imaging of calcium pyrophosphate deposition disease. Curr Rheumatol Rep. 2015;17(3):20.
85. Löffler C, Sattler H, Peters L, Löffler U, Uppenkamp M, Bergner R. Distinguishing gouty arthritis from calcium pyrophosphate disease and other arthritides. J Rheumatol. 2015;42(3):513–20.
86. Bianchi S, Della Santa D, Glauser T, Beaulieu JY, van Aaken J. Sonography of masses of the wrist and hand. AJR Am J Roentgenol. 2008;191(6):1767–75.
87. Bianchi S, Abdelwahab IF, Zwass A, Giacomello P. Ultrasonographic evaluation of wrist ganglia. Skelet Radiol. 1994;23(3):201–3.
88. Teh J. Ultrasound of soft tissue masses of the hand. J Ultrason. 2012;12(51):381–401.
89. Zhang A, Falkowski AL, Jacobson JA, Kim SM, Koh SH, Gaetke-Udager K. Sonography of wrist ganglion cysts: which location is most common? J Ultrasound Med. 2019;38(8):2155–60.
90. Teefey SA, Dahiya N, Middleton WD, Gelberman RH, Boyer MI. Ganglia of the hand and wrist: a sonographic analysis. AJR Am J Roentgenol. 2008;191(3):716–20.
91. De Keyser F, Isaac Z, Curtis MR. Ganglion cysts of the wrist and hand. UpToDate. 2021.
92. Zeiss J, Guilliam-Haidet L. MR demonstration of anomalous muscles about the volar aspect of the wrist and forearm. Clin Imaging. 1996;20(3):219–21.
93. Harvie P, Patel N, Ostlere SJ. Prevalence and epidemiological variation of anomalous muscles at guyon's canal. J Hand Surg Br. 2004;29(1):26–9.
94. Timins ME. Muscular anatomic variants of the wrist and hand: findings on MR imaging. AJR Am J Roentgenol. 1999;172(5):1397–401.
95. Chiavaras MM, Jacobson JA, Yablon CM, Brigido MK, Girish G. Pitfalls in wrist and hand ultrasound. AJR Am J Roentgenol. 2014;203(3):531–40.
96. Reynolds DL Jr, Jacobson JA, Inampudi P, Jamadar DA, Ebrahim FS, Hayes CW. Sonographic characteristics of peripheral nerve sheath tumors. AJR Am J Roentgenol. 2004;182(3):741–4.
97. Inampudi P, Jacobson JA, Fessell DP, Carlos RC, Patel SV, Delaney-Sathy LO, et al. Soft-tissue lipomas: accuracy of sonography in diagnosis with pathologic correlation. Radiology. 2004;233(3):763–7.
98. Anderson SE, De Monaco D, Buechler U, Triller J, Gerich U, Dalinka M, et al. Imaging features of pseudoaneurysms of the hand in children and adults. AJR Am J Roentgenol. 2003;180(3):659–64.
99. Plant MA, Panchapakesan V. Digital artery pseudoaneurysm in a patient with previous radial artery harvest. Can J Plast Surg. 2011;19(4):148–50.
100. Bianchi S, Zwass A, Abdelwahab IF, Merello A, Damiani A. Sonographic evaluation of pseudoaneurysm of a digital artery: a case report. J Hand Surg Am. 1993;18(4):638–40.
101. Alerhand S, Apakama D, Nevel A, Nelson BP. Radial artery pseudoaneurysm diagnosed by point-of-care ultrasound five days after transradial catheterization: a case report. World J Emerg Med. 2018;9(3):223–6.
102. Álvarez-Chinchilla PJ, Poveda Montoyo I, Illán F, Bañuls RJ. Rheumatoid nodulosis in an adult patient negative for rheumatoid factor. Actas Dermosifiliogr (Engl Ed). 2019;110(10):865–7.
103. Nalbant S, Corominas H, Hsu B, Chen LX, Schumacher HR, Kitumnuaypong T. Ultrasonography for assessment of subcutaneous nodules. J Rheumatol. 2003;30(6):1191–5.
104. Oo WM, Deveza LA, Duong V, Fu K, Linklater JM, Riordan EA, et al. Musculoskeletal ultrasound in symptomatic thumb-base osteoarthritis: clinical, functional, radiological and muscle strength associations. BMC Musculoskelet Disord. 2019;20(1):220.
105. Endo Y, Sivakumaran T, Lee SC, Lin B, Fufa D. Ultrasound features of traumatic digital nerve injuries of the hand with surgical confirmation. Skelet Radiol. 2021;50(9):1791–800.
106. Bøyesen P, Haavardsholm EA, van der Heijde D, Østergaard M, Hammer HB, Sesseng S, et al. Prediction of MRI erosive progression: a comparison of modern imaging modalities in early rheumatoid arthritis patients. Ann Rheum Dis. 2011;70(1):176–9.

107. Tang H, Liu Y, Liu Y, Zhao H. Comparison of role of hand and wrist ultrasound in diagnosis of subclinical synovitis in patients with systemic lupus erythematosus and rheumatoid arthritis: a retrospective, single-center study. Med Sci Monit. 2020;26:e926436.
108. Iagnocco A, Ceccarelli F, Rizzo C, Truglia S, Massaro L, Spinelli FR, et al. Ultrasound evaluation of hand, wrist and foot joint synovitis in systemic lupus erythematosus. Rheumatology (Oxford). 2014;53(3):465–72.
109. Fjellstad CM, Mathiessen A, Slatkowsky-Christensen B, Kvien TK, Hammer HB, Haugen IK. Associations between ultrasound-detected synovitis, pain, and function in interphalangeal and thumb base osteoarthritis: data from the nor-hand cohort. Arthritis Care Res (Hoboken). 2020;72(11):1530–5.
110. Ceponis A, Kissin E. Ultrasound of the hand and wrist. In: Kohler MJ, editor. Musculoskeletal ultrasound in rheumatology review. 2nd ed. Springer; 2021. p. 53–82.
111. Ohrndorf S, Halbauer B, Martus P, Reiche B, Backhaus TM, Burmester GR, et al. Detailed joint region analysis of the 7-joint ultrasound score: evaluation of an arthritis patient cohort over one year. Int J Rheumatol. 2013;2013:493848.
112. Navalho M, Resende C, Rodrigues AM, Ramos F, Gaspar A, Pereira da Silva JA, et al. Bilateral MR imaging of the hand and wrist in early and very early inflammatory arthritis: tenosynovitis is associated with progression to rheumatoid arthritis. Radiology. 2012;264(3):823–33.
113. Kissin E, DeMarco PJ, Malone DG, Bakewell C. Ultrasound training tips & pitfalls. The Rheumatologist. 2018.

Chapter 6
Anterior Elbow

Reasons to Do the Study
1. Antecubital elbow pain
2. Lateral elbow pain
3. Paresthesia or weakness in the hand in the distribution of the radial nerve or the median nerve
4. Weakness of elbow flexion or supination

Questions We Want Answered
1. Is there a joint effusion? If so, what is the cause?
2. Is there compression of the median nerve or the radial nerve?
3. Is there an injury to the distal biceps brachii tendon?
4. Is there bicipitoradial bursitis?

Anatomy

Bones and Joints (Fig. 6.1)

The terminology of the bones of the anterior elbow is confusing. The distal humeral ends are the capitellum (radial side) and the trochlea (ulnar side). The radial head is a "process," and the ulnar head is also a "process." The latter is also known as the

---

The contents do not represent the views of the U.S. Department of Veterans Affairs or the United States Government.

M. H. Greenberg et al., *Manual of Musculoskeletal Ultrasound*,
https://doi.org/10.1007/978-3-031-37416-6_6

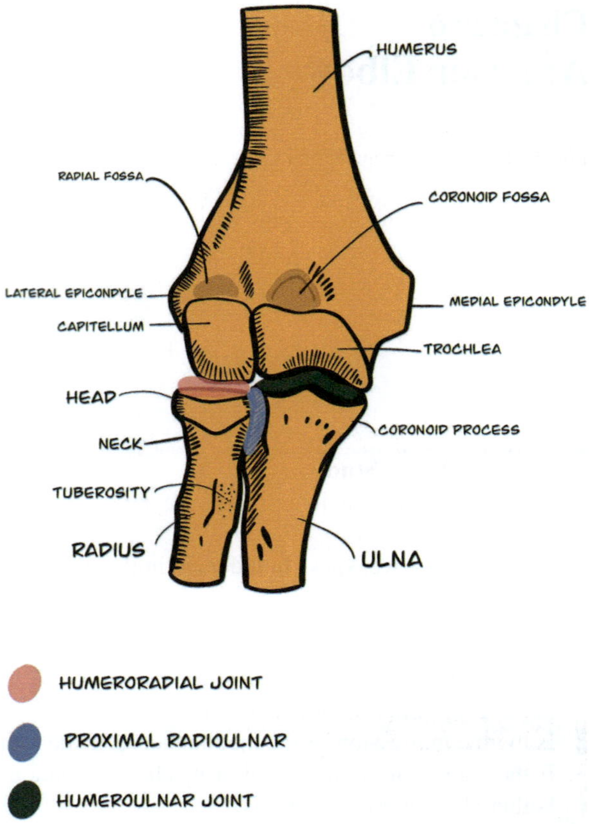

Fig. 6.1 Bony anatomy of the anterior elbow

coronoid process. The anterior recess has two fossae on the anterior distal humerus: the coronoid (ulnar) and radial. So, the bony landmarks, as we go from proximal to distal on the anterior elbow, are as follows:

A. Ulnar (medial) aspect:

Coronoid fossa (of the humerus) to the humeral trochlea to the ulnar head (coronoid process).

B. Radial (lateral) aspect:

Radial fossa (of the humerus) to the humeral capitellum to the radial head.

The three elbow joints are as follows:

A. Humeroulnar joint
B. Humeroradial joint
C. Radioulnar joint

Anterior Joint Recess

The coronoid and smaller radial fossa are concavities at the distal humerus within the anterior joint recess [1]. Each fossa contains a hyperechoic triangular fat pad sandwiched between the deeper synovium and the more superficial joint capsule. Fat pads will displace anteriorly by a joint effusion. An anechoic joint effusion is easy to visualize at the anterior recess and is accentuated on a dynamic ultrasound (US) exam with forearm supination/pronation. However, be aware that you may find a small amount of physiologic fluid posterior (deep) to the anterior fat pad.

Synovitis may occur in the anterior joint recesses due to chronic osteoarthritis, inflammatory arthropathy (rheumatoid arthritis [RA], gout, pseudogout, etc.), or infection [1]. Thickened synovium and hyperemia may be noted with RA. If synovitis is present, also consider synovial proliferative disorders such as pigmented villonodular synovitis or synovial osteochondromatosis (the latter may be seen as multiple hyperechoic foci within the synovial fluid). The examiner may also discover intra-articular bodies in the anterior recess with movement of the bodies activated by a gently rocking motion of the patient's elbow [2]. The capitellum may demonstrate a defect in the articular hyaline cartilage, the donor site for the intra-articular body, or bodies [1].

Muscles, Nerves, and Tendons

So, at the anterior elbow crease, the soft tissue structures from lateral to medial are the **br**achioradialis muscle, the **r**adial nerve, the **b**rachialis muscle, the **b**iceps brachii muscle, the **m**edian nerve, and the **p**ronator teres (Fig. 6.2). A suggested mnemonic is as follows:

Bring **R**eal **B**eer **B**y **M**y **P**orch

The radial nerve traverses the space between the brachialis and the brachioradialis. More distally, the radial nerve splits into superficial and deep branches; the latter is called the posterior interosseous nerve (PIN). The PIN most commonly passes between the superficial and deep portions of the supinator muscles [3]. The median nerve similarly traverses between the two heads of the pronator teres (PT) muscle.

Muscle Insertions [4]:

1. Brachioradialis muscle: distal radius.
2. Supinator muscle: proximal radius.
3. Brachialis muscle: coronoid process of the ulna.
4. Biceps brachii: radial tuberosity (via the distal biceps tendon) and deep forearm fascia (via the lacertus fibrosus).

Fig. 6.2 Simplified soft tissues of the anterior elbow

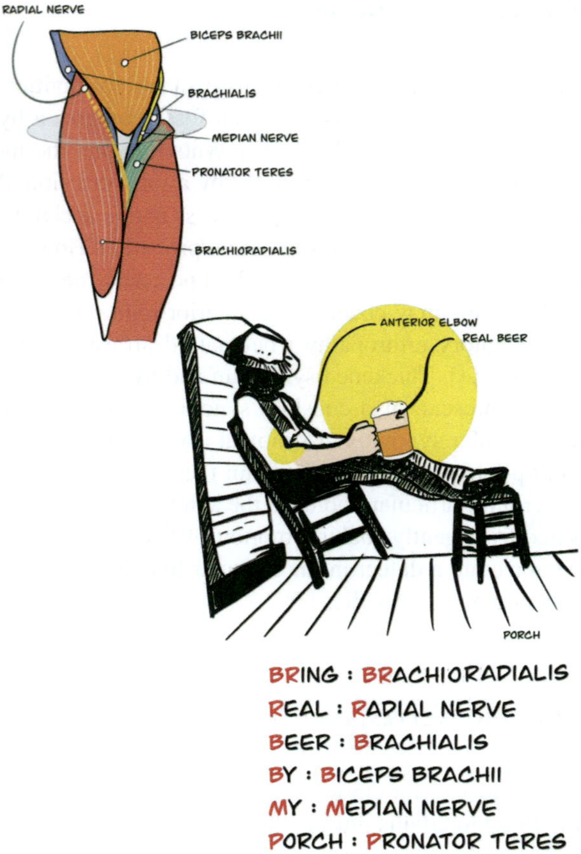

The distal biceps tendon (DBT) has a paratenon and not a true tendon sheath; therefore, no true tenosynovitis occurs in this location. The cubital bursa, interposed between the DBT and radial tuberosity, facilitates smooth gliding during pronation and supination [3]. The lacertus fibrosis is a flat aponeurosis that expands out from the biceps brachii tendon and inserts into the deep forearm fascia. It helps maintain the DBT in the correct position [1, 3].

5. Pronator teres: middle lateral radius.

Arteries

The brachial artery (BA) divides into the ulnar and radial arteries near or proximal to the elbow [4].

Clinical Comments

The complaint of antecubital pain prompts an assessment. The examiner can evaluate the anterior joint recesses, muscles tendons, nerves, arteries, and the bicipital bursa. Dynamic examination (pronation/supination and flexion/extension) may elucidate pathology undetected by a static MR study.

Effusion, Synovitis, Loose Bodies, and Erosions

The humeroulnar and humeroradial joints are evaluated. The anterior recesses have fat pads that are displaced anteriorly in the presence of synovial effusion or synovitis. An **effusion** may be caused by trauma, joint inflammation, degenerative changes, or infection. Ultrasound reveals a hypoechoic or anechoic effusion in the anterior recess with an anterior (superficial) displacement of the fat pad [3]. If you detect an effusion, evaluate it for **loose bodies** which may be associated with osteochondromatosis, osteochondral fractures, and osteochondritis dissecans, three entities with a proclivity to occur in the elbow. **Synovitis** may demonstrate power Doppler activity in inflammatory conditions such as rheumatoid arthritis [3]. **Cartilage or bone erosion** may be evident at the humeroradial joint or the humeroulnar joint. **Crystal disease** (gout and calcium pyrophosphate deposition disease [CPPD]) prefers other joints but may occur in the elbow. On US, examine for urate deposits on the cartilage surface (double contour sign of gout), gouty tophi, and CPPD crystals within the cartilage. Sonography may reveal gout and CPPD crystals within the joint, bursa, and tendons.

Fractures

Conventional radiographs may reveal a conspicuous fracture. However, both well-defined and occult fractures may produce a radiographic "sail sign," in which a secondary effusion elevates the fat pad, mimicking the appearance of a spinnaker sail [5]. If the fat pad is elevated on US or radiographs, a careful sonographic inspection may reveal cortical bone irregularities [3]. If a fracture is still suspected after radiographic and sonographic evaluations, further imaging is warranted.

Distal Biceps Tendon Pathology

Distal biceps tendon tears typically occur in middle-aged men, associated with activities such as weightlifting [6]. Distal biceps tendon ruptures are much less common than proximal biceps tendon tears [3]. A complete DBT tear often presents with sudden anterior elbow pain and a popping sensation after lifting a heavy object [1]. There may be a soft tissue mass or a palpable defect in the antecubital fossa accompanied by weakness in flexion and supination. A round lump along the anterior upper arm is described as a "Popeye sign" and is formed by the proximally retracted tendon and muscle [7].

Since treatment of a complete tear is surgical, better results require a prompt diagnosis [6]. If the diagnosis is clinically obvious, no imaging is required. However, the diagnosis of nonretracted tears may be challenging [1, 6]. Partial or nonretracted tears may occur due to intact lacertus fibrosis. The US image may be obscured by significant edema and hemorrhage [1]. Again, it is essential to distinguish between complete and partial tears since complete tears require surgery.

Partial tears of the DBT are best seen in longitudinal (LAX) view on US and appear as anechoic or hypoechoic discontinuities of tendon fibers without retraction; there may be surrounding hypoechoic fluid. If the lacertus fibrosis is ruptured, the tendon may retract, resulting in nonvisualization of the distal tendon in the expected location [1, 3]. The finding of a posterior acoustic shadow deep to the retracted tendon may increase diagnostic accuracy [8].

Incomplete ruptures may also appear as intratendinous hypoechogenicity with a thickened or thinned tendon [3]. The slanting course of the DBT and attendant hypoechogenicity create a sonographic imaging challenge, particularly at its attachment to the radial tuberosity. Multiple sonographic approaches have been offered [3]. Due to these imaging obstacles, US is unreliable to visualize tendon tears, although dynamic US may help differentiate complete from incomplete tears [1]. Evaluating the contralateral elbow may be of help.

Thus, a pitfall is missing a DBT tear or tendinosis. To mitigate the impediment of anisotropy, evaluate the tendon in transverse (TAX) and LAX, perform heel-to-toe maneuvers, and use full forearm supination. Further, evaluate the tendon in LAX with a *segmental approach* and assess any potential abnormality with varying degrees of elbow flexion and extension [1]. In one study comparing US to surgical findings in evaluating complete versus partial DBT tears, US demonstrated 95% sensitivity, 71% specificity, and 91% accuracy [6]. *As always, the diagnostic results of US studies rely on the experience of sonographers performing a particular US examination.*

A disadvantage of magnetic resonance imaging (MRI) is the typical delay in obtaining approval for the study since diagnostic delay affects surgical outcomes with complete tears. The disadvantage of US, however, is the technical challenge of DBT evaluation [6]. Some recommend that US be the first-line study to evaluate DBT injuries if a trained and experienced sonographer is available. Nevertheless, MRI is still considered the gold standard for the evaluation of this tendon, particularly at the distal attachment [9].

Bicipitoradial Bursitis

The DBT attaches to the radial tuberosity, located on the *posteromedial* aspect of the radius [10]. Repetitive supination–pronation creates wear and tear on the DBT. Since the DBT lacks a tenosynovium, contact between the tendon and the radius is mitigated by the bicipitoradial bursa (BRB) [10]. The BRB surrounds the DBT during forearm supination and is compressed against the radial tuberosity during forearm pronation. This bursa does not communicate with the joint space [10]. The BRB may distend due to tears of the DBT, repetitive microtrauma, infection, and inflammatory arthropathies such as RA and gout [3, 10].

Bursitis of the BRB, known as cubital (bicipitoradial) bursitis, may present as an antecubital mass or with symptoms of compression of the nearby radial nerve [11]. If neurologic symptoms are present, there may also be local pain. Cubital bursitis, while the most common bursitis in this area, may be challenging to diagnose since the cubital bursa is in a deep location [3]. Thus, cubital bursitis may be misdiagnosed as biceps tendinitis.

Ultrasound can detect anechoic bursal distension surrounding the distal DBT, best seen in TAX view with respect to the long axis of the tendon during forced forearm supination [10]. Ultrasound reveals a thickened bursal wall associated with distention and an anechoic effusion [3]. Bicipitoradial bursitis mimics synovial ganglion cysts and other soft tissue masses [1]. Additionally, US may help guide aspiration or a therapeutic bursal injection [10].

Median Nerve Pathology

Median nerve (MN) entrapment may occur at several locations in the anteromedial elbow [1]. The PT muscle has a humeral head and an ulnar head, which combine to form a muscle belly that inserts distally onto the middle third of the radius. The PT mainly pronates the forearm but also flexes the elbow. In more than 80% of individuals, the MN runs between the two heads and is less commonly posterior to the ulnar (deeper) head [1]. The MN on US has a speckled appearance in the short axis due to the hypoechoic nerve fascicles and intervening hyperechoic epineurium.

Median nerve (MN) compression between the two heads of the PT muscle causes pronator teres syndrome (PTS) [12]. This produces paresthesia in the distal extremity, mimicking carpal tunnel syndrome (CTS). It occurs with repetitive forearm pronation, an occupational hazard for carpenters, hairdressers, baseball pitchers, violinists, and mechanics [12]. Tinel's sign may be positive in the forearm. However, PTS does not produce nocturnal paresthesia as occurs with CTS, and EPS studies are often normal with PTS [12].

Struthers' ligament, present in 1% of individuals, is a fibrous or muscular band originating from the apex of the distal humeral supracondylar process that inserts on the medial epicondyle [12]. This creates a "tunnel" inside which the MN or BA may become entrapped. Radiographs of the humerus will detect this supracondylar process which is a clue to the possibility of Struthers' ligament compressing the MN and thus mimicking CTS [12]. Reduced radial artery pulsation may occur with BA compression in supination and forearm extension.

Other causes of proximal compression of the MN include pseudoaneurysms of the brachial or axillary arteries, dialysis fistula, post-puncture hematoma, anatomic variations, and compression by orthopedic material [12]. The lacertus fibrosis, located between the bicipital tendon and PT fascia, may also compress the MN. The PT muscle itself may also be injured with throwing sports, forearm fractures or dislocations, and during surgery [12].

Radial Nerve Pathology

The radial nerve (RN) at the elbow divides into the superficial radial nerve (SRN) and the deep branch, also known as the PIN [13]. This division may occur anywhere from 5 cm above to 1 cm below the radiohumeral joint line. The PIN may be compressed at multiple locations: at the medial edge of the extensor carpi radialis brevis, by fibrous bands near the head of the radius, by branches from the recurrent radial artery (Leash of Henry), under the Arcade of Frohse (a fibrous arch at the proximal edge of the superficial head of the supinator), and at the distal edge of the superficial supinator muscle [13]. Other causes of PIN compression include ganglion cysts, tumors, radial head and neck fractures, or posttraumatic scars [1, 14]. Compression of the PIN may cause the following:

1. Pure motor dysfunction distally, with weakness of finger extension, local pain, radial nerve palsy, and difficulty in extending the wrist [3].
2. The so-called radial tunnel syndrome may mimic lateral epicondylitis by causing lateral elbow pain. Sensory fibers carried by the PIN may be the source of pain. A variety of opinions in the literature makes this syndrome somewhat controversial [13].

Use US to follow the PIN distally with TAX images and compare with the contralateral unaffected side. Longitudinal views are more challenging due to the winding course of the PIN [13]. Compression of the PIN appears sonographically as hypoechoic enlargement proximal to the point of compression [1, 13]. Doppler ultrasound may reveal atypical and prominent pulsating vessels near the nerve in the case of PIN compression at the Leash of Henry [13]. PIN compression is best determined if there is enlargement and hypoechogenicity of the nerve near these vessels.

There may be *normal* PIN flattening at the Arcade of Frohse (just before the PIN enters the plane between the supinator heads), which may simulate entrapment, particularly on TAX view [14]. This pitfall is avoided by measuring the cross-sectional area (CSA) of the nerve 1 cm proximal to and 1 cm distal from the arcade. Another pitfall is that PIN appearance usually changes with pronation and supination [13].

The PIN is best followed between the superficial and deep heads of the supinator muscle while slowly pronating the forearm [13]. Magnetic resonance imaging is a complementary modality when diagnosing PIN compression. Bicipitoradial bursitis may uncommonly compress the SRN and the PIN [15]. Distal biceps tendon repair may injure these nerve branches as well [13].

Pitfalls

1. Tendinosis and tears of the DBT are easy to miss due to imaging challenges. To minimize this error, first evaluate the tendon in the transverse axis (TAX) and then evaluate the tendon in a longitudinal (LAX) segmental approach using heel-to-toe maneuvers employing maximal forearm supination. Also, learn the "crab" technique described in the **Method** section to delineate the DBT insertion on the radial tuberosity [1, 16].
2. Median nerve entrapment may occur at several locations in the anteromedial elbow and may mimic CTS. Nocturnal paresthesia, positive Tinel's and Phalen's signs at the volar wrist, as well as an enlarged MN proximal to the transverse carpal ligament, favors CTS. The diagnosis of more proximal MN entrapment requires careful US examination of the entire course of the MN in TAX view.
3. Do not mistake the normal flattening of the PIN as it enters the Arcade of Frohse (AF) for true compression. Always measure the CSA of the nerve 1 cm proximal to and 1 cm distal from the arcade [13].
4. Do not mistake the change in appearance of the PIN with pronation and supination for true compression of the PIN.
5. Do not mistake radial nerve compression for lateral epicondylitis since both cause pain near the lateral epicondyle.
6. Due to its deep location, cubital bursitis may be challenging to diagnose and may be mistaken for biceps tendinitis, synovial ganglion cysts, and other soft tissue masses.

Method

Based on the clinical history and location of the discomfort, try to determine what structure is the most likely cause of the problem. The patient is recumbent with an extended forearm resting on the examination table and palm up (supinated). Alternatively, the patient may be seated, with a supinated, extended arm on the lap. Use the middle depth or elbow preset.

Protocol Image 1: Anterior Distal Humerus, Transverse ± Power Doppler (Fig. 6.3)

Start with the probe in TAX on the ulnar (inner or medial) side of the elbow, at a point approximately 5 cm proximal to the elbow crease. Turn on the PD to visualize the BA and any of its branches on the anterior elbow's medial aspect. The MN is just medial to the BA.

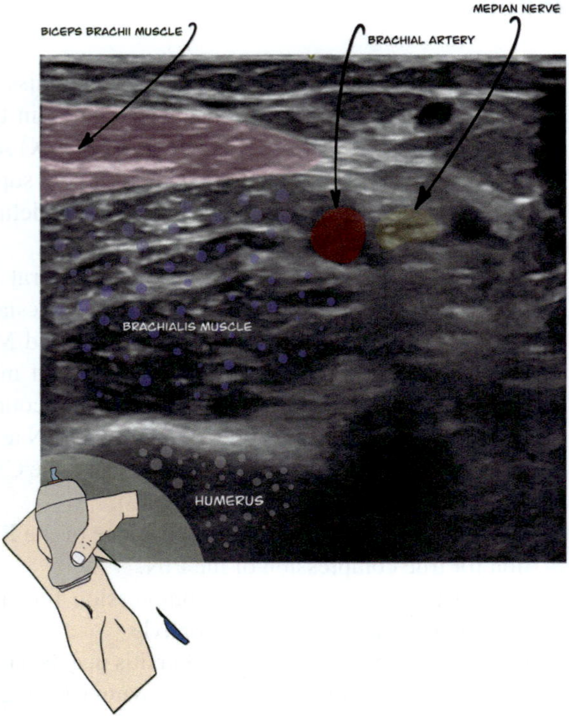

Fig. 6.3 Protocol Image 1: Anterior distal humerus, transverse ± power Doppler

Protocol Image 2: Anterior Distal Humeral Cartilage Ulnar, Transverse (Fig. 6.4)

To assess the cartilage of the distal humerus on the ulnar side (the trochlea), perform a distal TAX slide (move closer to the elbow crease) to visualize the wavy cartilage of the distal humerus. The transducer should be at or just proximal to the elbow crease. Note that this is *not* the true joint space, but the cartilage covering the distal humerus. From medial to lateral, locate the PT, BA, MN, and the brachialis muscle. The coronoid fossa is seen in the TAX view. The biceps brachii may be seen transitioning to the smaller biceps tendon at its myotendinous junction.

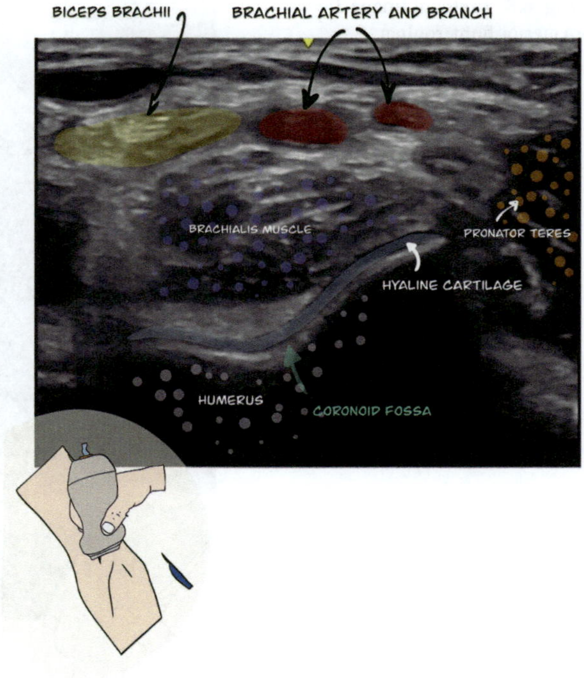

Fig. 6.4 Protocol Image 2: Anterior distal humeral cartilage ulnar, transverse

Protocol Image 3: Anterior Humeroulnar Joint, Longitudinal (Fig. 6.5)

Visualize the humeroulnar joint (HUJ) (trochlea-ulnar joint) in TAX by performing a slight distal TAX slide while still in TAX. The joint space, located immediately distal to the wavy line (distal humeral cartilage), is reached when the bone "drops out." The probe is now centered over the actual joint space. Then, rotate the transducer 90° to LAX orientation to visualize the HUJ in LAX. Recognize the curved humeral trochlea and the sharp, pointed ulnar head (coronoid process). Imaging may require slight TAX slides in the lateral or medial directions. If necessary, perform a proximal LAX slide to see the coronoid fossa of the distal humerus in LAX. The fat pad lies over the fossa; the anterior elbow fat pads are contained within the joint capsule but *outside* the synovium.

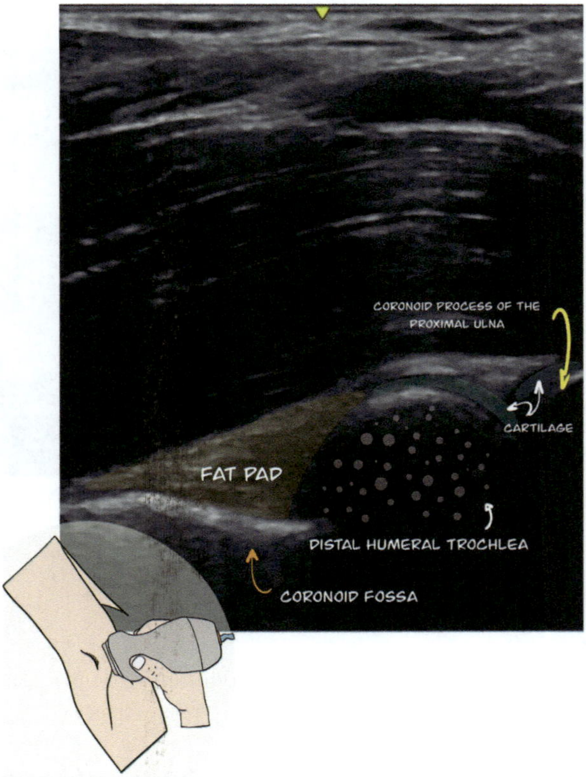

Fig. 6.5 Protocol Image 3: Anterior humeroulnar joint, longitudinal

Protocol Image 4: Anterior Humeroradial Joint, Longitudinal (Fig. 6.6)

With the transducer still in LAX orientation to the forearm, perform a TAX slide in the radial direction until the shallow radial fossa is seen at the distal humerus. The fat pad overlies the fossa. A slight distal LAX slide reveals the humeral-radial (radiocapitellar) joint with the curved hyperechoic humeral capitellum and the radial head, which has a characteristic hyperechoic "flat top" appearance.

Fig. 6.6 Protocol Image 4: Anterior humeroradial joint, longitudinal

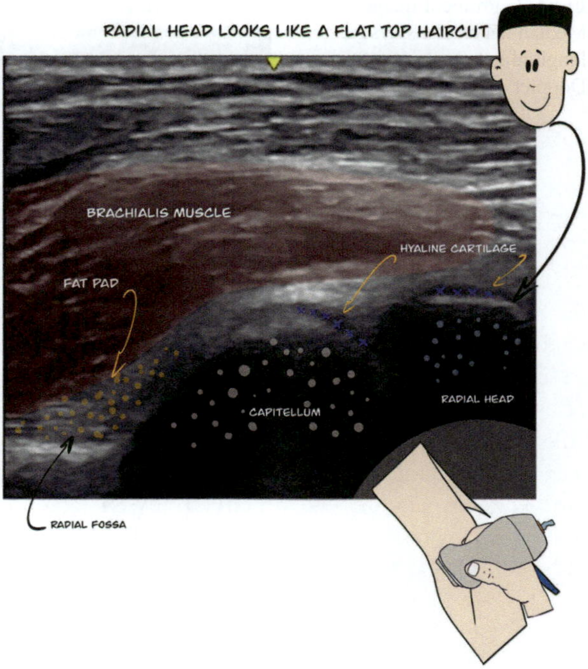

Protocol Image 5: Median Nerve, Transverse ± Longitudinal ± Power Doppler (Fig. 6.7)

Rotate the probe 90° into TAX, and then perform a proximal TAX slide back to Protocol Image 2. The elbow remains extended and supinated. Move the transducer in an LAX slide medially along the crease of the elbow to center it on the MN, which is just medial to the BA. The artery can be located with power Doppler. Then perform a distal TAX slide to observe the BA or one of its branches slip under the MN. This verifies the identity of the MN. At this level, the BA has split off into the ulnar and radial arteries, and the MN is between the two heads (ulnar and humeral) of the PT muscle. Compression of the MN between the two PT muscles may cause paresthesia, pain, or weakness and is called the *pronator teres syndrome.* See the **Clinical Comments** section.

If necessary, dynamic assessment for MN compression between the deeper ulnar head and the more superficial humeral head can be performed as follows. The patient's elbow is flexed, and the examiner actively resists forearm pronation. This

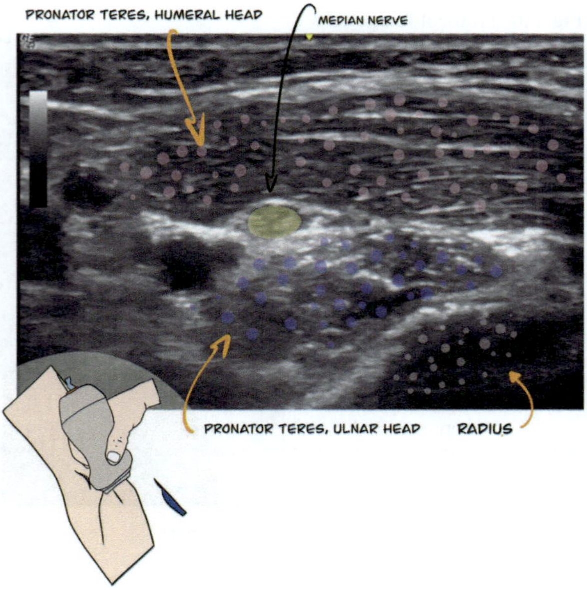

Fig. 6.7 Protocol Image 5: Median nerve, transverse ± longitudinal ± power Doppler

is followed by persistent, continued resistance to flexion, while the examiner passively extends the forearm [17]. Comparison to the contralateral, presumably unaffected side, may also be helpful. If desired, the MN can be examined in LAX by centering the image on the MN, rotating the transducer 90°, and performing a heel-to-toe maneuver.

Protocol Image 6: Anterior Distal Humeral Cartilage Radial, Transverse ± Power Doppler (Fig. 6.8)

Next, turn the probe back to TAX and return to Protocol Image 2, that is, TAX at the ulnar aspect of the distal humerus, just proximal to the joint line. The "wavy line" of the distal humeral cartilage is again visible. Perform an LAX slide in a radial (lateral) direction across the anterior elbow to see the distal humeral cartilage over the humeral capitellum. The radial fossa, shallower than the coronoid fossa, is also found here. The prominent hyperechoic diagonal line is the fascial plane (the lateral intermuscular septum) between the brachialis and the more lateral brachioradialis muscles. Sandwiched in between the two muscles is the radial nerve, which is starting to split into superficial and deep branches.

A distal TAX slide past the humeroradial joint shows the radial nerve further dividing into the deeper PIN and the SRN. The PIN is also known as the deep branch of the radial nerve. As you move the probe distally in a TAX slide, the fascicle that will become the PIN will dive deep between the two heads of the supinator

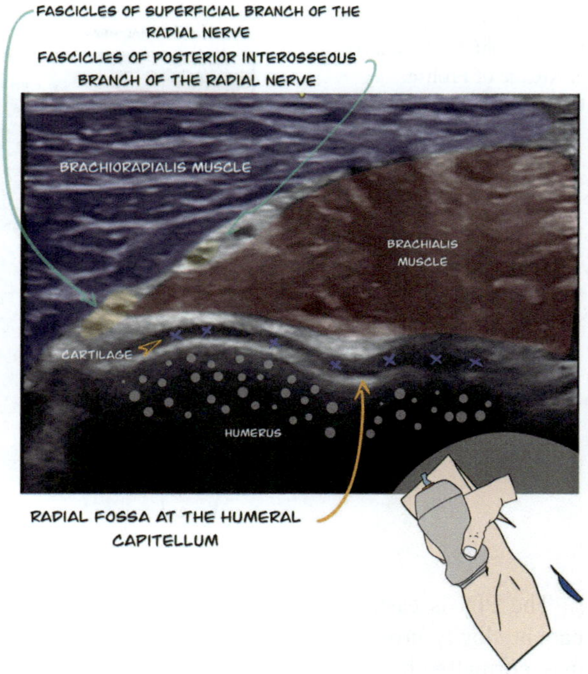

Fig. 6.8 Protocol Image 6: Anterior distal humeral cartilage radial, transverse ± power Doppler

muscle. This maneuver reveals which bundles in the lateral intermuscular septum become the PIN and the SRN.

Turn on the PD to look for pulsating blood vessels surrounding the PIN fascicle, which resembles a dog leash and is called the Leash of Henry and may be associated with PIN compression. See the **Clinical Comments** section. If necessary, turn the probe 90° to try to capture an optional image of the PIN in LAX, although this may be challenging.

Protocol Image 7: Posterior Interosseus Nerve at Arcade of Frohse, Transverse ± Longitudinal (Fig. 6.9)

Now, continue to move the probe distally in a TAX slide, following the PIN as it enters the AF between the deep and superficial heads of the supinator muscle. This is where the PIN may become symptomatically compressed, producing the *supinator syndrome*. However, the PIN may *normally* appear flattened here. Measure the CSA in TAX under the AF as well as 1 cm proximal to and distal from the AF to see if there is an actual enlargement of the PIN. Verified proximal enlargement might indicate actual nerve compression. The CSA of the PIN is normally about 1 mm^2. It may be entrapped if it has swollen to 2–3 mm^2 [13].

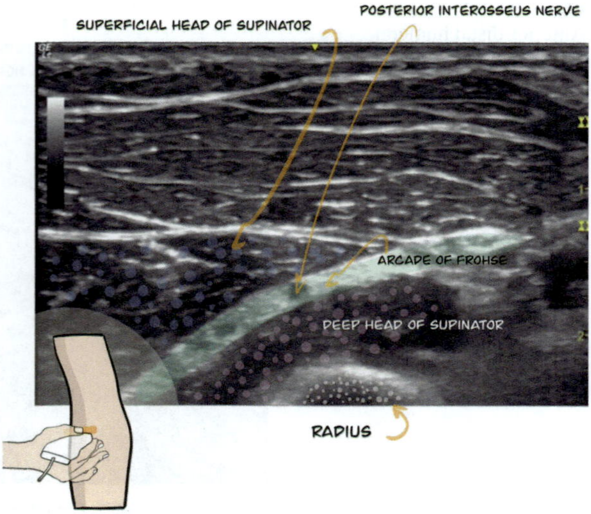

Fig. 6.9 Protocol Image 7: Posterior interosseus nerve at Arcade of Frohse, transverse ± longitudinal

The PIN is easily seen with a probe fixed on the supinator muscle while the patient slowly pronates and supinates the forearm, although the *shape* of the PIN may normally change with this movement. If necessary, turn the probe 90° to capture an optional image of the PIN in LAX. Image acquisition may be aided by slowly pronating the forearm.

Protocol Image 8: Biceps Tendon, Longitudinal (Composite Extended View) (Fig. 6.10)

Next, return to Image 1, that is, a TAX view 5 cm proximal to the elbow crease. Locate and center the probe on the cross section of the ovoid biceps muscle, just superficial to the larger brachialis muscle. Follow the biceps muscle distally by performing a TAX slide. The muscle decreases in size as it transitions to its myotendinous junction and then to the anisotropic circular DBT Center the probe over the DBT and rotate it 90°. The DBT is seen plunging downward at about a 30° angle toward the radial tuberosity, a bony protrusion on the *posteromedial* aspect of the radius just distal to the radial head.

Angle the probe about 10–15° in the radial direction and perform a heel-to-toe maneuver while performing a slow LAX slide. Image acquisition is challenging. Due to anisotropy, the bulk of the DBT is visualized as an anechoic or hypoechoic longitudinal structure. Make sure the forearm is in complete supination. The DBT needs to be *examined in segments* to mitigate the anisotropic mimicry of a tendon tear. By tilting, slightly rotating, and using heel-to-toe maneuvers, you can transform the appearance of several hypoechoic tendon segments into the hyperechoic fibrillar echotexture necessary to assess the tendon. Beam steering, if available, may be of assistance.

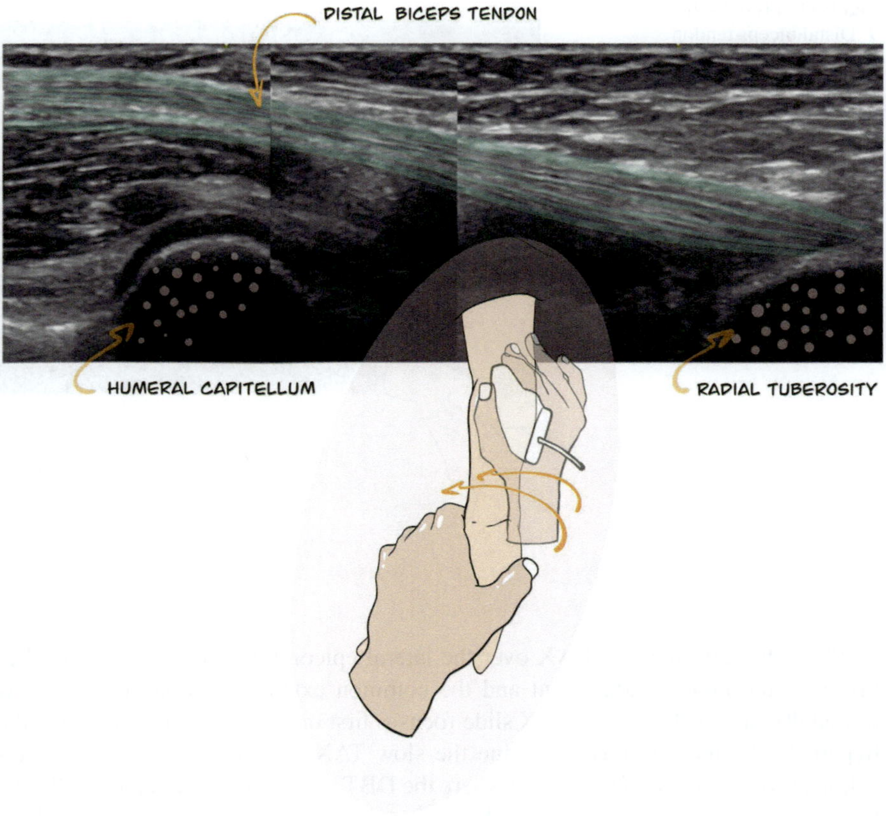

Fig. 6.10 Protocol Image 8: Biceps tendon, longitudinal [composite extended view]

Segmental analysis reveals a composite picture of this long tendon. The bicipitoradial bursa, if distended, may be seen surrounding the DBT in this view. The insertion of the DBT onto the radial tuberosity is difficult to visualize in this view, however.

Protocol Image 9: Distal Biceps Tendon Insertion, Longitudinal (Fig. 6.11)

Visualizing the DBT insertion onto the radial tuberosity is challenging. However, an innovative technique has proven useful [16]. The patient assumes the "crab position," in which the upper arm is extended backward at the shoulder, the elbow is flexed down at about 90°, and the palm is flat on the table with the fingers facing forward. This position exposes the posterior aspect of the radial tuberosity.

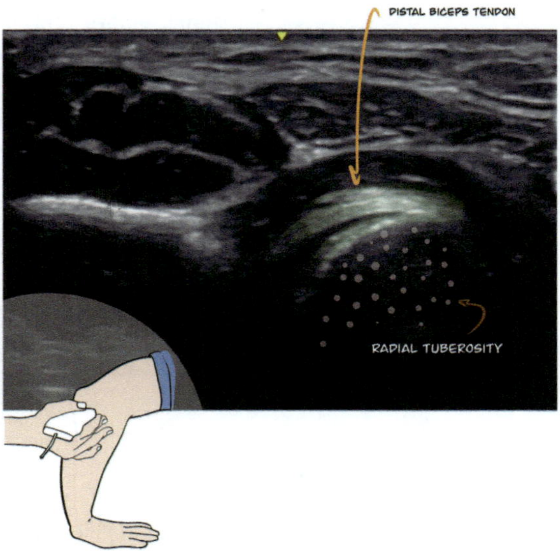

Fig. 6.11 Protocol Image 9: Distal biceps tendon insertion, longitudinal

Place the transducer in LAX over the lateral epicondyle and the radial head to visualize the humeroradial joint and the common extensor tendon. The probe is turned 90°, and a slow distal TAX slide focuses first on the radial head and then the hyperechoic curved radius. Continue the slow TAX slide to visualize the hyperechoic, prominent radial tuberosity where the DBT insertion can be seen as a "bird's beak" without anisotropy. Note that this view is longitudinal with respect to the long axis of the tendon at its insertion.

Complete Anterior Elbow Ultrasonic Examination Checklist

☐ Protocol Image 1: Anterior distal humerus, transverse ± power Doppler.
☐ Protocol Image 2: Anterior distal humeral cartilage ulnar, transverse.
☐ Protocol Image 3: Anterior humeroulnar joint, longitudinal.
☐ Protocol Image 4: Anterior humeroradial joint, longitudinal.
☐ Protocol Image 5: Median nerve, transverse ± longitudinal ± power Doppler.
☐ Protocol Image 6: Anterior distal humeral cartilage radial, transverse ± power Doppler.
☐ Protocol Image 7: Posterior interosseus nerve at Arcade of Frohse, transverse ± longitudinal.
☐ Protocol Image 8: Biceps tendon, longitudinal (Composite Extended View).
☐ Protocol Image 9: Distal biceps tendon insertion, longitudinal.

References

1. Konin GP, Nazarian LN, Walz DM. US of the elbow: indications, technique, normal anatomy, and pathologic conditions. Radiographics. 2013;33(4):E125–47.
2. Bianchi S, Martinoli C. Detection of loose bodies in joints. Radiol Clin N Am. 1999;37(4):679–90.
3. Draghi F, Danesino GM, de Gautard R, Bianchi S. Ultrasound of the elbow: Examination techniques and US appearance of the normal and pathologic joint. J Ultrasound. 2007;10(2):76–84.
4. Cleland JA, Koppenhaver S, Su J. Elbow and forearm. In: Netter's orthopaedic clinical examination. Elsevier; 2022. p. 513–38.
5. Chapman S. The sail sign. Br J Hosp Med. 1991;46(6):399–400.
6. Lobo Lda G, Fessell DP, Miller BS, Kelly A, Lee JY, Brandon C, et al. The role of sonography in differentiating full versus partial distal biceps tendon tears: correlation with surgical findings. AJR Am J Roentgenol. 2013;200(1):158–62.
7. Lozano V, Alonso P. Sonographic detection of the distal biceps tendon rupture. J Ultrasound Med. 1995;14(5):389–91.
8. de la Fuente J, Blasi M, Martínez S, Barceló P, Cachán C, Miguel M, et al. Ultrasound classification of traumatic distal biceps brachii tendon injuries. Skelet Radiol. 2018;47(4):519–32.
9. Chew ML, Giuffrè BM. Disorders of the distal biceps brachii tendon. Radiographics. 2005;25(5):1227–37.
10. Draghi F, Gregoli B, Sileo C. Sonography of the bicipitoradial bursa: a short pictorial essay. J Ultrasound. 2012;15(1):39–41.
11. Sofka CM, Adler RS. Sonography of cubital bursitis. AJR Am J Roentgenol. 2004;183(1):51–3.
12. Créteur V, Madani A, Sattari A, Bianchi S. Sonography of the pronator teres: normal and pathologic appearances. J Ultrasound Med. 2017;36(12):2585–97.
13. Becciolini M, Pivec C, Raspanti A, Riegler G. Ultrasound of the radial nerve: a pictorial review. J Ultrasound Med. 2021;40(12):2751–71.
14. Dong Q, Jamadar DA, Robertson BL, Jacobson JA, Caoili EM, Gest T, et al. Posterior interosseous nerve of the elbow: normal appearances simulating entrapment. J Ultrasound Med. 2010;29(5):691–6.
15. Tsz-Lung C, Tun-Hing L. Bicipitoradial bursitis: a review of clinical presentation and treatment. J Orthop Trauma Rehabil. 2014;18(1):7–11.
16. Draghi F, Bortolotto C, Ferrozzi G. Distal biceps brachii tendon insertion: a simple method of ultrasound evaluation. J Ultrasound Med. 2021;40(4):811–3.
17. Hsu SH, Moen TC, Levine WN, Ahmad CS. Physical examination of the athlete's elbow. Am J Sports Med. 2012;40(3):699–708.

Chapter 7
Posterior Elbow

Reasons to Do the Study

1. Posterior elbow pain
2. Posterior elbow swelling

Questions We Want Answered

1. Is there olecranon bursitis?
2. Is there a joint effusion or synovitis?
3. Are there any intra-articular bodies?
4. Is there evidence for cartilage or bone erosion?
5. Is there a distal triceps tendon tear?

Anatomy

The principal elbow joint is the humeroulnar joint (Fig. 7.1). The olecranon process of the ulna, shaped like a bottle opener, glides over the humeral trochlea during elbow flexion and extension. The humeral trochlea and capitellum are scroll-like, rounded bony structures best appreciated anteriorly. Proximal to the posterior aspect of the humeral trochlea, the concave olecranon fossa contains the posterior joint recess, including the hyperechoic posterior fat pad [1]. Bending the elbow enables sonographic visualization of the posterior recess. A second joint, the humeroradial joint, is also visible on ultrasound (US) from the posterior direction.

Superficial to the bones is the triceps muscle, which consists of three heads (the long, lateral, and medial). The medial head is the deepest (Fig. 7.2). The three heads form the triceps tendon, which inserts into the olecranon process [1]. Superficial to

The contents do not represent the views of the U.S. Department of Veterans Affairs or the United States Government.

M. H. Greenberg et al., *Manual of Musculoskeletal Ultrasound*,
https://doi.org/10.1007/978-3-031-37416-6_7

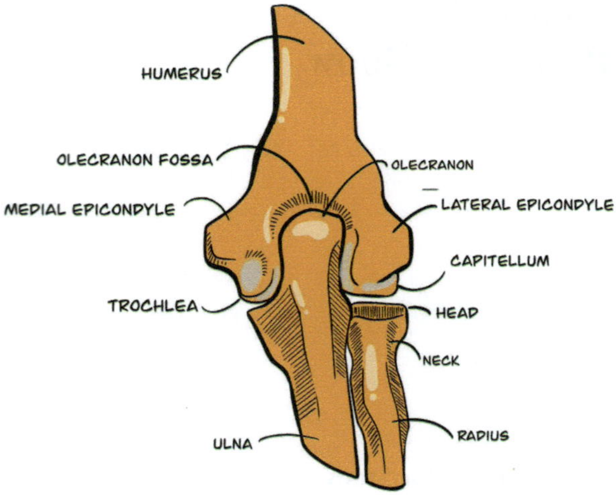

Fig. 7.1 Bony anatomy of the posterior elbow

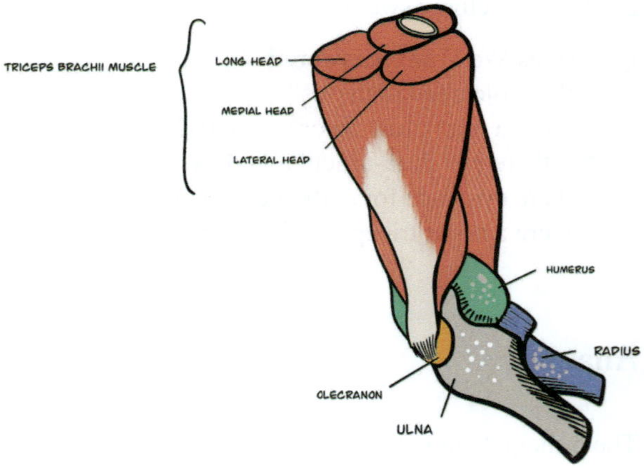

Fig. 7.2 Muscles of the posterior elbow

the olecranon process and the triceps tendon is the subcutaneous olecranon bursa, evident only when distended with fluid. The bursa is best visualized on US in elbow extension using abundant transmission gel and a light transducer touch [1].

Clinical Comments

Joint Effusion

The posterior recess is the preferred location to view excessive joint fluid (effusion), intra-articular loose bodies, and signs of inflammatory/erosive arthropathy [2]. Usually, fluid is undetectable within the recess, and the synovial membrane is invisible unless pathologically thickened [3]. The posterior joint recess is normally filled with an isoechoic posterior fat pad [1, 4]. An elbow effusion produces echogenically variable fluid deep within the recess that dislocates the fat pad in a superficial and proximal direction. The effusion is displaceable with transducer compression [1]. A posterior elbow effusion is easily aspirated using US guidance.

If synovitis or synovial thickening is present, US may also detect this [1, 4]. Synovial hypertrophy is implied by local hyperemia on power Doppler (PD). If synovial hypertrophy (synovitis) is present, consider inflammatory arthropathy (e.g., rheumatoid arthritis, spondyloarthropathy, or gout), infection, or chronic osteoarthritis [1]. Other causes of synovial hypertrophy include pigmented villonodular synovitis and synovial chondromatosis, the latter associated with calcified hyperechoic foci. Evaluate the hyaline cartilage over the trochlea and capitellum for erosion. If an intra-articular body is recognized on US or radiograph, you may discover the cartilage donor site [1].

Humeroradial Joint

Bone/Cartilage Erosion

Bone erosion may occur with inflammatory arthropathy.

Capitellum Osteochondritis dissecans

Capitellum osteochondritis dissecans (COCD) is a disorder of articular cartilage and subchondral bone of the capitellum associated with repetitive trauma, most commonly in teenage athletes who continually throw [5]. Conservative versus surgical treatment depends on the stability of the osteochondral fragment. Ultrasound may help evaluate fragment instability in COCD [6].

Triceps Brachii

Ultrasound can demonstrate tears and tendinosis of the distal triceps tendon (DTT) [4]. Degenerative, noninflammatory triceps tendon enthesopathy may be associated with localized pain [3]. The DTT may exhibit enthesopathy with spondyloarthropathy, with or without PD activity [7].

Frank DTT tears are uncommon, but US can differentiate complete from partial triceps tendon tears [3, 8]. Distal triceps tendon rupture may present as pain and swelling at the posterior elbow after falling on an outstretched hand [8]. There may be a bone avulsion from the tendon insertion on the olecranon. The surgical approach depends upon whether the tear is partial or complete [8]. Factors associated with a weakened triceps tendon include local steroid injections, attrition from degenerative arthritis, and olecranon bursitis.

The diagnosis of DTT tear is clinically challenging. The initial evaluation may be complicated by swelling and pain that limit strength and range of motion assessments. Magnetic resonance imaging has been helpful in preoperative planning, but US is gaining traction, particularly in the emergency department setting [8]. If a complete DTT tear occurs, US may demonstrate a retracted, wavy tendon [1].

Snapping triceps syndrome is described in Chap. 9.

Olecranon Bursa

The subcutaneous olecranon bursa, located over the posterior aspect of the olecranon, facilitates the gliding of the skin superficial to the joint and triceps tendon during elbow flexion and extension [3]. The olecranon bursa becomes visible when pathologically distended with effusion or when the bursal walls are significantly thickened, as occurs with synovial hypertrophy [3, 4]. Power Doppler activity within the bursa indicates local active inflammation [4]. Hyperechoic spots within the bursa may accompany crystal deposition. Since the olecranon bursa has a shallow location, it is crucial to use copious transmission gel and light touch to avoid displacing the intrabursal effusion.

Ulnar Nerve Compression

Although the ulnar nerve (UN) is technically part of the posterior elbow, UN evaluation at the elbow has been described in Chap. 9 since UN neuropathy often produces medial elbow symptoms.

Pitfalls

1. Avoid mistaking the anisotropy of the DTT for an insertional tear. Use sufficient transmission gel and a heel-to-toe maneuver to compensate.
2. When evaluating for olecranon bursitis, use ample transmission gel and a light touch to avoid displacing bursal fluid and missing the diagnosis. Extending one finger along the side of the transducer will prevent gel dispersion.
3. If posterior elbow discomfort is present after a fall on an outstretched hand injury, consider olecranon bone avulsion and triceps tendon tear since these diagnoses are clinically challenging.
4. In young athletes who repetitively throw, evaluate the posterior and anterior surfaces of the capitellum to look for COCD.
5. The posterior recess of the elbow is the ideal location to look for an effusion. Positioning the elbow facing downward enables gravity to assist the evaluation.

Method

The patient is recumbent with the hand placed over the umbilicus, elbow flexed at 90°, and pointed downward at least 45°. In this position, the posterior recess is gravity-dependent, which enhances effusion detection. Patients in pain usually tolerate this position. The examiner sets the transducer frequency at the midrange preset. Some examiners advocate an alternative position called the "crab" position, in which the forearm is internally rotated with the palm resting on the table [1].

Protocol Image 1: Posterior Joint Recess, Longitudinal ± Power Doppler (Fig. 7.3)

Place the transducer parallel to the long axis of the humerus at the center of the distal posterior elbow. Visualize the hyperechoic valley, the deep olecranon fossa, and the more superficial triceps muscle with its tendon insertion on the bony olecranon process of the ulna. Visualize the hyperechoic, curved trochlea. The posterior joint recess's hypoechoic or hyperechoic fat pad is within the olecranon fossa. The fat pad is just deep to the joint capsule.

Small transverse axis (TAX) slides medially and laterally reveal the full extent of the posterior recess. The hyperechoic, hill-like trochlea is covered by hypoechoic to anechoic cartilage, as is the less prominent capitellum, located radially. Evaluate for joint effusion, synovial hypertrophy, bone or joint erosion, loose bodies, and bone fracture. If a fracture is suspected, look for a discontinuity or step-off deformity. If there is a pathology question, rotate the probe 90° to verify in TAX.

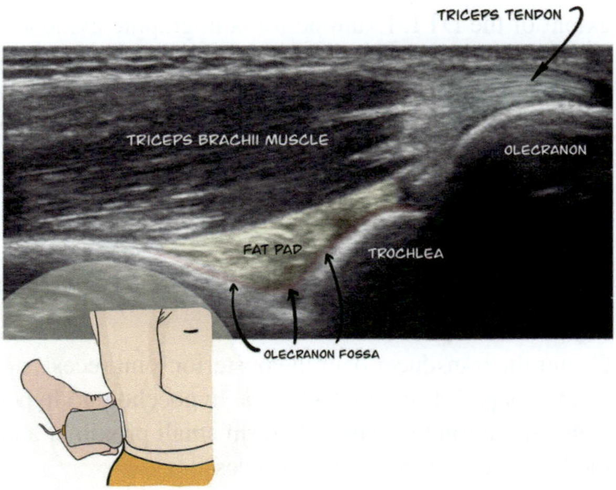

Fig. 7.3 Protocol Image 1: Posterior joint recess, longitudinal ± power Doppler

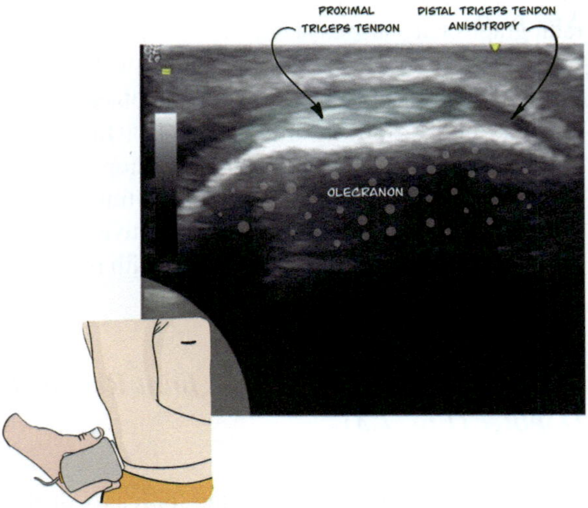

Fig. 7.4 Protocol Image 2: Distal triceps tendon, longitudinal ± power Doppler

Protocol Image 2: Distal Triceps Tendon, Longitudinal ± Power Doppler (Fig. 7.4)

Perform a slight distal longitudinal axis (LAX) slide halfway around the elbow bend to visualize the bird's beak configuration of the DTT insertion on the bony olecranon. Using adequate gel, perform a heel-to-toe maneuver, and tilt the probe to decrease anisotropy. Small medial and lateral TAX slides help evaluate the full extent of the DTT. Examine for sonographic evidence of enthesopathy, tendinosis, partial or complete tears, and bony avulsion. Slightly extend the elbow to avoid extreme tension on the tendon, which may ablate PD activity. Power Doppler activity might indicate active enthesitis. Rotate the probe 90° for a TAX view to verify any pathology.

Protocol Image 3: Posterior Joint Recess, Transverse (Fig. 7.5)

Center the transducer over the posterior joint recess, and rotate it 90° to evaluate the valley-shaped recess in TAX. The hypoechoic to hyperechoic fat pad is recognized beneath the joint capsule. Perform small proximal and distal TAX slides to take in the full extent of the posterior recess.

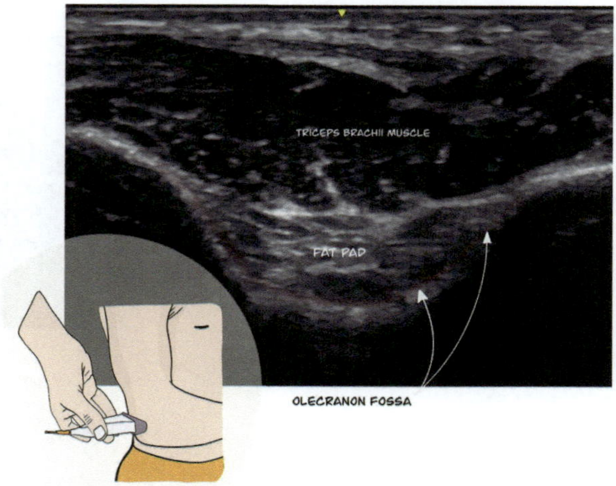

Fig. 7.5 Protocol Image 3: Posterior joint recess, transverse

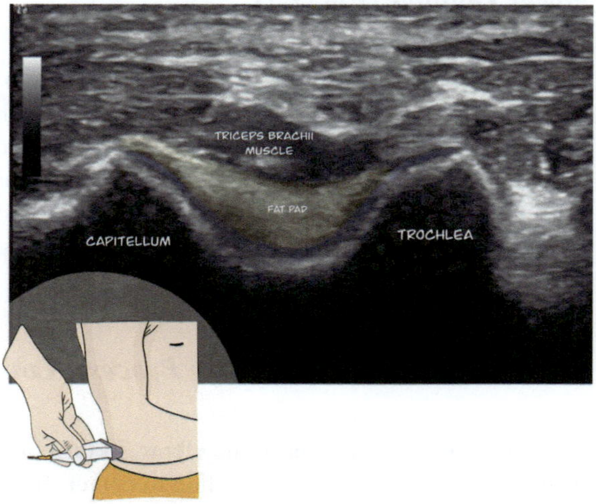

Fig. 7.6 Protocol Image 4: Distal trochlea and capitellum, transverse

Protocol Image 4: Distal Trochlea and Capitellum, Transverse (Fig. 7.6)

Perform a TAX slide distally to visualize the hyperechoic hill-like humeral trochlea and capitellum. The trochlea is medial (ulnar), and the capitellum is lateral (radial). Between these two hyperechoic "hills" are the distal portion of the olecranon fossa and its fat pad. Hyaline cartilage covers this distal portion of the humerus.

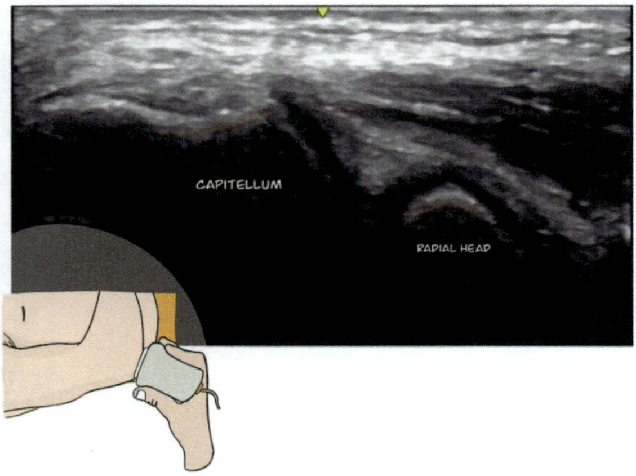

Fig. 7.7 Protocol Image 5: Posterior humeroradial joint, longitudinal ± power Doppler

Protocol Image 5: Posterior Humeroradial Joint, Longitudinal ± Power Doppler (Fig. 7.7)

Rotate the probe 90° back to the LAX orientation to the extremity. The transducer should be at the elbow bend, angled about 30° with respect to the forearm axis. Perform a TAX slide in a radial direction to visualize the humeroradial joint with the hyperechoic large curved humeral capitellum and the smaller flat-to-curved radial head. Verify any pathology in the TAX view. Turn on the PD, but remember that extreme elbow flexion may diminish Doppler activity.

Protocol Image 6: Olecranon Process, Longitudinal (Fig. 7.8)

The patient then fully extends the elbow. Return to the probe location for Protocol Image 1. Perform a distal LAX slide to center the probe over the bony olecranon fossa. Using a light touch and a liberal amount of transmission gel, evaluate for effusion or synovitis in the olecranon bursa (Fig. 7.9). Excessive pressure by the transducer will inadvertently disperse small amounts of olecranon bursal fluid, so float the transducer. An extended finger along the long axis of the transducer will help to keep the gel from dispersing.

Synovial hypertrophy may occur within this bursa due to inflammatory processes. Particle movement from transducer compression or a lack of PD activity favors effusion rather than synovial hypertrophy. Effusion within the bursa may occur with infection, inflammatory arthropathy, or trauma. Chronic inflammatory processes may cause bony erosion, depicted as cortical irregularities. Utilizing a higher frequency may improve the detection of more superficial pathology. Note that with the elbow extended, the insertion of the triceps tendon is not well-visualized.

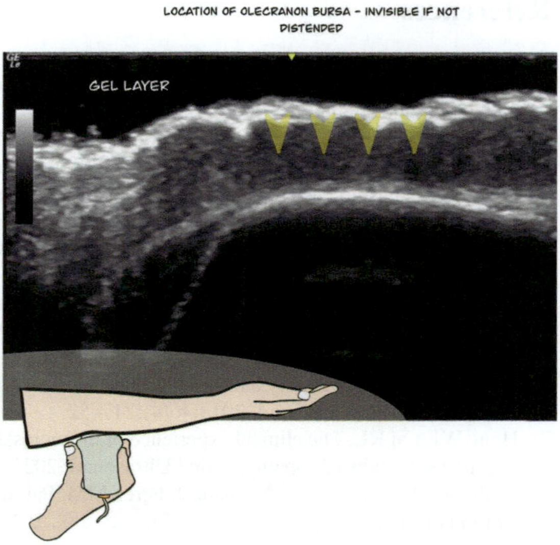

Fig. 7.8 Protocol Image 6: Olecranon process, longitudinal

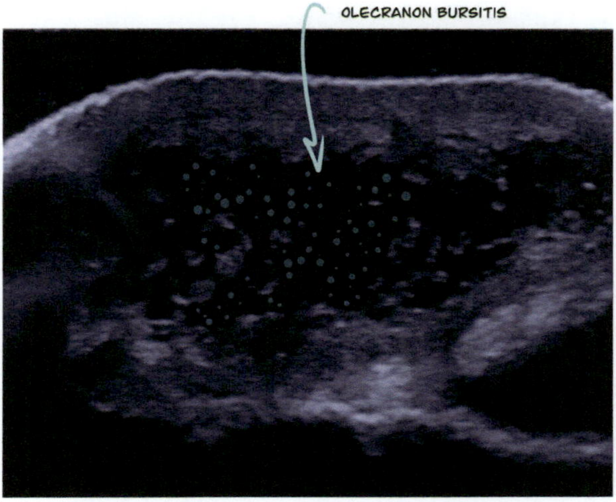

Fig. 7.9 Olecranon bursitis

Complete Posterior Elbow Ultrasonic Examination Checklist

☐ Protocol Image 1: Posterior joint recess, longitudinal ± power Doppler.
☐ Protocol Image 2: Distal triceps tendon, longitudinal ± power Doppler.
☐ Protocol Image 3: Posterior joint recess, transverse.
☐ Protocol Image 4: Distal trochlea and capitellum, transverse.
☐ Protocol Image 5: Posterior humeroradial joint, longitudinal ± power Doppler.
☐ Protocol Image 6: Olecranon process, longitudinal.

References

1. Konin GP, Nazarian LN, Walz DM. US of the elbow: indications, technique, normal anatomy, and pathologic conditions. Radiographics. 2013;33(4):E125–47.
2. Fessell DP, Jacobson JA, Craig J, Habra G, Prasad A, Radliff A, et al. Using sonography to reveal and aspirate joint effusions. AJR Am J Roentgenol. 2000;174(5):1353–62.
3. Draghi F, Danesino GM, de Gautard R, Bianchi S. Ultrasound of the elbow: examination techniques and US appearance of the normal and pathologic joint. J Ultrasound. 2007;10(2):76–84.
4. Radunovic G, Vlad V, Micu MC, Nestorova R, Petranova T, Porta F, et al. Ultrasound assessment of the elbow. Med Ultrason. 2012;14(2):141–6.
5. van Bergen CJ, van den Ende KI, Ten Brinke B, Eygendaal D. Osteochondritis dissecans of the capitellum in adolescents. World J Orthop. 2016;7(2):102–8.
6. Yoshizuka M, Sunagawa T, Nakashima Y, Shinomiya R, Masuda T, Makitsubo M, et al. Comparison of sonography and MRI in the evaluation of stability of capitellar osteochondritis dissecans. J Clin Ultrasound. 2018;46(4):247–52.
7. Hong WJ, Lai KL. The clinical experience of musculoskeletal ultrasound for enthesitis in seronegative spondyloarthropathy. J Med Ultrasound. 2021;29(4):237–8.
8. Tagliafico A, Gandolfo N, Michaud J, Perez MM, Palmieri F, Martinoli C. Ultrasound demonstration of distal triceps tendon tears. Eur J Radiol. 2012;81(6):1207–10.

Chapter 8
Lateral Elbow

Reasons to Do the Exam
1. Pain or swelling in the lateral elbow
2. Lateral epicondylar pain
3. Suspicion of "tennis elbow" (common extensor tendon tear or tendinosis)
4. Suspicion of lateral collateral ligament tear
5. Suspicion of elbow arthritis
6. Suspicion of elbow synovitis
7. Suspicion of elbow cartilage or bone erosion

Questions We Want Answered
1. What is causing the elbow pain?
2. Is the extensor tendon involved, and is the problem acute or chronic?
3. Is there enthesopathy at the lateral epicondyle?
4. Is there pathology of the lateral collateral ligament?
5. Is the pathology in the joint?
6. Is the visualized pathology the actual cause of the pain?
7. What is the mechanism of the condition?
8. Is impingement of a branch of the radial nerve mimicking tennis elbow pain?

Anatomy

Bones (Fig. 8.1)

The contents do not represent the views of the U.S. Department of Veterans Affairs or the United States Government.

M. H. Greenberg et al., *Manual of Musculoskeletal Ultrasound*,
https://doi.org/10.1007/978-3-031-37416-6_8

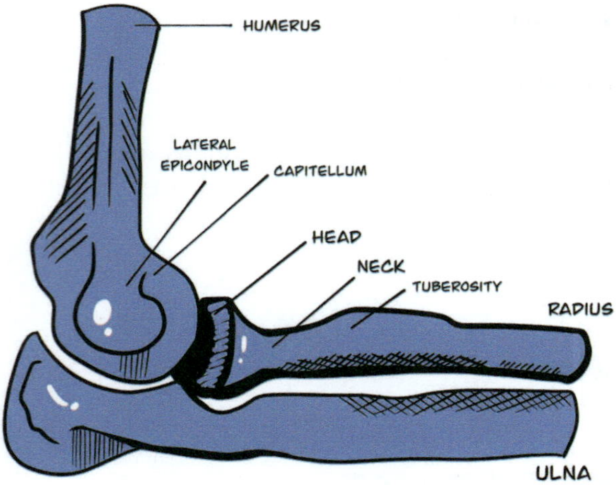

Fig. 8.1 Bony anatomy of the lateral elbow

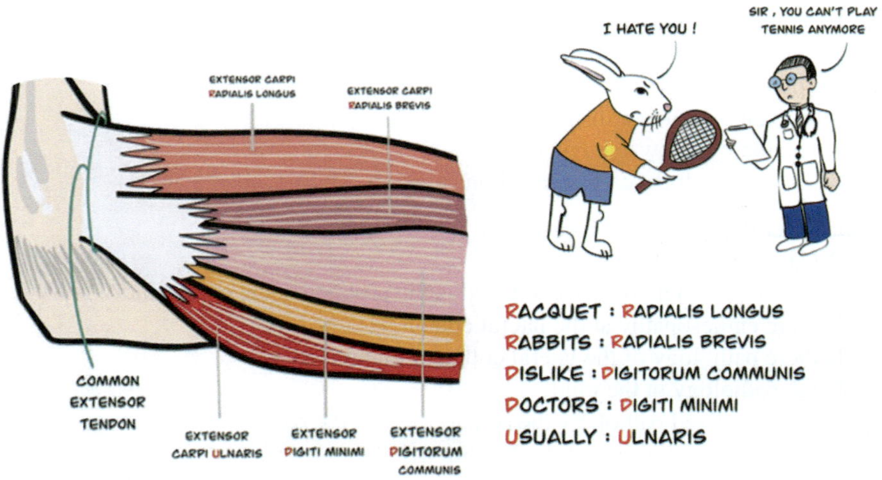

Fig. 8.2 Muscles and tendons of the lateral elbow

Muscles and Tendons (Fig. 8.2)

The **common extensor tendon (CET)**, a single large tendon derived from multiple extensor muscles, inserts onto the lateral epicondyle (LE) in a "bird-beak" configuration. The CET arises from these muscles [1], as follows:

1. Extensor carpi **r**adialis brevis (ECRB)

    The ECRB is the deepest and largest muscle of the CET.

2. Extensor carpi **r**adials longus (ECRL)

3. Extensor **d**igiti minimi (EDM)
4. Extensor **d**igitorum communis (EDC)

   The EDC is the most superficial muscle.

5. Extensor carpi **u**lnaris (ECU)

   A mnemonic derived from the bolded letters above is as follows:

   Racquet **R**abbits **D**islike **D**octors **U**sually

Joints (Fig. 8.3)

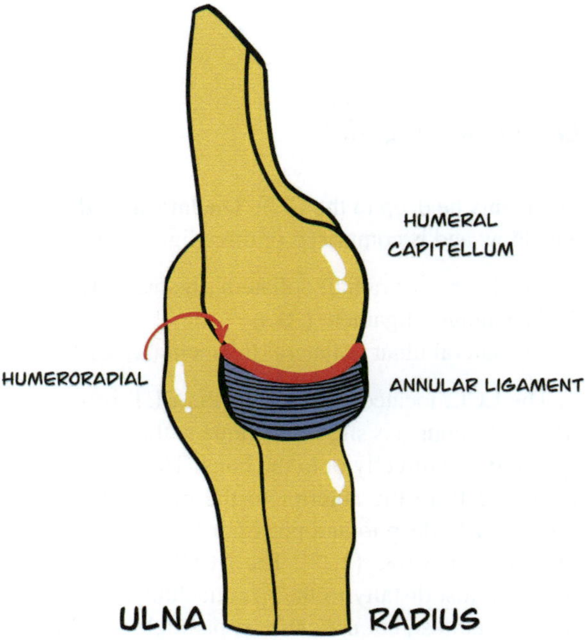

Fig. 8.3 Joints of the lateral elbow

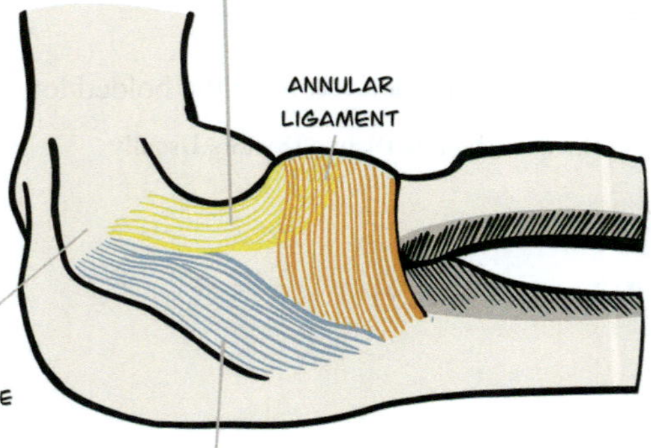

Fig. 8.4 Ligaments of the lateral elbow

Ligaments (Fig. 8.4)

Ligaments lie deep to the CET. The **lateral collateral ligament complex** stabilizes the elbow and is comprised of three ligaments:

1. The lateral (or radial) collateral ligament (LCL).
2. The annular ligament (AL).
3. The lateral ulnar collateral ligament (LUCL).

The LCL, located just deep to the CET, originates at the LE and attaches to the AL [2]. Its course is slightly oblique to the CET. The LCL inserts fibers into *another ligament,* not directly onto the bone. The AL stabilizes the radial head and neck by stretching from the anterior to the posterolateral ulna [2]. The LUCL originates from the LE, deep to and posterior to the LCL; the LUCL is an effective stabilizer against varus stress [1, 2]. The LUCL extends to the supinator crest of the ulna, attaching just distally to the AL attachment. Note that the LCL is deep to the CET and runs almost, but not entirely, parallel to it. Similarly, the LUCL lies deep to the LCL and runs virtually, but not entirely, parallel to the LCL.

Clinical Comments

Lateral Epicondylitis

Also known as "tennis elbow," this common disorder is usually due to repetitive overuse causing microtears and chronic secondary degeneration of the tendon near its insertion. Lateral epicondylitis is a misnomer since acute inflammation is generally lacking. Thus, lateral epicondylitis is more accurately described as CET tendinosis [1, 3]. However, microtears may sometimes progress to partial or even full-thickness tears. Lateral epicondylitis most often involves tendon fibers derived from the ECRB, but other CET layers may also be involved [1, 2].

Lateral epicondylitis clinically presents as localized tenderness with weakened strength of handgrip and elbow supination or extension [1]. Ultrasound (US) is helpful to confirm the diagnosis, evaluate disease severity and response to treatment, and guide injections. Ultrasound is also beneficial to exclude other causes of elbow symptoms. Sonography may demonstrate tendinosis with hypoechoic swelling at the tendon insertion. Anechoic tendon clefts or discontinuities suggest tendon tears. Doppler activity more commonly indicates neovascularization of an injured tendon than inflammation [1–6]. The clinical circumstance affects the interpretation of Doppler activity, for example, repetitive injury history versus spondyloarthropathy suspicion. Calcium deposits or enthesopathic changes at the bony attachment imply chronic tendinosis [1].

Regarding the diagnosis of lateral epicondylitis, US is specific but not sensitive [7]. A pitfall is mistaking anisotropy of the ECRB origin for hypoechogenicity and inferring tendinopathy or partial tear [1]. Be aware that CET tears may result from repeated intratendinous corticosteroid injections (CSIs) used to treat tennis elbow [8]. Long-term results after CSIs are less favorable than conservative treatment [9]. An entrapped posterior interosseous branch of the radial nerve may mimic tennis elbow [6].

(See Chap. 6 for more information.)

Ligament Pathology

With an injury to the LCL, US may demonstrate loss of ligament fibers and, if acute, perhaps a hematoma close to the humeral capitellum [2]. Consider and evaluate for concomitant LUCL and LCL injuries before treatment of lateral epicondylitis since conservative treatment failure occurs if an occult ligament tear is present [1, 2, 10, 11]. Distinguishing the CET from the LCL on the lateral epicondyle is challenging with US. A cadaver study described that the LCL footprint comprises 54% of the combined attachment of the CET and LCL [10]. Therefore, in most cases, the CET and the LCL each occupy roughly half the bony expanse of the LE. Injury to the LUCL may cause posterior rotatory joint instability with varus stress [10]. Perform dynamic imaging with hand supination and pronation to evaluate LUCL injury and abnormal radial head movement. However, magnetic resonance imaging is better suited for LUCL evaluation than US.

Joint Pathology

Ultrasound may detect joint effusion or synovitis at the humeroradial joint (HRJ). Complex effusion suggests infection, bleeding, intra-articular bodies, or inflammation [1]. Abnormal power Doppler (PD) activity argues for synovitis rather than a simple joint effusion; such synovitis may be due to inflammatory arthropathy, infection, or osteoarthritis. If synovitis is detected, consider pigmented villonodular synovitis or synovial osteochondromatosis, the latter displaying hyperechoic foci within the synovial fluid [1]. Additionally, US may guide joint aspiration.

Bone Pathology

1. **Bone erosion** may be noted at the humeroradial joint.
2. **Occult fractures** may be seen as step-off deformities at the epicondyle, the capitellum, or the radial head and neck.

Nerves

(See Chap. 6 for further information.)

Pitfalls

1. Do not mistake normal anisotropy of the ECRB origin for hypoechogenicity of tendinopathy or a partial tear [1].
2. Evaluate for occult ligament tears before conservative treatment of lateral epicondylitis since such treatment is likely to fail if there is a ligament tear [1].
3. Tennis elbow symptoms may be mimicked by entrapment of the posterior interosseous branch of the radial nerve [6].
4. Interpret Doppler activity at the CET insertion in light of the clinical situation. Doppler activity may indicate neovascularization of tendon injuries from repetitive trauma. Alternatively, Doppler activity may suggest active enthesitis if spondyloarthropathy is being considered.

Method

The patient is recumbent with the forearm in pronation. The elbow is bent to 90° with the palm down and the hand resting on the abdomen or examination table. Delineate any tender areas with a ballpoint pen. Palpate the LE and start with a moderate depth preset.

Protocol Image 1: Common Extensor Tendon Origin, Longitudinal ± Power Doppler (Fig. 8.5)

Place one end of the transducer over the bony, hyperechoic LE and the other over the small, slightly rounded radial head. The LE may be a bit more posterior than you think. The probe should roughly parallel the long axis of the forearm. Aim the distal transducer end toward the distal radius at the wrist. Visualize the hyperechoic gentle slope of the LE, the flat to slightly rounded distal humeral capitellum, and the small rounded radial head. Also, note the triangular synovial fold in the humeroradial joint and the CET, stretching distally from the most superficial portion of the LE. The triangular synovial fold is also called the meniscal homologue or lateral synovial fringe.

Visualize the fibrillar CET and its insertion on the LE by slightly rotating the distal portion of the transducer in either direction. Individual CET tendon components intermingle and cannot be separately delineated. While still in the longitudinal axis (LAX) orientation, perform minor transverse axis (TAX) slides medially and laterally to encompass the entire width of the CET. Look for CET tendinosis and tears and LE cortical irregularities and enthesophytes.

Turn on the PD to look for active enthesitis at the CET insertion on the LE. However, positive PD activity within the CET often indicates neovascularization from chronic tendinosis rather than an inflammatory process. Again, the clinical setting influences the interpretation of PD activity (e.g., repetitive trauma vs. possible spondyloarthropathy).

The curved hyperechoic structure just superficial to the radial head is the AL. The annular recess, located at the radial neck, may be distended with effusion or synovitis [8]. The brachioradialis muscle is superficial to the CET. The LCL lies deep to the CET. Since the CET and the LCL are not entirely parallel, only one of these two longitudinal structures may be in focus at a time. It is challenging to demarcate the CET from the LCL.

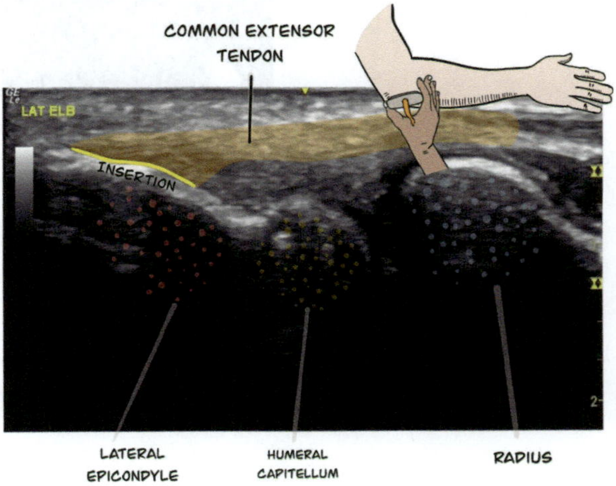

Fig. 8.5 Protocol Image 1: Common extensor tendon origin, longitudinal ± power Doppler

Tips to help distinguish the CET from the LCL:

1. Look distally to see the hyperechoic CET transform into a hypoechoic muscle.
2. The LCL is the deeper structure and more distally coalesces into the AL.
3. Extension of the fingers and wrist will elicit movement or tightening of the CET, helping to discern this structure from the underlying LCL.
4. Realize that the LCL and CET footprints each take up about 50% of the span of the LE.
5. After viewing the CET, rotate the distal end of the transducer 5–10° in a posterolateral direction to help delineate the LCL as a separate structure from the CET (see Protocol Image 2).

If you detect an abnormality, rotate the probe 90° to verify in a TAX view.

Protocol Image 2: Lateral Collateral Ligament, Longitudinal (Fig. 8.6)

The LCL lies deep to the CET, but the two structures are not quite parallel. The LCL may best be appreciated by rotating the transducer 5–10° in a posterolateral direction and tilting the transducer to visualize the hypoechoic to hyperechoic LCL shaped like a deep bowl.

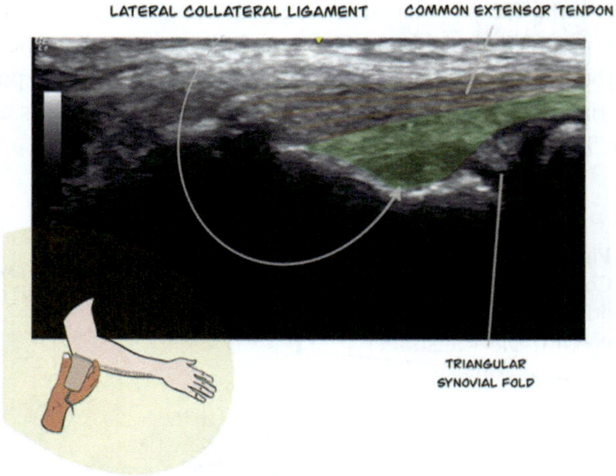

Fig. 8.6 Protocol Image 2: Lateral collateral ligament, longitudinal

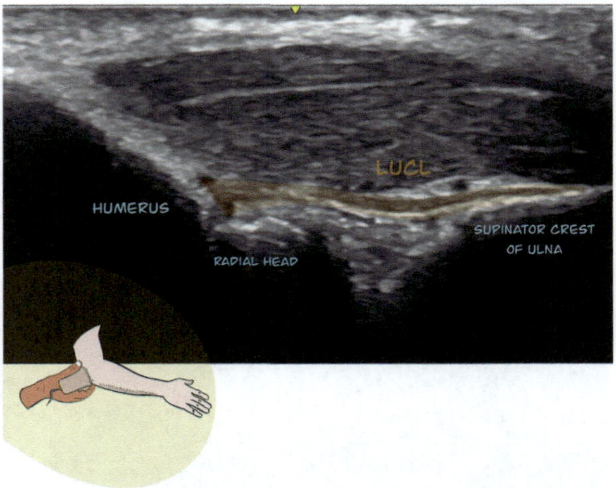

Fig. 8.7 Protocol Image 3: Lateral ulnar collateral ligament, longitudinal

Protocol Image 3: Lateral Ulnar Collateral Ligament, Longitudinal (Fig. 8.7)

To complete the sonographic evaluation of the lateral collateral ligamentous complex, evaluate the third ligament, the LUCL. This ligament is also deep and not quite parallel to the CET. To demonstrate the LUCL, move the probe in a TAX slide posterolaterally until the LE becomes a straight hyperechoic line. Then rotate the distal portion of the transducer posterolaterally to about a 45° angle to the long axis of the humerus. The curved radial head will fade out and be replaced by the ulna's hyperechoic supinator crest (SC). The hypoechoic longitudinal structure bridging the gap between the inferior portion of the LE and the SC is the LUCL. Remember, the LUCL is a primary elbow stabilizer.

Protocol Image 4: Humeroradial Joint, Longitudinal + Dynamic (Fig. 8.8)

Return the transducer to the position in Protocol Image 1. Perform a distal LAX slide to center the probe over the HRJ. Examine the triangular synovial fold within the joint, the cartilage covering the superficial surface of the humeral capitellum, and the radial head. The radial head has a slightly rounded shape. A curved hyperechoic structure, the AL is just superficial to the radial head cartilage. The AL extends distally to the radial neck.

At the radial neck, just distal to the radial head, is the location of the annular recess; the annular recess is detectable if distended by an effusion [8]. Look for synovitis (with or without PD activity), cartilage erosion, and effusion in the annular recess. To examine these structures more fully, perform tiny TAX slides slightly in

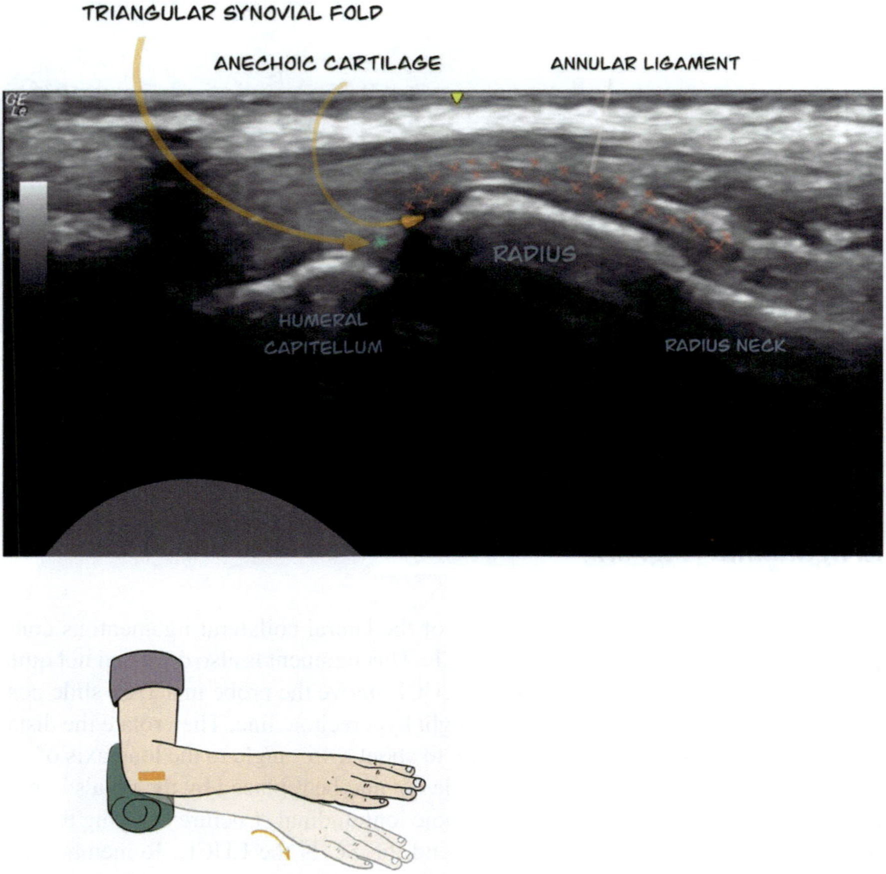

Fig. 8.8 Protocol Image 4: Humeroradial joint, longitudinal + dynamic

lateral and medial directions. Discontinuity in the smooth cortex of the radial head and neck suggests an occult fracture in the appropriate clinical setting. To perform a varus dynamic stress test for LCL complex laxity, place a bolster under the elbow and manually press the distal forearm toward the table.

Protocol Image 5: Radial Head, Transverse + Dynamic (Fig. 8.9)

Center the probe over the radial head and rotate it 90° to see the round hyperechoic bony cortex of the radial head in a transverse view. Perform tiny TAX slides proximally and distally. You can see the overlying anechoic cartilage and, superficially,

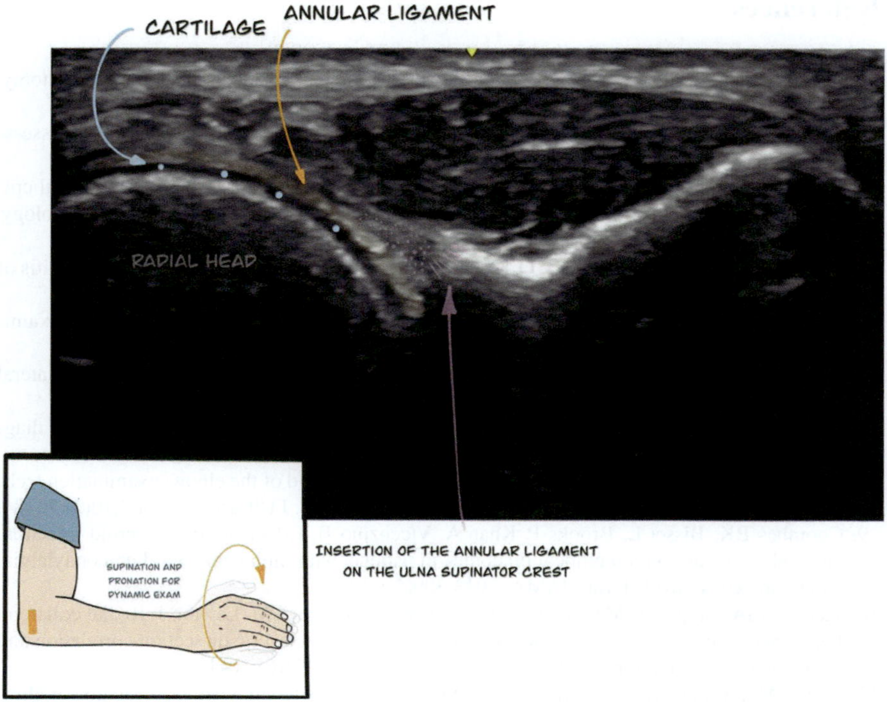

Fig. 8.9 Protocol Image 5: Radial head, transverse + dynamic

the hyperechoic AL. Small bony cortical irregularities may be present. The ECRB muscle has, at this point, become a tendon, but the EDC and ECRL are still muscles in this location. The AL may be seen inserting posterolaterally on the supinator crest of the ulna.

Manually pronate and then supinate the forearm to verify the margins and integrity of the AL. Portions of the AL may exhibit anisotropy. This maneuver reveals a fuller view of the radial head cortex and its cartilage cover when evaluating for erosive disease. A distal TAX slide will also examine the radial neck and annular recess in cross-section.

Complete Lateral Elbow Ultrasonic Examination Checklist

☐ Protocol Image 1: Common extensor tendon origin, longitudinal ± power Doppler.
☐ Protocol Image 2: Lateral collateral ligament, longitudinal.
☐ Protocol Image 3: Lateral ulnar collateral ligament, longitudinal.
☐ Protocol Image 4: Humeroradial joint, longitudinal + dynamic.
☐ Protocol Image 5: Radial head, transverse + dynamic.

References

1. Konin GP, Nazarian LN, Walz DM. US of the elbow: indications, technique, normal anatomy, and pathologic conditions. Radiographics. 2013;33(4):E125–47.
2. Radunovic G, Vlad V, Micu MC, Nestorova R, Petranova T, Porta F, et al. Ultrasound assessment of the elbow. Med Ultrason. 2012;14(2):141–6.
3. Potter HG, Hannafin JA, Morwessel RM, DiCarlo EF, O'Brien SJ, Altchek DW. Lateral epicondylitis: correlation of MR imaging, surgical, and histopathologic findings. Radiology. 1995;196(1):43–6.
4. Levin D, Nazarian LN, Miller TT, O'Kane PL, Feld RI, Parker L, et al. Lateral epicondylitis of the elbow: US findings. Radiology. 2005;237(1):230–4.
5. Connell D, Burke F, Coombes P, McNealy S, Freeman D, Pryde D, et al. Sonographic examination of lateral epicondylitis. AJR Am J Roentgenol. 2001;176(3):777–82.
6. Kotnis NA, Chiavaras MM, Harish S. Lateral epicondylitis and beyond: imaging of lateral elbow pain with clinical-radiologic correlation. Skelet Radiol. 2012;41(4):369–86.
7. Miller TT, Shapiro MA, Schultz E, Kalish PE. Comparison of sonography and MRI for diagnosing epicondylitis. J Clin Ultrasound. 2002;30(4):193–202.
8. Draghi F, Danesino GM, de Gautard R, Bianchi S. Ultrasound of the elbow: examination techniques and US appearance of the normal and pathologic joint. J Ultrasound. 2007;10(2):76–84.
9. Coombes BK, Bisset L, Brooks P, Khan A, Vicenzino B. Effect of corticosteroid injection, physiotherapy, or both on clinical outcomes in patients with unilateral lateral epicondylalgia: a randomized controlled trial. JAMA. 2013;309(5):461–9.
10. Jacobson JA, Chiavaras MM, Lawton JM, Downie B, Yablon CM, Lawton J. Radial collateral ligament of the elbow: sonographic characterization with cadaveric dissection correlation and magnetic resonance arthrography. J Ultrasound Med. 2014;33(6):1041–8.
11. Clarke AW, Ahmad M, Curtis M, Connell DA. Lateral elbow tendinopathy: correlation of ultrasound findings with pain and functional disability. Am J Sports Med. 2010;38(6):1209–14.

Chapter 9
Medial Elbow

Reasons to Do the Study

1. Evaluation of elbow pain
2. Dynamic evaluation of the ulnar nerve (UN)
3. Dynamic evaluation of the medial head of the triceps muscle (MHTr)
4. Possible compression of the UN in, or proximal to, the cubital tunnel
5. Evaluation of soft tissue swelling
6. Evaluation of the common flexor tendon (CFT) for tears, tendinosis, and tenosynovitis
7. Evaluation of the ulnar collateral ligament (UCL) (also known as the medial collateral ligament or MCL)
8. Evaluation for inflammatory arthritis

Questions We Want Answered

1. What is causing the elbow pain?
2. Is the CFT involved, and is the problem acute or chronic?
3. Is there enthesopathy at the insertion at the medial epicondyle (ME)?
4. Is there pathology of the UCL?
5. Is the pathology within the joint itself?
6. Is the visualized pathology the actual cause of the pain?
7. What is the mechanism of the condition?
8. Is the UN or the MHTr creating the problem dynamically?
9. Is the UN compressed proximal to or within the true cubital tunnel (TCuT)?
10. What is the next step to help the problem?

The contents do not represent the views of the U.S. Department of Veterans Affairs or the United States Government.

M. H. Greenberg et al., *Manual of Musculoskeletal Ultrasound*,
https://doi.org/10.1007/978-3-031-37416-6_9

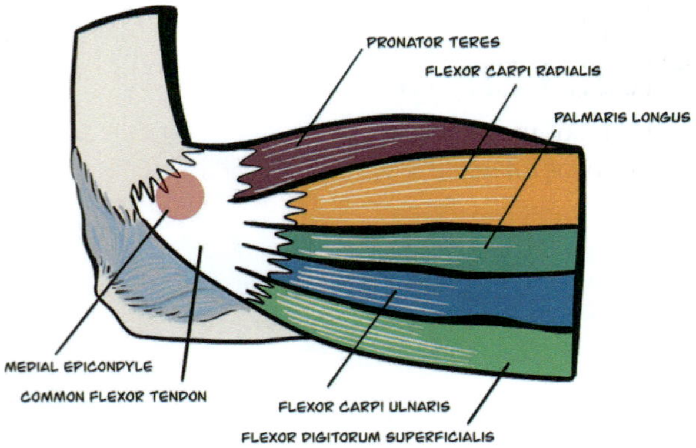

Fig. 9.1 Muscles and common flexor tendon

Basic Anatomy

The Medial Epicondyle and Common Flexor Tendon (Fig. 9.1)

The ME is a smooth bony ridge to which the CFT and the UCL attach, with the CFT being more superficial. The CFT is the confluence of four or five tendons that insert on the most superficial part of the ME. The tendons include the **f**lexor carpi radialis, **p**almaris longus (PL), **f**lexor digitorum superficialis, **p**ronator teres, and **f**lexor carpi ulnaris (FCU). The PL may not be present in up to 15% of normal individuals [1]. A mnemonic to remember the five tendons that form the CFT is as follows:

Five People Found Peace Forever

The CFT is mostly muscular and has a smaller tendon than the CET. The examiner may evaluate this tendon for enthesitis on power Doppler (PD).

Ulnar Collateral Ligament (Fig. 9.2)

Deep to the CFT is the UCL. The UCL also inserts on the ME, but further (deeper) down the slope. The mnemonic "**CU down the slope**" reminds us of the locations. The UCL stretches from the ME to the ulna and may exhibit enthesitis at its attachment to the ME. The UCL comprises three bands: the anterior oblique, posterior oblique, and transverse. The anterior band is the most important of the three in terms of elbow joint stabilization and is a common location of pathology [2]. This band of the UCL is triangular-shaped, with the apex attaching distally to the ulnar coronoid (sublime) tubercle. The anterior band is most often injured in athletes, particularly baseball pitchers [2]. "Tommy John surgery" is a popular surgery in which an autologous graft replaces the torn anterior band of the UCL. On a sonographic longitudinal (LAX) view, the anterior band is often hypoechoic due to anisotropy but will appear hyperechoic with fibrillar echotexture if the angle of insonation is precisely 90°.

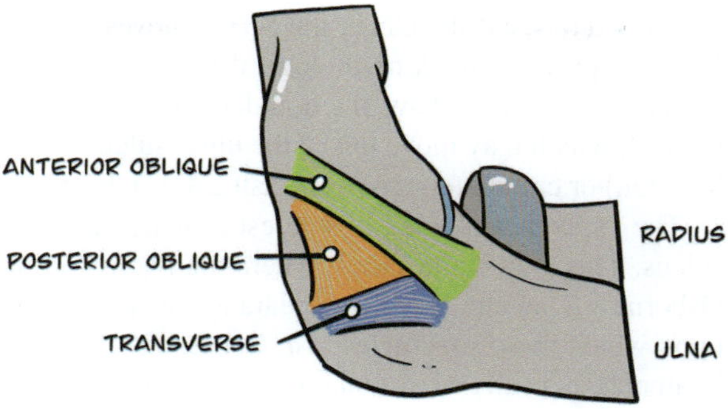

Fig. 9.2 Ulnar collateral ligament bands

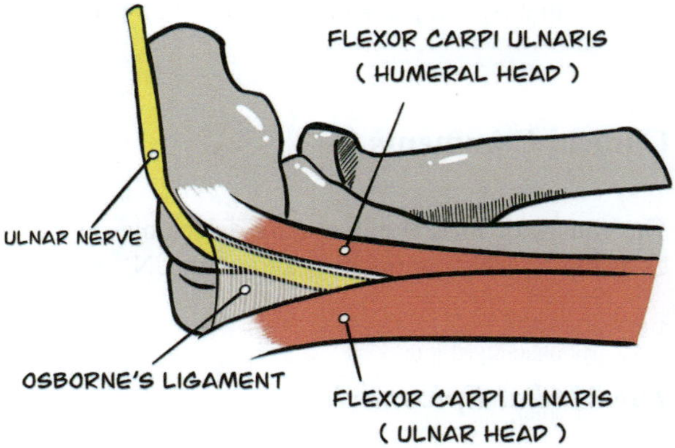

Fig. 9.3 Posteromedial elbow simplified

Posteromedial Elbow (Fig. 9.3)

We include posteromedial elbow anatomy in the medial elbow discussion since UN compromise may cause medial elbow pain. The UN lies within the sulcus formed by the ME and the olecranon. Here, the UN is covered with Osborne's *fascia*. More distally, within the TCuT, the nerve runs deep to Osborne's *ligament*, the latter uniting the two heads of the flexor carpi ulnaris muscle. The terminology is confusing. Osborne's ligament and Osborne's fascia go by different names in the literature, including the arcuate ligament of Osborne, the cubital tunnel retinaculum, and Osborne's band, among others. Our convention is to call the connective tissue over the sulcus "Osborne's fascia" and that forming the roof of the TCuT "Osborne's ligament."

Sonographic Anatomy

We use US to evaluate the areas we label **the three "S's": ski slope, Sisyphus, and spectacles**. The **"ski slope"** is the LAX view of the ulnar aspect of the bony ME, where the CFT and the UCL attach. The **"Sisyphus"** area is the posterior and radial aspect of the ME seen in sonographic transverse (TAX), where a dynamic exam is performed to see if the UN or the MHTr moves up and perhaps over the apex of the ME. **Sisyphus**, a Greek mythological figure, was doomed to eternally push a boulder up a hill, only to have the boulder roll back down each time. Our "boulder" is the UN, which may move out of the ulnar sulcus (groove) and partially climb (subluxation) or completely go over (dislocation) the peak of the ME.

The **"spectacles"** area of interest is where the UN travels distally to the ulnar sulcus. This area is the TCuT, where the two ovoid heads of the FCU, tethered by Osborne's ligament, give the appearance of **spectacles**. The UN is the round "nose" upon which these **spectacles** rest. Osborne's ligament is a bit unusual in that most ligaments typically bind bone to bone; however, this ligament attaches muscle to muscle and is the roof of the TCuT [3]. By examining this area in TAX and LAX, we can sonographically confirm or refute clinical suspicion of CuT syndrome (CuTS), that is, compression of the UN in the TCuT.

Clinical Comments

The causes of medial elbow pain detectable on US include medial epicondylitis, UCL injury, UN compression, snapping UN, and snapping MHTr [4].

The Medial Epicondyle

Medial epicondylitis is also called "golfer's elbow." Medial epicondylitis is a misleading term since, in most cases, it is caused by tendinosis of the CFT rather than actual inflammation. Doppler activity in this tendon may be from tendon injury or active enthesitis, such as in spondyloarthropathy. The clinical setting is critical.

When evaluating the UCL, the goal is to delineate a partial or complete tear in this oft-injured ligament [5]. Extend the elbow to look at the UCL fibers crossing the humeroulnar joint [6]. However, the more distal portion of the UCL is best seen with the elbow flexed so that the ligament is taut. Visualization of both UCL tears and laxity is enhanced on US when the examiner does a dynamic valgus stress test while observing if the humeroulnar joint enlarges when stressed [7].

Ulnar Nerve Entrapment

Ulnar nerve compromise at the elbow is included in the medial elbow despite the nerve's location on the posterior elbow, as pain from ulnar neuropathy is often experienced in the medial elbow. Ulnar neuritis may be caused by entrapment in different locations, but it most commonly occurs under Osborne's ligament in the TCuT [4]. Again, such entrapment is termed CuTS [8]. In ulnar neuritis of any origin, paresthesia may radiate to the fourth and fifth fingers.

Entrapment of the UN occurs at the entrance to the TCuT, which lies beneath Osborne's ligament and the humeral and ulnar heads of the FCU. The TCuT entrance is our **"spectacles"** location and reminds us to look carefully at this site. If compressed within the TCuT, the UN may enlarge proximally, with swelling apparent near the ME. In this scenario, the UN will have a smaller caliber within the TCuT, that is, under Osborne's ligament.

It is important to note that nerves tend to enlarge proximally to compression when squeezed within tight fibro-osseous canals such as the TCuT and the carpal tunnel [9, 10]. One criterion for CuTS is a cross-sectional area (CSA) of the UN greater than 9 mm^2 [11]. An alternative criterion is a ratio greater than 2.8 when comparing the CSA of the UN at the site of maximal swelling to a proximal site [12]. It is essential to recognize that some normal (physiologic) UN enlargement may occur with asymptomatic UN subluxation/dislocation out of the ulnar sulcus [13]. However, in the proper clinical setting, a de facto diagnosis of CuTS can best be made on US in LAX view if the UN is enlarged proximal to the entrance of the TCuT [9].

It is noteworthy that at the elbow, the UN typically does not have the familiar honeycombing that we expect to see in TAX. This appearance may be due to anisotropy or mild edema due to daily microtrauma. The usual UN in the elbow has a hypoechoic or anechoic center surrounded by a thicker, hyperechoic epineurium. UN neuropathy at the elbow may occur in three circumstances:

1. Compromise within the TCuT [10].
2. Repetitive subluxation of the UN out of the sulcus may cause chronic pressure damage and UN swelling [10, 13].
3. Compression by a space-occupying lesion, including ganglia, lipomas, anconeus epitrochlearis (AE), osteophytes from the humeroulnar joint (HUJ), HUJ synovitis, intra-articular bodies, hemorrhage, and MHTr dislocation [10].

The AE is an accessory muscle present in up to 23% of the population (Fig. 9.4) [14]. It connects the ME to the olecranon process. It is proximal to the TCuT and appears as a hypoechoic structure superficial to the UN in the ulnar sulcus, in the expected location of Osborne's fascia. The AE and MHTr are both adjacent to the UN at elbow flexion, but the MHTr separates from the UN in elbow extension.

Fig. 9.4 Anconeus epitrochlearis

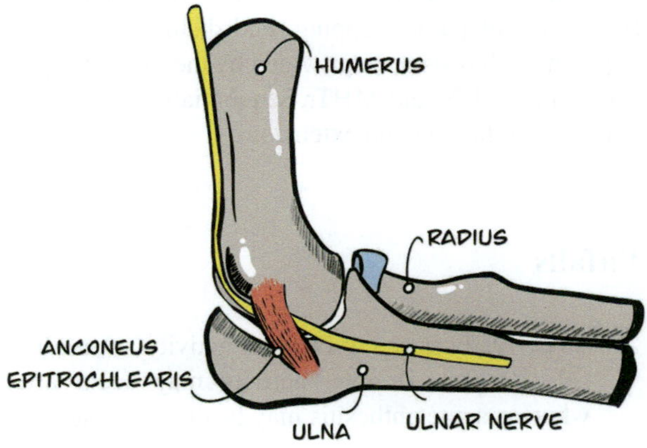

Avoid confusing the AE with a ganglion. The AE compression of the UN may occur in an extended elbow. Not only can an AE mimic CuTS by compressing the UN, but the AE may also mimic a snapping MHTr by subluxing over the ME [9].

Other Ulnar Nerve Syndromes

Dynamic US of the posteromedial elbow is a priceless diagnostic modality when focusing on two potentially mobile structures: the UN and the MHTr. In **ulnar nerve instability (UNI)**, the UN chronically subluxes and relocates with elbow flexion and extension [15]. The UN fully dislocates in 20% of normal elbows and is usually without symptoms [16]. However, UNI may increase UN vulnerability to trauma, mainly when the UN lies superficially on the ME [17]. The most probable cause of UNI is ligamentous laxity. Ulnar nerve instability may occur with or without radiating paresthesia or a snapping UN or MHTr [17, 18]. It is separate from UN compression, although UNI may cause friction neuritis, producing radiating paresthesia mimicking CuTS [17].

Snapping ulnar nerve occurs in elbow flexion (between 70° and 90°) when the UN dislocates over the apex of the ME. This condition is typically painful and often accompanied by a palpable and audible snap [19, 20]. Neurologic symptoms from the snapping UN are rarely reported, with the most common symptom being localized pain [19]. Another potential cause of elbow pain in a throwing athlete with UN dislocation is pain upon extension of the elbow when the UN rapidly relocates over the apex of the ME [21].

Snapping triceps may or may not be painful. Snapping triceps occurs with about 115° of elbow flexion when the distal MHTr displaces over the apex of the ME. There may be audible/palpable snapping, and the MHTr may push the UN out of the groove (subluxation) and perhaps even further to a point over the apex of the ME (dislocation). Hypertrophy of the triceps may predispose to UN subluxation, dislocation, and snapping triceps syndrome [4]. Snapping MHTr itself is associated with UN dislocation [20]. If both snapping UN and MHTr are present, there may be two separate palpable or audible sequential snaps [9].

Again, dislocation of the UN and MHTr may occur without pain or snapping. Symptoms of pain, snapping, and dislocation of the UN or MHTr may only be apparent with resistance provided by the examiner [22]. When examining the elbow for snapping UN and MHTr, sequentially evaluate the elbow with passive, active, and resisted flexion and extension.

Pitfalls

1. To correctly interpret Doppler activity at the CFT insertion, consider the clinical situation. Doppler activity may occur after repetitive tendon injuries, whereas acute enthesitis may be due to spondyloarthropathy.

2. When examining the elbow for snapping UN and MHTr, scan the elbow with passive, active, and resisted flexion and extension. Do not forget to do the dynamic exam with resistance; this maneuver is critical for snapping UN/MHTr.
3. Cutoff values and CSA ratios of the UN will help diagnose CuTS, but UN compression in LAX view is a strong sign of CuTS in the right clinical setting.
4. Do not mistake the appearance of the UN as necessarily abnormal. The UN typically has some edema, possibly due to TCuT compression with repetitive elbow flexion. In addition, it does not always have a honeycomb appearance.
5. UN swelling may also be due to repetitive dislocation trauma over the ME and may not necessarily indicate UN compression in the TCuT.
6. Examine the UCL thoroughly to look for UCL tears. The UCL is hypoechoic in elbow extension and becomes more hyperechoic as it tightens with elbow flexion.
7. Valgus stress testing of the UCL with only gravity is sufficient.
8. Remember that a dislocating UN or MHTr may not cause pain.
9. Evaluate the contralateral elbow when evaluating a putative US abnormality.
10. Nerves tend to enlarge proximally to compression when squeezed within tight fibro-osseous canals such as the TCuT and the carpal tunnel.
11. Avoid immediately assuming that ulnar neuropathy is due to CuTS, even if you note a swollen UN. Look for more proximal causes of ulnar neuritis, such as UN trauma or compression near the ME. Also, do not forget about more distal causes of ulnar neuropathy, including compression within the ulnar (Guyon's) canal in the wrist with possible retrograde paresthesia.
12. The AE will appear as a hypoechoic structure superficial to the UN in the ulnar sulcus, proximal to the TCuT. Avoid confusing the AE with a ganglion.
13. The AE may mimic either CuTS by compressing the UN or a snapping MHTr by subluxing over the ME.
14. The elbow offers an extraordinary diagnostic window. Do not neglect to carefully examine the elbow for bone/cartilage erosion, step-off deformity of an occult fracture, rheumatoid nodules, double contour sign of gout, and calcium deposition in cartilage.

Method

The patient is recumbent with a pillow supporting the elbow, which is externally rotated, abducted, and extended. Slight flexion of the elbow will likely occur with this maneuver and is expected. Use a transducer with a frequency of at least 12 MHz.

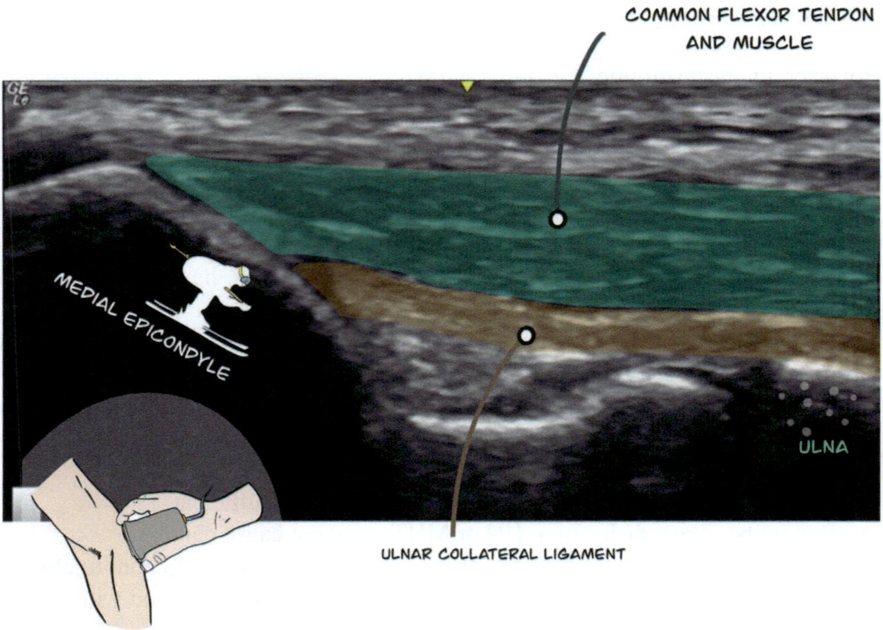

Fig. 9.5 Protocol Image 1: Medial epicondyle, longitudinal, extension ± power Doppler

Protocol Image 1: Medial Epicondyle, Longitudinal, Extension ± Power Doppler (Fig. 9.5)

Palpate the ME and place the probe in LAX over the joint along the anterior medial elbow to visualize the HUJ. See the bony hyperechoic curved "**ski slope**" of the ME leading down the curved valley, the coronoid recess, and then a smaller "hill," the trochlea of the humerus. The proximal ulna is distal to the trochlea and separated by the joint space. Look at the CFT, its muscle, and the insertion on the ME. Elbow extension enhances CFT visualization. Look for signs of enthesopathy, muscle tears, and bony avulsions. Next, turn on the PD to look for active enthesitis. Deep to the CFT and muscles is the UCL's anterior bundle, which we will refer to simply as the "UCL." To delineate the separation of the CFT from the UCL, look at the more hypoechoic CFT muscle just superficial to the ligament. In elbow extension, the UCL is hypoechoic. Look for gaps that may indicate partial or complete UCL tears. Turn on the PD to look for Doppler activity suggestive of active enthesitis or an acute ligament tear.

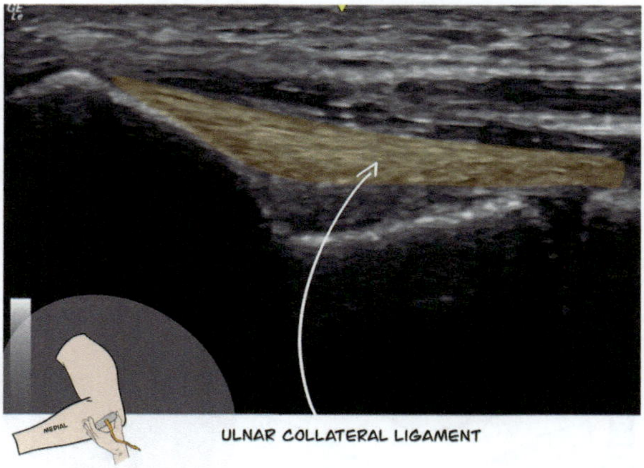

Fig. 9.6 Protocol Image 2: Medial epicondyle, longitudinal, flexion ± power Doppler

Protocol Image 2: Medial Epicondyle, Longitudinal, Flexion ± Power Doppler (Fig. 9.6)

Flexing the elbow to 90° will create tension in the UCL. This results in a more hyperechoic echotexture and improves UCL visualization. Rotate the probe to a position nearly parallel to the forearm to better align the transducer with the ligament. Look for alteration of the normal fibrillar echostructure, hypoechoic areas, and ligament swelling. In males, the average ligament thickness ranges from 2.6 to 4 mm [23]. However, comparing the (unaffected) contralateral side might be more revealing. Irregularity or protrusion of a portion of the medial epicondyle may point to a prior UCL avulsion injury [24]. Again, look for interruptions in the integrity of the ligament, which may indicate partial or complete ligament tears. Look at the enthesis on the ME for evidence of enthesopathy. Turn on the PD to evaluate for active enthesitis or an acute ligament tear.

Protocol Image 3: Medial Epicondyle, Valgus Maneuver, Longitudinal (Fig. 9.7)

Then, perform a valgus stress maneuver to test the structural integrity of the UCL, mainly the anterior band. The elbow should be flexed to 30° or more during the valgus maneuver. Place a roll of paper towels beneath the elbow. Stabilize the elbow with the probe and let gravity perform the valgus maneuver [25]. Note that the UCL is even better defined when it is taut during the valgus stress maneuver. During the stress maneuver, measure the width of the HUJ space to see if it enlarges. Comparison with the contralateral (unaffected) side is helpful. The expected change in the UCL space with stress valgus is <1.3 mm. If the difference in UCL joint space with valgus stress on the affected side is at least 1 mm more than on the unaffected side, suspect a UCL tear [26].

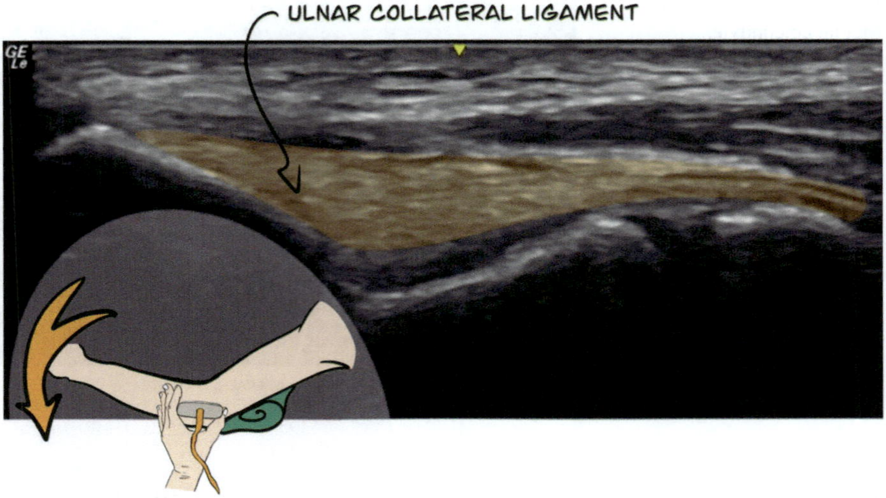

Fig. 9.7 Protocol Image 3: Humeroulnar joint, valgus maneuver, longitudinal

Protocol Image 4: Ulnar Nerve at the Medial Epicondyle, Transverse + Measurement (Fig. 9.8)

Next, extend the elbow off the examination table while maximally supinating the forearm. Rotate the probe 90° to a TAX position and then center the transducer over the ME. Slide the probe posteriorly, still in TAX, to see the UN in TAX. The two bony landmarks on opposite sides of the picture are the ME and the olecranon process. The probe should bridge these two bones. The posterolateral portion of the ME is hyperechoic and looks like a hill. The UN appears to have a central hypoechoic area with a thick hyperechoic rim. The UN seems round, like **Sisyphus's boulder**, sitting at the bottom of the hill. Note that the UN, in this view, may not exhibit a classic honeycomb appearance due to anisotropy. The UN lies close to and may abut the ME.

In this location, there may or may not be a retinaculum called Osborne's fascia, which stretches between the ME and the olecranon superficial to the UN. Note again that Osborne's fascia is not the same as Osborne's ligament; the fascia (if present) becomes Osborne's ligament more distally, the latter connecting the two heads of the FCU. Verify the identity of the UN by moving the probe 1–2 cm distally to follow this circular structure. Once you have confirmed that the visualized circular form is indeed the UN, reverse course and move the probe proximally back to the ME. Be aware that the UN may split into two separate or connected rounded sections for some individuals. The AE is an accessory muscle that may be present in place of Osborne's fascia and reside just superficially to the UN. It may have clinical

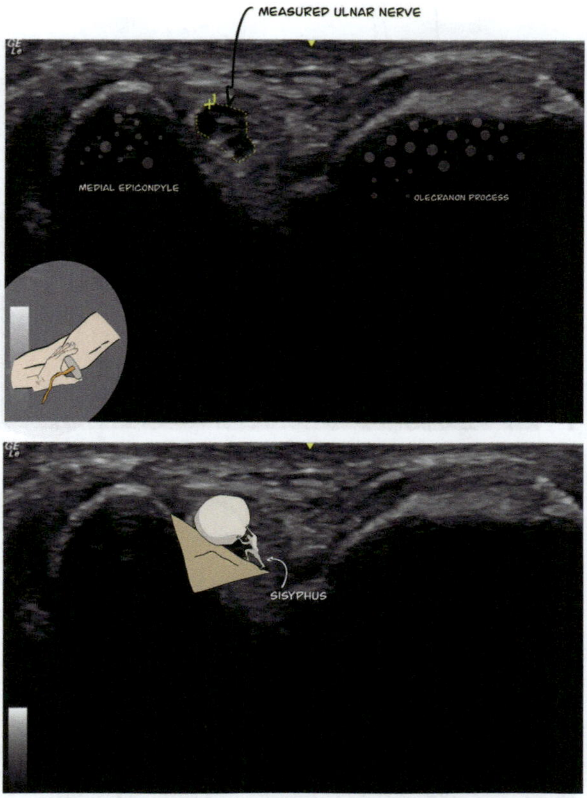

Fig. 9.8 Protocol Image 4: Ulnar nerve at the medial epicondyle, transverse + measurement

significance since it may cause UN compression. Measure the CSA of the UN in TAX. The typical CSA of the UN is 8–12 mm^2; however, a CSA of the UN >9 mm^2 may indicate pathology [11].

Protocol Image 5: Ulnar Nerve at the Medial Epicondyle, Transverse, Dynamic Exam, Passive Flexion (Fig. 9.9)

With a light touch of the probe, grasp the forearm and *very slowly* flex the elbow to look for UN subluxation. Try to keep the ME in focus. We are watching for the UN to ascend the ME with *passive* elbow bending. In flexion, the UN may normally climb the ME (subluxation), like in the **Sisyphus** allegory. The UN may even go over the peak of ME (dislocation). However, these UN movements may be asymptomatic. Alternatively, the UN subluxation or dislocation may be associated with pain or an audible snap. The examiner may see the snap or feel a "popping" sensation through the transducer.

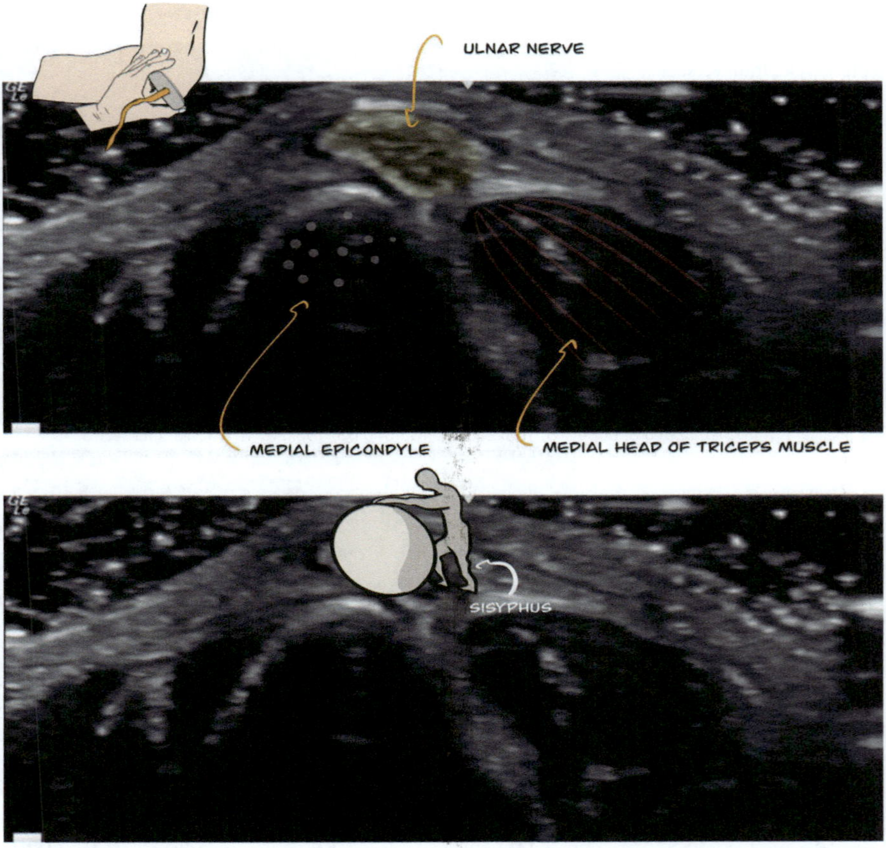

Fig. 9.9 Protocol Image 5: Ulnar nerve at the medial epicondyle, transverse, dynamic exam, passive flexion

Recurrent subluxation or dislocation may predispose to UN injury. With repetitive trauma from dislocating and relocating, the UN itself may occasionally swell to the point that it can mimic the swelling associated with nerve compression. The MHTr may also move over the ME in elbow flexion, producing a snapping sensation known as the snapping triceps syndrome. The UN's snapping over the ME's peak occurs at about 80° of elbow flexion, while MHTr snapping takes place at about 115° of elbow flexion. When both the UN and the MHTr dislocate over the ME, the MHTr may appear to be propelling the UN.

Protocol Image 6: Ulnar Nerve at the Medial Epicondyle, Transverse, Dynamic Exam, Active Flexion

The patient *actively* but slowly flexes the elbow while the examiner observes for UN and MHTr subluxations.

Protocol Image 7: Ulnar Nerve at the Medial Epicondyle, Transverse, Dynamic Exam, Resisted Flexion

The patient flexes the elbow *against resistance* provided by the examiner or an assistant while the examiner observes for UN and MHTr subluxations. Resistance may best evoke subluxation.

Protocol Image 8: Ulnar Nerve in the True Cubital Tunnel, Transverse (Fig. 9.10)

With the transducer still in TAX, perform a distal TAX slide toward the TCuT, focusing on the UN. You have reached the TCuT when the UN looks like a round nose under a pair of oval eyeglass lenses. The "**spectacle lenses**" are the two heads of the FCU; the "bridge" or "nosepiece" is Osborne's ligament connecting the two heads of the FCU. The humeral head of the FCU is on the same side as the ME. Again, this is the TCuT, the most common site of UN entrapment. As you move more distally away from the TCuT, the two FCU heads coalesce, and the UN is just beneath the single FCU. Just proximal to the TCuT, the UN may be enlarged (pseudoneuroma) if there is chronic compression within the TCuT. If you suspect compression, measure the CSA of the UN at its largest circumference and compare it with the CSA more proximally. Comparison with the unaffected side may often help. Normal UN CSA is 8–12 sq. mm. in this location [11].

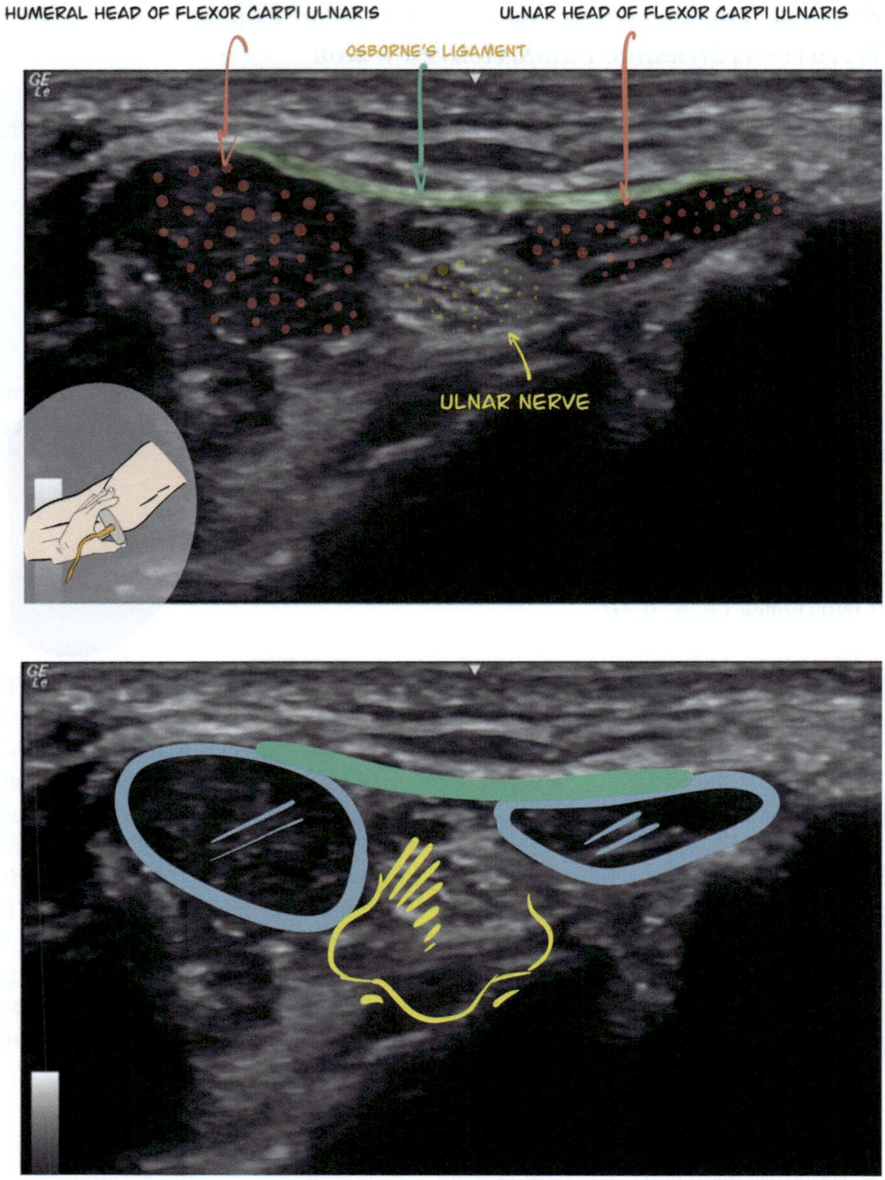

Fig. 9.10 Protocol Image 8: Ulnar nerve in the true cubital tunnel, transverse

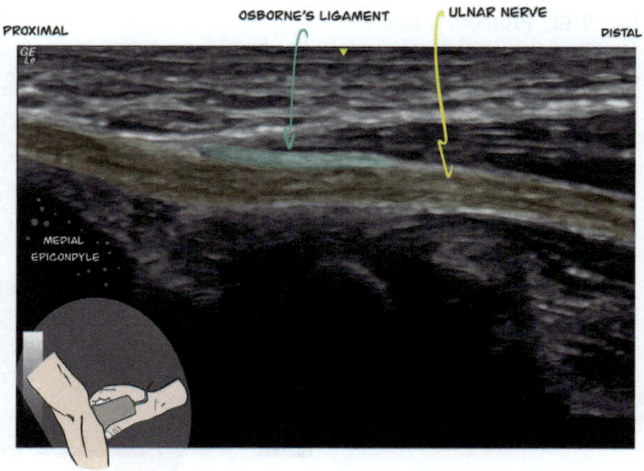

Fig. 9.11 Protocol Image 9: Ulnar nerve in true cubital tunnel, longitudinal

Protocol Image 9: Ulnar Nerve in True Cubital Tunnel, Longitudinal (Fig. 9.11)

Visualize and center the UN under Osborne's ligament within the TCuT, and slowly rotate the probe 90° to visualize the UN in LAX. Note the caliber of the UN proximal to, within, and distal to the TCuT. If there is a discrepancy, measure the CSAs in TAX in these three areas.

In the LAX view, and in the appropriate clinical setting, the finding of a compressed UN within the TCuT with clear-cut proximal swelling often defines CuTS, irrespective of the measurements. Again, be aware that the UN may swell with repetitive trauma, such as traumatic dislocation and relocation over the ME, mimicking swelling from chronic compression either within or proximal to the TCuT.

Protocol Image 10: Ulnar Nerve in Forearm, Transverse (Fig. 9.12)

Next, rotate the probe back to TAX and attempt to follow UN distally to the forearm. You will need to straighten the elbow. Remember, as you move more distally, the two FCU heads coalesce, and the UN is just deep into the single FCU. You can follow the UN to the ulnar canal (Guyon's) in the wrist.

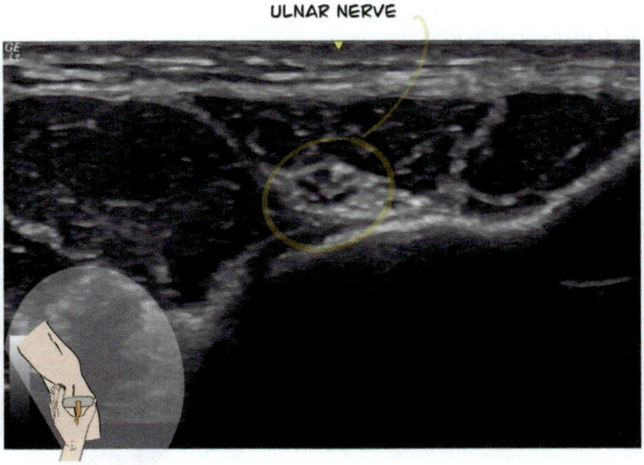

Fig. 9.12 Protocol Image 10: Ulnar nerve in forearm, transverse

Protocol Image 11 (Optional): Ulnar Nerve Proximal to the Medial Epicondyle

Starting at the ME, examine the proximal UN in TAX as much as possible. If there is a question of UN enlargement, measure the CSA and compare it with that in a more distal location of potential compression, such as may occur with AE compression of the UN in the ulnar groove.

Complete Medial Elbow Ultrasonic Examination Checklist

☐ Protocol Image 1: Medial epicondyle, longitudinal, extension ± power Doppler
☐ Protocol Image 2: Medial epicondyle, longitudinal, flexion ± power Doppler
☐ Protocol Image 3: Medial epicondyle, valgus maneuver, longitudinal
☐ Protocol Image 4: Ulnar nerve at medial epicondyle, transverse + measurement
☐ Protocol Image 5: Ulnar nerve at medial epicondyle, transverse, dynamic exam, passive flexion
☐ Protocol Image 6: Ulnar nerve at medial epicondyle, transverse, dynamic exam, active flexion
☐ Protocol Image 7: Ulnar nerve at medial epicondyle, transverse, dynamic exam, resisted flexion
☐ Protocol Image 8: Ulnar nerve in true cubital tunnel, transverse
☐ Protocol Image 9: Ulnar nerve in true cubital tunnel, longitudinal
☐ Protocol Image 10: Ulnar nerve in forearm, transverse

References

1. Thompson NW, Mockford BJ, Cran GW. Absence of the palmaris longus muscle: a population study. Ulster Med J. 2001;70(1):22–4.

2. Erickson BJ, Harris JD, Chalmers PN, Bach BR Jr, Verma NN, Bush-Joseph CA, et al. Ulnar collateral ligament reconstruction: anatomy, indications, techniques, and outcomes. Sports Health. 2015;7(6):511–7.
3. Granger A, Sardi JP, Iwanaga J, Wilson TJ, Yang L, Loukas M, et al. Osborne's ligament: a review of its history, anatomy, and surgical importance. Cureus. 2017;9(3):e1080-e.
4. Barco R, Antuna SA. Medial elbow pain. EFORT Open Rev. 2017;2(8):362–71.
5. Miller TT, Adler RS, Friedman L. Sonography of injury of the ulnar collateral ligament of the elbow-initial experience. Skelet Radiol. 2004;33(7):386–91.
6. Bianchi S, Martinoli C. Elbow. In: Ultrasound of the musculoskeletal system. Springer; 2007.
7. De Smet AA, Winter TC, Best TM, Bernhardt DT. Dynamic sonography with valgus stress to assess elbow ulnar collateral ligament injury in baseball pitchers. Skelet Radiol. 2002;31(11):671–6.
8. Cutts S. Cubital tunnel syndrome. Postgrad Med J. 2007;83(975):28–31.
9. Jacobson JA. Elbow. In: Fundamentals of musculoskeletal ultrasound. 3rd ed. Elsevier; 2018.
10. Kele H. Ultrasonography of the peripheral nervous system. Pers Med. 2012;1(1):417–21.
11. Thoirs K, Williams MA, Phillips M. Ultrasonographic measurements of the ulnar nerve at the elbow: role of confounders. J Ultrasound Med. 2008;27(5):737–43.
12. Yoon JS, Walker FO, Cartwright MS. Ultrasonographic swelling ratio in the diagnosis of ulnar neuropathy at the elbow. Muscle Nerve. 2008;38(4):1231–5.
13. Kopf H, Loizides A, Mostbeck GH, Gruber H. Diagnostic sonography of peripheral nerves: indications, examination techniques and pathological findings. Ultraschall Med. 2011;32(3):242–63; quiz 64–6.
14. Husarik DB, Saupe N, Pfirrmann CW, Jost B, Hodler J, Zanetti M. Elbow nerves: MR findings in 60 asymptomatic subjects—normal anatomy, variants, and pitfalls. Radiology. 2009;252(1):148–56.
15. Lazaro L 3rd. Ulnar nerve instability: ulnar nerve injury due to elbow flexion. South Med J. 1977;70(1):36–40.
16. Okamoto M, Abe M, Shirai H, Ueda N. Morphology and dynamics of the ulnar nerve in the cubital tunnel. Observation by ultrasonography. J Hand Surg Br. 2000;25(1):85–9.
17. Xarchas KC, Psillakis I, Koukou O, Kazakos KJ, Ververidis A, Verettas DA. Ulnar nerve dislocation at the elbow: review of the literature and report of three cases. Open Orthop J. 2007;1:1–3.
18. Watts AC, McEachan J, Reid J, Rymaszewski L. The snapping elbow: a diagnostic pitfall. J Shoulder Elb Surg. 2009;18(1):e9–10.
19. Bjerre JJ, Johannsen FE, Rathcke M, Krogsgaard MR. Snapping elbow—a guide to diagnosis and treatment. World J Orthop. 2018;9(4):65–71.
20. Jacobson JA, Jebson PJ, Jeffers AW, Fessell DP, Hayes CW. Ulnar nerve dislocation and snapping triceps syndrome: diagnosis with dynamic sonography—report of three cases. Radiology. 2001;220(3):601–5.
21. Greenberg MH, Day AL, Fant JW, Mazoue CG. Recurrent medial elbow pain after successful Tommy John surgery: ultrasound helps to define an ulnar nerve relocation syndrome. The Rheumatologist. 2020.
22. Spinner RJ, Goldner RD, Lee RA. Diagnosis of snapping triceps with US. Radiology. 2002;224(3):933–4; author reply 4.
23. Ward SI, Teefey SA, Paletta GA Jr, Middleton WD, Hildebolt CF, Rubin DA, et al. Sonography of the medial collateral ligament of the elbow: a study of cadavers and healthy adult male volunteers. AJR Am J Roentgenol. 2003;180(2):389–94.
24. Minagawa H, Wong KH. Musculoskeletal ultrasound: echo anatomy & scan technique. Amazon Digital Services; 2017.
25. Harada M, Takahara M, Maruyama M, Nemoto T, Koseki K, Kato Y. Assessment of medial elbow laxity by gravity stress radiography: comparison of valgus stress radiography with gravity and a Telos stress device. J Shoulder Elb Surg. 2014;23(4):561–6.
26. Roedl JB, Gonzalez FM, Zoga AC, Morrison WB, Nevalainen MT, Ciccotti MG, et al. Potential utility of a combined approach with US and MR arthrography to image medial elbow pain in baseball players. Radiology. 2016;279(3):827–37.

Chapter 10
Shoulder

Reasons to Do the Study
1. Shoulder pain [1]

Questions We Want Answered
1. What is causing the shoulder pain?
2. Is the pain directly from the shoulder, or is it referred from elsewhere?
3. Is the shoulder pain from a structural problem or an inflammatory condition such as rheumatoid arthritis (RA) or spondylarthritis (SpA)?
4. Does the pathology on the image represent the actual cause of the pain?
5. Is the pain caused by movement?
6. Is a "frozen shoulder" (adhesive capsulitis [AC]) present, and if so, is it the source of pain?

Basic Anatomy

The **bony shoulder architecture** is depicted in Fig. 10.1. There are four shoulder joints: glenohumeral (GHJ), acromioclavicular (ACJ), scapulothoracic, and sternoclavicular (SCJ) [2].

Shoulder soft tissue anatomy is complicated; however, a few simple anatomic concepts are worth remembering. Also, bear in mind that the long head of the biceps tendon (LHBT) is *not* part of the rotator cuff and is just "passing through," straight up the middle. However, as you will see, impairment of other shoulder structures may negatively influence the LHBT. The key to understanding the shoulder is the **rotator cuff tendons** and their insertions (Fig. 10.2). The time-honored mnemonic

The contents do not represent the views of the U.S. Department of Veterans Affairs or the United States Government.

M. H. Greenberg et al., *Manual of Musculoskeletal Ultrasound*,
https://doi.org/10.1007/978-3-031-37416-6_10

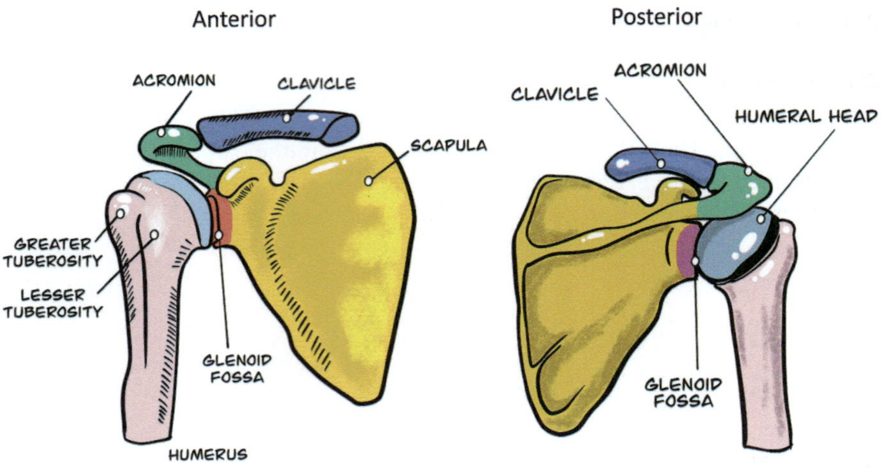

Fig. 10.1 Shoulder bony anatomy

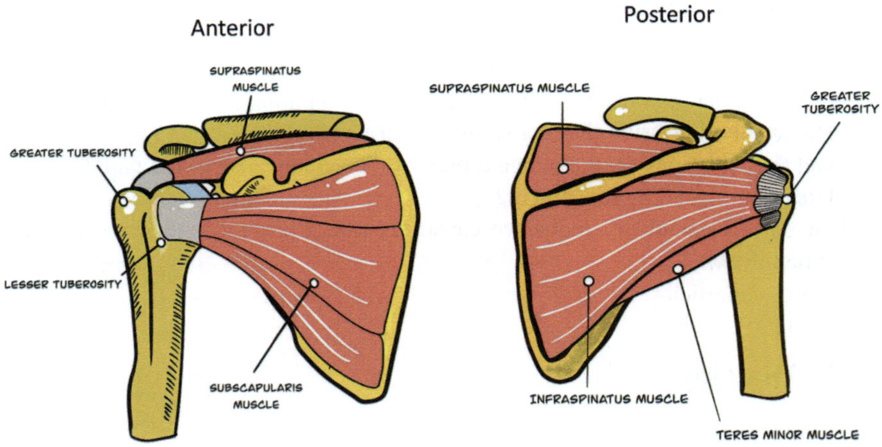

Fig. 10.2 Rotator cuff tendons

acronym for the *names* of the rotator cuff tendons and muscles is SITS: **supra**spinatus (SST), **inf**raspinatus (IST), **ter**es minor (TMT), and **sub**scapularis (SSCT) [2]. To remember the *function* of the rotator cuff muscles, think of the physical actions of the muscles that produce specific tennis strokes:

> Tennis players with a **sub**par forehand and a **ter**ribly **inf**erior backhand must rely on a **supr**eme serve.

The four tendons insert on two tuberosities, namely the greater tuberosity (GT) and the lesser tuberosity (LT). The subscapularis tendon inserts on the LT. The GT has three facets: superior (SF), middle (MF), and inferior (IF) [3]. To remember the GT tendon insertion locations, consider this story (Fig. 10.3):

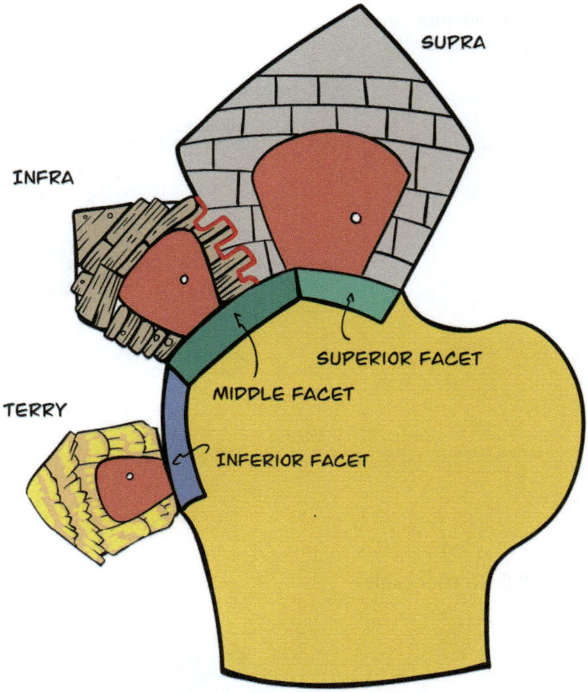

Fig. 10.3 Greater tuberosity attachments cartoon

There were three Sit brothers, Supra, Infra, and Terry, who lived in a great town (GT). Firstborn Supra lived high and mighty in a superior house (SF). Infra, the middle child (MF), lived nearby. Terry, the youngest, lived far away from his older siblings since he felt inferior (IF) to them. Supra and Infra were close and spent time at each other's houses (GT insertional overlap). Supra, however, was rambunctious and tended to get "tore up," sometimes extending his damage to nearby Infra. Terry stayed out of the way and was rarely injured.

Fluid accumulation detectable by US occurs in three recesses (anterior, posterior, and axillary), two bursae (subcoracoid and subacromial-subdeltoid [SASDB]), and the LHBT sheath. The posterior recess is most accessible on US when the shoulder is in external rotation (Fig. 10.4) [4].

The term **rotator cable** (RC) (Fig. 10.5) refers to a semicircular ovoid thickening of the GHJ capsule stretching from the lateral aspect of the LT to the GT. This cable runs just deep to the SST and IST and anchors them to the tubercles, functioning like a suspension bridge cable [5]. The RC is superficial to the LHBT. On US, the RC may be seen as a thickened hyperechoic curved structure underlining the SST and may be confused with calcium deposition of the deeper aspect of the tendon [6]. The RC fibers are perpendicular to the tendon fibril orientation.

The **rotator cuff interval** (RCI) is the triangular space bordered by the SST, the SCCT, and the coracoid process [7]. On US, we look at the base of this triangle, delineated as the space between the insertion areas between the SST and the SSCT insertions on the anterior-superior GT and superior LT, respectively (Fig. 10.6).

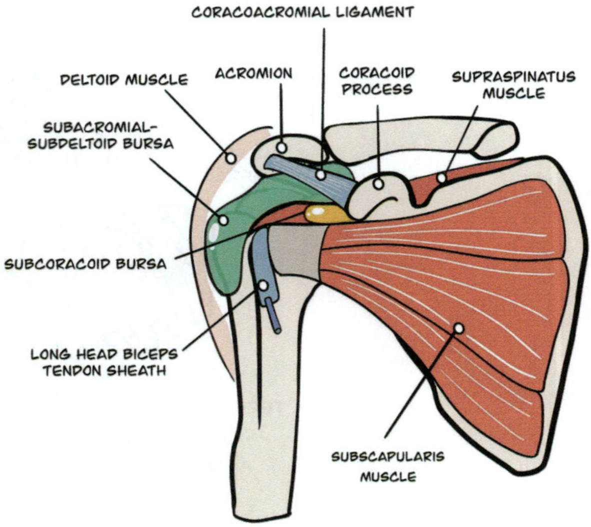

Fig. 10.4 Shoulder bursae and tendon sheath

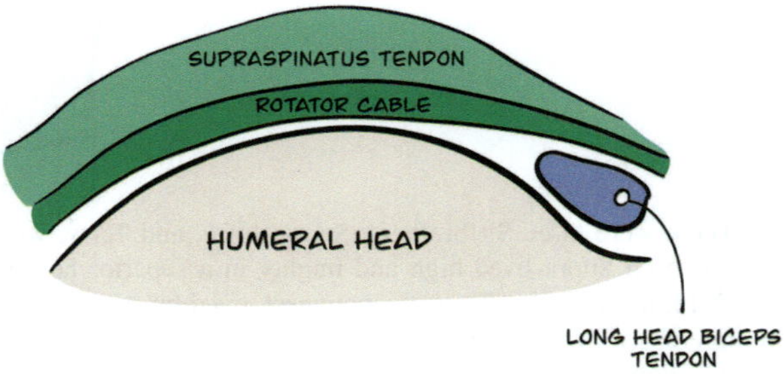

Fig. 10.5 Rotator cable

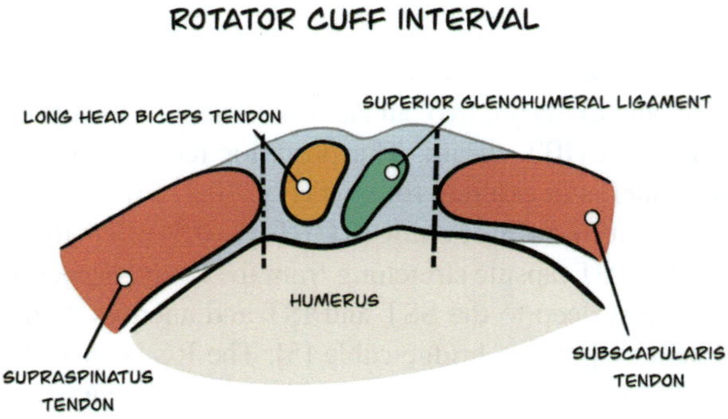

Fig. 10.6 (Intra-articular) Rotator cuff interval

The proximal portion of the RCI is where the LHBT is ovoid (intra-articular). In this location, the LHBT does not touch bone since it is supported by two ligaments that form a sling, the **biceps pulley system.** These ligaments are the more superficial coracohumeral ligament (CHL) and the deeper superior glenohumeral ligament (SGHL). More distal in the RCI, the LHBT becomes more circular and touches the base of the biceps groove, tethered in place by the transverse humeral ligament (THL). The RCI encompasses the path of the LHBT as it traverses this area and is best seen in the transverse (TAX) view in a modified Crass position.

Clinical Comments

Shoulder pain is common. It is impractical to order a magnetic resonance imaging (MRI) study for everyone with shoulder pain. Point-of-care ultrasound (US) excels in delineating many causes of shoulder pain and is complementary to MRI [8]. However, shoulder US can be challenging, humbling, and loaded with potential pitfalls [9]. We call it a shoulder exam, but US assesses soft tissue more than actual shoulder joints. A shoulder US exam typically visualizes the posterior GHJ but not the scapulothoracic joint. The ACJ is the shoulder joint best evaluated by US. The SCJ joint can also be assessed by US, although it is not part of our standard protocol.

Since clinical evaluation imperfectly pinpoints the source of shoulder pain, it is essential to perform a complete shoulder US examination in almost every case [10]. If the exam is unremarkable, the pain may have been referred to the shoulder from elsewhere. The radiograph is a valuable roadmap for the US exam. It will demonstrate the shape of the acromion, the humeral head (HH) position with respect to the glenoid fossa, the presence or absence of an os acromiale variant, GT cortical irregularity, osteophytes that may cause impingement, calcium deposition in the soft tissue, and the status of the GHJ and ACJ [11].

Rotator Cuff Impingement

Anterosuperior Impingement

Anterosuperior impingement describes compression of the SASDB (and sometimes even the SST itself) under the coracoacromial arch (CAA) (Fig. 10.4). The CAA consists of the distal acromion, the coracoid process, and the coracoacromial ligament (CAL). Shoulder abduction provokes compression. Impingement catalysts are structural (downsloping distal acromion or osteophytes, os acromiale, large SST calcium deposits, SASDB thickening) or functional (SST tendinosis or tears, SST weakness from neurologic disease) [12]. However, functional impingement may occur in healthy over-head throwing athletes due to humeral head instability while throwing [13].

In severe impingement, the bony GT can abut the undersurface of the distal acromion in abduction. Impingement may result from a complete SST tear since this tendon keeps the HH tamped down. So, impingement of the SST under the CAA may cause an SST tear, and a large SST tear itself may cause or exacerbate impingement due to an ascending HH, creating a vicious cycle. Impingement syndrome can lead to chronic SASD bursitis/bursal thickening, SST tendinosis, and ultimately tearing. Some suggest that impingement under the CAA initiates the bulk of rotator cuff tears [14]. The sonographer tests for impingement using passive shoulder abduction. Positive findings include bursal material or fluid accumulating at the point of impingement ("rug-under-the-door sign") or simply sluggish gliding of the bursa [8, 15]. Also, be aware that SASDB impingement may occur under the CAL and not just under the distal acromion [4].

Subcoracoid Impingement

Only rarely is the SSCT truly impinged under the coracoid. See the **Dynamic Views** section below.

Posterosuperior Impingement Syndrome

Posterosuperior impingement syndrome, also called *internal impingement,* occurs most often in over-head throwing athletes. The posterior rotator cuff is pinched at the junction of the SST and IST between the GT and the posterosuperior glenoid rim in maximal abduction and external shoulder rotation (the cocking phase for throwing). It causes SST, undersurface tendinosis, and tears [16].

Rotator Cuff Tears

Rotator cuff tears most commonly occur in the SST, but isolated tears may also occur in the IST and the SSCT, usually as an extension of an SST tear [17, 18]. The most common partial-thickness SST tear is articular-sided on the anterior portion of the SST near its insertion on the GT [19]. Such tears may account for 70% of all partial-thickness rotator cuff tears. Critical zone (approximately 1 cm proximal to the GT insertion) and bursal-sided tears are less common [19]. *Thus, the "critical zone" of the SST is not as vulnerable to tears as previously thought.*

The rotator cuff tear you detect on US may not be the source of pain. Rotator cuff tears are common, and their prevalence increases with age. In one study, full-thickness tears were present in about 11% of males in their 50s and 37% in their 80s [20]. The tears were asymptomatic in half of the males in their 50s and in 2/3 of the males in their 80s. Comparing US to MRI and magnetic resonance imaging arthrography (MRA) for the detection of rotator cuff tears, MRA was shown to be more sensitive and specific than either US or MRI [21]. For all tears, the sensitivity of US, MRI, and MRA, respectively, is about 85%, 85%, and 92%, respectively. Specificity is about 82%, 90%, and 97% [21]. Therefore, US and MRI are comparable in both sensitivity and specificity. US and MRI detect about 2/3 of partial-thickness tears, whereas MRA detects 86% [21]. All three modalities were greater than 90% specific for partial-thickness tears. *Thus, even if you do not see a tear on US, one can still be present.*

Description of Rotator Cuff Tears

To avoid confusion, remember that the tendons are three-dimensional, so tears vary in width, depth, location, and extent. Adding to the bewilderment are descriptions of tendon tears such as a rim-rent tear, which is also called a PASTA (partial articular side tear). Tears can be defined as being either complete or incomplete. A complete tear means that the tendon has torn to the point of having no interconnecting fibers on either side of the tear. In an incomplete tendon tear, communicating remnants persist. Furthermore, incomplete tears may be partial, intrasubstance, or even full-thickness. However, incomplete full-thickness tears do not encompass both the tendon's depth *and* width.

Verifying tears in two orthogonal planes is essential. The US report description should include the location of the tear in the cross-sectional plane (e.g., anterior, middle, or posterior) or longitudinal plane. Also, depict the tear shape, length, and note if retraction (partial or complete) is present [8]. In our institution, we use descriptive terms such as full- or partial-thickness and full- or partial-width to describe tears. We further describe the extent of partial tears by percent. For instance, "partial articular-side tear involving 50% of the tendon thickness" [9].

Specific Types of Tears (Fig. 10.7)

Partial-Thickness Tears

Partial-thickness or incomplete tears are tears that do not completely penetrate the full thickness of the tendon. The partial-thickness tear should have a well-delineated anechoic to hypoechoic discontinuity, to differentiate it from tendinosis, which is verified in two orthogonal planes [9]. However, tendinosis may coexist with and be confused with an intrasubstance tear. *Cortical irregularity* at the GT footprint may be a clue to the presence of intrasubstance tears [9]. Measure the depth in mm or a fraction of the tendon depth involved, the location in the tendon, and add a location descriptor:

(a) **Bursal**: The tear communicates with SASDB but may be difficult to delineate due to SASDB synovial fluid or synovitis filling the gap. Transducer pressure may accentuate the migration of the SASDB into the tear defect in bursal-sided tears [8].

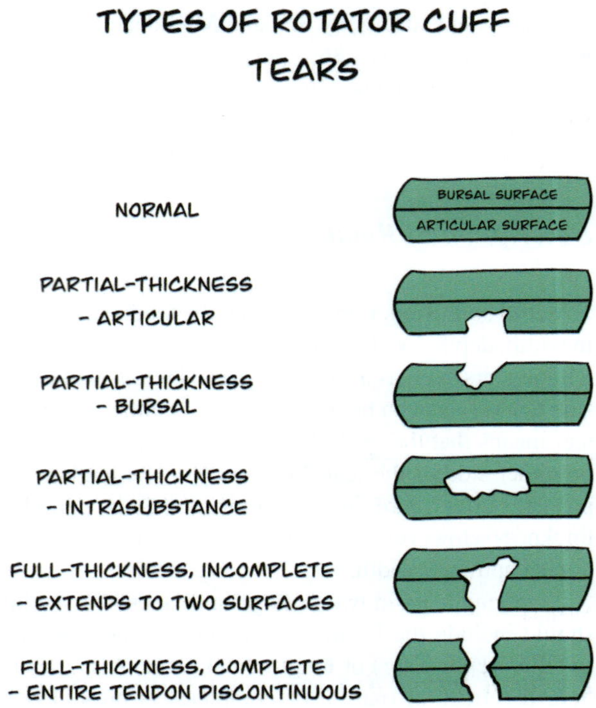

Fig. 10.7 Types of rotator cuff tears

(b) **Articular**: The tear communicates with GHJ and may cause a *cartilage interface sign* (CIS) (see Chap. 1 for more information). A CIS is a detection clue for articular-sided tears and occurs when fluid in the tear reaches the cartilage, producing a thin hyperechoic interface on the surface of the cartilage [22].

(c) **Intrasubstance**: The tear occurs in the middle of the tendon and does not communicate with the bursa or the joint. Intrasubstance tears may involve the insertional footprint (as long as they do not concomitantly involve the bursa or articular side) [3]. Logically, intrasubstance tears noted on US or MRI are invisible to arthroscopy.

Focal/Incomplete Full-Thickness Tears

This tear goes through the complete depth of the tendon, thereby communicating with the GHJ and the SASDB; however, the tendon width is not entirely involved. Like a partial-thickness tear, a full-thickness tear is delineated on US as an anechoic or hypoechoic tendon defect. Partial tendon retraction may be seen; granulation tissue and scarring may fill the space where the tendon once existed [8]. Effusion of the SASDB or the GHJ is found in about 2/3 of acute rotator cuff tears and about 1/3 of chronic rotator cuff tears [23].

Complete Tear

A complete tear is a full-thickness (depth) tear that involves the tendon's full width; this is a 100% total tendon tear. There may be no residual tendon remnant, and the stumps may not be appreciated. Debris or other structures may shift into the area. A classic example is that after a massive SST tear, the deltoid muscle shifts into and fills the tendon defect [24]. Alternatively, as may occur with incomplete full-thickness tears, granulation tissue and scarring may fill the space where the SST previously resided [8]. Look for at least some evidence of fibrillar echotexture, or else you may not be looking at a tendon.

Clues That Point Toward a Full-Thickness Rotator Cuff Tear

1. **Loss of volume of the tendon**
   With a SST tear, the normal convex contour of the tendon may become flat or even slightly concave. The SASDB, the peribursal fat pad, or even the deltoid muscle may outline the new, abnormal SST silhouette; the latter is called the *deltoid herniation sign* [25]. Loss of volume is often more noticeable in bursal-sided tears but also occurs with full-thickness tears. External probe pressure may highlight this "*flat-tire sign*." Small tears will not cause significant volume loss.
2. **A total absence of the tendon** [26]
3. **Superior facet GT cortical irregularity** (*for supraspinatus tears only*)
   This bone abnormality is present in 86% of full-thickness tears and 50% of articular-sided partial-thickness tears [26]. However, there are two caveats to consider: (1) Ccortical irregularity may occur in 21% of patients *without* a tear, and (2) Do not mistake a step-off deformity of a GT fracture for cortical irregularity [26].
4. **Effusion in the GHJ, SASDB, and LHBT sheath** [26]
   Recall that the GHJ and LHBT sheaths usually communicate with each other. Also, recall that the SASDB may or may not communicate with the GHJ, but will often do so in the presence of a rotator cuff tear [27]. However, effusion in the LHBT sheath or the SASDB is not unambiguous evidence of a rotator cuff tear [28].
5. **The CIS**
   The CIS is a hyperechoic line covering articular cartilage, it is magnified by fluid within a tear that reaches the articular side. Thus, the CIS is an acoustic enhancement artifact due to the tendon tear extending to the articular surface, whether a partial- or full-thickness tear [26].

Supraspinatus Tendon Tears

The etiology of SST tears is often multifactorial [29]. Degenerative cuff defects are common, occurring in more than 65% of people, often with no symptoms [30]. This degenerative (intrinsic) change, perhaps starting in middle life, coupled with extrinsic factors such as subacromial impingement, glenohumeral instability, and internal impingement, creates a perfect storm for rotator cuff tears. Younger patients may likewise develop rotator cuff tears, often from trauma [30]. Indeed, trauma may also play a role in middle-aged athletes with preexisting degenerative rotator cuff changes. In those individuals, clues to chronic degenerative change are tendinosis or GT cortical irregularity [31].

1. **Bursal-sided tears** of the SST may be obscured by the encroachment of SASDB material, synovial fluid, peribursal fat, or even the deltoid muscle upon the damaged SST. You may only see a flattening or reversal of the normal SST convexity accentuated by probe pressure ("*flat tire sign*"). If the tear progresses to a complete tear, the deltoid muscle may migrate into the former position of the SST and mislead the sonographer into believing that the SST is present.
2. **Articular-sided tears** are more common; look for fluid in the gap or a CIS.
3. **Intrasubstance tears** are longitudinal and do not break through to the bursal or articular surface. Remember, an SST tear extending to the GT footprint but *not* reaching either the articular or bursal sides is *still* called an intrasubstance tear. Tilt the probe and evaluate in transverse (TAX) and longitudinal (LAX) views to verify that the hypoechoic area you identify as a putative intrasubstance tear is not simply anisotropy.
4. **Full-thickness tears** are *full-depth*. They are termed "**complete tears**" when the *full width* is also involved. The larger the tear size, the greater the chance for tendon stump retraction. Measure the distance between the two stumps. If you see a complete SST tear, look carefully for an extension to the IST and check for IST insertion integrity with dynamic maneuvers [18].
5. **Rim-rent tears** are distal partial-thickness articular-sided tears adjacent to the GT. If a rim-rent tear has an additional bony avulsion from the GT, this is called a PASTA (partial articular-sided supraspinatus tear *with avulsion*) lesion.

Subscapularis Tendon Tears

The SSCT stabilizes the anterior shoulder [8]. Isolated tears may be sports-related; however, in many cases, an SSCT tear may be caused by an SST tear extending anteriorly right across the RCI. The SSCT fibers blend into the CHL and the THL, and, thus, SSCT tears may be associated with LHBT instability due to the loss of integrity of the THL [8]. The LHBT will subluxate or even dislocate medially if this occurs; this is most apparent during a dynamic US examination.

Infraspinatus Tendon Tears

Infraspinatus tendon tears are rarely isolated to the IST but often result from an SST tear extending posteriorly [6]. If an SST tear extends posterolaterally by more than 2.5 cm, then there is a high likelihood that this tear has spread to the IST. Like SST tears, these tears may be partial-thickness, intrasubstance, full-thickness, or complete. One special mention is a partial-thickness articular-side tear of the IST, which may occur with posterosuperior impingement syndrome. Impingement between the posterior HH and glenoid during external rotation can cause a posterior labral tear and significant posterior cortical irregularity of the posterior GT. However, this "bare area" may normally have some irregularity since it is not covered by cartilage.

Teres Minor Tendon Tears

Teres minor tendon tears are rare, unrelated to IST tear extension, and may be due to acute shoulder trauma. Measure the length of the tear or describe the percent of the tendon involved. An acute tear may demonstrate neovascularization with power Doppler (PD).

Other Findings

Tendinosis: Tendinosis is sonographically represented initially by tendon thickening and enlargement and later by focal or diffuse hypoechogenicity with loss of the normal fibrillar echotexture [8]. Tendinosis may occur in the SST, the LHBT, the IST, and the SSCT. Doppler activity may or may not be seen. Any abnormality should be defined as a tear only when a genuine defect is documented in both TAX and LAX and an anisotropy effect is ruled out. Tendinosis is a precursor to a tear but may be challenging to separate from an actual tear [8]. Tendinosis, often referred to as tendinopathy, may be caused by age-related degeneration, overactivity, trauma, or chronic impingement, such as may occur in the SST under the CAA [14, 32, 33]. Statin medications may increase the incidence of tendinopathy [34]. Likewise, quinolone antibiotics may facilitate supraspinatus tendinopathy [35]. Sonopalpation may be diagnostically helpful in specific areas:

1. SST impingement at the distal acromion
2. Tenderness over the LHBT
3. Tenderness over an acute tendon tear
4. Tenderness over tenosynovitis, such as along the LHBT sheath

Bursitis: Bursitis, an inflamed and distended bursa, may be detected in the SASDB and the subcoracoid bursa. Bursitis may be caused by repetitive pressure irritation, infection, crystal disease, or inflammatory conditions.

Glenohumeral joint effusion: Excess synovial fluid can reliably be demonstrated in the posterior GHJ recess (in external rotation) and the LHBT sheath since the latter communicates with the GHJ [36]. In the posterior GHJ recess, an anechoic or hypoechoic effusion may be located deep to the IST but superficial to the junction of the HH and the glenoid labrum [4].

Synovitis: Shoulder US may detect synovitis in the LHBT sheath, the GHJ posterior recess, the SASDB, the ACJ, and the subcoracoid bursa. One study described more significant bursal synovitis with PD activity in elderly onset RA compared with polymyalgia rheumatica (PMR) [37].

Enthesitis: Entheses can be evaluated at all three GT facets (SST, IST, and TMT) and the LT's SSCT insertion.

Dynamic Views

A significant strength of US is dynamic evaluation; MRI cannot duplicate this. Dynamic US uses:

1. Abduction detects SST impingement under the CAA.
2. External rotation detects small to moderate posterior GHJ recess effusions.
3. External rotation detects SSCT impingement under the coracoid due to coracoid process anatomic variations, chunks of calcium within the SSCT, and ganglion cysts [4, 38, 39].
4. External rotation discloses medial LHBT subluxation if there is a loss of integrity of the biceps pulley system or the THL.

Adhesive Capsulitis

Adhesive capsulitis results in markedly decreased shoulder range of motion (ROM), often with pain [4]. Although often idiopathic, causes include diabetes, certain medications, and prolonged immobilization after surgery or trauma. It starts with synovial proliferation, followed by collagen deposition, and ultimately articular volume contraction, resulting in severe stiffening and decreased GHJ ROM. Also known as "frozen shoulder," it progresses in three stages: freezing, frozen, and thawing. Doppler activity may occur before it becomes frozen. There are four sonographic signs of AC on US. Using the first 3 (a–c) parameters, US has high sensitivity and specificity (100% and 87%, respectively) for the diagnosis of AC of the shoulder [40]. Power Doppler activity (d) is also diagnostically helpful [41].

(a) Decreased dynamic movement of the SSCT under the coracoid.
(b) Decreased SST dynamic movement under the acromion.
(c) Increased RCI size due to swelling of the ligaments surrounding the LHBT.
(d) ± PD activity in the RCI.

Rotator Cuff Muscle Atrophy

Atrophic muscles are hyperechoic (whiter), but be aware that diabetes may produce bright-appearing muscles with increased echogenicity. Compare the muscle size and echogenicity to normal contiguous and contralateral muscles. Additionally, look for a loss of normal muscle echotexture and contour [3, 8]. Ultrasound compares favorably to MRI for detecting rotator cuff atrophy, but is sonographer-dependent [42, 43]. One important use of US is to evaluate the infraspinatus muscle (ISM) for atrophy by comparing it to the teres minor muscle (TMM) in TAX, since an atrophic IST predicts poor outcomes of rotator cuff repair [44–46]. In the TAX view at the scapula ridge, the cross-sectional area of the normal ISM is approximately twice that of the TMM [3].

Postoperative Rotator Cuff Repair

Ultrasound is useful for assessing postoperative rotator cuff repair if the shoulder has had sufficient time to heal. Most rotator cuff repairs are simply decompression procedures, done arthroscopically to alleviate impingement on the SST [47]. The surgeon resects any structure encroaching upon the SST. Tendon reattachment procedures can be done for complete rotator cuff tears. Even large tears may be amenable to arthroscopic repair [48].

First, look at plain films as a roadmap to evaluate a post-op cuff repair. The reattachment site appears as a trough, and you may see hyperechoic sutures [49]. If a bursectomy was done, then the SASDB would not be identified. Distorted tendon sonoanatomy complicates perioperative interpretation due to variable thickness and echogenicity. Ultrasound image clarity improves with time, perhaps 6–9 months post-op, as the tendon gradually resumes a more "normal" hyperechoic and fibrillar echotexture and shape [49]. Recurrent rotator cuff tears may be noted, ranging from focal to complete tears, sometimes with complete retraction. Free-floating metal sutures may indicate a recurrent tear [49].

After tendon repair, US can diagnose recurrent tears with an accuracy of 89% [50]. A recurrent tear may not correlate with pain or dysfunction; however, an intact rotator cuff correlates with greater strength [51]. Thus, ultrasonic detection of a recurrent tear does not necessarily warrant repeat surgery. Dynamic scans help determine the repair integrity or the emergence of adhesive capsulitis.

Postoperative Arthroplasty

Total shoulder replacements (TSR) and hemiarthroplasties are done with an intact rotator cuff. Both procedures preserve the rotator cuff and the tuberosities into which the rotator cuff inserts. A "reverse shoulder" prosthesis is performed for patients who lack an intact rotator cuff [52]. Look at plain films for the position of the prosthesis, fracture, hardware aberrancy, and periprosthetic lucency >2 mm; the latter suggesting prosthesis loosening [49]. On US, the prosthesis is echogenic, smooth, and has posterior reverberation artifacts. Evaluate the overlying rotator cuff tendons and their insertions to see if they remain intact. Since the prosthetic metal humeral head artifact is deep to the rotator cuff, US assesses the integrity of the rotator cuff and the position of surgical screws. Normal, intact rotator cuff tendons appear hypoechoic and heterogeneous immediately after surgery. The diagnosis of tendinosis or even a tear is challenging unless you see a clear-cut tendon defect with anechoic fluid [49]. A dynamic US exam may demonstrate hardware instability.

Calcium in the Shoulder

Calcium deposits are common, consist primarily of **hydroxyapatite**, and occur mainly in the SST, the LHBT, and the SADSB [8]. Calcium deposition is a dynamic process of unknown cause, typically occurring in people aged 30–50 [53]. The calcium deposits may be appreciated on plain films. On US, calcium is hyperechoic, perhaps with an acoustic shadowing artifact [8]. The three stages are: precalcific, calcific (formative and resorptive phases), and postcalcific [54]. Pain, and often Doppler activity, is most prominent in the resorptive phase [8]. Treatment is not needed if asymptomatic, but if symptomatic, conservative treatment is tried first.

Calcific Tendinosis Stages

1. **Thin/linear stage**
   This is a degenerative stage. Do not aspirate, and do not confuse this with a CIS.
2. **Well-defined or globular stage**
   This may cause inflammation and may respond to aspiration. Determine if a block of calcium is causing impingement. If so, needling the calcium (barbotage) and saline lavage and aspiration may help [55].
3. **Resorptive or "slurry" stage**
   The deposits are in liquid or semiliquid ("toothpaste") form and can simply be aspirated if necessary [56]. Positive Doppler and pain reproduction with sonopalpation help determine if the calcium deposit is symptomatic.

Aside from hydroxyapatite, there are other causes of calcium deposition in the shoulder. Calcium deposition may occur with calcific arthropathy, such as calcium pyrophosphate deposition (CPPD) disease. Calcium in the SASDB may migrate from calcium deposition in the SST, perhaps resulting in bursitis of the SASDB [17, 57, 58]. Loose bodies may occur when cartilage pieces detach and calcify in the synovial fluid [59]. This process may occur in the joint recess, the LHBT sheath, and the bursae. Loose bodies are benign, occasionally painful, and are best demonstrated on dynamic US.

Other Shoulder Structures

Biceps Tendon

Although not technically part of the rotator cuff, LHBT pathology may be associated with disorders of the rotator cuff, the glenoid labrum, and LHBT impingement [8]. The biceps' brachii muscle, the classic symbol of strength, is interesting. The LHBT originates from the spinoglenoid tubercle, but 50% of the time, there is a dual attachment to the glenoid labrum [60]. Consequently, this tendon may be partially attached to a cartilage structure. The proximal LHBT drapes over the HH, acting as a secondary stabilizer of the HH. The tendon curves downward about 40° as it enters the intertubercular sulcus or groove. Like many tendons that rub against bone, friction is reduced by a surrounding protective sheath, the tenosynovium. More distally, the LHBT transitions into the myotendinous junction. Beyond the muscle, the distal biceps tendon inserts on the radial tuberosity and the bicipital aponeurosis.

Many presume the biceps to be the ultimate elbow flexor, but it is subordinate to the underlying brachialis muscle regarding elbow flexion. However, the biceps muscle is the primary forearm supinator. The origin and course of the LHBT expose the tendon to bystander damage from impairment of the GHJ, the glenoid labrum, the SASDB, and particularly the SST. A completely torn SST encourages the HH to migrate upward; the proximal biceps tendon will mitigate such movement to its own detriment. Another example of collateral damage would be the effect of a glenoid labrum tear in a person whose LHBT is partially anchored to the labrum. Thus, the LHBT is, in a sense, the barometer of the shoulder. In the case of a complete LHBT tear, hyperechoic tissue, including hematoma and synovitis, may mimic an intact LHBT within the biceps groove on the TAX view. Again, look for the familiar fibrillar echostructure of the tendon to verify that the LHBT is in the groove. The LHBT can be associated with several pathologic findings:

1. **Tendinosis**

   The LHBT is a bit different from many tendons. With tendinosis, the tendon swells, *but fibrillar echotexture may persist*, barring a frank tear. However, the presence of some hypoechoic areas may indicate tendinosis [8]. In the tendon groove, the LHBT with tendinosis appears enlarged. Like other disorders of the LHBT, tendinosis is often a secondary event [61]. Occasionally, isolated LHBT pathology may occur in younger athletes. Additionally, the LHBT may be subject to adverse effects from medications (e.g., statins, quinolone antibiotics, or injected or systemic corticosteroids) [62].

2. **Tears**

   Tears of the LHBT may be associated with a complete SST tear; in this case, the BT becomes the HH's main depressor, resulting in excess stress on the tendon. The LHBT may even split. An LHBT tear will cause distal tendon sheath effusion and loss of fibrillary echotexture, which may occur at the intra-articular or extra-articular level. If a proximal LHBT tear occurs, suspect a glenoid labrum (SLAP) tear [63]. However, with such a proximal LHBT tear, the detached LHBT may remain in the biceps groove. Consider a labral tear with appropriate signs and symptoms, particularly in an overhead throwing athlete. Again, the best test for a labral tear is an MRA of the shoulder [63].

Complete LHBT tears result in Popeye's sign due to a balling up of the muscle. Acutely, this is painful and may have a dramatic appearance, but the pain subsides eventually, and elbow flexion strength is minimally affected. These tears are often spontaneous and tend to occur in people older than 50 [64]. It may be challenging to visualize partial tears of the LHBT on US. The strength of sonography lies in confirming an intact LHBT or full-thickness tear, but it is less accurate in diagnosing partial-thickness tears [65, 66].

Tracking the tendon from proximal and distal locations is imperative since debris may mimic an intact LHBT. There may be visible, retracted tendon stumps. If the rupture occurs at the myotendinous junction, then the LHBT may still be in the groove proximally. Most LHBT tears are associated with SST or SSCT tears. Treatment is usually conservative, but surgery may be needed for a young, high-level athlete. It is not uncommon for anomalous biceps muscle heads to be present and have separate tendons in the biceps groove [67]. Such a variant may mimic a split LHBT [68].

3. **Tenosynovial effusion**

   The LHBT communicates with the GHJ, so pathology in the GHJ may be reflected by increased synovial fluid, perhaps with calcified intra-articular bodies originating from the GHJ. In TAX views of an LHBT tenosynovial effusion, synovial fluid almost wholly surrounds the tendon. If there is increased synovial fluid in the LHBT sheath, look posteriorly at the GHJ in external rotation for effusion and consider potential causes of increased GHJ synovial fluid, including a rotator cuff tear.

4. **Tenosynovitis**

   The tendon sheath may demonstrate anechoic fluid, synovial hypertrophy, or PD activity [8]. Such findings should prompt a search for an underlying inflammatory condition such as RA. Power or color flow Doppler may produce spectacular diagnostic images. The SASDB, if distended, may mimic an LHBT sheath effusion or tenosynovitis in TAX view at the biceps groove but will *not completely* surround the LHBT; the SASDB will be located superficially to the LHBT.

5. **Instability**

   The accuracy of US in detecting LHBT instability is high [66]. The LHBT is held in place by the bicipital groove, the biceps pulley at the intra-articular level, and, more distally, by the THL at the extra-articular level. The SSCT gives supporting fibers to the THL, so if the SSCT tears, the THL may also be impaired. A compromise of the tethering effect of either the biceps pulley or the THL may result in LHBT subluxation or dislocation.

During the external rotation dynamic exam, the LHBT may medially sublux-ate/dislocate superficial to or even into an SSCT tear. The LHBT may even dislocate under the coracoid, thus becoming invisible to the US beam. So, start distally and follow the LHBT proximally to look for dislocation. In summary, if dynamic US reveals LBHT subluxation, consider an injury to the SSCT, the THL, or the biceps pulley [8, 69].

6. **Entrapment**

   The LHBT may rarely become impinged between the HH and the glenoid during forward arm elevation [4]. This may result in an "hourglass" configuration of the LHBT [70]. There may be buckling of the LHBT as it is squeezed between the HH and the glenoid [71].

7. **Acromial impingement**

   If the SST completely tears, the HH may rise toward the acromion. By default, the LHBT becomes a tether of last resort, ineffectually impeding the HH's ascension. The LHBT itself may become impinged under the acromion, often damaging the tendon [72].

Subacromial-Subdeltoid Bursa

This synovial-lined bursa, the largest in the body, separates the rotator cuff from the CAA and the deltoid muscle [27]. It extends from the coracoid process to beyond the GT and decreases tendon friction [8, 73]. It is surrounded by a hyperechoic peribursal fat layer, the latter producing a "tram-track" appearance [8]. The SASDB is hypoechoic, ordinarily 1–2 mm deep, and may typically contain a slight amount of fluid [8]. It can thicken with underlying SST pathology or recurrent impingement under the CAA [27]. A clue to a bursal-sided SST tear is the downward dipping of the SASDB into the usually convex outer tendon layer. In the event of a complete SST tear, a thickened SASDB may move into the vacant area to mimic an intact SST [9]. With an effusion, the distended SASDB may be seen in TAX to overlie the LHBT, resembling the beam and fulcrum of a teeter-totter or seesaw. Use probe pressure to verify whether an effusion is present or not.

The bursa may become distended due to inflammation, even if it does not communicate with the GHJ [8]. It can harbor effusions, synovitis, and calcium deposits [8]. Power Doppler activity may be present with synovitis or an infection. Causes of noncommunicating SASDB distension include direct bursal trauma, repetitive trauma, and inflammatory disease affecting the synovium [27]. The major causes of SASDB distension are [8]:

1. Effusion
2. Synovitis (RA, spondyloarthropathy [SpA], PMR [74])
3. Crystal arthropathy (gout, CPPD, calcium hydroxyapatite)
4. Hemorrhage
5. Impingement
6. Infection (septic bursitis)

7. Rotator cuff tear
8. Amyloidosis
9. Synovial proliferative disorder (pigmented villonodular synovitis, synovial osteochondromas)

The SASDB may be isoechoic with the deltoid muscle and thus challenging to delineate [8]. However, since the distended SASDB extends beyond the SST, the sonographer can detect the tip of a dependent SASDB effusion *lateral* to the insertion of the SST on the GT [27]. Additionally, PD activity within the bursa may indicate acute inflammation [8]. The SASDB does not usually communicate with the GHJ but often does so with a rotator cuff tear; more than 90% of rotator cuff tears are associated with SASDB distension [27].

Acromioclavicular Joint

The most commonly seen ACJ abnormality is osteoarthritis. The fibrous capsule and coracoacromial and coracoclavicular ligaments (CAL and CCL, respectively) stabilize the joint [75]. The os acromiale (an anomalous nonfused accessory bone in about 8% of people) may be seen on the superior acromion [76]. An os acromiale has a distinct US appearance, with the extra bony fragment being readily apparent adjacent to the distal acromion [77]. An os acromiale can cause SST impingement syndrome if the deltoid inserts on it, tugging it down and compressing the SST. Alternatively, the os acromiale may develop a downward osteophyte that impinges on the SST [78]. Cysts over the ACJ are painless or slightly tender masses associated with ACJ degenerative joint disease and full-thickness rotator cuff tears [79]. Other ACJ findings include:

1. **Geyser sign**: Fluid tracks into the ACJ from an underlying SST tear. This fluid accumulates superficially to the ACJ and is known as a geyser sign. Probe pressure shows debris moving back and forth across the ACJ [80].
2. **Synovitis and bony erosion**: This may occur with RA. The ACJ may also be affected by other inflammatory polyarthropathies, including CPPD.
3. **Effusion**: In this case, the ACJ capsule can be distended with effusion, perhaps with PD activity. Effusion causes include degenerative changes, infections, and inflammatory conditions [8].
4. **Instability**: This may be painful and caused by trauma or osteoarthritis; it is classified according to the ACJ space width and degree of ligament injury [81]. A dynamic exam of an injured ACJ will show a narrowing of the joint space in cross-arm adduction [4]. A comparison with the contralateral, uninjured joint is useful. A 2–3 mm discrepancy between sides is abnormal if this correlates with the clinical situation [75]. Tossy type I–III instability grades can be determined by US [82]. Significant tears of the acromioclavicular ligament may result in misalignment of the acromion and clavicle, resulting in an elevated clavicle.
5. **Widening of the ACJ**: This may occur with effusion, hematoma, surgical resection, post-traumatic osteolysis, distal clavicle resorption from RA, systemic sclerosis, or hyperparathyroidism.

Sternoclavicular and Costochondral Joints

Trauma may result in SCJ instability or dislocation, which US can diagnose [83]. Synovitis is quite detectable by US [84]. Ultrasound may also reveal bony erosion, osteoarthritis, and even bone destruction from SCJ infection [85, 86].

Glenoid Labrum

This fibrocartilage rim augments the bony glenoid "socket" into which the "ball" or humeral head fits. Ultrasound best visualizes the more superficial posterior labrum, but other quadrants can be viewed [87]. The acronym "SLAP" refers to a superior labrum tear from anterior to posterior. This injury may occur after a fall on an outstretched hand or repetitive overhead throwing by a younger patient. Symptoms range from a dull ache to a loss of throwing strength or a catching sensation.

Posteriorly, the labrum has a triangular shape. Tears are anechoic or hypoechoic fissures, but this may be a difficult call to make on US. However, US excels in determining if the posterior labrum is *normal* [88]. The presence of a paralabral cyst implies a torn labrum. The posterior labrum is best visualized with the shoulder in external rotation since this maneuver augments posterior recess distension, thereby enhancing labral tear detection. However, external rotation may also enlarge a suprascapular vein, which may be mistaken for a paralabral cyst (see the **Pitfalls** section below). Paralabral cysts may also be seen in the spinoglenoid notch. With the improved resolution of newer US equipment, it is possible to create images of all four quadrants of the glenoid labrum [87]. Nonetheless, MRA is the best imaging modality to evaluate the glenoid labrum [89, 90].

Glenohumeral Joint

Visualization of the GHJ is best in the posterior view. Even so, US is not a premier technique for GHJ evaluation [8]. Simple radiographs are superior for the assessment of arthritic changes. However, US may demonstrate cartilage loss, osteophytes, humeral head bony cortex irregularity, or joint recess synovitis [8].

Glenohumeral Joint Instability and Dislocation

Humeral head instability or dislocation out of the glenoid fossa causes pain, dysfunction, and accelerated osteoarthritis. Radiographs are the gold standard for the evaluation of dislocation. However, in experienced hands, US can be confirmatory and accurate for diagnosing dislocation as a stand-alone modality [91]. Ultrasound is also a valid tool for evaluating GHJ instability [92, 93]. Ultrasound also helps to look for related pathology, such as rotator cuff tears. The cause of GHJ dislocation may be a single significant trauma, repetitive lower-grade injury (e.g., throwing athlete), ligamentous laxity, or neuromuscular causes such as axillary nerve injury [94].

1. **Anterior dislocation**: About 95–97% of shoulder dislocations are anterior [95, 96]. Radiographs, including axillary views, are diagnostic [96]. Anterior dislocation may produce a *Hill-Sachs deformity*, a depressed area at the posterolateral aspect of the HH from recurrent impact against the edge of the anterior glenoid rim. Ultrasound may diagnose this deformity reliably [97].
2. **Posterior dislocation**: This accounts for 1–4% of shoulder dislocations, may be challenging to identify, and is easily missed [98, 99]. The posterior glenoid may impact the anterior HH, causing a fracture, which is referred to as a "reverse Hill-Sachs lesion" or "*McLaughlin lesion*" [100, 101]. Additional axillary radiographs may be needed to diagnose a posterior dislocation. An undetected posterior dislocation may lead to chronic pain and disability [102]. In contrast to Hill-Sachs lesions, McLaughlin lesions are not readily evaluated by US; this requires a CT or MRI [101].
3. **Superior dislocation**: The HH tends to migrate superiorly with a complete SST tear since the latter structure depresses the HH. With a substantial SST tear, the HH may ascend to abut the inferior acromial surface, accelerating the development of osteoarthritis [103, 104]. The plain radiograph and MRI can delineate superior migration of the HH [105].
4. **Inferior dislocation**: Dynamic US has been favorably compared with stress radiography for assessing and quantifying inferior glenohumeral laxity [93].
5. **Multidirectional instability**: This is due to excessive laxity in more than one direction and may be due to recurrent shoulder injuries. Such instability may be asymptomatic in elite swimmers [106]. Multidirectional instability may have neurologic causes, including cerebrovascular accidents [107].

Suprascapular Nerve

With a SLAP tear, a paralabral cyst may develop and compress the nerve in the spinoglenoid notch (SGN), causing weakness of the infraspinatus muscle. If you see a cyst in the SGN, be concerned about a SLAP tear, and consider an MRA of the shoulder to look for a labrum tear [108]. However, varicose veins in the SGN may imitate a paralabral cyst. Varicosities may also compress the nerve, but varicosities collapse with internal shoulder rotation (cross-arm maneuver) and expand with external rotation [4]. Be cautious about placing a needle into a cystic structure in this area since this may be a vein. Even if you successfully decompress a paralabral cyst, the cyst will often recur unless you address the underlying cause, which is the labral tear [8]. Since the SGN is a relatively deep structure, a cyst in the SGN may be hypoechoic and not as well-demarcated as a more superficial cyst.

Inflammatory Disease

Rheumatoid arthritis and other inflammatory arthropathies may cause effusion or synovitis in the bursae, joint recesses, and the LHBT sheath, as well as bone erosion of the humeral head [109]. For polymyalgia rheumatica, US detection of bursitis, effusion, and tenosynovitis is incorporated in the 2012 EULAR/ACR classification criteria to improve sensitivity and specificity [110, 111].

Pitfalls

1. Always request and directly review baseline shoulder radiographs before the US. Review the history and do a focused physical exam. Have the patient indicate the pain location and shoulder positions that exacerbate the pain [112].
2. Use a ballpoint pen to draw topical bony landmarks before the exam.
3. Rotator cuff tears are common. The unambiguous rotator cuff tear you detect may *not* be the source of the patient's pain.
4. Based on the limits of the technology, if a rotator cuff tear is not observed, it may still be present. It is best described in reports as "no tears were visualized." Be aware that the former location of a completely torn SST may fill in with hyperechoic material or the SASDB, fooling even experienced sonographers.
5. Angle the probe downward (heel-to-toe maneuver) to look at the footprint of the SST in LAX to minimize anisotropy [112].
6. The heel-to-toe maneuver is also critical in evaluating the LHBT in LAX to correct for anisotropy since the tendon angles downward more distally [113].
7. Tilting the probe back and forth is critical when evaluating the LHBT in TAX within the bicipital groove to avoid the effect of anisotropy [112].
8. In LAX view, don't mistake a horizontal hypoechoic band in the anterior SST for an intrasubstance tear. This line separates the superficial from the deeper, flatter fibers of the SST. On TAX, the superficial fibers may appear as a hypoechoic ovoid defect due to anisotropy, so don't mistake this for a tear or tendinosis [17].
9. Do not forget to externally rotate the shoulder to enhance the detection of small to moderate-sized GHJ effusions in the posterior recess [36].
10. The suprascapular vein may distend in external rotation, mimicking a paralabral cyst. It will collapse in internal rotation, whereas a cyst will not [4]. Color flow Doppler will help to delineate this.
11. Completely evaluate the entire width and insertion of the SST and the SSCT [112].
12. Do not diagnose tendinosis or tears at the junction of the SST and IST on the GT since this is where the two tendons intermingle at different angles, resulting in a hypoechoic insertional overlap due to anisotropy [112].
13. Do not misconstrue the normal echotexture of the cross-section of the SSCT as having tears [112, 113].
14. Ultrasound can look at the glenoid labrum to a degree, but not nearly as accurately as an MRA.
15. In the LAX view, the SASDB and the SST sometimes blend and may be challenging to differentiate. To discern the two, follow the SASDB distally since it extends beyond the GT [113]. Remember, the SASB lacks fibrillar echotexture.

16. A dislocated or completely torn LHBT is a challenge to detect. Look for the fibrillar tendon and search for a dislocated LHBT and retracted tendon stumps. If you are uncertain, describe that this structure is not visualized [10, 113]. If you do not locate the LHBT in the groove, track the LHBT from distal to proximal, starting with the musculotendinous junction.
17. Verify tendon tears in both TAX and LAX views; double-check for anisotropy mimicking a tear.
18. Don't mistake a CIS for calcific tendinopathy or a double contour sign (see Chap. 8).
19. Don't mistake the SASDB or the deltoid muscle for an intact SST with complete SST tears. In this situation, the deltoid muscle, or the SASDB, occupies the absent tendon position. Verify that the putative SST inserts on the GT and has fibrillar echotexture.
20. Tendinosis of the SST may be confused with a tear. Tendinosis is hypoechoic with a *swollen tendon*. Tears may also be hypoechoic, but they are often anechoic and have a *thin tendon* [3].
21. Look very carefully at the anterior SST footprint, a common location for tears.
22. It is difficult to determine if an intrasubstance tear at the SST footprint on the GT extends to the tendon's articular surface (and thus would be called a partial thickness, articular side tear, or "rim-rent" tear). Look for a CIS or simply fluid in the tear and apply probe pressure to see if the anechoic discontinuity reaches the articular side.
23. Look for clues to amplify your suspicion of SST tears. For instance, GT cortical irregularity increases the probability of a tear.
24. The SASDB, if distended, may mimic LHBT sheath effusion or tenosynovitis, but the SASDB, in this case, *will not surround* the LHBT; it will lie superficial to the LHBT.

Method

At our institution, we have adopted the four regions "ASAP" protocol (anterior, superior, anterolateral, and posterior) used by others [8]. Use a 12 MHz linear probe except for the GHJ, which may require an 8 MHz frequency to reach a 6 cm depth. Obtain static, moving, and dynamic imaging. Many views are labeled "A, B, C, etc.," which are standard views, but may not apply to every patient and are left to the examiner's discretion.

Position The patient sits with the elbow flexed to 90° and the forearm and hand on the lap. The palm is up. Palpate the shoulder and, with a ballpoint pen, mark off specific landmarks: the proximal biceps tendon (long head) in the biceps groove, the coracoid process (if palpable), the ACJ, and the most lateral portion of the acromion. These markings are beneficial for landmark identification during an examination. The examiner may be in front of or behind the patient, whichever is more comfortable.

Crass vs. Modified Crass (Fig. 10.8)

These are shoulder positions designed to bring the proximal SST out from under the acromion. The Crass position is more extreme; the patient's hand is placed in the contralateral back pocket or across the mid-back. This maneuver rotates the humerus, bringing the GT more anteriorly. The modified Crass (Middleton) position is less extreme; the patient's hand is placed in the ipsilateral back pocket. Most patients with shoulder pain will not tolerate the Crass position. but do well with the modified Crass position. For that reason, we favor the modified Crass position. The advantages and disadvantages of these two positions are summarized as follows:

Crass

- More accurate for SST tear size in both planes [114]

– Does not fully evaluate the anterior third of the SST
– Difficult for patients with shoulder pain to maintain

Modified Crass

- Overestimates SST tears in LAX but is accurate in TAX [114]

– Better demonstrates the intra-articular LBHT and the RCI
– Easier for the patient in pain to tolerate

Fig. 10.8 Crass vs. modified crass

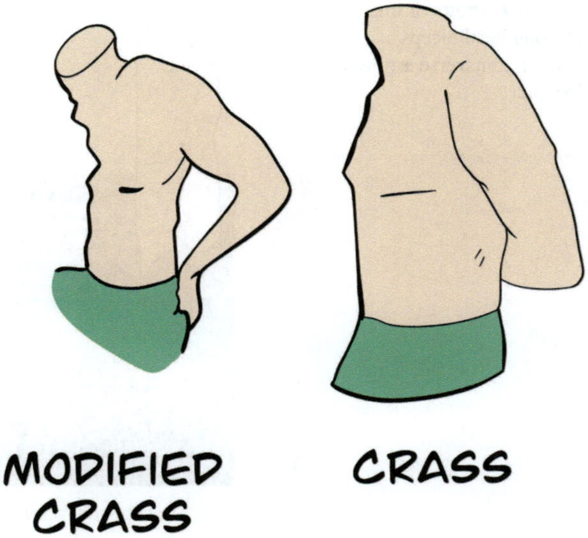

Protocol Image 1: Long Head Biceps Tendon, Transverse ± Power Doppler (Fig. 10.9)

Place the probe transverse to the upper arm, centered over the LHBT, which sits in the bony biceps groove, between the LT and the GT. The LT is medial, more pointed, and perhaps more superficial. Move the probe proximally until you see the tendon's ovoid, hyperechoic fibrillar cross-section. Observe for fiber hypertrophy, disruption, or loss of echogenicity. Tilt the probe to eliminate anisotropy. The oval appearance of the LHBT indicates its intraarticular location. Turn on PD if you suspect an acute tendon tear or synovitis. Just deep to the LHBT is the SGHL. Superficial to and surrounding the LBHT is the CHL. The CHL combines with the SGHL to form the biceps pulley system, or sling.

Next, move the probe in TAX more distally, where the LHBT assumes a round or semiround shape and the biceps groove is better delineated. The LHBT is now extra-articular. Look at the tendon sheath surrounding the LHBT for effusion or synovitis. A small amount of fluid is normal, but it should not surround the tendon [17]. Just superficial to the LHBT is the THL, which is composed of fibers from the SST and SSCT. If you do not see the tendon, it may be dislocated (over the LT superficial to or within the SSCT) or completely torn. Toggle the probe to sharpen the image. Look for LHBT swelling, which may indicate tendinosis. If you see excess fluid surrounding the LHBT, this is strong, albeit inconclusive, evidence for a rotator cuff tear or a GHJ effusion. Power Doppler may indicate tenosynovitis or perhaps an acute tendon tear. You will often visualize a branch of the anterior circumflex humeral artery on PD, just lateral to the LHBT. Avoid this vessel when injecting the LHBT sheath.

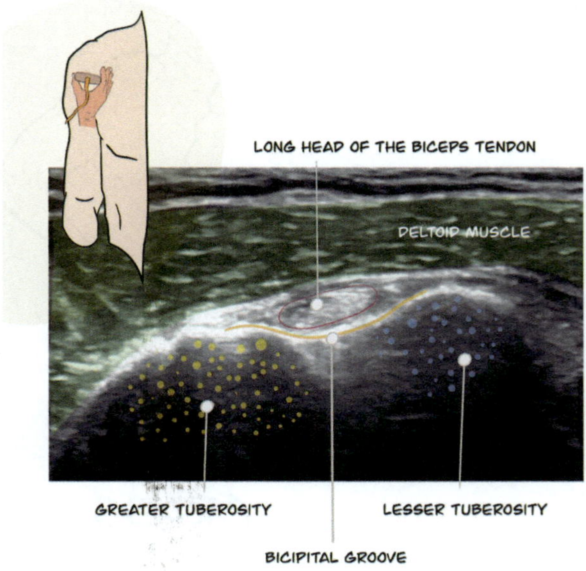

Fig. 10.9 Protocol Image 1: Long head biceps tendon, transverse ± power Doppler

With the patient's elbow tucked into the body, you can passively externally and internally rotate the forearm to see if the LHBT subluxes or even completely dislocates medially over the LT. In the case of an SSCT tear, the LHBT may move into the SSCT [4]. Resume the prior arm position and then move the probe in TAX distally until the tendon blends into the biceps muscle. Note that the biceps groove becomes shallower. Look for the tendon of the pectoralis major muscle, seen as a horizontal hyperechoic structure just superficial to the LHBT, near the tendon-muscle junction.

Protocol Image 2: Long Head Biceps Tendon, Longitudinal ± Power Doppler (Fig. 10.10)

Next, center the probe on the LHBT and turn it 90° to see the fibrillar LHBT in LAX. Slowly move the probe in a proximal LAX slide, repositioning the probe slightly medially and then laterally as you go to visualize the entire width of the LHBT. Make sure that the distal portion of the probe is angled downward in a "heel-to-toe" orientation that parallels the diving tendon. Look for loss of normal echotexture (tendinosis vs. tears), fluid in the distal tendon sheath (a small amount of fluid is normal), calcium deposits in the tendon sheath, and tenosynovitis. Power Doppler may help define acute tenosynovitis. Evaluate for partial or complete tears. If the tear is complete, there may be tendon retraction. The clinical finding of "Popeye's sign" occurs with a complete tendon tear; US may confirm a ball of muscle.

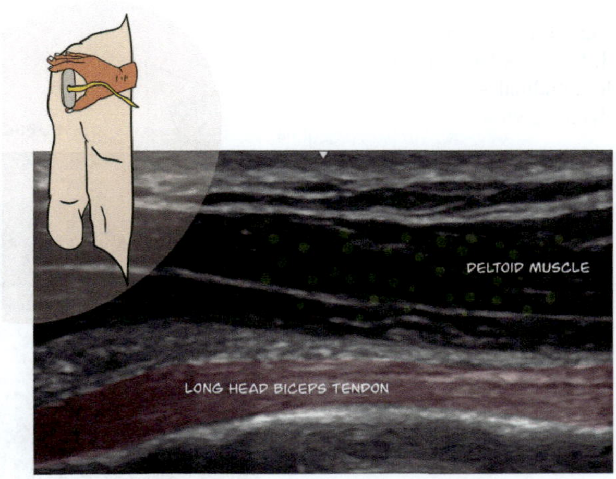

Fig. 10.10 Protocol Image 2: Long head biceps tendon, longitudinal ± power Doppler

Protocol Image 3: Subscapularis Tendon, Longitudinal + Dynamic View (Fig. 10.11)

For conceptual expediency, we break with convention and describe LAX and TAX views with respect to the SSCT itself. However, according to strict nomenclature, the views should be described with reference to the upper arm axis. Therefore, what we describe as our TAX view of the SSCT is the traditional LAX view, and vice versa.

To visualize the superior portion of the broad SSCT:

1. The patient keeps the elbow tucked in and then externally rotates the forearm to rest on a stable surface.
2. Move the probe back into a TAX position with respect to the upper arm, start at the biceps tendon groove, and perform an LAX slide medially (over the LT) to see the first bird's beak view, the SSCT insertion on the LT.
3. Medially, visualize the curved hyperechoic coracoid process.

Inspect the SSCT fibers for tears and the humeral head for bony irregularities. Look for subcoracoid bursal (SCB) fluid or thickening. The SCB is anterior to the SSC muscle and inferior to the coracoid process. It does not normally communicate with the GHJ, but may connect with the SASDB. If you do not see the curved, bony hyperechoic coracoid process, move the probe medially until it appears, since this verifies that you are looking at the superior portion of the SSCT.

Have the patient keep the elbow tucked into the body while you gently internally rotate the forearm and thereby the shoulder. Look for impingement of the SSCT, which normally glides smoothly under the coracoid process. Look for an enlarged SCB, which may be just superficial to the SSCT. If the SSCT does not slide under the coracoid and does not appear to impinge, this may indicate adhesive

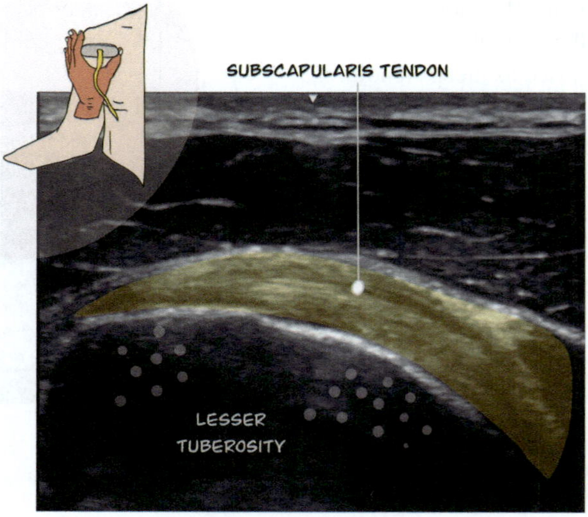

Fig. 10.11 Protocol Image 3: Subscapularis tendon, longitudinal + dynamic view

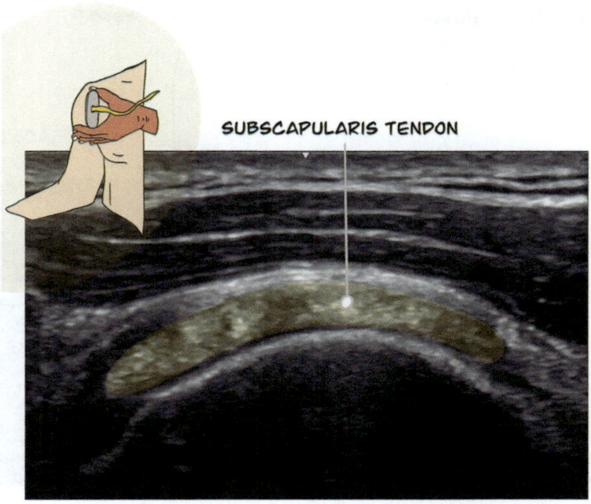

Fig. 10.12 Protocol Image 4: Subscapularis tendon, transverse

capsulitis. Only rarely is the SSCT truly impinged under the coracoid. Perform a slight distal TAX slide to see the inferior portion of the SSCT and examine those fibers for tears and loss of echotexture. The coracoid process disappears. A slight lateral LAX slide and toggling the probe will sharpen the insertion on the LT.

Protocol Image 4: Subscapularis Tendon, Transverse (Fig. 10.12)

Next, rotate the probe 90° to look at the SSCT in TAX (officially, LAX with respect to the humerus). This image reveals our first "wagon-wheel" view. Note that you see tendon (hyperechoic) alternating with muscle (hypoechoic). These are not tears! You often see three pure tendon sections of the SSCT in this view. Move the probe in a medial TAX slide to examine the proximal portion of the tendon. Then, move the transducer laterally to look at the more distal portion of the tendon and its insertion on the LT.

Protocol Image 5: Acromioclavicular Joint, Longitudinal ± Power Doppler (Fig. 10.13)

The patient then loosely hangs the arm downward. Place the probe parallel to and above the clavicle, over your ink marks delineating the ACJ in LAX. Look at the joint space and alignment. Look at the acromioclavicular ligament if it is visible. Ascertain if the joint capsule is abnormally tented or distended. Hyperechoic

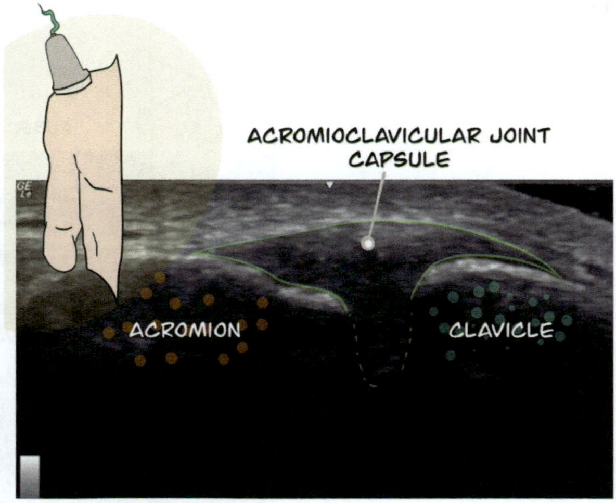

Fig. 10.13 Protocol Image 5: Acromioclavicular joint, longitudinal ± power Doppler

fibrocartilage may sometimes be seen within the joint space. Acromial or distal clavicle osteophytes may be noted superiorly. You may see synovitis and bone erosion here in cases of RA. Os acromiale may be seen on the superior acromion. It appears to be "a double ACJ." Use PD if there is joint capsule distension.

Perform a dynamic exam if there is a question of instability [8]. The patient places the hand of the extremity onto the contralateral shoulder to achieve cross-arm adduction, which may cause localized pain. Ultrasound may show significant joint narrowing. If there is a widening of the ACJ, compare it with the other, presumably unaffected side. Excessive widening may indicate an acromioclavicular ligament tear.

Protocol Image 6: Supraspinatus Tendon, Distal Acromion, Longitudinal + Dynamic View (Fig. 10.14)

With the probe in the same LAX position, perform a lateral LAX slide; the distal acromion is medial, and the bird's beak of the SST is lateral. The probe is at a 45° angle to the shoulder. Focus on the bird's beak and the bony cortex of the distal acromion. Each structure may be a bit out of focus. Inspect the location of the SASDB just superficial to the SST. Take the patient's forearm, bend it 90°, and gently grasp it near the elbow. The patient points the thumb downward [8]. Gradually, slowly, abduct the humerus to 90° as the forearm hangs downward. Watch the SST and SASDB glide under the acromion. You and the patient may need to practice this a bit since most patients with shoulder pain will guard and not initially permit the examiner to freely move the shoulder until it is clear that the passive movement is

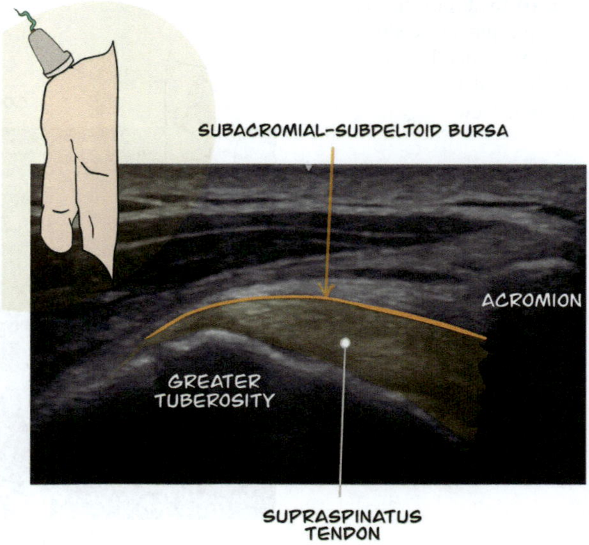

Fig. 10.14 Protocol Image 6: Supraspinatus tendon, distal acromion, longitudinal + dynamic view

slow and gentle. If there is impingement, note whether bursal fluid, the SASDB, or the tendon bunches up ("rug under the door sign") [115]. With impingement, the patient almost always feels pain under the probe. This positive sonopalpation sign is helpful if this maneuver reproduces the patient's pain since asymptomatic bunching is clinically irrelevant. Note that the GT may directly contact the acromion [4, 112]. If the SST does not slide under the acromion and does not impinge, this may indicate adhesive capsulitis [40].

Protocol Image 7: Supraspinatus Tendon, Coracoacromial Ligament, Longitudinal + Dynamic View (Fig. 10.15)

There may also be SST impingement under the CAL [4]. While focusing on the SST, perform a TAX slide inferiorly until the bony distal acromion disappears. The distal end of the transducer will need to be directed slightly downward to visualize the fibrillar SST. Passively abduct the arm as done for Protocol Image 6 and look for impingement under the CAL. You may see fluid within the SASDB fluid bunching up, accompanied by pain reproduction. The SASDB impingement under the CAL may occur with or without concomitant impingement under the distal acromion.

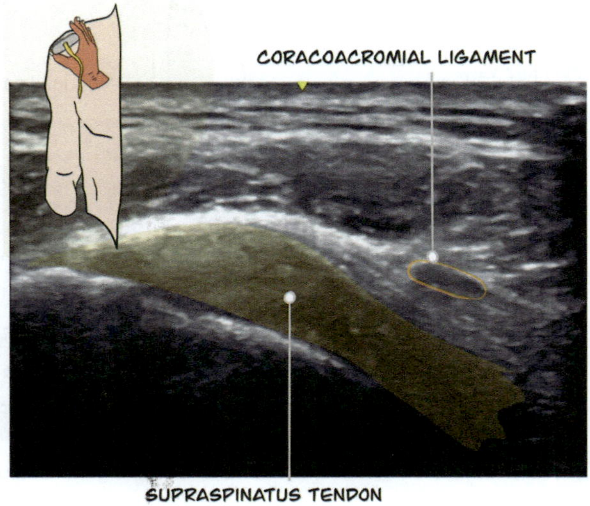

Fig. 10.15 Protocol Image 7: Supraspinatus tendon, coracoacromial ligament, longitudinal + dynamic view

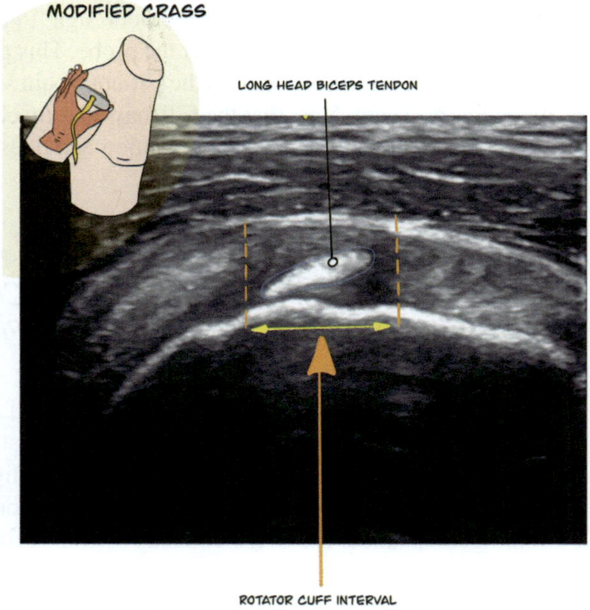

Fig. 10.16 Protocol Image 8: Rotator interval, modified Crass position

Protocol Image 8: Rotator Interval, Modified Crass Position (Fig. 10.16)

To visualize the RCI, the patient places the hand into the ipsilateral back pocket (modified Crass position) to bring the SST out from under the bony acromion. The elbow should be tucked into the body (adducted) and pointing backward. Place the probe in TAX to the humerus anteriorly over the ink marks of the LHBT and aim the medial end of the transducer toward the ipsilateral nipple [3]. Visualize the oval cross-section of the LHBT (intra-articular portion) again, along with the SSCT medially and the SST laterally. This view best defines the intra-articular portion of the RCI. A slight distal TAX slide reveals the LBHT with a more rounded appearance; this is the extra-articular portion of the RCI.

Protocol Image 9a: Supraspinatus Tendon, Distal Anterior, Longitudinal ± Power Doppler (Fig. 10.17)

With the patient still in the modified Crass position, center the probe over the LHBT and rotate the proximal end of the probe toward the distal end of the clavicle while keeping the center of the probe and your focus on the LHBT. Watch the LHBT elongate and visualize as much of the LHBT in LAX as possible. Next, move the probe in a lateral TAX slide to see the large bird's-beak of the SST insertion on the GT.

The anterior SST is a common area for tendon tears [112]. Carefully examine the distal anterior portion of the SST at its insertion on the SF of the GT. The hyperechoic bony SF is concave, with the distal portion of the concavity being somewhat

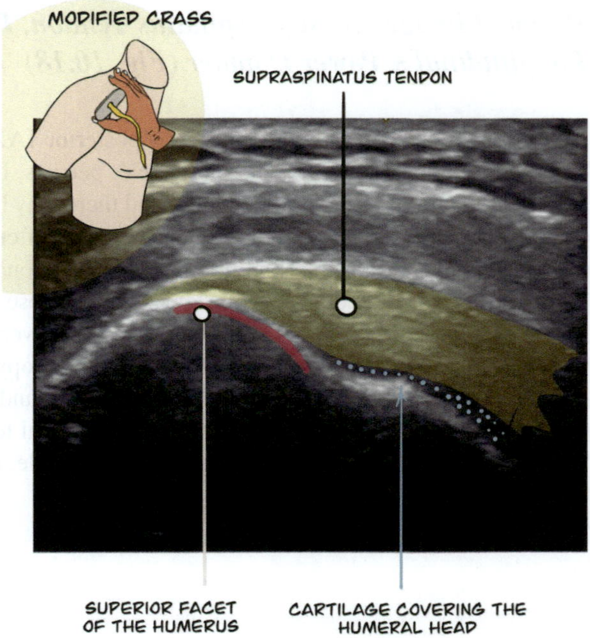

Fig. 10.17 Protocol Image 9a: Supraspinatus tendon, distal anterior, longitudinal ± power Doppler

elongated and a bit flattened. Look at the GT for irregularities in the bone since this may heighten suspicion of an SST tear. When looking at the SST in LAX, the goal is to sharpen and "fill in the blanks" of the fibrillar SST by alternatively performing heel-to-toe and tilting probe maneuvers.

At the insertion of the anterior portion of the SST on the SF, there is a very thin hypoechoic line due to the SST insertional fibers' downward curve; this causes anisotropy. If you see a *large* anechoic area, then heel-to-toe the probe downward to verify an SST footprint tear. A footprint tear is an *intrasubstance tear* unless it extends to the bursal or articular sides. Note the loss of fibrillar echotexture from tendinosis or tears. Fluid may fill a tear, making it more conspicuous. Look for a CIS if there appears to be an articular-sided tear [112]. Measure the length and location of any tear and observe for any calcium deposition. Look at the distal portion of the SASDB, which may contain effusion or be thickened, perhaps indicating synovitis. Look for any flattening or concavity of the substance of the normally convex SST, which may indicate SST volume loss from a tear. Sometimes the bursa, peribursal fat, or overlying deltoid muscle dips down into a torn area. The more medial hyperechoic bone is the HH, covered by anechoic hyaline cartilage. Turn on the PD to look for bursal synovitis or acute SST tears.

Avoid mistaking a horizontal hypoechoic band in the anterior SST for an intrasubstance tear. This line separates the superficial from the deeper, flatter fibers of the SST [17].

Protocol Image 9b: Supraspinatus Tendon, Distal Posterior, Longitudinal ± Power Doppler (Fig. 10.18)

Next, remain in LAX and perform a slow posterior TAX slide to look at the middle facet and the posterior SST. The bird's beak persists, but is smaller and sharper at the end. The middle facet will be flatter, and there may be some hypoechoic areas of the SST near the insertion on the GT. These areas, often mistaken for small tears or tendinosis, are normal interdigitation of the insertional fibers of the IST coming from a different direction, causing anisotropy. These thin hypoechoic lines may resemble zebra stripes [112]. Perform probe maneuvers (heel-to-toe and tilting) to "fill in the blanks" to avoid misinterpreting anisotropy for tendinosis or a tendon tear. Be aware that the SST is about 2.5 cm wide, and you need to scan its entire girth [112]. Again, the first 1.5 cm of the SST lateral to the biceps groove is "pure SST," but beyond that, the IST fibers may intermingle, causing anisotropy, simulating tendinosis, or tears [116].

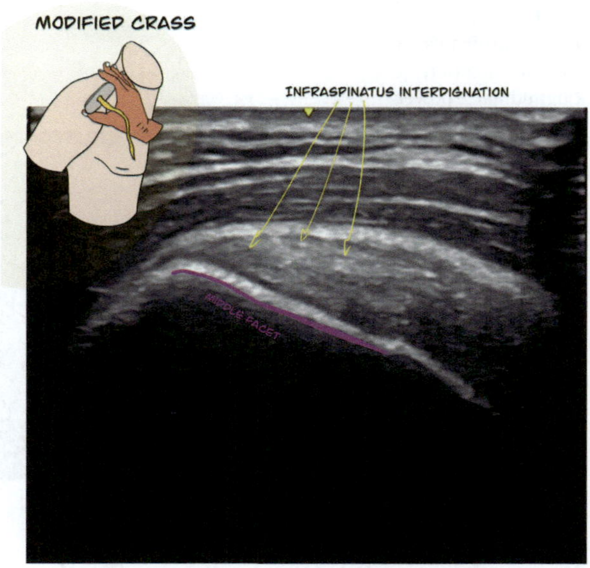

Fig. 10.18 Protocol Image 9b: Supraspinatus tendon, distal posterior, longitudinal ± power Doppler

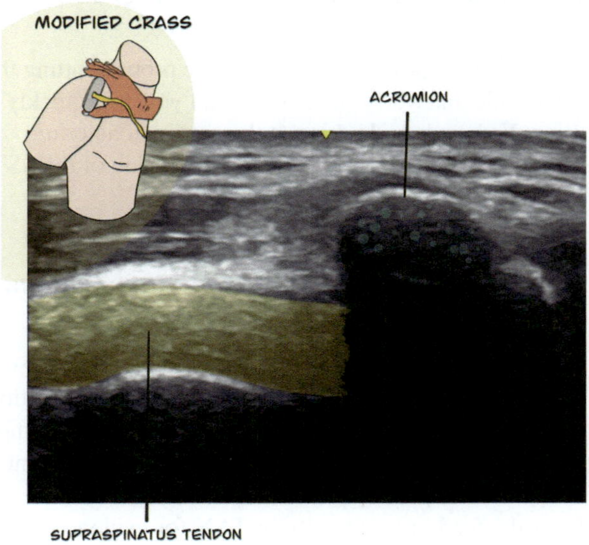

Fig. 10.19 Protocol Image 9c: Supraspinatus tendon, proximal posterior, longitudinal ± power Doppler

Protocol Image 9c: Supraspinatus Tendon, Proximal Posterior, Longitudinal ± Power Doppler (Fig. 10.19)

Next, move the probe proximally in an LAX slide until the SST fibers disappear beneath the bony distal acromion. Use the same probe maneuvers to "fill in the blanks" as before, and turn on the PD if necessary.

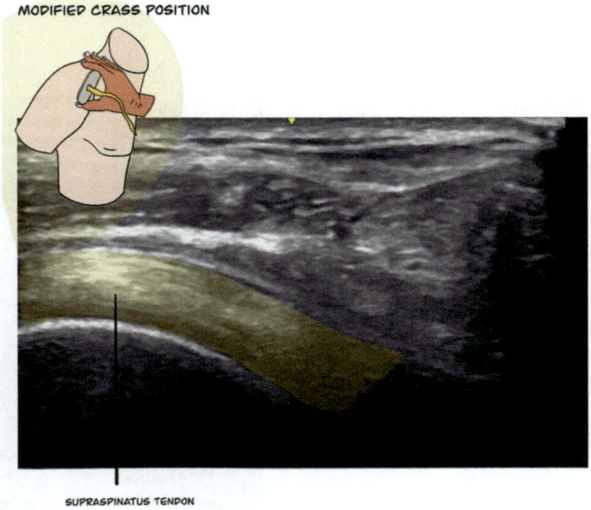

Fig. 10.20 Protocol Image 9d: Supraspinatus tendon, proximal anterior, longitudinal ± power Doppler

Protocol Image 9d: Supraspinatus Tendon, Proximal Anterior, Longitudinal ± Power Doppler (Fig. 10.20)

With the most proximal portion of the probe abutting the bony acromion, perform a TAX slide anteriorly (medially), and you will quickly lose sight of the bony acromion. You are now looking at the proximal anterior portion of the SST, which runs deep to the CAL. Carefully scan this portion of the SST.

Protocol Image 10a: Supraspinatus Tendon, Anterior, Transverse (Fig. 10.21)

Next, rotate the probe into TAX and place it in a position to once again visualize the RCI (Protocol Image 8). The patient remains in the modified Crass position. Move the transducer with a slight lateral LAX slide to center it on the SF of the GT to visualize the anterior portion of the SST; this is the second "wagon-wheel" view. Scan the tendon proximally in TAX for evidence of tears, splitting, or tendinosis. The average SST is about 6 mm in thickness. You may see the RC, which is a hyperechoic curved line running along the deepest portion of the SST. It may be mistaken for a calcium deposit. Also, remember that the superficial SST fibers are oriented differently from the deeper, flatter fibers and may normally appear as a hypoechoic ovoid defect due to anisotropy; don't mistake this area for a tear or tendinosis [17].

Check the SASDB for fluid, thickening, and peribursal fat indentation/flattening/bulging. Flattening or concavity of the usually convex SST itself may indicate

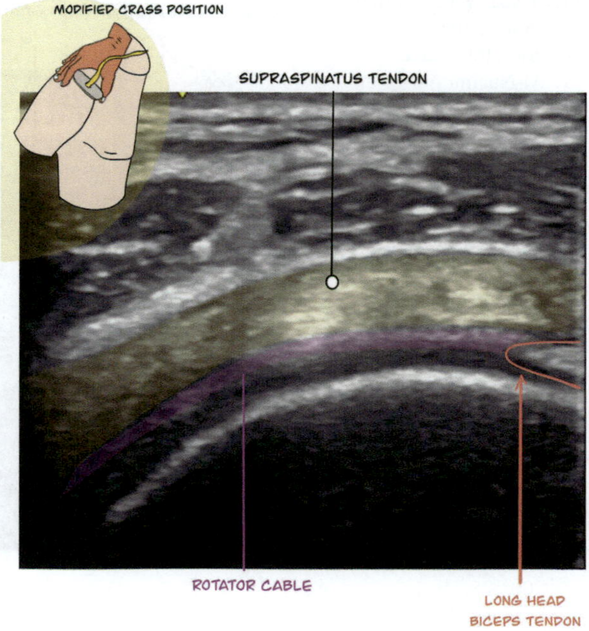

Fig. 10.21 Protocol Image 10a: Supraspinatus tendon, anterior, transverse, modified Crass position

volume loss, a good indicator of a possible SST tear. The SASDB does not usually communicate with the GHJ, but may do so after a full-thickness SST tear. Scan the HH and the covering hyaline cartilage for irregularities; note if there are any calcium deposits within the cartilage. The SST should be scanned proximally and then distally to the insertion on the GT. You have reached the insertion on the GT when tendon fibers drop off, replaced with bony cortex.

Protocol Image 10b: Supraspinatus Tendon, Posterior, Transverse (Fig. 10.22)

Refocus on the SF and perform a slight posterior LAX slide to visualize the flatter MF. Then start a slow TAX slide proximally to scan the entire posterior portion of the SST. The posterior SST fibers will disappear once you have reached the hyperechoic acromion. Remember that the posterior part of the SST inserts on the MF and is interspersed with fibers from the IST. The IST fibers are oriented differently from the SST, causing anisotropy that mimics a tear or tendinosis. As a general rule, starting at the intra-articular LHBT and moving in a posterolateral direction, the first 1.5 cm of the SST in TAX is considered "pure SST" without interdigitating IST fibers [116]. Again, if you recognize the flatter MF, you are looking at the SST, the IST, or an overlap area of both.

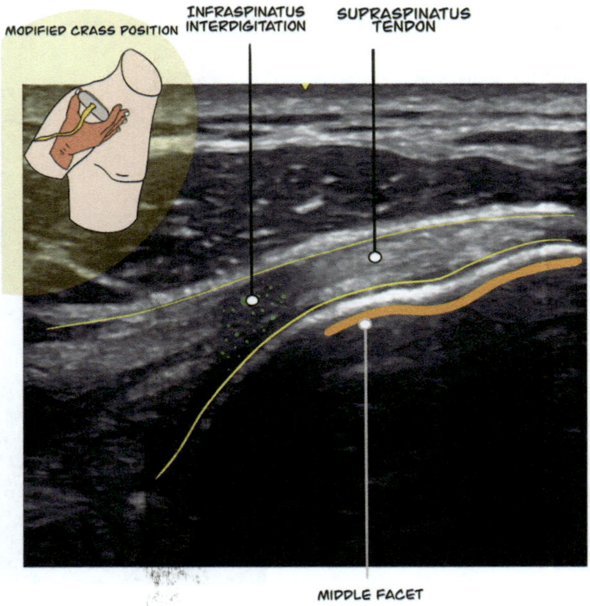

Fig. 10.22 Protocol Image 10b: Supraspinatus tendon, posterior, transverse

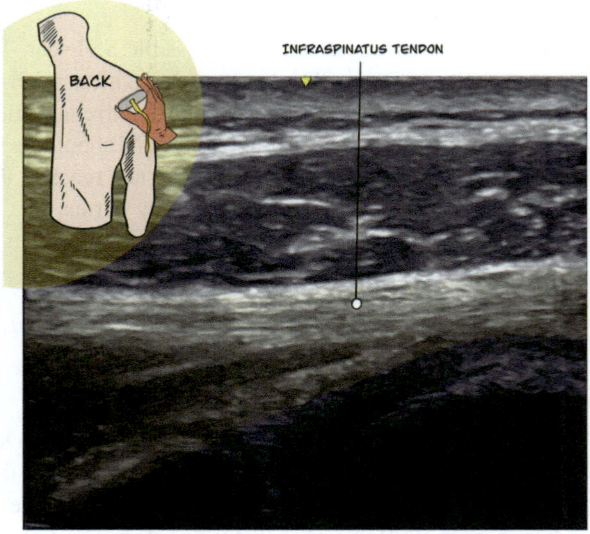

Fig. 10.23 Protocol Image 11a: Infraspinatus tendon and muscle, proximal, longitudinal

Protocol Image 11a: Infraspinatus Tendon and Muscle, Proximal, Longitudinal (Fig. 10.23)

The patient places the hand on the thigh in a neutral position with an upward palm as positioned for Protocol Image 1. Place the probe on the posterior shoulder about a 2-3 cm caudal to the posterior acromion. The medial aspect of the transducer is rotated downward, resulting in a 35-45° angle. You will need more depth to look at the posterior shoulder. Try to visualize the curved HH with a linear probe at its lowest frequency, but switch to a curvilinear transducer if necessary. For a moment, focus on the curved hyperechoic surfaces of the HH and the glenoid. Then concentrate more superficially on the IST in a longitudinal view. Inspect the IST fibrillar echotexture for tears, tendinosis, and calcium deposits. Perform a medial LAX slide to visualize the proximal portion of the IST, the myotendinous junction, and the ISM.

Protocol Image 11b: Infraspinatus Tendon, Distal, Longitudinal (Fig. 10.24)

With the transducer still at a 35–45° angle, perform a superolateral LAX slide to look at the IST insertion on the middle facet (posterior aspect) of the GT. Ask the patient to keep the elbow tucked in and externally rotate the forearm about 30° to visualize a bird's beak insertion. Again, look carefully for tendon tears, tendinosis, and calcium deposits. Slow internal and external rotation of the shoulder reveals insertional integrity.

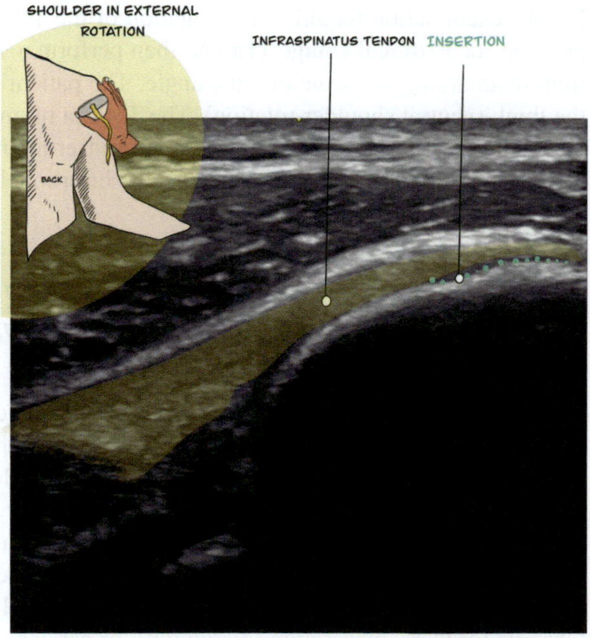

Fig. 10.24 Protocol Image 11b: Infraspinatus tendon, distal, longitudinal

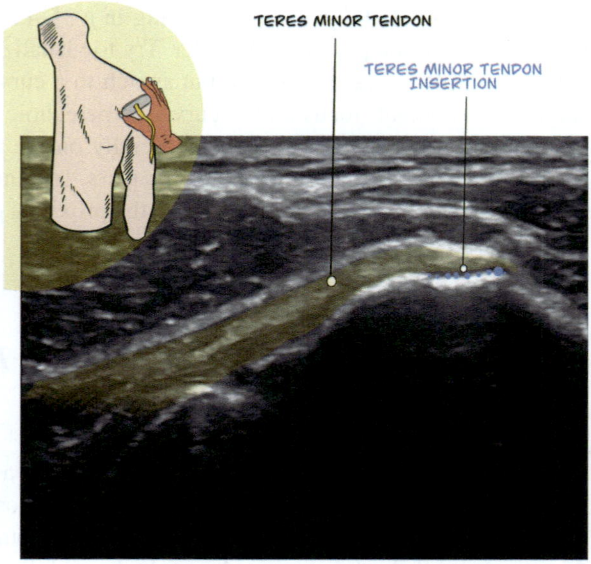

Fig. 10.25 Protocol Image 12: Teres minor tendon, longitudinal

Protocol Image 12: Teres Minor Tendon, Longitudinal (Fig. 10.25)

If your examination requires an evaluation of the TMT, go back to the transducer position for **Protocol Image 11a** and then perform a TAX slide in a caudal direction, maintaining the same oblique angle. The patient's hand should be resting on the thigh (neutral shoulder rotation). The TMT, a bit more superficial than the IST, inserts into the IF of the GT. The bird's beak insertion is small. Passive internal and external shoulder rotation will check for the insertional integrity of the TMT.

Protocol Image 13: Posterior Glenohumeral Joint, Longitudinal, Internal + External Rotation ± Power Doppler (Fig. 10.26)

The patient places the hand on the contralateral shoulder to produce internal rotation. Reestablish Image 11a, the IST in LAX. Rotate the transducer to a horizontal position. Perform small cephalad and caudal TAX slides until you visualize the hyperechoic curved cortex of the HH and the smaller hyperechoic bony cortex of the glenoid. Between these two bones lies the triangular, hyperechoic posterior glenoid labrum. Clarify the labrum borders by toggling the probe. The superficial edge of the labrum is the joint capsule, which extends laterally over the humeral cartilage. Observe for tears, maceration (disruption), or paralabral cysts. Use the highest frequency that will still enable you to visualize the labrum. The time-gain control may

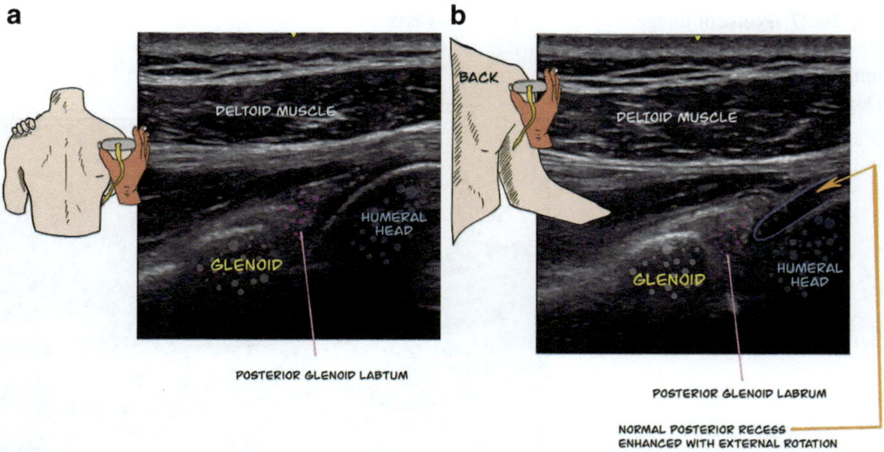

Fig. 10.26 (**a**, **b**) Protocol Image 13: Posterior glenohumeral joint, longitudinal, internal + external rotation ± power Doppler

be useful to elucidate posterior labrum echotexture. Report if you do not see the labrum; it may not be well visualized with a larger body habitus. If there is a tear, use the PD to determine if it is acute.

Next, perform a dynamic scan to evaluate for an effusion in the posterior recess of the GHJ. With the elbow tucked into the chest, the patient slowly externally rotates the forearm to see if an effusion distends the posterior recess joint capsule. Anechoic fluid will be seen superficially and anteriorly to the labrum in external rotation, but a large effusion will distend the posterior recess even in a neutral position. An effusion height (measured from the HH to the IST) >3 mm is abnormal [117]. External rotation may enhance posterior glenoid labrum visualization as well.

Protocol Image 14: Spinoglenoid Notch, Longitudinal + Power Doppler (Fig. 10.27)

Perform a medial LAX slide to assess the hyperechoic SGN. The SGN connects the supraspinatus and infraspinatus fossae [8]. The "U-shaped" notch is deep to the IST myotendinous junction. Slightly rotate the transducer to delineate the SGN sharply. The suprascapular artery and nerve are occasionally seen in this notch [4]. Power Doppler may assist in locating the artery; however, the nerve is challenging to see. A paralabral cyst in the notch implies a glenoid labrum tear. The cyst can entrap the suprascapular nerve [4]. An enlarged suprascapular vein may simulate a paralabral cyst in external rotation, so turn on the CF to delineate this. Do a dynamic maneuver; if this is varicosity, it should collapse when the shoulder is internally rotated [4].

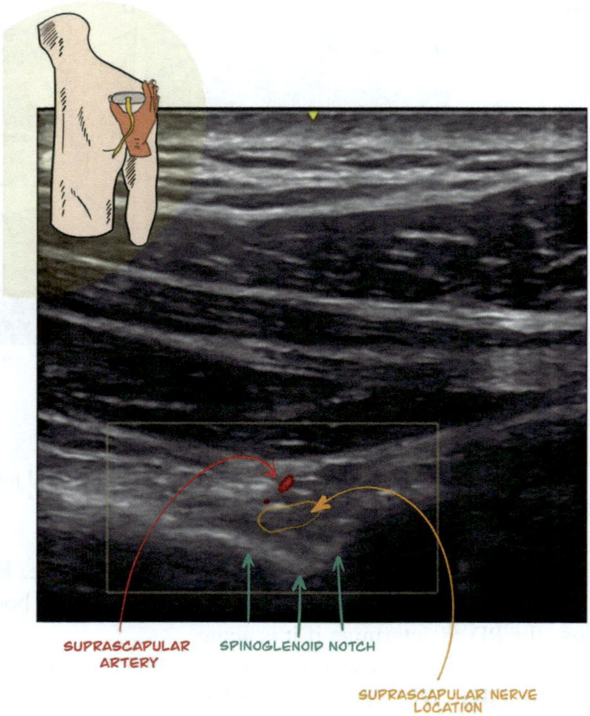

Fig. 10.27 Protocol Image 14: Spinoglenoid notch, longitudinal + power Doppler

Protocol Image 15 (Optional): Infraspinatus and Teres Minor Muscles, TAX (Fig. 10.28)

The patient's hand should be resting on the thigh (neutral shoulder rotation). Rotate the transducer so that it is parallel to the spine. Perform a medial TAX slide to see the bony cortex of the scapula edge. Then perform a slow caudal LAX slide to look for a bony protuberance (the posterior scapula ridge). At this protuberance, the ISM and TMM are adjacent and easily delineated, separated by a diagonal fascial plane [3]. When assessing for atrophy of the ISM, compare its thickness at the posterior scapula ridge to that of the TMM. The normal ISM should be about twice the size of the TMM [3]. A hyperechogenic ISM (compared with the echogenicity of the TMM) may also indicate atrophy.

Fig. 10.28 Protocol Image 15: Infraspinatus and teres minor muscles, TAX

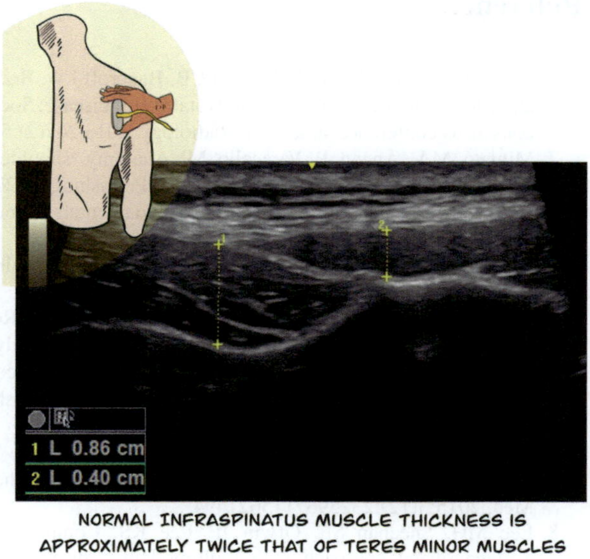

Complete Shoulder Ultrasonic Examination Checklist

☐ Protocol Image 1: Long head biceps tendon, transverse ± power Doppler.
☐ Protocol Image 2: Long head biceps tendon, longitudinal ± power Doppler.
☐ Protocol Image 3: Subscapularis tendon, longitudinal + dynamic view.
☐ Protocol Image 4: Subscapularis tendon, transverse.
☐ Protocol Image 5: Acromioclavicular joint, longitudinal ± power Doppler.
☐ Protocol Image 6: Supraspinatus tendon, distal acromion, longitudinal + dynamic view.
☐ Protocol Image 7: Supraspinatus tendon, coracoacromial ligament, longitudinal + dynamic view.
☐ Protocol Image 8: Rotator interval, modified Crass position.
☐ Protocol Image 9a: Supraspinatus tendon, distal anterior, longitudinal ± power Doppler.
☐ Protocol Image 9b: Supraspinatus tendon, distal posterior, longitudinal ± power Doppler.
☐ Protocol Image 9c: Supraspinatus tendon, proximal posterior, longitudinal ± power Doppler.
☐ Protocol Image 9d: Supraspinatus tendon, proximal anterior, longitudinal ± power Doppler.
☐ Protocol Image 10a: Supraspinatus tendon, anterior, transverse.
☐ Protocol Image 10b: Supraspinatus tendon, posterior, transverse.
☐ Protocol Image 11a: Infraspinatus tendon and muscle, proximal, longitudinal.
☐ Protocol Image 11b: Infraspinatus tendon, distal, longitudinal.
☐ Protocol Image 12: Teres minor tendon, longitudinal.
☐ Protocol Image 13: Posterior glenohumeral joint, longitudinal, internal + external rotation ± power Doppler.
☐ Protocol Image 14: Spinoglenoid notch, longitudinal + power Doppler.

References

1. Nazarian LN, Jacobson JA, Benson CB, Bancroft LW, Bedi A, McShane JM, et al. Imaging algorithms for evaluating suspected rotator cuff disease: Society of Radiologists in Ultrasound consensus conference statement. Radiology. 2013;267(2):589–95.
2. Miniato MA, Anand P, Varacallo M. Anatomy, shoulder and upper limb, shoulder. In: StatPearls. Treasure Island, FL: StatPearls Publishing; 2022.
3. Jacobson JA. Shoulder ultrasound. In: Fundamentals of musculoskeletal ultrasound. 3rd ed. Elsevier; 2018.
4. Park J, Chai JW, Kim DH, Cha SW. Dynamic ultrasonography of the shoulder. Ultrasonography. 2018;37(3):190–9.
5. Podgórski MT, Olewnik Ł, Grzelak P, Polguj M, Topol M. Rotator cable in pathological shoulders: comparison with normal anatomy in a cadaveric study. Anat Sci Int. 2019;94(1):53–7.
6. Morag Y, Jamadar DA, Boon TA, Bedi A, Caoili EM, Jacobson JA. Ultrasound of the rotator cable: prevalence and morphology in asymptomatic shoulders. AJR Am J Roentgenol. 2012;198(1):W27–30.
7. Frank RM, Taylor D, Verma NN, Romeo AA, Mologne TS, Provencher MT. The rotator interval of the shoulder: implications in the treatment of shoulder instability. Orthop J Sports Med. 2015;3(12):2325967115621494.
8. Lee MH, Sheehan SE, Orwin JF, Lee KS. Comprehensive shoulder US examination: a standardized approach with multimodality correlation for common shoulder disease. Radiographics. 2016;36(6):1606–27.
9. Rutten MJ, Jager GJ, Blickman JG. From the RSNA refresher courses: US of the rotator cuff: pitfalls, limitations, and artifacts. Radiographics. 2006;26(2):589–604.
10. Jamadar DA, Jacobson JA, Caoili EM, Boon TA, Dong Q, Morag Y, et al. Musculoskeletal sonography technique: focused versus comprehensive evaluation. AJR Am J Roentgenol. 2008;190(1):5–9.
11. Small KM, Adler RS, Shah SH, Roberts CC, Bencardino JT, Appel M, et al. ACR Appropriateness Criteria(®) shoulder pain-atraumatic. J Am Coll Radiol. 2018;15(11s):S388–s402.
12. Michener LA, McClure PW, Karduna AR. Anatomical and biomechanical mechanisms of subacromial impingement syndrome. Clin Biomech (Bristol, Avon). 2003;18(5):369–79.
13. Page P. Shoulder muscle imbalance and subacromial impingement syndrome in overhead athletes. Int J Sports Phys Ther. 2011;6(1):51–8.
14. Neer CS 2nd. Impingement lesions. Clin Orthop Relat Res. 1983;173:70–7.
15. Farin PU, Jaroma H, Harju A, Soimakallio S. Shoulder impingement syndrome: sonographic evaluation. Radiology. 1990;176(3):845–9.
16. Manske RC, Grant-Nierman M, Lucas B. Shoulder posterior internal impingement in the overhead athlete. Int J Sports Phys Ther. 2013;8(2):194–204.
17. Bianchi S, Martinoli C. Shoulder. In: Ultrasound of the musculoskeletal system. Springer; 2007.
18. Morag Y, Jacobson JA, Miller B, De Maeseneer M, Girish G, Jamadar D. MR imaging of rotator cuff injury: what the clinician needs to know. Radiographics. 2006;26(4):1045–65.
19. Vinson EN, Helms CA, Higgins LD. Rim-rent tear of the rotator cuff: a common and easily overlooked partial tear. AJR Am J Roentgenol. 2007;189(4):943–6.
20. Minagawa H, Yamamoto N, Abe H, Fukuda M, Seki N, Kikuchi K, et al. Prevalence of symptomatic and asymptomatic rotator cuff tears in the general population: From mass-screening in one village. J Orthop. 2013;10(1):8–12.
21. de Jesus JO, Parker L, Frangos AJ, Nazarian LN. Accuracy of MRI, MR arthrography, and ultrasound in the diagnosis of rotator cuff tears: a meta-analysis. AJR Am J Roentgenol. 2009;192(6):1701–7.
22. Smith CD, Corner T, Morgan D, Drew S. Partial thickness rotator cuff tears: what do we know? Shoulder Elb. 2010;2(2):77–82.

23. Teefey SA, Middleton WD, Bauer GS, Hildebolt CF, Yamaguchi K. Sonographic differences in the appearance of acute and chronic full-thickness rotator cuff tears. J Ultrasound Med. 2000;19(6):377–8; quiz 83.
24. Wiener SN, Seitz WH Jr. Sonography of the shoulder in patients with tears of the rotator cuff: accuracy and value for selecting surgical options. AJR Am J Roentgenol. 1993;160(1):103–7; discussion 9–10.
25. Gaitini D. Shoulder ultrasonography: performance and common findings. J Clin Imaging Sci. 2012;2:38.
26. Jacobson JA, Lancaster S, Prasad A, van Holsbeeck MT, Craig JG, Kolowich P. Full-thickness and partial-thickness supraspinatus tendon tears: value of US signs in diagnosis. Radiology. 2004;230(1):234–42.
27. van Holsbeeck M, Strouse PJ. Sonography of the shoulder: evaluation of the subacromial-subdeltoid bursa. AJR Am J Roentgenol. 1993;160(3):561–4.
28. Arslan G, Apaydin A, Kabaalioglu A, Sindel T, Lüleci E. Sonographically detected subacromial/subdeltoid bursal effusion and biceps tendon sheath fluid: reliable signs of rotator cuff tear? J Clin Ultrasound. 1999;27(6):335–9.
29. Matthewson G, Beach CJ, Nelson AA, Woodmass JM, Ono Y, Boorman RS, et al. Partial thickness rotator cuff tears: current concepts. Adv Orthop. 2015;2015:458786.
30. Matsen FA 3rd. Clinical practice. Rotator-cuff failure. N Engl J Med. 2008;358(20):2138–47.
31. Norwood LA, Barrack R, Jacobson KE. Clinical presentation of complete tears of the rotator cuff. J Bone Joint Surg Am. 1989;71(4):499–505.
32. Tempelhof S, Rupp S, Seil R. Age-related prevalence of rotator cuff tears in asymptomatic shoulders. J Shoulder Elb Surg. 1999;8(4):296–9.
33. Yamamoto A, Takagishi K, Osawa T, Yanagawa T, Nakajima D, Shitara H, et al. Prevalence and risk factors of a rotator cuff tear in the general population. J Shoulder Elb Surg. 2010;19(1):116–20.
34. Eliasson P, Dietrich-Zagonel F, Lundin AC, Aspenberg P, Wolk A, Michaëlsson K. Statin treatment increases the clinical risk of tendinopathy through matrix metalloproteinase release—a cohort study design combined with an experimental study. Sci Rep. 2019;9(1):17958.
35. Gopikrishnan K, Mariadoss A, Sambandam SN, Sakthivel A, Pratheep NS, Bakthavasthsalam GB. Quinolone induced supraspinatus tendinopathy. Scholars Acad J Biosci. 2017;5(1):32–5.
36. Zubler V, Mamisch-Saupe N, Pfirrmann CW, Jost B, Zanetti M. Detection and quantification of glenohumeral joint effusion: reliability of ultrasound. Eur Radiol. 2011;21(9):1858–64.
37. Suzuki T, Yoshida R, Hidaka Y, Seri Y. Proliferative synovitis of the shoulder bursae is a key feature for discriminating elderly onset rheumatoid arthritis mimicking polymyalgia rheumatica from polymyalgia rheumatica. Clin Med Insights Arthritis Musculoskelet Disord. 2017;10:1179544117745851.
38. Martetschläger F, Rios D, Boykin RE, Giphart JE, de Waha A, Millett PJ. Coracoid impingement: current concepts. Knee Surg Sports Traumatol Arthrosc. 2012;20(11):2148–55.
39. Drakes S, Thomas S, Kim S, Guerrero L, Lee SW. Ultrasonography of subcoracoid bursal impingement syndrome. PM R. 2015;7(3):329–33.
40. Tandon A, Dewan S, Bhatt S, Jain AK, Kumari R. Sonography in diagnosis of adhesive capsulitis of the shoulder: a case-control study. J Ultrasound. 2017;20(3):227–36.
41. Lee JC, Sykes C, Saifuddin A, Connell D. Adhesive capsulitis: sonographic changes in the rotator cuff interval with arthroscopic correlation. Skelet Radiol. 2005;34(9):522–7.
42. Strobel K, Hodler J, Meyer DC, Pfirrmann CW, Pirkl C, Zanetti M. Fatty atrophy of supraspinatus and infraspinatus muscles: accuracy of US. Radiology. 2005;237(2):584–9.
43. Wall LB, Teefey SA, Middleton WD, Dahiya N, Steger-May K, Kim HM, et al. Diagnostic performance and reliability of ultrasonography for fatty degeneration of the rotator cuff muscles. J Bone Joint Surg Am. 2012;94(12):e83.
44. Vad VB, Warren RF, Altchek DW, O'Brien SJ, Rose HA, Wickiewicz TL. Negative prognostic factors in managing massive rotator cuff tears. Clin J Sport Med. 2002;12(3):151–7.

45. Goutallier D, Postel JM, Gleyze P, Leguilloux P, Van Driessche S. Influence of cuff muscle fatty degeneration on anatomic and functional outcomes after simple suture of full-thickness tears. J Shoulder Elb Surg. 2003;12(6):550–4.
46. Oh JH, Kim SH, Ji HM, Jo KH, Bin SW, Gong HS. Prognostic factors affecting anatomic outcome of rotator cuff repair and correlation with functional outcome. Arthroscopy. 2009;25(1):30–9.
47. Mohana-Borges AV, Chung CB, Resnick D. MR imaging and MR arthrography of the postoperative shoulder: spectrum of normal and abnormal findings. Radiographics. 2004;24(1):69–85.
48. Gowd AK, Liu JN, Garcia GH, Cabarcas BC, Verma NN. Arthroscopic massive rotator cuff repair and techniques for mobilization. Arthrosc Tech. 2018;7(6):e633–8.
49. Jacobson JA, Miller B, Bedi A, Morag Y. Imaging of the postoperative shoulder. Semin Musculoskelet Radiol. 2011;15(4):320–39.
50. Prickett WD, Teefey SA, Galatz LM, Calfee RP, Middleton WD, Yamaguchi K. Accuracy of ultrasound imaging of the rotator cuff in shoulders that are painful postoperatively. J Bone Joint Surg Am. 2003;85(6):1084–9.
51. Russell RD, Knight JR, Mulligan E, Khazzam MS. Structural integrity after rotator cuff repair does not correlate with patient function and pain: a meta-analysis. J Bone Joint Surg Am. 2014;96(4):265–71.
52. Ha AS, Petscavage JM, Chew FS. Current concepts of shoulder arthroplasty for radiologists: part 2—anatomic and reverse total shoulder replacement and nonprosthetic resurfacing. AJR Am J Roentgenol. 2012;199(4):768–76.
53. Greis AC, Derrington SM, McAuliffe M. Evaluation and nonsurgical management of rotator cuff calcific tendinopathy. Orthop Clin North Am. 2015;46(2):293–302.
54. Uhthoff HK, Sarkar K, Maynard JA. Calcifying tendinitis: a new concept of its pathogenesis. Clin Orthop Relat Res. 1976;118:164–8.
55. de Witte PB, Selten JW, Navas A, Nagels J, Visser CP, Nelissen RG, et al. Calcific tendinitis of the rotator cuff: a randomized controlled trial of ultrasound-guided needling and lavage versus subacromial corticosteroids. Am J Sports Med. 2013;41(7):1665–73.
56. Bazzocchi A, Pelotti P, Serraino S, Battaglia M, Bettelli G, Fusaro I, et al. Ultrasound imaging-guided percutaneous treatment of rotator cuff calcific tendinitis: success in short-term outcome. Br J Radiol. 2016;89(1057):20150407.
57. Serafini G, Sconfienza LM, Lacelli F, Silvestri E, Aliprandi A, Sardanelli F. Rotator cuff calcific tendonitis: short-term and 10-year outcomes after two-needle us-guided percutaneous treatment—nonrandomized controlled trial. Radiology. 2009;252(1):157–64.
58. Della Valle V, Bassi EM, Calliada F. Migration of calcium deposits into subacromial-subdeltoid bursa and into humeral head as a rare complication of calcifying tendinitis: sonography and imaging. J Ultrasound. 2015;18(3):259–63.
59. Bianchi S, Martinoli C. Detection of loose bodies in joints. Radiol Clin N Am. 1999;37(4):679–90.
60. Vangsness CT Jr, Jorgenson SS, Watson T, Johnson DL. The origin of the long head of the biceps from the scapula and glenoid labrum. An anatomical study of 100 shoulders. J Bone Joint Surg Br. 1994;76(6):951–4.
61. Longo UG, Loppini M, Marineo G, Khan WS, Maffulli N, Denaro V. Tendinopathy of the tendon of the long head of the biceps. Sports Med Arthrosc Rev. 2011;19(4):321–32.
62. Knobloch K. Drug-induced tendon disorders. Adv Exp Med Biol. 2016;920:229–38.
63. Calcei JG, Boddapati V, Altchek DW, Camp CL, Dines JS. Diagnosis and treatment of injuries to the biceps and superior labral complex in overhead athletes. Curr Rev Musculoskelet Med. 2018;11(1):63–71.
64. Carter AN, Erickson SM. Proximal biceps tendon rupture: primarily an injury of middle age. Phys Sportsmed. 1999;27(6):95–101.
65. Armstrong A, Teefey SA, Wu T, Clark AM, Middleton WD, Yamaguchi K, et al. The efficacy of ultrasound in the diagnosis of long head of the biceps tendon pathology. J Shoulder Elb Surg. 2006;15(1):7–11.

66. Skendzel JG, Jacobson JA, Carpenter JE, Miller BS. Long head of biceps brachii tendon evaluation: accuracy of preoperative ultrasound. AJR Am J Roentgenol. 2011;197(4):942–8.
67. Gheno R, Zoner CS, Buck FM, Nico MA, Haghighi P, Trudell DJ, et al. Accessory head of biceps brachii muscle: anatomy, histology, and MRI in cadavers. AJR Am J Roentgenol. 2010;194(1):W80–3.
68. Brasseur JL. The biceps tendons: from the top and from the bottom. J Ultrasound. 2012;15(1):29–38.
69. Bennett WF. Subscapularis, medial, and lateral head coracohumeral ligament insertion anatomy. Arthroscopic appearance and incidence of "hidden" rotator interval lesions. Arthroscopy. 2001;17(2):173–80.
70. Boileau P, Ahrens PM, Hatzidakis AM. Entrapment of the long head of the biceps tendon: the hourglass biceps—a cause of pain and locking of the shoulder. J Shoulder Elb Surg. 2004;13(3):249–57.
71. Pujol N, Hargunani R, Gadikoppula S, Holloway B, Ahrens PM. Dynamic ultrasound assessment in the diagnosis of intra-articular entrapment of the biceps tendon (hourglass biceps): a preliminary investigation. Int J Shoulder Surg. 2009;3(4):80–4.
72. Nakata W, Katou S, Fujita A, Nakata M, Lefor AT, Sugimoto H. Biceps pulley: normal anatomy and associated lesions at MR arthrography. Radiographics. 2011;31(3):791–810.
73. Bureau NJ, Dussault RG, Keats TE. Imaging of bursae around the shoulder joint. Skelet Radiol. 1996;25(6):513–7.
74. Ruta S, Rosa J, Navarta DA, Saucedo C, Catoggio LJ, Monaco RG, et al. Ultrasound assessment of new onset bilateral painful shoulder in patients with polymyalgia rheumatica and rheumatoid arthritis. Clin Rheumatol. 2012;31(9):1383–7.
75. Mall NA, Foley E, Chalmers PN, Cole BJ, Romeo AA, Bach BR Jr. Degenerative joint disease of the acromioclavicular joint: a review. Am J Sports Med. 2013;41(11):2684–92.
76. Sammarco VJ. Os acromiale: frequency, anatomy, and clinical implications. J Bone Joint Surg Am. 2000;82(3):394–400.
77. Stieler MA. The use of sonography in the detection of bony and calcific disorders of the shoulder. J Diagn Med Sonogr. 2001;17(6):331–8.
78. Kurtz CA, Humble BJ, Rodosky MW, Sekiya JK. Symptomatic os acromiale. J Am Acad Orthop Surg. 2006;14(1):12–9.
79. Cruz JC, Kirby CL. Acromioclavicular joint cyst. Ultrasound Q. 2016;32(3):314–6.
80. Gilani SA, Mehboob R, Bacha R, Gilani A, Manzoor I. Sonographic presentation of the Geyser sign. Case Rep Med. 2019;2019:5623530.
81. Ferri M, Finlay K, Popowich T, Jurriaans E, Friedman L. Sonographic examination of the acromioclavicular and sternoclavicular joints. J Clin Ultrasound. 2005;33(7):345–55.
82. Kock HJ, Jurgens C, Hirche H, Hanke J, Schmit-Neuerburg KP. Standardized ultrasound examination for evaluation of instability of the acromioclavicular joint. Arch Orthop Trauma Surg. 1996;115(3–4):136–40.
83. Noh YM, Jeon SH, Yoon HM. Ultrasonography in sternoclavicular joint posterior dislocation in an adolescent—a case report. Clin Shoulder Elb. 2014;17:205.
84. Wisniewski SJ, Smith J. Synovitis of the sternoclavicular joint: the role of ultrasound. Am J Phys Med Rehabil. 2007;86(4):322–3.
85. Mondal S, Sinha D, Nag A, Chakraborty A, Ete T, Ghosh A. Sterno clavicular joint osteoarthritis in a patient with rheumatoid arthritis: a case report with classical sonographic features. J Diagn Med Sonogr. 2015;3:186–9.
86. Kawashiri SY, Edo Y, Kawakami A. Early detection of inflammation and joint destruction revealed by ultrasound in a patient with sternoclavicular septic arthritis. Intern Med. 2019;58(6):865–9.
87. Krzyżanowski W. The use of ultrasound in the assessment of the glenoid labrum of the glenohumeral joint. Part I: ultrasound anatomy and examination technique. J Ultrason. 2012;12(49):164–77.
88. Taljanovic MS, Carlson KL, Kuhn JE, Jacobson JA, Delaney-Sathy LO, Adler RS. Sonography of the glenoid labrum. Am J Roentgenol. 2000;174(6):1717–22.

89. Tirman PFJ, Smith ED, Stoller DW, Fritz RC. Shoulder imaging in athletes. Semin Musculoskelet Radiol. 2004;8(1):29–40.
90. Van der Woude HJ, Vanhoenacker FM. MR arthrography in glenohumeral instability. JBR-BTR. 2007;90(5):377–83.
91. Gottlieb M, Russell F. Diagnostic accuracy of ultrasound for identifying shoulder dislocations and reductions: a systematic review of the literature. West J Emerg Med. 2017;18(5):937–42.
92. Borsa PA, Jacobson JA, Scibek JS, Dover GC. Comparison of dynamic sonography to stress radiography for assessing glenohumeral laxity in asymptomatic shoulders. Am J Sports Med. 2005;33(5):734–41.
93. Cheng SC, Hulse D, Fairbairn KJ, Clarke M, Wallace WA. Comparison of dynamic ultrasound and stress radiology for assessment of inferior glenohumeral laxity in asymptomatic shoulders. Skelet Radiol. 2008;37(2):161–8.
94. Neer CS 2nd. Involuntary inferior and multidirectional instability of the shoulder: etiology, recognition, and treatment. Instr Course Lect. 1985;34:232–8.
95. John Marx M, Robert Hockberger M, Ron Walls M. Rosen's emergency medicine: concepts and clinical practice. Mosby; 2002.
96. Yuen CK, Mok KL, Kan PG, Wong YT. Ultrasound diagnosis of anterior shoulder dislocation. Hong Kong J Emerg Med. 2009;16(1):29–34.
97. Farin PU, Kaukanen E, Jaroma H, Harju A, Väätäinen U. Hill-Sachs lesion: sonographic detection. Skelet Radiol. 1996;25(6):559–62.
98. Kaar TK, Wirth MA, Rockwood CA Jr. Missed posterior fracture-dislocation of the humeral head. A case report with a fifteen-year follow-up after delayed open reduction and internal fixation. J Bone Joint Surg Am. 1999;81(5):708–10.
99. Perron AD, Jones RL. Posterior shoulder dislocation: avoiding a missed diagnosis. Am J Emerg Med. 2000;18(2):189–91.
100. McLaughlin HL. Preoperative and postoperative care of shoulder injuries. Clin Orthop Relat Res. 1965;38:58–64.
101. Bize P, Pugliese F, Bacigalupo L, Bianchi S. Unrecognized bilateral posterior shoulder dislocation diagnosed by ultrasound (2003:11b). Eur Radiol. 2004;14(2):350–2.
102. Hawkins RJ, Neer CS 2nd, Pianta RM, Mendoza FX. Locked posterior dislocation of the shoulder. J Bone Joint Surg Am. 1987;69(1):9–18.
103. Weiner DS, Macnab I. Superior migration of the humeral head. A radiological aid in the diagnosis of tears of the rotator cuff. J Bone Joint Surg Br. 1970;52(3):524–7.
104. Keener JD, Wei AS, Kim HM, Steger-May K, Yamaguchi K. Proximal humeral migration in shoulders with symptomatic and asymptomatic rotator cuff tears. J Bone Joint Surg Am. 2009;91(6):1405–13.
105. Park SH, Choi CH, Yoon HK, Ha JW, Lee C, Chung K. What can the radiological parameters of superior migration of the humeral head tell us about the reparability of massive rotator cuff tears? PLoS One. 2020;15(4):e0231843.
106. Zemek MJ, Magee DJ. Comparison of glenohumeral joint laxity in elite and recreational swimmers. Clin J Sport Med. 1996;6(1):40–7.
107. Arya KN, Pandian S, Puri V. Rehabilitation methods for reducing shoulder subluxation in post-stroke hemiparesis: a systematic review. Top Stroke Rehabil. 2018;25(1):68–81.
108. Tung GA, Entzian D, Stern JB, Green A. MR imaging and MR arthrography of paraglenoid labral cysts. AJR Am J Roentgenol. 2000;174(6):1707–15.
109. Alasaarela E, Suramo I, Tervonen O, Lähde S, Takalo R, Hakala M. Evaluation of humeral head erosions in rheumatoid arthritis: a comparison of ultrasonography, magnetic resonance imaging, computed tomography and plain radiography. Br J Rheumatol. 1998;37(11):1152–6.
110. Dasgupta B, Cimmino MA, Maradit-Kremers H, Schmidt WA, Schirmer M, Salvarani C, et al. 2012 provisional classification criteria for polymyalgia rheumatica: a European League Against Rheumatism/American College of Rheumatology collaborative initiative. Ann Rheum Dis. 2012;71(4):484.

111. Macchioni P, Boiardi L, Catanoso M, Pazzola G, Salvarani C. Performance of the new 2012 EULAR/ACR classification criteria for polymyalgia rheumatica: comparison with the previous criteria in a single-centre study. Ann Rheum Dis. 2014;73(6):1190–3.
112. Jacobson JA. Shoulder US: anatomy, technique, and scanning pitfalls. Radiology. 2011;260(1):6–16.
113. Jamadar DA, Robertson BL, Jacobson JA, Girish G, Sabb BJ, Jiang Y, et al. Musculoskeletal sonography: important imaging pitfalls. AJR Am J Roentgenol. 2010;194(1):216–25.
114. Ferri M, Finlay K, Popowich T, Stamp G, Schuringa P, Friedman L. Sonography of full-thickness supraspinatus tears: comparison of patient positioning technique with surgical correlation. AJR Am J Roentgenol. 2005;184(1):180–4.
115. Read JW, Perko M. Ultrasound diagnosis of subacromial impingement for lesions of the rotator cuff. Australas J Ultrasound Med. 2010;13(2):11–5.
116. Teefey SA, Middleton WD, Yamaguchi K. Shoulder sonography. State of the art. Radiol Clin N Am. 1999;37(4):767–85, ix.
117. Schmidt WA, Schmidt H, Schicke B, Gromnica-Ihle E. Standard reference values for musculoskeletal ultrasonography. Ann Rheum Dis. 2004;63(8):988–94.

Chapter 11
Anterior Ankle

Reasons to Do the Study
1. Anterior ankle pain
2. Dorsal foot pain
3. Foot or toe extension dysfunction
4. Evaluate for underlying systemic diseases such as rheumatoid arthritis (RA), spondyloarthropathy (SpA), or crystalline disease

Questions We Want Answered
1. What is causing foot dorsum or anterior ankle pain?
2. What is the cause of foot dorsum pain/swelling?

Basic Anatomy (Fig. 11.1)

Bones, Joints, Recesses of the Hindfoot and Midfoot

1. **Tibiotalar joint**
2. **Chopart joint (transverse tarsal joint)** is the talonavicular joint combined with the calcaneocuboid joint.
3. **Naviculocuneiform joint**
4. More distally, the three cuneiforms and the cuboid bone articulate with the metatarsal bones to form the **Lisfranc joint (tarsometatarsal joints)**.

The contents do not represent the views of the U.S. Department of Veterans Affairs or the United States Government.

M. H. Greenberg et al., *Manual of Musculoskeletal Ultrasound*,
https://doi.org/10.1007/978-3-031-37416-6_11

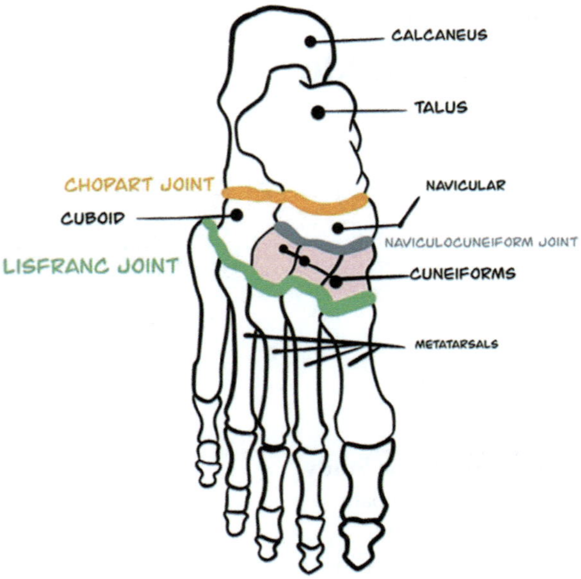

Fig. 11.1 Basic bony anatomy

Soft Tissue Structures (Fig. 11.2a) [1]

1. **Tibialis anterior tendon (TAT)**
   (a) Origin: proximal tibia and interosseous membrane
   (b) Insertion: base of the first MT and medial cuneiform
2. **Extensor hallucis longus (EHL) tendon**
   (a) Origin: fibula and interosseous membrane
   (b) Insertion: distal phalanx of the first toe
3. **Extensor digitorum longus (EDL) tendon**
   (a) Origin: tibia, fibula, interosseous membrane
   (b) Insertion: phalanges of the second-fifth toes
4. **Peroneus tertius**
   The peroneus tertius is an accessory tendon that, if present, may lie within the same tendon sheath as the EDL [2].
   (a) Origin: fibula and interosseous membrane
   (b) Insertion: base of the fifth MT or the cuboid
5. **Anterior tibial artery (ATA),** which becomes the **dorsalis pedis artery (DPA)** at the anterior ankle joint
6. **Deep peroneal nerve (DPN)**
7. **Superior extensor retinaculum (SER)**
8. **Inferior extensor retinaculum (IER)**
9. **Anterior recess superficial to the tibiotalar joint**

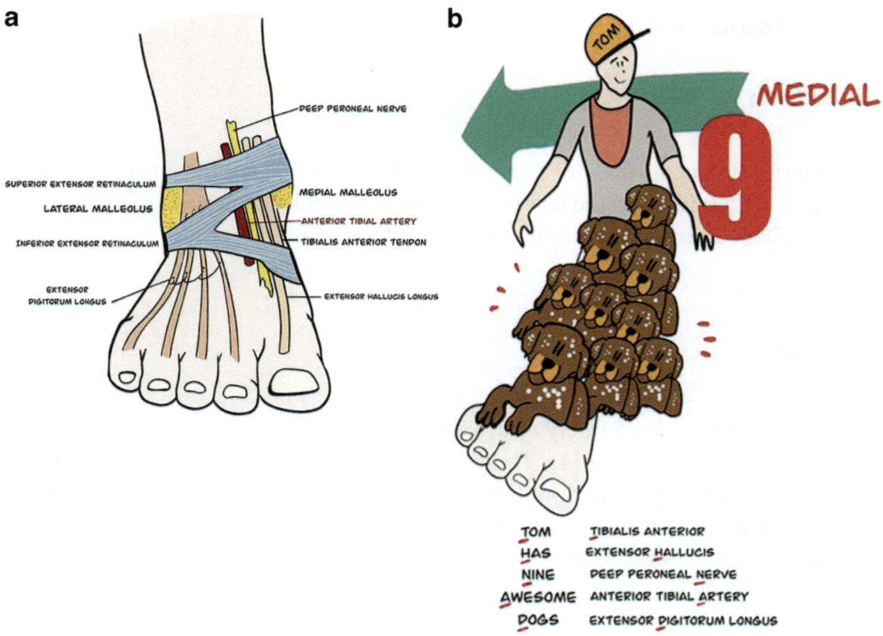

Fig. 11.2 (**a**) Soft tissue structures (**b**) Soft tissue structure mnemonic

Soft Tissue Information

A mnemonic for many of the soft tissue structures of the anterior ankle, from medial to lateral (Fig. 11.2b):

> "**Tom Has Nine Awesome Dogs**" (**T**AT, E**H**L, **n**erve, **a**rtery, E**D**L)

Additional Tips

1. The anterior tendons are held in place by the SER and IER.
2. The ATA goes beneath the SER to become the DPA between the EHL and EDL tendons.
3. The DPN medially follows the ATA/DPA, then may cross to the lateral aspect of the artery [3].

Clinical Comments

Ligaments

The anterior ligaments that are sonographically accessible are the anterior talofibular ligament (ATaFL) and the anterior inferior tibiofibular ligament (AITiFL).

(See Chap. 13 for more information.)

Tendons

The extensor tendons are rarely affected by tendinosis and tenosynovitis; complete tears are even less common [4]. These disorders arise from trauma, overuse, infections, or inflammatory conditions. Since the extensor tendons follow linear courses, uninjured tendons exhibit pristine fibrillar echotexture; however, of all the extensor tendons, the TAT is most often damaged [5].

Tendinosis

With tendinosis, the tendon swells with heterogeneous hypoechoic distortions within the ordinarily well-defined linear fibrillar echotexture, but it is impossible to delineate low-grade tears [6, 7]. Power Doppler (PD) activity may be appreciated, indicating local hypervascularity [6]. Comparison with the unaffected contralateral tendon is quite helpful. The TAT may develop tendinosis distally near its two insertions, which may present as medial midfoot burning [8]. Osteophytes might exert an abrasive effect on the undersurface of the distal TAT, causing hypoechoic swelling, sometimes with PD activity and pain during sonopalpation.

Tenosynovitis

Tenosynovitis is represented on ultrasound (US) as thickening of the tendon sheath with increased anechoic or hypoechoic synovial fluid within the sheath, perhaps with PD activity [3, 6]. Tenosynovitis may be due to repetitive trauma, infection, or inflammatory polyarthropathy such as RA or SpA. Tenosynovitis of the EDL occurs most frequently; next is the TAT [9]. The myotendinous junction of the EHL may be mistaken for tenosynovitis. The extensor tendons present with serous tenosynovitis, in contrast to the medial and lateral ankle tendons, which frequently show proliferative tenosynovitis. If you can visualize the TAT sheath, then tenosynovial pathology is likely [8].

Tears

Anterior tendon tears are uncommon and may be caused by impingement from osteophytes or orthopedic hardware, local steroid injection, penetrating trauma, or possibly a fracture [5, 6, 8, 10]. Dynamic US reveals the underlying impinging source if the tendon has not been torn completely. The TAT is the most frequently torn of the extensor tendons, with tears occurring beneath the extensor retinaculum [6]. Partial tears include longitudinal splitting, best delineated when filled with anechoic fluid; however, if debris fills the fissure, it may be impossible to differentiate them from tendinosis. Complete tears may show retraction, gap separation on dynamic US, or present as a dorsal foot mass [6].

Muscles

Herniation

Muscle herniation most often occurs in the lower leg, primarily affecting the tibialis anterior muscle [8]. Causes include athletic activities, occupational stresses, direct trauma, and chronic compartment syndrome. The muscle protrudes through a defect in the fascia and the subcutaneous fat, presenting as a soft tissue mass that may be mistaken for a tumor. Dynamic US, in which the patient contracts the muscle, will delineate defect margins and, thus, the true nature of the mass.

Atrophy

Tibialis anterior muscle atrophy may occur with direct trauma, common peroneal neuropathy, knee dislocation, fibula fracture, tumors, or a tight cast. Ultrasound objectively evaluates foreleg muscle atrophy by assessing thickness compared to the contralateral side [8]. Atrophic muscle tissue converts to fat and fibrosis, increasing reflectivity and rendering the echotexture hyperechoic.

Calcification

Foreleg calcification presenting as a soft tissue mass occurs with calcific myonecrosis, hematoma, synovial sarcoma, or osteosarcoma [8]. Calcific myonecrosis from blunt trauma may appear as a hypoechoic mass with internal irregular linear echoes and posterior acoustic shadowing. Other causes of soft tissue calcification in the foreleg are myositis ossificans, posttraumatic pseudoaneurysms, dermatomyositis or polymyositis, tumoral calcinosis, and diabetic myonecrosis [8].

Anterior Retinacula

The SER and the IER, composed of deep fascia, stabilize the extensor tendons during muscle contraction to prevent bowstringing [11]. The anterior retinacula (AR) may be injured playing sports, acutely, or by repetitive microtrauma. The AR is best seen sonographically with the probe transverse (TAX) to the foot so that the linear, thin hyperechoic fibrillar bands are displayed superficial to the tendons. The average normal AR thickness is approximately 1 mm [12]. Injured retinacula appear thickened and hypoechoic; hyperemia, delineated by PD, may be present in acute and subacute stages. Chronic injuries lead to retinacular loosening and secondary tendon instability. Retinacula integrity can be assessed with dynamic US by passively or actively dorsiflexing the foot or digits [11]. The AR may also play a role in stabilizing the subtalar joint since AR injuries are implicated in subtalar instability.

Anterior Tibiotalar Joint

Effusion

Joint effusion may result from infection, fractures, inflammatory polyarthritis, gout, hemophilia, sickle cell disease, and amyloid arthropathy [13]. Anechoic or hypoechoic fluid beneath the fat pad is best seen on a sagittal view with the ankle in maximal plantar flexion [4, 6, 13]. The effusion displaces or floats the fat pad anteriorly (superficial on the US image). Be aware that up to 3 mm of synovial fluid may be expected within the anterior recess, so do not mistake physiologic synovial fluid for effusion [4, 14]. Likewise, do not mistake anechoic cartilage for effusion.

To differentiate effusion from cartilage, note that compared to talar dome cartilage, fluid appears anechoic, and transducer pressure displaces fluid but not cartilage [4]. For effusion detection, US is more sensitive than lateral radiography, but less sensitive than magnetic resonance imaging. Clues about the cause of effusion may be in plain sight: osteophytes indicating osteoarthritis, intra-articular tophi, a double contour sign (DCS) of gout over the talar dome cartilage, or perhaps cartilage erosion from inflammatory polyarthritis.

Echogenic fluid can mimic synovial proliferation or active synovitis [15]. Synovitis may or may not have PD activity; however, effusion lacks PD. Hence, PD activity argues for synovitis. It is best to evaluate for effusion and PD activity at both the lateral and medial aspects of the tibiotalar joint; these areas are more superficial than the deeper central portion of the joint [9].

Dynamic imaging with dorsiflexion and plantar flexion may reveal movement of an effusion; synovitis will not reveal such action [15]. Plantar flexion facilitates synovial fluid flow from the posterior recess into the anterior tibiotalar joint and recess [3, 7, 13, 16]. Directly compressing an effusion with the probe may differentiate simple from complex effusions since there may be debris swirling with the latter [14]. Consider fluid aspiration under direct US guidance; however, synovial fluid aspiration is mandatory if an infection is of concern.

Synovitis

Synovitis of the tibiotalar joint is located deep to the fat pad in a similar location to an effusion; however, US reveals synovitis to be slightly hypoechoic compared to anechoic effusion. Synovitis appears as soft tissue thickening and has minimal to no compressibility to probe pressure [16]. Tibiotalar joint synovitis in RA appears as a hypoechoic area superficial to the talar dome [9]. Synovitis causes include infection, inflammatory joint disease, osteoarthritis, and noninflammatory synovial proliferation [7]. The latter can be seen with pigmented villonodular synovitis (PVNS) and synovial osteochondromatosis. PD may indicate acute inflammation.

Erosion

Erosion of the cartilage and bone of the tibiotalar joint occurs less commonly than one might expect.

Arthritis

Here, we refer to causes of tibiotalar arthritis, *not* periarticular mimics of ankle arthritis.

1. **Osteoarthritis**
   Osteoarthritis occurs much less commonly in the ankle than in the knee and hip [17]. When it does happen, it is not usually primary but secondary to preceding physical trauma or the end-stage result of RA, hemochromatosis, recurrent bleeding, avascular necrosis, or postinfectious arthritis. The low prevalence of primary osteoarthritis of the ankle is surprising considering its weight-bearing burden. A putative biomechanical protective factor may be the difference in cartilage density and stiffness over the talar dome compared with other joints [18].
2. **Spondyloarthropathy**
   This includes axial and peripheral SpA, psoriatic arthritis, and SpA associated with inflammatory bowel disease [17].
3. **Rheumatoid arthritis**
   Please see the section below on RA.
4. **Other systemic rheumatic diseases**
   While both lupus and systemic sclerosis may cause ankle arthralgia, occasionally, true ankle synovitis is noted with systemic sclerosis [17].

5. **Infectious arthritis**
   (a) **Bacterial**
   The ankle is the third most infected joint after the knee and hip [19]. Gonococcal disease is also included in this category [17].
   (b) **Viral**
   This includes hepatitis B and C, chikungunya, and human immunodeficiency virus infections [17].
   (c) **Mycobacterial** [17]
6. **Crystal disease**
   Gout frequently involves the ankle [17]. Calcium pyrophosphate deposition disease (CPPD), although more common in the wrist and knee, may sometimes involve the ankle.
7. **Sarcoidosis**
   Older literature describes frequent ankle involvement in sarcoidosis; however, closer inspection reveals that sarcoidosis more commonly *surrounds* soft tissue with panniculitis and tenosynovitis instead of directly involving the ankle [17].
8. **Intra-articular bodies**
   These bodies, comprised of cartilage or bone, are caused by damage to the articular surface. They may be asymptomatic or painful. Causes include osteochondral fractures, osteochondritis dissecans (see below), PVNS (see below), osteoarthritis (see above), and synovial osteochondromatosis [20]. Loose bodies are located within the joint capsule or the tendon sheath [4]. Dynamic US confirms loose body movement.

Rheumatoid Arthritis

Rheumatoid arthritis usually starts in the hands, but 20% of patients first develop symptoms in the foot or ankle [21]. Nearly 90% of RA patients develop foot or ankle involvement during the disease course [22]. Sonographic evaluation of tenosynovitis, erosion, and PD activity informs diagnosis, prognosis, and treatment selection [23]. When evaluating RA in the ankle, the regions of interest are the tibiotalar and talonavicular joints, the tendons, and the tenosynovial sheaths. Fluid distending the recesses by more than 3 mm in the tibiotalar joint or 2.6 mm in the talonavicular joint is regarded as an effusion [24]. An inflammatory effusion may also present in the subtalar joint, best viewed laterally at the sinus tarsi (see Chap. 13 for more information). Definitions of tendinosis, erosions, synovitis, and tenosynovitis, as well as scoring of PD vascularity, synovitis, tenosynovitis, and erosions, have been described in Chaps. 1 and 5.

In one study of 63 patients with active disease, evaluation of the tibiotalar and talonavicular joints, ankle tendons, and tenosynovial sheaths revealed RA ankle involvement in 44% [22]. The right (dominant) ankle had more findings than the left ankle, indicating the need to evaluate both ankles. Tenosynovitis was the earliest sign of RA and correlated with disease activity scores. However, bone erosion was associated with disease duration and the presence of rheumatoid factor. A second study compared 100 symptomatic ankles in 74 patients with early RA (<6 months duration) and established RA (>6 months duration) [25]. Nearly 70% of the early RA patients had tenosynovitis, which most often occurred in the medial flexor tendons. Thus, the shorter the disease duration (and perhaps less exposure to treatment), the greater the chance of detecting ankle tenosynovitis.

If RA is suspected in a patient with ankle pain, US examination of ankle joints and tendon sheaths may aid in diagnosis. When a patient presents with a clinically inflamed ankle, evaluate the contralateral ankle to look for subclinical signs of synovitis or tenosynovitis.

Masses and Cysts

Ultrasound will locate a mass in relation to other structures and will define the internal echogenic character (e.g., cystic, solid), compressibility, vascularity, and the presence of nearby nerves and vessels [6]. Serial US examinations can follow masses over time to verify stability or changes in size. However, other modalities, or ultimately, a biopsy may be necessary to determine the precise diagnosis. Ultrasound can help safely guide the biopsy [6]. Neoplasms are rare on the foot dorsum. Several other entities may present as a foot mass.

Ganglion Cyst

The familiar, benign ganglion cyst may be uni- or multiloculated and has a narrow stalk connecting it to a nearby joint or tendon sheath [6]. Ganglia appear anechoic or hypoechoic on US, but may be so tense as to mimic a solid mass or bone on palpation [26].

Nerve Sheath Tumors

These may occur on the DPN in the anterior compartment or the superficial peroneal nerve at the distal fibula [16]. Schwannomas and neurofibromas are the most common types of peripheral nerve sheath tumors. Usually benign, they rarely undergo malignant transformation. Most are homogeneous and hypoechoic on US. Other sonographic features include posterior acoustic enhancement, target appearance (hyperechoic center with hypoechoic periphery), and PD activity.

These tumors may mimic ganglion cysts; however, if PD activity is present, a ganglion cyst is ruled out [6, 16]. Benign and malignant peripheral nerve sheath tumors have variable sonographic features; thus, sonography cannot reliably differentiate the tumor type or presence of malignancy [27]. Sonographic imaging of the nerve entering the mass (peripheral nerve continuity) supports the diagnosis of a peripheral nerve sheath tumor [6, 27].

Gouty Tophus

A tophus may appear as a localized mass on an extensor tendon [6]. Tophaceous material within the tendon may cause it to lose its typical fibrillar pattern; there may also be associated tenosynovitis and surrounding soft tissue swelling. There may or may not be a classic sonographic picture of a tophus. Clinical history and physical examination, as well as the presence of a DCS in the ankle or other joints, support the diagnosis. Fluid aspiration and sometimes direct aspiration of the mass with polarized microscopic analysis may be diagnostic.

(See Chap. 24 for more information.)

Retained Foreign Body

A foreign body may incite a reactive inflammatory mass depicted on US as a hyperechoic foreign body surrounded by a hypoechoic rim [6]. The foreign body itself may produce reverberation artifacts if it is smooth and flat [28]. Acoustic shadowing may be seen if the body is irregular with a small, curved radius.

(See Chap. 15 for more information.)

Complete Tendon Tear

A complete tendon tear with retraction may produce a mass on the foot dorsum [6].

Extensor Muscle Herniation

See Muscle Disorders, Herniation, above.

Calcium-Containing Masses

See Muscle Disorders, Calcifications, above.

Pigmented Villonodular Synovitis

Also known as a tenosynovial giant cell tumor or a giant cell tumor of the tendon sheath, PVNS is a rare, localized, potentially aggressive lesion involving the synovial lining of joints, tendon sheaths, or bursae [29]. PVNS occurs more often in the knee, hand flexor tendons, and hip than in the foot or ankle. Due to the high recurrence rate (up to 50%), some authors refer to this condition as a semi-malignant tumor [30]. In the foot, PVNS may occur in the ankle joint, the tenosynovium, or other areas [31]. The lesion may present as monoarticular arthritis or as a palpable, painful or painless mass. Magnetic resonance imaging (MRI) is the most useful noninvasive means of diagnosis; however, it is sensitive but nonspecific. A biopsy is often necessary. Radiography may reveal localized bony erosion [32]. Ultrasound shows the relationship of the nodule to the adjacent tendon during dynamic evaluation. The mass is usually homogeneous, lobulated, and hypoechoic, with internal vascularity on PD. Maintain a high index of suspicion of any painless or painful mass in the foot and ankle, despite the rarity of this condition [31].

Aneurysm/Pseudoaneurysm

Vascular injury may follow direct trauma, ankle arthroscopy, or extreme plantar flexion or inversion [33]. Such damage may present as swelling, hemarthrosis, vascular insufficiency, a pulsatile mass, or a compartment syndrome and may result in severe pain, ulceration, or ischemia. Doppler US accurately detects injury to the anterior tibial artery; however, a transfemoral arteriogram is the gold standard for diagnosis [6, 33, 34].

Abscess

The sonographic appearance of an abscess may be hyper- or hypoechoic, often with an echogenic rim and perhaps hyperemia [6]. The abscess lacks internal Doppler flow. Transducer pressure may produce swirls of complex fluid.

Lipoma

These are well-defined, mildly hyperechoic, ovoid masses parallel to the skin surface that lack internal blood flow on Doppler evaluation [5]. The appearance may overlap with other masses, and MRI is typically diagnostic.

Subcutaneous Tissue

Edema

On US, hypoechoic fluid separates soft tissue and creates a marbled appearance [6].

Subcutaneous Gas

Due to infection or invasive intervention, subcutaneous gas produces reflective foci with shadowing [6].

Anterior Ankle Impingement Syndrome

A common clinical syndrome due to synovial tissue crowding the anterolateral recess and tibiotalar joint space, anterior ankle impingement syndrome (AAIS), is associated with bone and soft tissue abnormalities [35, 36]. This condition is often due to repetitive microtrauma injuries with dorsiflexion, occurring in soccer players, gymnasts, runners, and ballet dancers [35]. There is often previous joint instability [37]. Direct ankle trauma may also be the trigger. Cartilage or osteochondral lesions may result in the abnormal growth of scar tissue and synovial abnormalities [38]. Anterior ankle impingement syndrome may culminate in chronic anterior ankle pain, soft tissue swelling, pain with motion (especially dorsiflexion), or decreased range of motion [35, 36]. Dorsiflexion may increase pain since it reduces articular joint space volume [39]. Premature ankle osteoarthritis and spur formation are also associated with AAIS, although cause and effect are not always clear.

Imaging delineates associated ligament injuries, articular cartilage damage to the talar dome, intra-articular foreign bodies, and peroneal tendon disorders [36]. Radiographs may show spurs at the distal tibia or talus; however, the presence of bony spurs is insufficient for the diagnosis. MRI shows synovial hypertrophy and talofibular ligament damage, while MR or CT arthrography further increases diagnostic accuracy. Ultrasound is less accurate than the above contrast studies, but may be helpful [36]. Ultrasound may detect a preexisting anterior talofibular ligament injury. Sonopalpation causing pain over a synovial mass may be diagnostically useful. PD activity within a synovial mass implies an underlying pathological origin [40]. Initial conservative treatment includes rest, physical therapy, and shoe modification [35]. Injection guided by US may be beneficial [36]. Surgical procedures are often necessary, including arthroscopic procedures [41].

Osteochondritis Dissecans

Persistent pain after an ankle sprain should prompt consideration for AAIS, instability, and osteochondral lesions [42]. Osteochondritis dissecans (OCD) of the dome of the talus may cause chronic posttraumatic ankle pain [43]. Delay in diagnosis and treatment of OCD may cause loose body arthritis and long-term morbidity [44]. OCD results from sudden impacts causing bone contusions or prolonged repetitive microtrauma such as excessive weight-bearing [43]. Such bony trauma causes a secondary microscopic fracture, bone marrow edema, and eventual detachment of the osteochondral fragment with intra-articular loose body formation. The patient complains of vague pain exacerbated by weight-bearing, swelling, stiffness, and the sensation of ankle instability and giving way. The diagnostic gold standard is the MRI [45, 46]. If traumatic ankle pain becomes chronic, the US finding of a large ankle effusion should heighten suspicion of OCD [43].

Pitfalls

1. Evaluate the anterior joint recess over the tibiotalar joint with the foot in plantar flexion since this maximizes detection of joint effusion.
2. In a patient with persistent pain after an ankle sprain, consider impingement syndrome, instability, and osteochondral lesions [42].
3. Don't overlook OCD in a patient with a history of an ankle sprain or other ankle trauma who has persistent ankle pain, but presents with only a tibiotalar effusion on US.
4. Do not mistake cartilage for effusion of the tibiotalar joint.
5. When evaluating a tibiotalar joint effusion, do not neglect to perform PD, compress the effusion with the probe, and compare it with the contralateral, normal side.
6. There may be a cartilage interface sign (CIS) over the talus with a tibiotalar effusion [47]. CIS may imitate a DCS. In contrast to a DCS, the CIS is very thin and will disappear with a slight change in the angle of insonation.
7. Clues about the cause of effusion may be in plain sight: osteophytes indicating osteoarthritis, intra-articular tophi, a DCS over the talar dome cartilage, or perhaps cartilage erosion from inflammatory arthritis.
8. If RA is suspected in a patient with ankle pain, do not neglect to perform an US examination of the ankle joints and tendon sheaths to evaluate for synovitis or tenosynovitis to support the diagnosis.
9. If RA is suspected of causing ankle monoarthritis, the lack of inflammatory findings in the contralateral ankle should raise doubts about the diagnosis.

Method

Although located anteriorly, the anterior talofibular ligament and the anterior inferior tibiofibular ligament are included on the lateral foot US examination.

(See Chap. 13 for more information.)

The patient is supine with the knee flexed at 90°, the foot flat on the exam table, and slightly rotated inward. Choose a linear probe with an ankle or mid-depth preset.

Protocol Image 1: Tibialis Anterior Tendon + Extensor Hallucis Longus, Transverse ± Power Doppler (Fig. 11.3)

Place the probe in TAX at the medial aspect of the distal anterior tibia. Move the probe distally in a TAX slide. The hyperechoic tibial surface will drop off to reveal another hyperechoic bone that looks like a mountain with a gentle slope and a steep cliff on the medial aspect. This is the medial process of the proximal

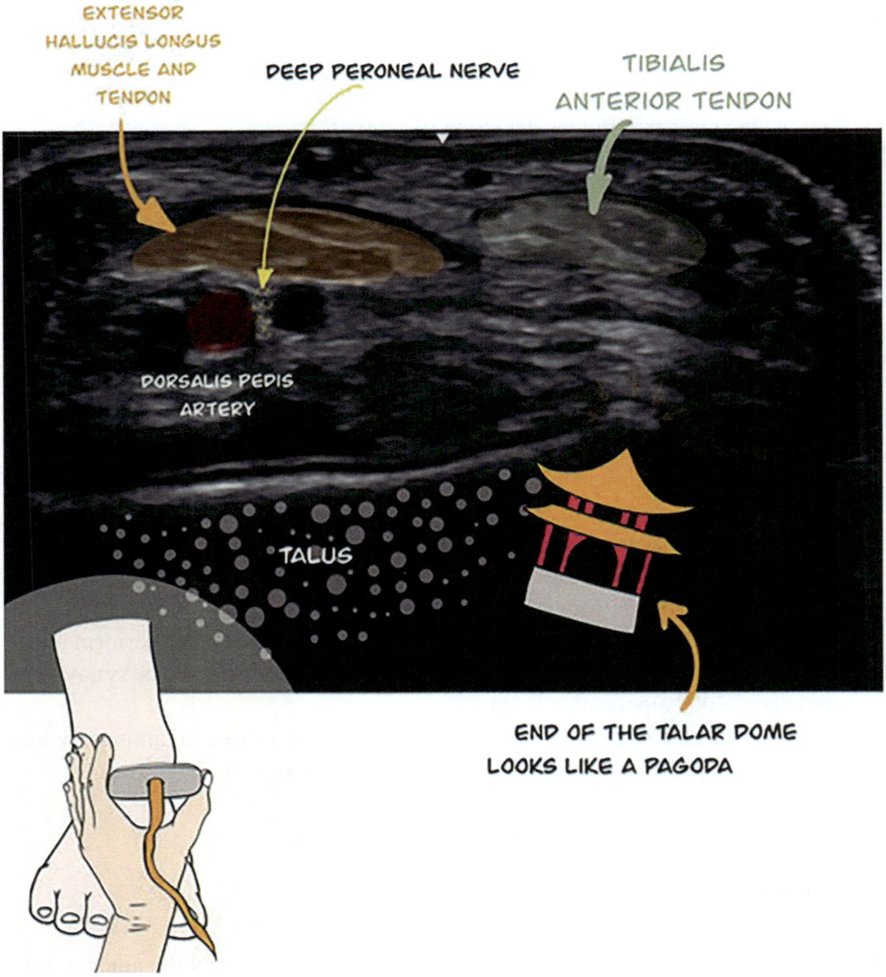

Fig. 11.3 Protocol Image 1: Tibialis anterior tendon + extensor hallucis longus, transverse ± power Doppler

talus, the talar dome. The appearance on US has been likened to the corner of the roof of a pagoda.

Look at the soft tissue structures superficial to the bone. To verify the identity of the structures, turn on the PD to see the more central DPA. Superficial and slightly medial to the DPA is the EHL; medial to the EHL is the TAT. This is the "home base" position, and if the probe is sufficiently long, you should be able to simultaneously see the TAT, the EHL, the neurovascular bundle, and at least part of the EDL. The SER is superficial to all three tendons, is hyperechoic, and exhibits anisotropy. If present, the accessory peroneus tertius tendon is lateral to the EDL and often lies within the same synovial sheath [2]. Tilting the probe will diminish tendon anisotropy.

Center the probe on the large TAT. Manually dorsiflex and invert the foot to verify tendon movement and that this tendon is, in fact, the TAT [16]. Look for tendon sheath fluid or tenosynovitis surrounding the TAT. Examine the hypoechoic or anechoic cartilage covering the surface of the talar dome for damage. Verify the identity of the EHL by passively flexing and extending the large toe. The musculotendinous junction of the EHL is somewhat distal; thus, the EHL muscle may sometimes be confused with tenosynovitis. The DPA is deep and slightly lateral to the EHL. Contiguous to the DPA, the small DPN may at first be medial to and then cross over to the lateral aspect of the artery. Use PD to confirm the DPA location since the DPN may be difficult to discern. The DPN displays a honeycomb appearance; tilting the probe delineates the nerve by exploiting its anisotropy. If necessary, the DPN and the DPA can each be followed distally in TAX view.

Hints:

1. The TAT is double the size of the other extensor tendons. Normal ankle tendon synovial sheaths cannot be delineated.
2. The DPN is adjacent to, or "hugs," the medial aspect of the DPA. At this level, it is often medial to the DPA, but sometimes it is on the lateral aspect of the DPA.

Protocol Image 2: Extensor Digitorum Longus, Transverse ± Power Doppler (Fig. 11.4)

Next, move the probe laterally to center it over the EDL. This tendon or tendon group is lateral to the EHL. Verify the tendon identity by observing tendon movement while manually dorsiflexing the second through fifth toes [16]. Note several tendons on their course to the digits and examine them for tenosynovitis.

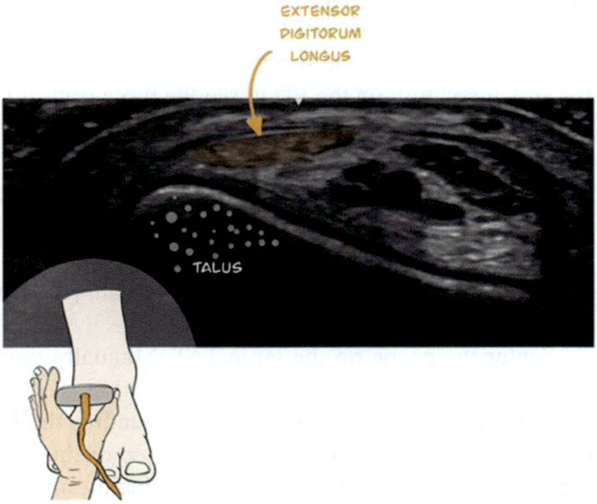

Fig. 11.4 Protocol Image 2: Extensor digitorum longus, transverse ± power Doppler

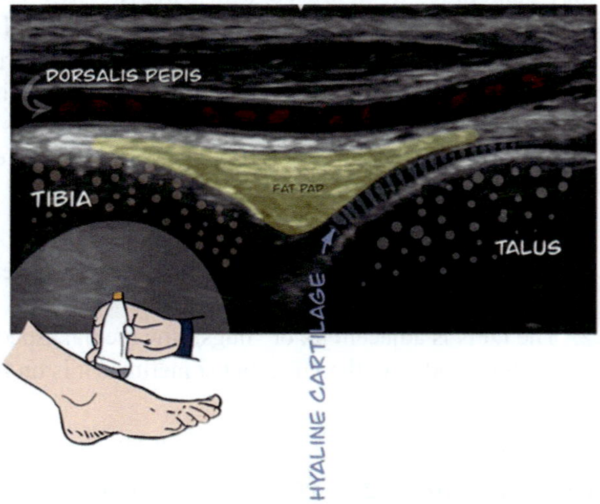

Fig. 11.5 Protocol Image 3: Tibiotalar joint, dorsalis pedis artery, longitudinal ± power Doppler

Protocol Image 3: Tibiotalar Joint, Dorsalis Pedis Artery, Longitudinal ± Power Doppler (Fig. 11.5)

Once again, go back to "home base" and center the probe on the DPA. Rotate the probe 90° to see the DPA in a longitudinal (LAX) view. This is the first look at the tibiotalar joint. Note the two curved hyperechoic bones, the hypoechoic or anechoic cartilage covering the more distal talus, and the mildly hyperechoic triangular or hammock-shaped fat pad within the joint space. Superficial to the fat pad is the hyperechoic joint capsule, which is sometimes challenging to discern. The joint recess is deep to the fat pad and may typically contain a small amount of anechoic synovial fluid. An abnormal effusion is noted when the fat pad is elevated or "floats" on the excess liquid. This image may demonstrate joint effusion, synovitis, DCS, or cartilage erosion. This joint may typically contain up to 3 mm of synovial fluid [4, 14]. Next, perform a TAX slide medially and laterally to examine the entire width of the tibiotalar joint in LAX. The cartilage covering the talar dome is usually thinner laterally [48].

Hints:

1. Maximal plantar flexion will expose more of the talar dome cartilage and accentuate an abnormal effusion.
2. There may be osteophytes on the distal tibia or the talus, which may be associated with AAIS.
3. The shallower lateral and medial aspects of the tibiotalar joint may best demonstrate effusion.

Protocol Image 4: Talonavicular Joint, Longitudinal ± Power Doppler (Fig. 11.6)

Move the probe more distally, still in LAX, and center on the talonavicular joint. If necessary, examine the full breadth of this joint by performing a slow medial and then lateral TAX slide, similar to what was done for the tibiotalar joint. Turn on PD to look for active synovitis.

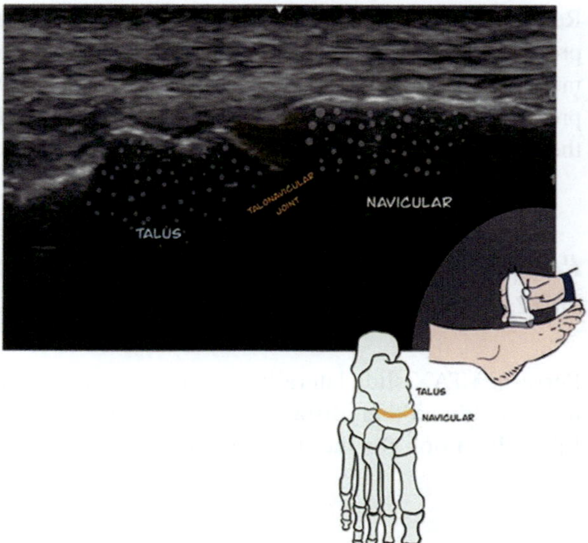

Fig. 11.6 Protocol Image 4: Talonavicular Joint, longitudinal ± power Doppler

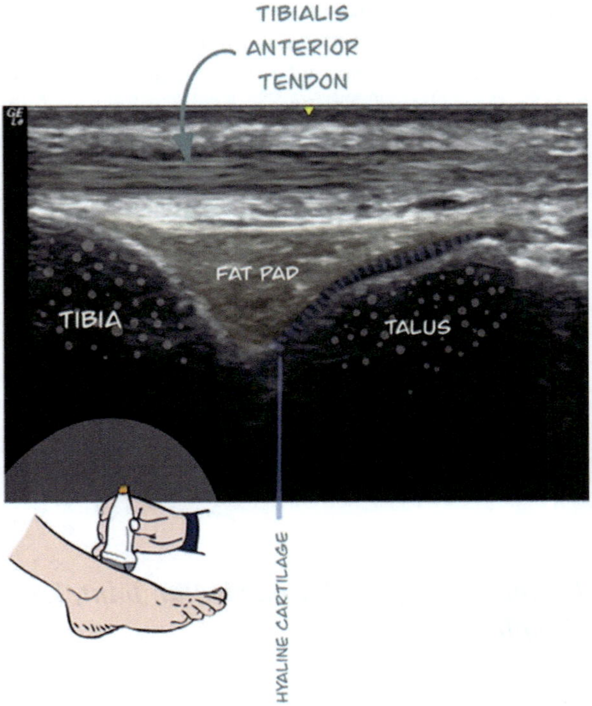

Fig. 11.7 Protocol Image 5: Tibialis anterior tendon over tibitotalar joint, longitudinal ± power Doppler

Protocol Image 5: Tibialis Anterior Tendon over Tibiotalar Joint, Longitudinal ± Power Doppler (Fig. 11.7)

Return to the "home base" view in TAX and center the probe on the TAT. Turn the probe 90° into LAX. Move the transducer distally, moving it slightly laterally and medially as you go to see the complete TAT in LAX. As you look at the talus from proximal to distal, you will sequentially visualize the tibiotalar joint, the talar dome, the talar neck, the talar head, and finally, the talonavicular joint.

Protocol Image 6: Extensor Hallucis Longus over Tibiotalar Joint, Longitudinal ± Power Doppler (Fig. 11.8)

Perform a TAX slide laterally to center the probe over the EHL. Move the probe proximally and then distally, as well as laterally and medially, to visualize the full EHL. Look for the fibrillar echotexture of the EHL, but be aware that the

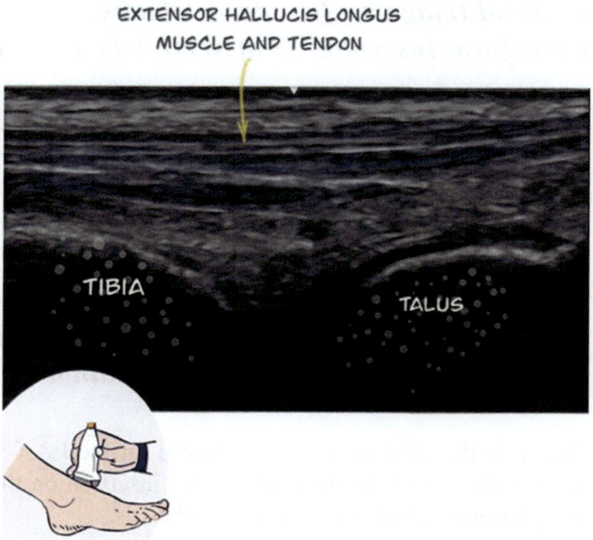

Fig. 11.8 Protocol Image 6: Extensor hallucis longus over the tibiotalar joint, longitudinal ± power Doppler

EHL muscle may be present at the tibiotalar joint; the muscle may mimic tenosynovitis.

Protocol Image 7 (Optional): Tibialis Anterior Tendon at Naviculocuneiform Joint, Longitudinal ± Power Doppler

Return to the view in Protocol Image 5 for the LAX view of the TAT. Continue to move the probe in LAX distally, following the TAT. The next joint distal to the talo-navicular joint is the naviculocuneiform joint. Technically, you are looking at the navicular-medial cuneiform joint.

Protocol Image 8 (Optional): Tibialis Anterior Tendon at Medial Cuneiform Insertion, Longitudinal ± Power Doppler

Continue to move the probe distally, following the TAT, to see its insertion on the distal medial cuneiform. Turn on PD at the insertion if there is a question of active enthesitis.

Protocol Image 9 (Optional): Tibialis Anterior Tendon at Medial Cuneiform Insertion, Transverse ± Power Doppler

Turn the probe 90° into the TAX position. Perform small proximal and distal TAX slides to examine the tendon and to see the part of the TAT that inserts on the medial cuneiform. This image may require tilting the probe to minimize anisotropy. Confirm any PD findings in transverse view.

Protocol Image 10 (Optional): Extensor Hallucis Longus at Insertion on First Phalanx, Longitudinal ± Power Doppler

Return to the view in Protocol Image 6 for the LAX view of the EHL. Move the probe distally in LAX to visualize the insertion on the first digit phalanx. Power Doppler may help delineate active enthesitis.

Protocol Image 11 (Optional): Extensor Hallucis Longus at Insertion on First Phalanx, Transverse ± Power Doppler

Turn the probe 90° into TAX to examine the tendon insertion on the dorsum of the first toe's distal phalanx. Again, confirm any PD findings.

Protocol Image 12 (Optional): Extensor Digitorum Longus at Tibiotalar Joint, Longitudinal ± Power Doppler

Return to the "home base" view of Protocol Image 1 in TAX and center the probe on the EDL. Note how this flat tendon divides into four slips, each inserted separately on the bases of digits 2–5. Lift each toe manually to ensure that you look at the correct tendon. Rotate the probe 90° to see the EDL in LAX. Since this is often the area of the myotendinous junction of the EDL, you may see muscle, tendon, or a combination.

Protocol Image 13 (Optional): Extensor Digitorum Longus at Insertion on Digits 2–5, Longitudinal ± Power Doppler

Next, move the probe distally to view the EDL insertions on digits 2–5.

Protocol Image 14 (Optional): Extensor Digitorum Longus Insertion on the Base of Digits 2–5, Transverse ± Power Doppler

Rotate the probe 90° into TAX to examine the tendon insertion on the dorsum of each digit 2–5.

Complete Anterior Ankle Ultrasonic Examination Checklist

☐ Protocol Image 1: Tibialis anterior tendon + extensor hallucis longus, transverse ± power Doppler.
☐ Protocol Image 2: Extensor digitorum longus, transverse ± power Doppler.
☐ Protocol Image 3: Tibiotalar joint, dorsalis pedis artery, longitudinal ± power Doppler.
☐ Protocol Image 4: Talonavicular joint, longitudinal ± power Doppler.
☐ Protocol Image 5: Tibialis anterior tendon over tibiotalar joint, longitudinal ± power Doppler.
☐ Protocol Image 6: Extensor hallucis longus over tibiotalar joint, longitudinal ± power Doppler.
☐ Protocol Image 7 (optional): Tibialis anterior tendon at naviculocuneiform joint, longitudinal ± power Doppler.
☐ Protocol Image 8 (optional): Tibialis anterior tendon at medial cuneiform insertion, longitudinal ± power Doppler.
☐ Protocol Image 9 (optional): Tibialis anterior tendon at medial cuneiform insertion, transverse ± power Doppler.
☐ Protocol Image 10 (optional): Extensor hallucis longus at insertion on first phalanx, longitudinal ± power Doppler.
☐ Protocol Image 11 (optional): Extensor hallucis longus at insertion on first phalanx, transverse ± power Doppler.
☐ Protocol Image 12 (optional): Extensor digitorum longus at tibiotalar joint, longitudinal ± power Doppler.
☐ Protocol Image 13 (optional): Extensor digitorum longus at insertion on digits 2–5, longitudinal ± power Doppler.
☐ Protocol Image 14 (optional): Extensor digitorum longus insertion on the base of digits 2–5, transverse ± power Doppler.

References

1. De Maeseneer M, Madani H, Lenchik L, De Mey J, Provyn S, Shahabpour M. Ultrasound of the distal insertions of the ankle and foot tendons with anatomical correlation: a review. Can Assoc Radiol J. 2018;69(3):282–92.
2. Precerutti M, Bonardi M, Ferrozzi G, Draghi F. Sonographic anatomy of the ankle. J Ultrasound. 2013;17(2):79–87.
3. Park JW, Lee SJ, Choo HJ, Kim SK, Gwak HC, Lee SM. Ultrasonography of the ankle joint. Ultrasonography. 2017;36(4):321–35.

4. Fessell DP, Vanderschueren GM, Jacobson JA, Ceulemans RY, Prasad A, Craig JG, et al. US of the ankle: technique, anatomy, and diagnosis of pathologic conditions. Radiographics. 1998;18(2):325–40.
5. Fessell DP, Jacobson JA. Ultrasound of the hindfoot and midfoot. Radiol Clin N Am. 2008;46(6):1027–43, vi.
6. Fessell DP, Jamadar DA, Jacobson JA, Caoili EM, Dong Q, Pai SS, et al. Sonography of dorsal ankle and foot abnormalities. AJR Am J Roentgenol. 2003;181(6):1573–81.
7. Sahu A, Rath P, Aggarwal B. Ultrasound of ankle and foot in rheumatology. Indian J Rheumatol. 2018;13(5):43–7.
8. Varghese A, Bianchi S. Ultrasound of tibialis anterior muscle and tendon: anatomy, technique of examination, normal and pathologic appearance. J Ultrasound. 2014;17(2):113–23.
9. Suzuki T. Power Doppler ultrasonographic assessment of the ankle in patients with inflammatory rheumatic diseases. World J Orthop. 2014;5(5):574–84.
10. Shetty M, Fessell DP, Femino JE, Jacobson JA, Lin J, Jamadar D. Sonography of ankle tendon impingement with surgical correlation. AJR Am J Roentgenol. 2002;179(4):949–53.
11. Bianchi S, Becciolini M. Ultrasound features of ankle retinacula: normal appearance and pathologic findings. J Ultrasound Med. 2019;38(12):3321–34.
12. Numkarunarunrote N, Malik A, Aguiar RO, Trudell DJ, Resnick D. Retinacula of the foot and ankle: MRI with anatomic correlation in cadavers. AJR Am J Roentgenol. 2007;188(4):W348–54.
13. Jacobson JA, Andresen R, Jaovisidha S, De Maeseneer M, Foldes K, Trudell DR, et al. Detection of ankle effusions: comparison study in cadavers using radiography, sonography, and MR imaging. AJR Am J Roentgenol. 1998;170(5):1231–8.
14. Nazarian LN, Rawool NM, Martin CE, Schweitzer ME. Synovial fluid in the hindfoot and ankle: detection of amount and distribution with US. Radiology. 1995;197(1):275–8.
15. Patel S, Fessell DP, Jacobson JA, Hayes CW, van Holsbeeck MT. Artifacts, anatomic variants, and pitfalls in sonography of the foot and ankle. AJR Am J Roentgenol. 2002;178(5):1247–54.
16. Micu MC, Nestorova R, Petranova T, Porta F, Radunovic G, Vlad V, et al. Ultrasound of the ankle and foot in rheumatology. Med Ultrason. 2012;14(1):34–41.
17. Kiely PDW, Lloyd ME. Ankle arthritis - an important signpost in rheumatologic practice. Rheumatology (Oxford). 2021;60(1):23–33.
18. Cole AA, Kuettner KE. Molecular basis for differences between human joints. Cell Mol Life Sci. 2002;59(1):19–26.
19. Kaandorp CJ, Dinant HJ, van de Laar MA, Moens HJ, Prins AP, Dijkmans BA. Incidence and sources of native and prosthetic joint infection: a community based prospective survey. Ann Rheum Dis. 1997;56(8):470–5.
20. Sedeek SM, Choudry Q, Garg S. Synovial chondromatosis of the ankle joint: clinical, radiological, and intraoperative findings. Case Rep Orthop. 2015;2015:359024.
21. Matricali GA, Boonen A, Verduyckt J, Taelman V, Verschueren P, Sileghem A, et al. The presence of forefoot problems and the role of surgery in patients with rheumatoid arthritis. Ann Rheum Dis. 2006;65(9):1254–5.
22. Elsaman AM, Mostafa ES, Radwan AR. Ankle evaluation in active rheumatoid arthritis by ultrasound: a cross-sectional study. Ultrasound Med Biol. 2017;43(12):2806–13.
23. McKie S. Imaging of the foot and ankle rheumatoid arthritis. In: Helliwell PS, editor. The foot and ankle in rheumatoid arthritis. Churchill Livingstone Elsevier; 2007. p. 99–112.
24. Machado FS, Natour J, Takahashi RD, de Buosi AL, Furtado RN. Sonographic assessment of healthy peripheral joints: evaluation according to demographic parameters. J Ultrasound Med. 2014;33(12):2087–98.
25. Suzuki T, Okamoto A. Ultrasound examination of symptomatic ankles in shorter-duration rheumatoid arthritis patients often reveals tenosynovitis. Clin Exp Rheumatol. 2013;31(2):281–4.
26. Ortega R, Fessell DP, Jacobson JA, Lin J, Van Holsbeeck MT, Hayes CW. Sonography of ankle Ganglia with pathologic correlation in 10 pediatric and adult patients. AJR Am J Roentgenol. 2002;178(6):1445–9.

27. Reynolds DL Jr, Jacobson JA, Inampudi P, Jamadar DA, Ebrahim FS, Hayes CW. Sonographic characteristics of peripheral nerve sheath tumors. AJR Am J Roentgenol. 2004;182(3):741–4.
28. Jacobson JA, Powell A, Craig JG, Bouffard JA, van Holsbeeck MT. Wooden foreign bodies in soft tissue: detection at US. Radiology. 1998;206(1):45–8.
29. Lee M, Mahroof S, Pringle J, Short SC, Briggs TW, Cannon SR. Diffuse pigmented villonodular synovitis of the foot and ankle treated with surgery and radiotherapy. Int Orthop. 2005;29(6):403–5.
30. Illian C, Kortmann HR, Künstler HO, Poll LW, Schofer M. Tenosynovial giant cell tumors as accidental findings after episodes of distortion of the ankle: two case reports. J Med Case Rep. 2009;3:9331.
31. Sharma H, Jane MJ, Reid R. Pigmented villonodular synovitis of the foot and ankle: Forty years of experience from the Scottish bone tumor registry. J Foot Ankle Surg. 2006;45(5):329–36.
32. Sirlyn Q. Pigmented villonodular synovitis. Sonography. 2014;1(1):19–24.
33. Marron CD, McKay D, Johnston R, McAteer E, Stirling WJ. Pseudo-aneurysm of the anterior tibial artery, a rare cause of ankle swelling following a sports injury. BMC Emerg Med. 2005;5:9.
34. Rubin JM, Adler RS, Bude RO, Fowlkes JB, Carson PL. Clean and dirty shadowing at US: a reappraisal. Radiology. 1991;181(1):231–6.
35. Vaseenon T, Amendola A. Update on anterior ankle impingement. Curr Rev Musculoskelet Med. 2012;5(2):145–50.
36. Pesquer L, Guillo S, Meyer P, Hauger O. US in ankle impingement syndrome. J Ultrasound. 2014;17(2):89–97.
37. Robinson P, White LM. Soft-tissue and osseous impingement syndromes of the ankle: role of imaging in diagnosis and management. Radiographics. 2002;22(6):1457–69; discussion 70–1.
38. Mardani-Kivi M, Mirbolook A, Khajeh-Jahromi S, Hassanzadeh R, Hashemi-Motlagh K, Saheb-Ekhtiari K. Arthroscopic treatment of patients with anterolateral impingement of the ankle with and without chondral lesions. J Foot Ankle Surg. 2013;52(2):188–91.
39. Ray RG, Gusman DN, Christensen JC. Anatomical variation of the tibial plafond: the anteromedial tibial notch. J Foot Ankle Surg. 1994;33(4):419–26.
40. Cochet H, Pelé E, Amoretti N, Brunot S, Lafenêtre O, Hauger O. Anterolateral ankle impingement: diagnostic performance of MDCT arthrography and sonography. AJR Am J Roentgenol. 2010;194(6):1575–80.
41. Berman Z, Tafur M, Ahmed SS, Huang BK, Chang EY. Ankle impingement syndromes: an imaging review. Br J Radiol. 2017;90(1070):20160735.
42. Ogilvie-Harris DJ, Gilbart MK, Chorney K. Chronic pain following ankle sprains in athletes: the role of arthroscopic surgery. Arthroscopy. 1997;13(5):564–74.
43. Zanon G, Di Vico G, Marullo M. Osteochondritis dissecans of the talus. Joints. 2014;2(3):115–23.
44. Schenck RC Jr, Goodnight JM. Osteochondritis dissecans. J Bone Joint Surg Am. 1996;78(3):439–56.
45. De Smet AA, Fisher DR, Burnstein MI, Graf BK, Lange RH. Value of MR imaging in staging osteochondral lesions of the talus (osteochondritis dissecans): results in 14 patients. AJR Am J Roentgenol. 1990;154(3):555–8.
46. Mintz DN, Tashjian GS, Connell DA, Deland JT, O'Malley M, Potter HG. Osteochondral lesions of the talus: a new magnetic resonance grading system with arthroscopic correlation. Arthroscopy. 2003;19(4):353–9.
47. De Maeseneer M, Marcelis S, Jager T, Shahabpour M, Van Roy P, Weaver J, et al. Sonography of the normal ankle: a target approach using skeletal reference points. AJR Am J Roentgenol. 2009;192(2):487–95.
48. Minagawa H, Wong KH. Musculoskeletal ultrasound: echo anatomy & scan technique. Amazon Digital Services; 2017.

Chapter 12
Posterior Ankle and Heel

Reasons to Do the Study

1. Posterior or inferior heel pain

Questions We Want Answered

1. What is causing posterior ankle pain?

   (a) Is there evidence for an Achilles tendon tear or tendinopathy?
   (b) Is there evidence for a systemic disease such as crystalline arthritis (gout, pseudogout), spondyloarthropathy (SpA), rheumatoid arthritis (RA), etc.?
   (c) Is there evidence for bursitis?
   (d) Is there evidence for Haglund's syndrome?
   (e) Is there evidence for Achilles paratenonitis, masses, or calcium deposits?

2. What is causing posterior calf pain?

   (a) Is there evidence of a muscle or tendon tear?

3. What is causing inferior heel pain?

   (a) Is there an inferior calcaneal spur?
   (b) Is there evidence for SpA?
   (c) Is there evidence for plantar fascial tears, plantar fasciitis, or fibromatosis?
   (d) Is there compression of Baxter's nerve (first branch of the lateral plantar nerve) [1]?
   (e) Is there a mass?
   (f) Is the ultrasound (US) exam normal?

The contents do not represent the views of the U.S. Department of Veterans Affairs or the United States Government.

M. H. Greenberg et al., *Manual of Musculoskeletal Ultrasound*,
https://doi.org/10.1007/978-3-031-37416-6_12

Bony Anatomy

The ankle structure evokes an ancient Egyptian pyramid; the base is the *heel*, or *calcaneus*, which contacts the ground. "*Calx*" is Latin for *chalk or limestone*. The ancient Egyptians constructed pyramids with limestone, a material that could support a great deal of weight. Stubborn people who are re*calci*trant are always *digging in their heels*. Perched on the calcaneus is the *talus*, a curved bone similar to an eagle's *talon*. On the back of a United States dollar bill are a pyramid and an eagle, a convenient mnemonic device.

Posterior Heel Anatomy (Fig. 12.1)

The *Achilles tendon* (AT), the chief ankle plantar flexor, is formed proximally by the soleus and medial and lateral gastrocnemius muscles and inserts on the posterior calcaneus. The mean AT thickness is 4.3 mm when measured 2 cm proximal to the calcaneus [2]. Deep (anterior) to the AT and abutting the proximal posterior calcaneus is the *retrocalcaneal* (subtendinous) *bursa.* It may normally contain fluid measuring up to 3.4 mm in thickness [2, 3]. The normal retrocalcaneal bursa is comma-shaped [4]. More proximal, *Kager's fat pad* (KFP), rests on the calcaneus. Superficial (posterior) to the AT is the *subcutaneous calcaneal bursa* (also called the *retro-Achilles*, *pre-Achilles*, *precalcaneal*, *superficial*, or *Achilles bursa*), usually undetectable unless pathologically distended [3]. The thin *plantaris tendon (PT)* along the medial aspect of the AT may be absent in up to 20% of people [3, 5]. The PT may mimic intact AT fibers in the event of a complete AT rupture [3]. There may be a normal variant *accessory soleus muscle* in approximately 3% of people, which runs parallel and adjacent to the anteromedial aspect of the AT near Kager's fat pad [6]. The tendon inserts into the AT, or the calcaneus, and appears as a mass displaying muscle echotexture. It may be asymptomatic, but may also cause pain and swelling after exercise.

Proximal/Posterior Calf Anatomy (Fig. 12.2)

The AT is formed by the *gastrocnemius muscles* (*medial and lateral heads*) (MHG and LHG, respectively) and the deeper *soleus muscle* [7]. The hyperechoic muscle fascia separates these unipennate muscles. The long, thin, ovoid PT, variably absent

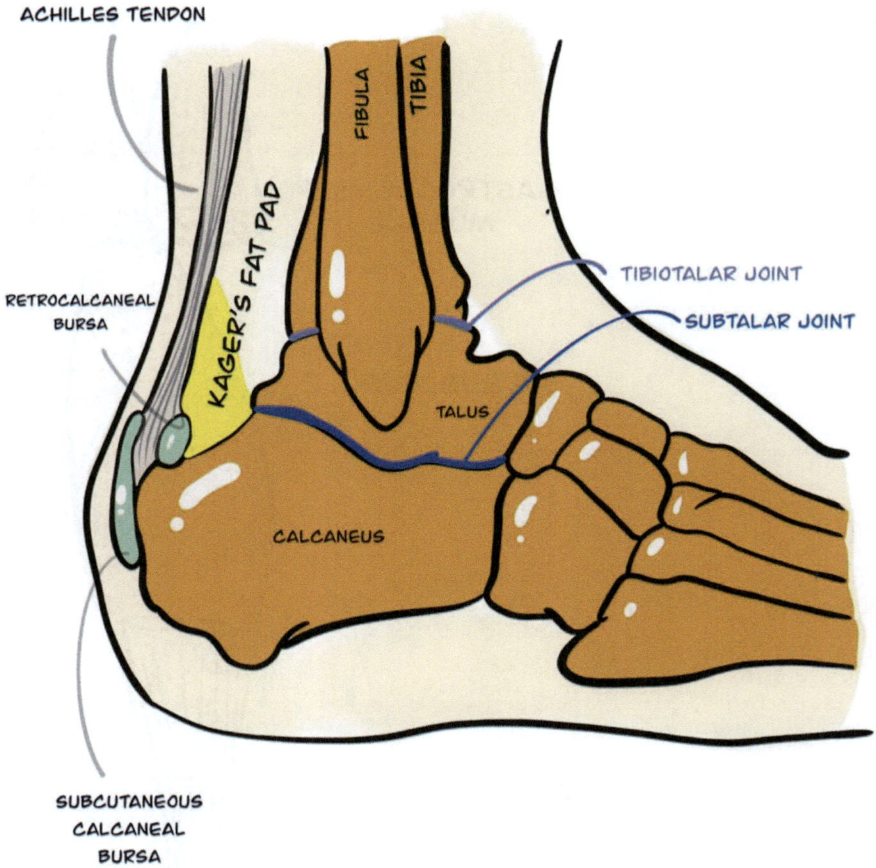

Fig. 12.1 Posterior heel bone and soft tissue anatomy

in normal people, originates from the distal lateral femur [8]. The PT moves distally from lateral to medial, sandwiched between the MHG muscle and the deeper soleus muscle. The PT has various insertion points, including the calcaneus or the AT [8]. Again, the PT can be mistaken for intact AT fibers or perhaps even a nerve. The biomechanical properties of the PT are unclear, but it may be used as a graft in orthopedic surgery [9].

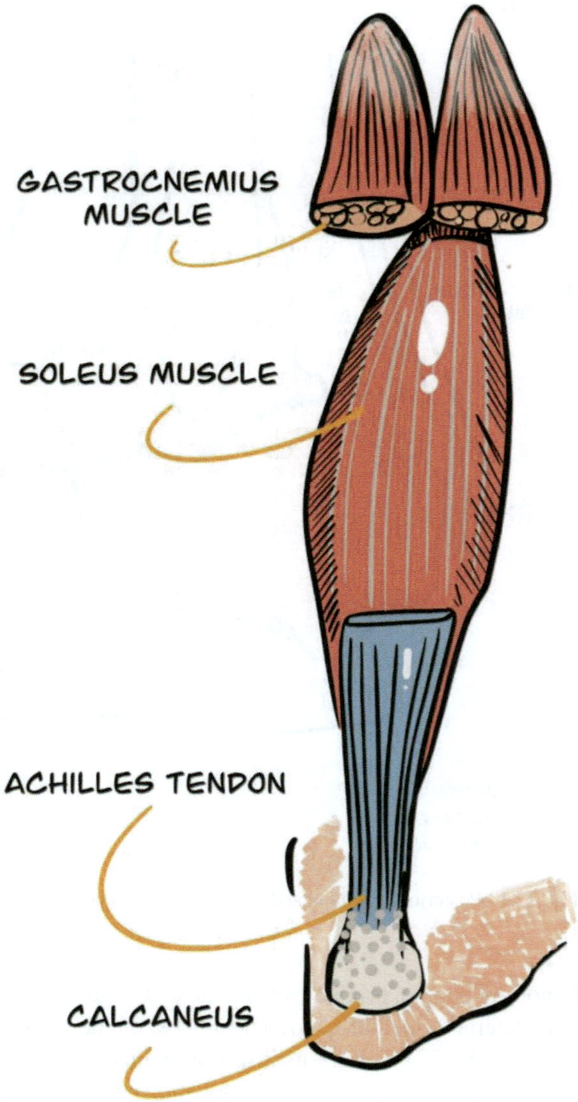

Fig. 12.2 Proximal calf anatomy

Inferior Heel Anatomy (Fig. 12.3)

The calcaneus is the primary bony landmark of the inferior hindfoot. The normally fibrillar *plantar fascia* (PF) is a ligamentous structure connecting the inferior calcaneus to the toes and maintaining the foot's arch. It has three bundles (central, lateral, and medial), with the maximal normal thickness of the central bundle being 4.0 mm [10].

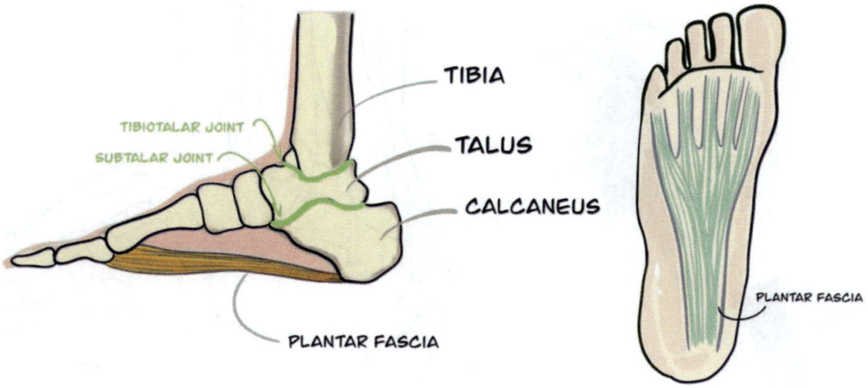

Fig. 12.3 Inferior heel anatomy

Clinical Comments

Posterior and inferior heel pain complaints are common. History, a physical exam, and radiographs are the first lines of evaluation. Radiographs may help delineate fractures, Haglund's *deformity* (superior calcaneal prominence), posterior or inferior bony spurs (enthesophytes where ligaments or tendons attach to bone), foreign bodies, or osteomyelitis. Ultrasound should be the next step if further workup is needed.

Posterior Heel Potential Ultrasound Findings

1. **Haglund's syndrome** consists of *calcaneal cortical irregularity* with *adjacent subcutaneous and retrocalcaneal bursitis*, an *abnormally prominent posterosuperior calcaneal corner, and associated AT thickening or swelling (tendinosis)* [3, 11]. The calcaneal prominence is often called a "pump bump" (Fig. 12.4). Radiographs reveal a prominent posterosuperior calcaneus. The US may demonstrate swelling of the distal AT, perhaps with power Doppler (PD) activity [11, 12]. One or both posterior heel bursae may reveal an anechoic fluid collection. Ill-fitting shoes produce bony enlargement of the posterior calcaneus and subsequent soft tissue inflammation, resulting in Haglund's syndrome [12].
2. **Bony spurs (enthesophytes)** at the insertion of the AT on the posterior calcaneus may be associated with osteoarthritis, RA, or SpA [13–15].
3. **Crystal disease**. Calcium deposits from calcium pyrophosphate deposition disease (CPPD) may appear within the AT, as can monosodium urate (MSU) deposits, the latter suggesting gout. Gouty tophi may develop within or outside the AT or within the two posterior bursae [16–18].

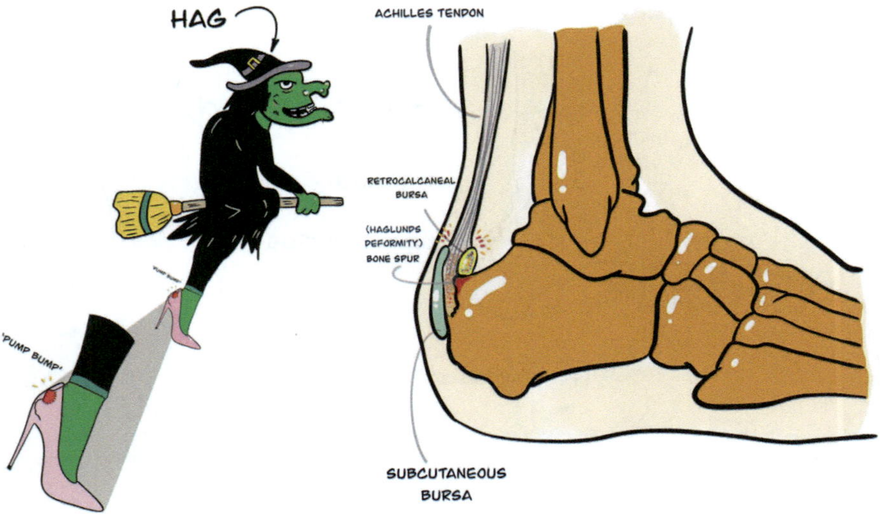

Fig. 12.4 Elements of Haglund's syndrome

4. **Rheumatoid arthritis** may produce paratenonitis and retrocalcaneal bursitis [19].
5. **Enthesopathy** with or without PD activity may be seen at the AT insertion on the posterior calcaneus or the PF insertion on the inferior calcaneus.
6. **Achilles tendon tear** (complete or partial). Tears of the AT, often caused by forceful dorsiflexion, most frequently occur 2–6 cm from the calcaneal insertion, also known as the "critical zone" [20]. For a complete AT rupture, the decision to treat it surgically or conservatively may be determined by how closely the two stump ends approximate each other in plantar flexion [21]. Partial tears appear as anechoic or hypoechoic areas or perhaps as clefts with a fibrillar interruption [3]. An AT thickness greater than 10 mm suggests a partial tear [3]. Other sonographic signs of an AT tear include distortion of the normal fibrillar echotexture, PD activity at the edge of a tear, hematoma, and possibly a gap in the tendon [22]. A healed AT tear may show residual fibrillar echotexture alteration, anterior local bulging, or a hypoechoic area at the site of the tear. Partial tears may be associated with Haglund's syndrome.
7. **Tendinosis of the Achilles tendon** may appear as focal or diffuse hypoechoic swelling with or without fibrillar disruption; there may also be loss of the anterior concavity of the AT in the transverse plane [23]. Tendinosis most often occurs in the proximal two-thirds of the AT and is best measured in the transverse axis (TAX) view [24, 25]. Tendinosis may be associated with pain; however, it is challenging to differentiate tendinosis from a partial thickness AT tear [12].

8. **Power Doppler activity** in the AT is abnormal, represents neovascularization, and may occur with inflammation, tendinosis, paratenonitis, an acute AT tear, or infection [26]. Keep the foot in a neutral position to avoid extreme tension on the AT, which may falsely obliterate PD activity [27].
9. **Paratenonitis**. The AT has no proper tenosynovial sheath, so swelling bordering the outside of the AT may be due to paratenonitis. Paratenonitis may appear as isoechoic/hypoechoic soft tissue thickening or hypoechoic fluid, perhaps with PD activity [3, 26]. Paratenonitis is best appreciated at the posterior aspect of the AT [19]. Paratendinopathy is often due to overuse and may occur with AT degeneration [28].
10. **Retrocalcaneal bursitis** can be caused by mechanical compression (shoes), direct or repetitive trauma, gout, RA, SpA, or Haglund's deformity [29]. Haglund's deformity impinges on the bursa during ankle dorsiflexion [30]. Another cause of retrocalcaneal bursitis is a misalignment of the subtalar joint axis [31].
11. **Subcutaneous calcaneal bursitis** may be seen on US superficial to the AT and can be due to direct pressure from the back of a shoe [29].
12. **Xanthomas** are painless soft tissue masses due to fat accumulation. Usually occurring in the distal AT, xanthomas may be present bilaterally. Ultrasound reveals a thickened and speckled (reticular) pattern within the AT due to hypoechoic xanthoma foci [12, 32].
13. **Integrity of Achilles tendon repair**. After repair, the AT may have heterogeneous echogenicity; however, if the fibers are continuous and there is no retraction on passive maneuvers, the repair is likely intact [33].
14. **Calcium deposits within the Achilles tendon** may be due to prior trauma (dystrophic), CPPD, or hydroxyapatite deposition disease.

(See Chap. 24 for more information.)

Proximal Calf Potential Ultrasound Findings

1. **"Tennis leg"** or MHG muscle tear occurs near the aponeurosis, resulting in disrupted fibers and possible tendon retraction [3].
2. **Plantaris tendon tears** can occur in conjunction with the MHG muscle tear or more distally associated with an AT tear. In addition, isolated PT tears rarely occur but can mimic a partial AT tear, thrombophlebitis, or a ruptured popliteal cyst [34]. Isolated PT tears are usually benign and do not require surgery.
3. **Other calf injuries**, such as those to the soleus, LHG contusion or tear, deep venous thrombosis, or ruptured Baker's cyst, may be delineated by US [35, 36].

Inferior Heel Potential Ultrasound Findings

1. **Enthesopathy** (with or without PD activity) may be seen at the PF insertion on the inferior calcaneus [37].
   (See Chap. 25 for more information.)
2. **Hypoechoic thickening (>4 mm) of the proximal PF near the calcaneus** may be due to enthesitis, plantar fasciitis, PF tears, degeneration, or edema. Tears of the PF may demonstrate partial or complete fiber disruption. However, edema may obscure these tears on US [38]. Look for PF tears in track and field athletes and patients previously treated with local corticosteroid injections [39, 40]. Confirm tears with a dynamic US exam to demonstrate the widening of any gap [1].
3. **Plantar fasciitis** is suggested by painful enlargement and hypoechogenicity of the PF near the calcaneal insertion, often with calcaneal cortical irregularities [38] (Fig. 12.5). Plantar fasciitis is frequently a result of repetitive stress in the presence of biomechanical risk factors (foot deformities, excess body mass, improper footwear, etc.). There is a loss of fibrillar echotexture and possibly increased perifascial soft tissue, calcifications, and hyperemia [38]. The origin is often degenerative; hence, a better term is plantar fasciopathy. However, SpA and RA may be associated conditions [41]. The presence of inferior calcaneal spurs (enthesophytes) may be present, but is not specific for plantar fasciitis since they occur in asymptomatic people. In one study, the mean PF thickness was 5.9 mm in patients with plantar fasciitis and 3.3 mm in asymptomatic controls [42]. Thus, PF thickening greater than 3–4.5 mm is considered abnormal, although comparison with the contralateral asymptomatic side should be taken into consideration [19]. Thickening of the proximal plantar fascia at the calcaneal origin is best seen in the LAX view [43].

Fig. 12.5 Plantar fasciitis

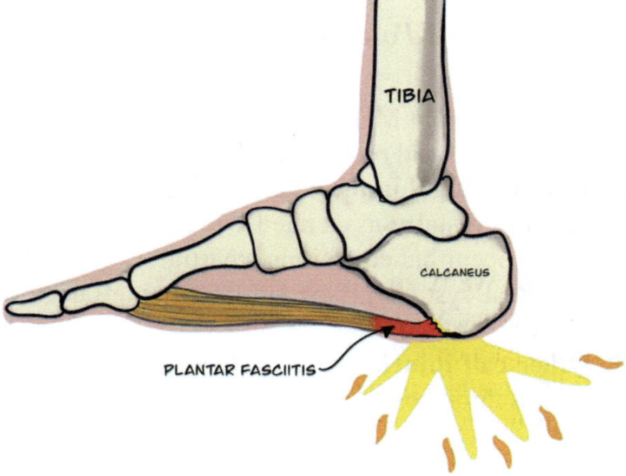

4. **Plantar fibromatosis (Ledderhose's disease)** is a benign condition that may or may not cause ambulatory pain in the central, medial, or more distal aspect of the PF [38, 44] (Fig. 12.6). The proliferation of fibroblasts in the PF may appear as **fusiform nodules** of varied echogenicity, often isoechoic or hypoechoic. The subcutaneous nodules are usually solitary, less than 3 cm in diameter, and well-demarcated, perhaps with calcium or fluid collection [38]. Ledderhose's disease may be associated with Dupuytren's contracture [45]. The nodules may be superficial and clinically visible distal to the calcaneus. The nodules are often bilateral and may demonstrate increased through-transmission and vascularity on PD.
5. **Inferior calcaneal spurs (enthesophytes)** at the insertion of the PF may be a primary pain source. Again, inferior calcaneal spurs are not specific to plantar fasciitis, but may be associated with chronic heel pain in the presence of fat pad abnormalities [38, 46].

   (See Chap. 25 for more information.)

Fig. 12.6 Ledderhose's disease

6. **Other lesions**: Xanthomas, usually asymptomatic nodules with a speckled pattern on US, may occasionally be seen near and within the PF [38, 47]. Likewise, foreign bodies, inclusion cysts, and adventitial bursae may be sonographically visualized at the inferior heel.
   (See Chap. 15 for more information.)

Lack of Ultrasound Findings with Heel Pain

Medial plantar fascial pain with an unremarkable US examination suggests **flexor digitorum brevis tendinosis** (the muscle deep to the plantar fascia) [48]. Another mimic of plantar fasciitis pain is **entrapment of Baxter's nerve** deep to the PF or between a calcaneal spur and a muscle [49] (see Chap. 14 for more information). Other conditions such as occult calcaneal stress fracture, vascular disease, heel fat pad atrophy, or necrosis may mimic plantar fasciitis [50–52].

Pitfalls

1. When examining posterior calcaneal radiographs, look for spurs (enthesophytes), since these bony protrusions may form a valley mimicking erosion on US. Also, remember to look for Haglund's deformity.
2. Scan the entire surface and verify posterior calcaneal and AT pathology in two orthogonal views [53].
3. Use a heel-to-toe maneuver at the AT calcaneal insertion to minimize anisotropy, which can mimic an insertional tear.
4. After a complete AT tear, the medially located PT may migrate posteriorly into the space previously occupied by the AT. This wandering tendon may mimic intact AT fibers and mislead the examiner into thinking there is only a partial AT tear [3].
5. The normal retrocalcaneal bursa may be challenging to visualize, but it is more detectable when distended by a pathologic process [54]. If this bursa is >2–3 mm, it may be abnormal.
6. Extreme tension of the AT (foot dorsiflexion) may falsely eliminate PD activity.
7. Ultrasound guidance enhances injection accuracy into the PF. However, for a procedure such as fenestration of the PF, a tibial nerve block at the medial ankle will improve patient comfort.
8. The AT and PF may be thickened with diabetes mellitus; PF thickness is particularly affected by body mass [55].
9. When investigating enthesopathy, remember that magnetic resonance imaging is a valuable adjunct to US since it can detect insertional bone marrow edema [56].

Method

The patient is preferentially prone, with the foot hanging off the table and slightly dorsiflexed. Choose a linear probe with an ankle or mid-depth preset. The virtual convex setting will extend the viewing field for the Achilles tendon, the plantar fascia, and the posterior ankle joints if needed.

Protocol Image 1: Achilles Tendon, Longitudinal ± Power Doppler (Fig. 12.7)

Place the probe along the longitudinal axis (LAX) with the distal end over the superior posterior calcaneus. This position is the "home base." Move the probe laterally and medially to encompass the entire width of the AT. Look for the AT insertion on the posterior calcaneus. The insertion point is approximately 1 cm long and is hypoechoic or anechoic due to the anisotropy of AT fibers. Use the heel-to-toe maneuver and slightly increase foot dorsiflexion to diminish anisotropy.

Look for AT tears, swelling, hypoechogenicity, calcifications, MSU aggregates, or tophi adjacent to or within the AT. Outside the AT, look for bony posterior calcaneal spurs, bursitis, signs of Haglund's syndrome, KFP location, calcaneal spurs,

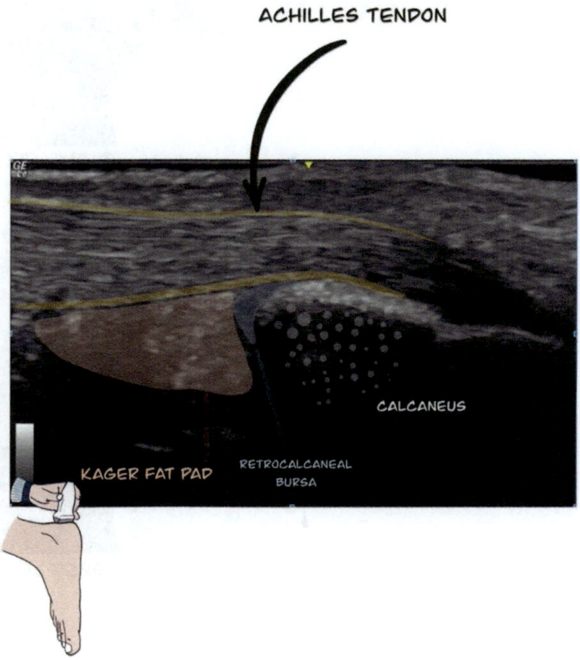

Fig. 12.7 Protocol Image 1: Achilles tendon, longitudinal ± power Doppler

and paratenonitis. Move the probe proximally to evaluate the entire length of the AT. The AT will merge into the gastrocnemius-soleus complex at the myotendinous junction. Recognize the gastrocnemius muscles (medial and lateral heads) and the deeper soleus muscle. These unipennate muscles are separated by hyperechoic muscle fascia. The MHG is more prominent and is subject to tears in middle-aged athletes (tennis leg). Such tears most often occur at the distal medial head.

Protocol Image 2: Achilles Tendon, Dynamic Scan, Longitudinal (Fig. 12.8)

Return to the "home base" position of Protocol Image 1. Keep the probe in place and slowly, passively plantar and dorsiflex the foot to look for AT tears and the relationship of movement to the enthesis, the KFP, and the two bursae. If there is a suggestion of more proximal pathology, perform this dynamic scan at that location.

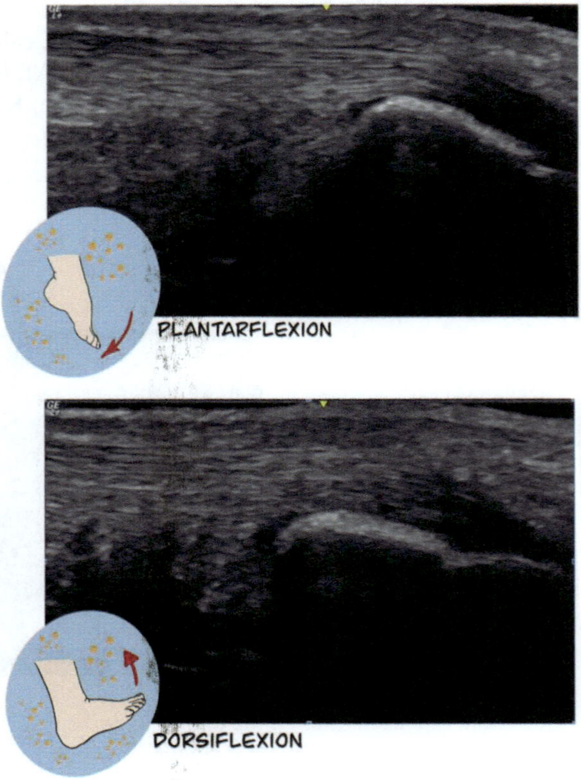

Fig. 12.8 Protocol Image 2: Achilles tendon, dynamic scan, longitudinal

Protocol Image 3: Achilles Tendon at the Enthesis, Longitudinal + Power Doppler (Fig. 12.9)

Turn on PD and scan the AT calcaneal insertion with TAX slides medially and laterally to look for enthesitis. Again, avoid extreme foot dorsiflexion since this may falsely abate the PD signal.

Protocol Image 4: Achilles Tendon, Transverse ± Power Doppler (Fig. 12.10)

Return to the calcaneus and rotate the probe 90° to TAX. Look for spurs, but be aware that some degree of posterior calcaneal irregularity is common and is not truly a spur. Consider PD if indicated. Locate the hyperechoic ovoid AT in cross-section, tilting the probe back and forth to enhance the hyperechogenicity. A trick to verify that the visualized ovoid structure is truly the AT is to visualize the AT in LAX, move the AT to the center, and then rotate the probe 90°.

Next, move the probe proximally to reach the mid tendon. The normal anterior (ventral) AT should be flat or slightly concave but not convex. Achilles tendinopathy may be supported by focal thickening, disruption of fibrils, and loss of anterior concavity. If needed, compare this image with the contralateral AT at the same distance from the calcaneal insertion. A discrepancy in the volume of the AT between two sides may indicate tendinosis of the enlarged side. Now continue to perform a proximal TAX slide while following the AT to inspect the myotendinous junction. Appreciate the emergence of the AT from the deeper lateral and medial

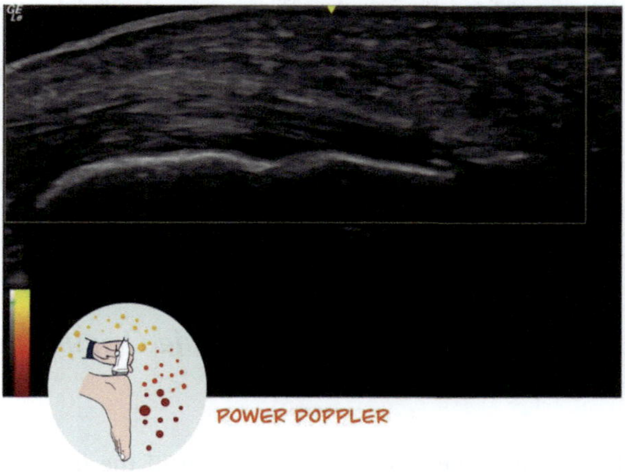

Fig. 12.9 Protocol Image 3: Achilles tendon at the enthesis, longitudinal + power Doppler

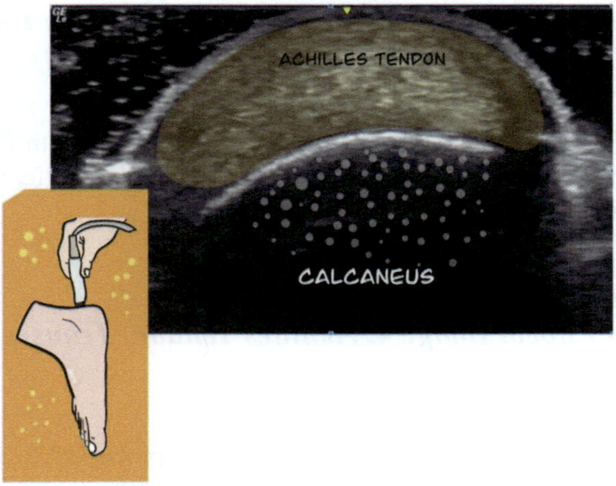

Fig. 12.10 Protocol Image 4: Achilles tendon, transverse ± power Doppler

gastrocnemius muscles and the even deeper soleus muscle. Medial to the AT, the thin, ovoid plantaris tendon may be seen deep to the gastrocnemius but superficial to the soleus.

Protocol Image 5: Posterior Tibiotalar Recess, Longitudinal ± Power Doppler (Fig. 12.11)

Now, go back to the "home base" position of Protocol Image 1 in LAX, but increase the depth to look at the deeper posterior tibiotalar and subtalar joints. The lower frequency enhances bone visualization, and the virtual convex setting extends the field of view. Passively dorsiflex the foot to open the joint space; this facilitates detection of an effusion. An effusion may distend the posterior tibiotalar joint recess, which normally contains a small amount of synovial fluid. The transducer is placed just lateral to the midline of the posterior heel and tilted to aim the US beam toward the midline to visualize posterior recess distension [57]. The bulk of the deep muscle just superficial to the joint is the flexor hallucis longus (FHL), not the soleus. Remember, as you move proximally in the calf, the soleus becomes dominant, and the FHL muscle mass decreases.

Protocol Image 6: Plantar Fascia Origin, Longitudinal (Fig. 12.12)

Place the probe in LAX at the center of the inferior calcaneus. The probe frequency may need to be further reduced to visualize the calcaneus. Perform TAX slides medially and laterally to examine the full extent of the fibrillar PF; rocking the probe will help overcome the significant anisotropy of the PF. The

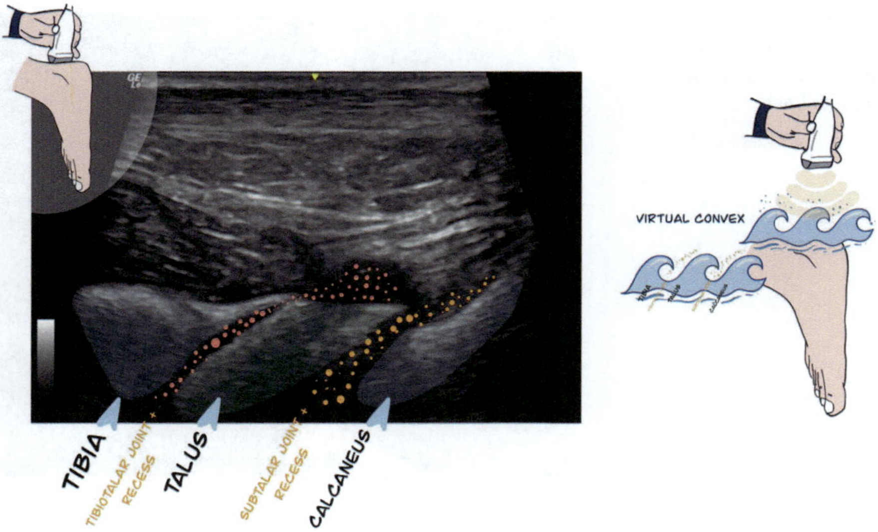

Fig. 12.11 Protocol Image 5: Posterior tibiotalar recess, longitudinal ± power Doppler

Fig. 12.12 Protocol Image 6: Plantar fascia origin, longitudinal

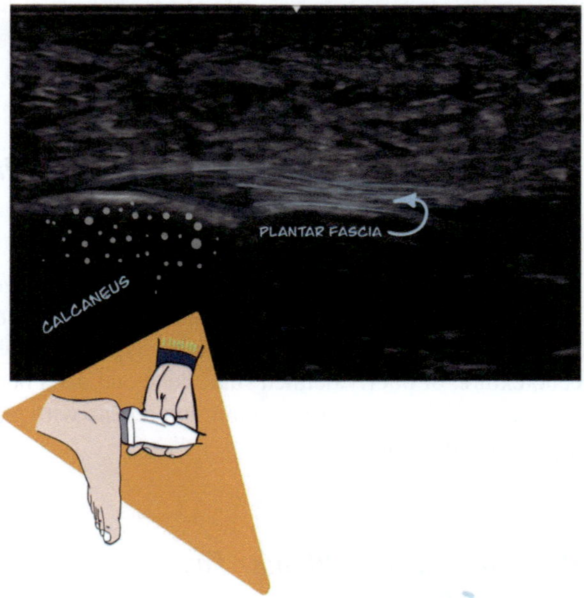

central cord of the PF is just superficial to the flexor digitorum brevis muscle (FDB). Manually dorsiflexing the ankle may clarify the PF image. Use a heel-to-toe maneuver (with plenty of gel) to diminish anisotropy, which might mimic an insertional tear of either the PF or the FDB on the calcaneus. The PF will lose fibrillar definition near the calcaneal origin, but this is normal if there is no attendant PF thickening [58].

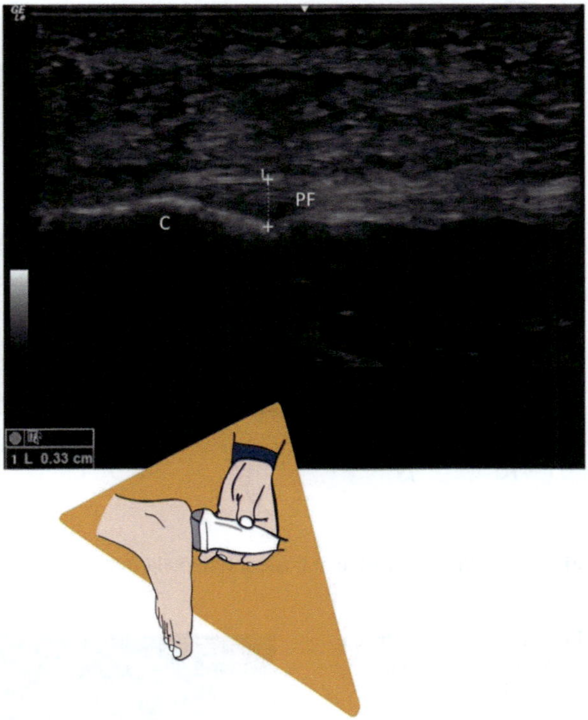

Fig. 12.13 Protocol Image 7: Plantar fascia origin, thickness measurement, longitudinal

Protocol Image 7: Plantar Fascia Origin, Thickness Measurement, Longitudinal (Fig. 12.13)

Measure the thickness of the plantar fascia. The average PF thickness is 3–4 mm [4, 58]. A comparison with the contralateral foot might be helpful. With plantar fasciitis, there may be increased thickness, hypoechoic changes with loss of fibrillar echotexture, or poorly defined superficial and deep borders of the PF. In one study, the mean PF thickness was 5.9 mm in patients with plantar fasciitis and 3.3 mm in asymptomatic controls [42].

Protocol Image 8: Plantar Fascia, Longitudinal + Power Doppler (Fig. 12.14)

Turn on PD to look for active enthesopathy at the origin, which may be from SpA, but may also occur with the more acute phase of plantar fasciitis [59]. The ankle should not be maximally dorsiflexed since this may ablate PD activity.

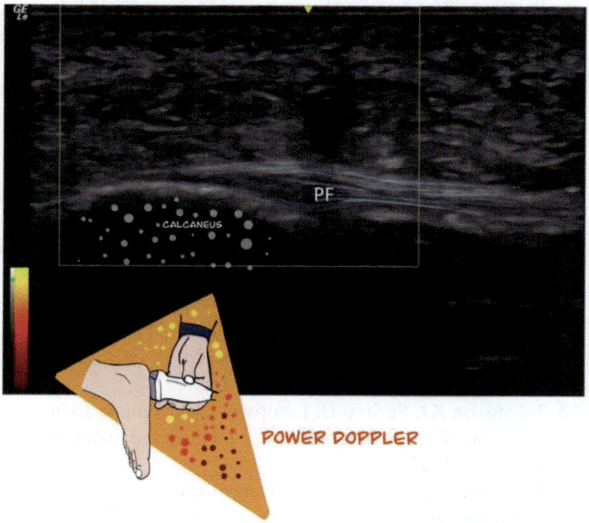

Fig. 12.14 Protocol Image 8: Plantar fascia origin, longitudinal + power Doppler

Complete Posterior Ankle Ultrasound Checklist

□ Protocol Image 1: Achilles tendon, longitudinal ± power Doppler
□ Protocol Image 2: Achilles tendon, dynamic scan, longitudinal
□ Protocol Image 3: Achilles tendon at the enthesis, longitudinal + power Doppler
□ Protocol Image 4: Achilles tendon, transverse ± power Doppler
□ Protocol Image 5: Posterior tibiotalar recess, longitudinal ± power Doppler
□ Protocol Image 6: Plantar fascia origin, longitudinal
□ Protocol Image 7: Plantar fascia origin, thickness measurement, longitudinal
□ Protocol Image 8: Plantar fascia origin, longitudinal + power Doppler

References

1. Hoffman DF, Grothe HL, Bianchi S. Sonographic evaluation of hindfoot disorders. J Ultrasound. 2014;17(2):141–50.
2. Schmidt WA, Schmidt H, Schicke B, Gromnica-Ihle E. Standard reference values for musculoskeletal ultrasonography. Ann Rheum Dis. 2004;63(8):988–94.
3. Dong Q, Fessell DP. Achilles tendon ultrasound technique. AJR Am J Roentgenol. 2009;193(3):W173.
4. Fessell DP, Vanderschueren GM, Jacobson JA, Ceulemans RY, Prasad A, Craig JG, et al. US of the ankle: technique, anatomy, and diagnosis of pathologic conditions. Radiographics. 1998;18(2):325–40.
5. Simpson SL, Hertzog MS, Barja RH. The plantaris tendon graft: an ultrasound study. J Hand Surg Am. 1991;16(4):708–11.

6. Plečko M, Knežević I, Dimnjaković D, Josipović M, Bojanić I. Accessory soleus muscle: two case reports with a completely different presentation caused by the same entity. Case Rep Orthop. 2020;2020:8851920.
7. O'Brien M. The anatomy of the Achilles tendon. Foot Ankle Clin. 2005;10(2):225–38.
8. Spang C, Alfredson H, Docking SI, Masci L, Andersson G. The plantaris tendon. Bone Joint J. 2016;98-B(10):1312–9.
9. Vlaic J, Josipovic M, Bohacek I, Jelic M. The plantaris muscle: too important to be forgotten. A review of evolution, anatomy, clinical implications and biomechanical properties. J Sports Med Phys Fitness. 2019;59(5):839–45.
10. Ehrmann C, Maier M, Mengiardi B, Pfirrmann CW, Sutter R. Calcaneal attachment of the plantar fascia: MR findings in asymptomatic volunteers. Radiology. 2014;272(3):807–14.
11. Sofka CM, Adler RS, Positano R, Pavlov H, Luchs JS. Haglund's syndrome: diagnosis and treatment using sonography. HSS J. 2006;2(1):27–9.
12. Park JW, Lee SJ, Choo HJ, Kim SK, Gwak HC, Lee SM. Ultrasonography of the ankle joint. Ultrasonography. 2017;36(4):321–35.
13. Johansson KJ, Sarimo JJ, Lempainen LL, Laitala-Leinonen T, Orava SY. Calcific spurs at the insertion of the Achilles tendon: a clinical and histological study. Muscles Ligaments Tendons J. 2012;2(4):273–7.
14. Bassiouni M. Incidence of calcaneal spurs in osteo-arthrosis and rheumatoid arthritis, and in control patients. Ann Rheum Dis. 1965;24(5):490–3.
15. Resnick D, Feingold ML, Curd J, Niwayama G, Goergen TG. Calcaneal abnormalities in articular disorders. Rheumatoid arthritis, ankylosing spondylitis, psoriatic arthritis, and Reiter syndrome. Radiology. 1977;125(2):355–66.
16. Greenberg MH. Is the patient's arthritis from crystal disease. In: Borneman PH, editor. Ultrasound for primary care. Wolters; 2020.
17. Gutierrez M, Di Geso L, Filippucci E, Grassi W. Calcium pyrophosphate crystals detected by ultrasound in patients without radiographic evidence of cartilage calcifications. J Rheumatol. 2010;37(12):2602–3.
18. Carroll M, Dalbeth N, Allen B, Stewart S, House T, Boocock M, et al. Ultrasound characteristics of the Achilles tendon in tophaceous gout: a comparison with age- and sex-matched controls. J Rheumatol. 2017;44(10):1487–92.
19. Higgs JB. Ultrasound of the ankle and foot. In: Kohler MJ, editor. Musculoskeletal ultrasound in rheumatology review. 2nd ed. Springer; 2021.
20. Scheller AD, Kasser JR, Quigley TB. Tendon injuries about the ankle. Orthop Clin North Am. 1980;11(4):801–11.
21. Thermann H, Hoffmann R, Zwipp H, Tscherne H. The use of ultrasonography in the foot and ankle. Foot Ankle. 1992;13(7):386–90.
22. Artul S, Habib G. Ultrasound findings of the painful ankle and foot. J Clin Imaging Sci. 2014;4:25.
23. Hartgerink P, Fessell DP, Jacobson JA, van Holsbeeck MT. Full- versus partial-thickness Achilles tendon tears: sonographic accuracy and characterization in 26 cases with surgical correlation. Radiology. 2001;220(2):406–12.
24. Gibbon WW, Cooper JR, Radcliffe GS. Distribution of sonographically detected tendon abnormalities in patients with a clinical diagnosis of chronic Achilles tendinosis. J Clin Ultrasound. 2000;28(2):61–6.
25. Fornage BD. Achilles tendon: US examination. Radiology. 1986;159(3):759–64.
26. Harris CA, Peduto AJ. Achilles tendon imaging. Australas Radiol. 2006;50(6):513–25.
27. Smith E, Azzopardi C, Thaker S, Botchu R, Gupta H. Power Doppler in musculoskeletal ultrasound: uses, pitfalls and principles to overcome its shortcomings. J Ultrasound. 2021;24(2):151–6.
28. Almekinders LC, Temple JD. Etiology, diagnosis, and treatment of tendonitis: an analysis of the literature. Med Sci Sports Exerc. 1998;30(8):1183–90.

29. van Dijk CN, van Sterkenburg MN, Wiegerinck JI, Karlsson J, Maffulli N. Terminology for Achilles tendon related disorders. Knee Surg Sports Traumatol Arthrosc. 2011;19(5):835–41.
30. Brinker MR, Miller MD. The adult foot. Fundamentals of orthopaedics. Philadelphia, PA: WB Saunders Co; 1999.
31. Reule CA, Alt WW, Lohrer H, Hochwald H. Spatial orientation of the subtalar joint axis is different in subjects with and without Achilles tendon disorders. Br J Sports Med. 2011;45(13):1029–34.
32. Rodriguez CP, Goyal M, Wasdahl DA. Best cases from the AFIP: atypical imaging features of bilateral Achilles tendon xanthomatosis. Radiographics. 2008;28(7):2064–8.
33. Gitto S, Draghi AG, Bortolotto C, Draghi F. Sonography of the Achilles tendon after complete rupture repair: what the radiologist should know. J Ultrasound Med. 2016;35(12):2529–36.
34. Bianchi S, Sailly M, Molini L. Isolated tear of the plantaris tendon: ultrasound and MRI appearance. Skelet Radiol. 2011;40(7):891–5.
35. Bright JM, Fields KB, Draper R. Ultrasound diagnosis of calf injuries. Sports Health. 2017;9(4):352–5.
36. Jamadar DA, Jacobson JA, Theisen SE, Marcantonio DR, Fessell DP, Patel SV, et al. Sonography of the painful calf: differential considerations. AJR Am J Roentgenol. 2002;179(3):709–16.
37. Borman P, Koparal S, Babaoğlu S, Bodur H. Ultrasound detection of entheseal insertions in the foot of patients with spondyloarthropathy. Clin Rheumatol. 2006;25(3):373–7.
38. Draghi F, Gitto S, Bortolotto C, Draghi AG, Ori Belometti G. Imaging of plantar fascia disorders: findings on plain radiography, ultrasound and magnetic resonance imaging. Insights Imaging. 2017;8(1):69–78.
39. Saxena A, Fullem B. Plantar fascia ruptures in athletes. Am J Sports Med. 2004;32(3):662–5.
40. Lee HS, Choi YR, Kim SW, Lee JY, Seo JH, Jeong JJ. Risk factors affecting chronic rupture of the plantar fascia. Foot Ankle Int. 2014;35(3):258–63.
41. Falsetti P, Frediani B, Fioravanti A, Acciai C, Baldi F, Filippou G, et al. Sonographic study of calcaneal entheses in erosive osteoarthritis, nodal osteoarthritis, rheumatoid arthritis and psoriatic arthritis. Scand J Rheumatol. 2003;32(4):229–34.
42. Gibbon W, Long G. Plantar fasciitis: US evaluation. Radiology. 1997;203(1):290.
43. Jacobson JA. Ankle, foot, and lower leg ultrasound. In: Fundamentals of musculoskeletal ultrasound. Elsevier; 2018. p. 328–406.
44. Adib O, Noizet E, Croue A, Aubé C. Ledderhose's disease: radiologic/pathologic correlation of superficial plantar fibromatosis. Diagn Interv Imaging. 2014;95(9):893–6.
45. Gudmundsson KG, Jónsson T, Arngrímsson R. Association of Morbus Ledderhose with Dupuytren's contracture. Foot Ankle Int. 2013;34(6):841–5.
46. McMillan AM, Landorf KB, Barrett JT, Menz HB, Bird AR. Diagnostic imaging for chronic plantar heel pain: a systematic review and meta-analysis. J Foot Ankle Res. 2009;2:32.
47. Filippini C, Teh J. Ultrasound features of sole of foot pathology: a review. J Ultrason. 2019;19(77):145–51.
48. Christie S, Styn G Jr, Ford G, Terryberry K. Proximal plantar intrinsic tendinopathy: anatomical and biomechanical considerations in plantar heel pain. J Am Podiatr Med Assoc. 2019;109(5):412–5.
49. Lareau CR, Sawyer GA, Wang JH, DiGiovanni CW. Plantar and medial heel pain: diagnosis and management. J Am Acad Orthop Surg. 2014;22(6):372–80.
50. Berger FH, de Jonge MC, Maas M. Stress fractures in the lower extremity. The importance of increasing awareness amongst radiologists. Eur J Radiol. 2007;62(1):16–26.
51. Siegal DS, Wu JS, Brennan DD, Challies T, Hochman MG. Plantar vein thrombosis: a rare cause of plantar foot pain. Skelet Radiol. 2008;37(3):267–9.
52. Theodorou DJ, Theodorou SJ, Resnick D. MR imaging of abnormalities of the plantar fascia. Semin Musculoskelet Radiol. 2002;6(2):105–18.
53. Kissin E, DeMarco PJ, Malone DG, Bakewell C. Ultrasound training tips & pitfalls. Rheumatologist. 2018.

54. Nazarian LN, Rawool NM, Martin CE, Schweitzer ME. Synovial fluid in the hindfoot and ankle: detection of amount and distribution with US. Radiology. 1995;197(1):275–8.
55. Abate M, Schiavone C, Di Carlo L, Salini V. Achilles tendon and plantar fascia in recently diagnosed type II diabetes: role of body mass index. Clin Rheumatol. 2012;31(7):1109–13.
56. McNally EG, Shetty S. Plantar fascia: imaging diagnosis and guided treatment. Semin Musculoskelet Radiol. 2010;14(3):334–43.
57. Smith J, Maida E, Murthy NS, Kissin EY, Jacobson JA. Sonographically guided posterior subtalar joint injections via the sinus tarsi approach. J Ultrasound Med. 2015;34(1):83–93.
58. Cardinal E, Chhem RK, Beauregard CG, Aubin B, Pelletier M. Plantar fasciitis: sonographic evaluation. Radiology. 1996;201(1):257–9.
59. Walther M, Radke S, Kirschner S, Ettl V, Gohlke F. Power Doppler findings in plantar fasciitis. Ultrasound Med Biol. 2004;30(4):435–40.

Chapter 13
Lateral Ankle

Reasons to Do the Study

1. Lateral ankle pain or swelling
2. Ankle instability, particularly after sports trauma
3. Loss of ankle eversion and plantar flexion

Questions We Want Answered

1. What is the cause of lateral ankle pain?
2. Is an underlying inflammatory condition such as spondyloarthropathy (SpA), rheumatoid arthritis (RA), or crystal disease causing the pain?
3. Is there a structural, functional, traumatic, or repetitive injury problem causing lateral ankle pain, such as a ligament or tendon tear, retinaculum tear, or painful os peroneum syndrome (POPS)?
4. Is tenosynovitis present, and if so, what is the cause?
5. What is causing soft tissue swelling?

Basic Anatomy

Bones (Fig. 13.1)

We primarily deal with the fibula, talus, and calcaneus of the lateral ankle.

The contents do not represent the views of the U.S. Department of Veterans Affairs or the United States Government.

M. H. Greenberg et al., *Manual of Musculoskeletal Ultrasound*,
https://doi.org/10.1007/978-3-031-37416-6_13

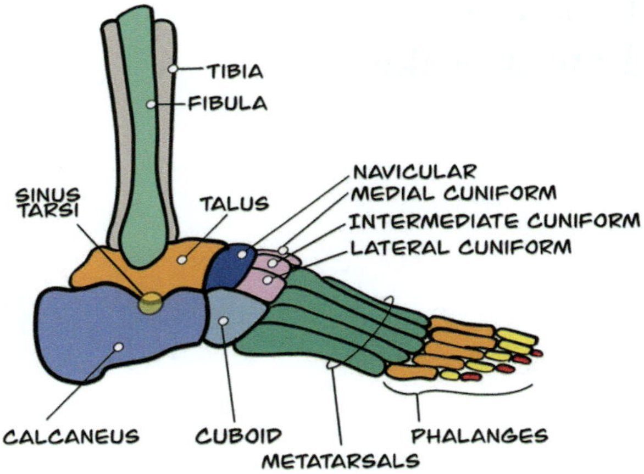

Fig. 13.1 Basic bony anatomy

Ligaments (Fig. 13.2)

Ligament names are daunting unless we remember two simple rules:

1. Ligament names start with the more prominent bone.
2. Ligaments that bind adjacent bones are logically deeper than superficial ligaments that bind nonadjacent bones.

The ligaments of the lateral ankle are divided into high ankle and low ankle ligaments [1].

High Ankle Ligaments

1. The **anterior and posterior inferior tibiofibular ligaments (AITiFL and PITiFL)** extend from the tibia to the fibula.
2. The **accessory anterior inferior tibiofibular ligament (Bassett's Ligament)** extends from the tibia to the fibula.
3. The **interosseous membrane (IOM)** (**interosseous** or **middle tibiofibular ligament**) stabilizes the tibia and tibia in relation to each other.

Low Ankle Ligaments

1. The **anterior and posterior talofibular ligaments (ATaFL and PTaFL)** extend from the fibula to the talus.
2. The **calcaneofibular ligament (CFL)** extends from the fibula to the calcaneus and is deep to the peroneal tendons.

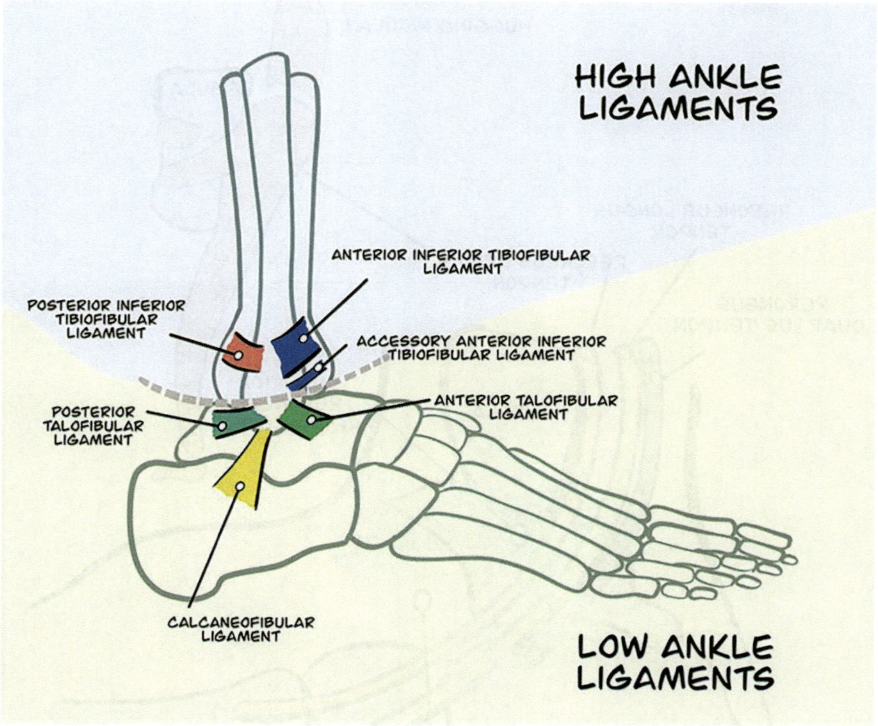

Fig. 13.2 Ligaments of the lateral foot

Muscles and Tendons (Fig. 13.3)

1. The **Peroneus brevis tendon (PBT)** inserts on the fifth metatarsal (MT) base and enables ankle eversion and plantar flexion [2].
2. The **Peroneus longus tendon (PLT)** inserts on the first MT and medial cuneiform and facilitates ankle eversion and plantar flexion.
3. An accessory tendon variant, the **peroneus quartus tendon (PQT)**, has a variable insertion, most often onto the lateral calcaneus. The PQT assists ankle eversion, depending on the insertion.

Peroneus means "related to the fibula" [3]. The fibula is shaped like a colossal hair brooch with a long stem and an oval head. The PBT and PLT tendons usually arise from their foreleg muscles before curving around the posterior distal fibula; however, the PLT myotendinous junction is more cephalad than the PBT.

We begin our sonographic focus primarily on the PBT and PLT at a point posterior to the distal fibula. The PBT is more anterior in this location and typically contacts the posterior fibula or retromalleolar groove. The PBT and the PLT curve around the groove and extend distally on each side of the peroneal tubercle (PT) on

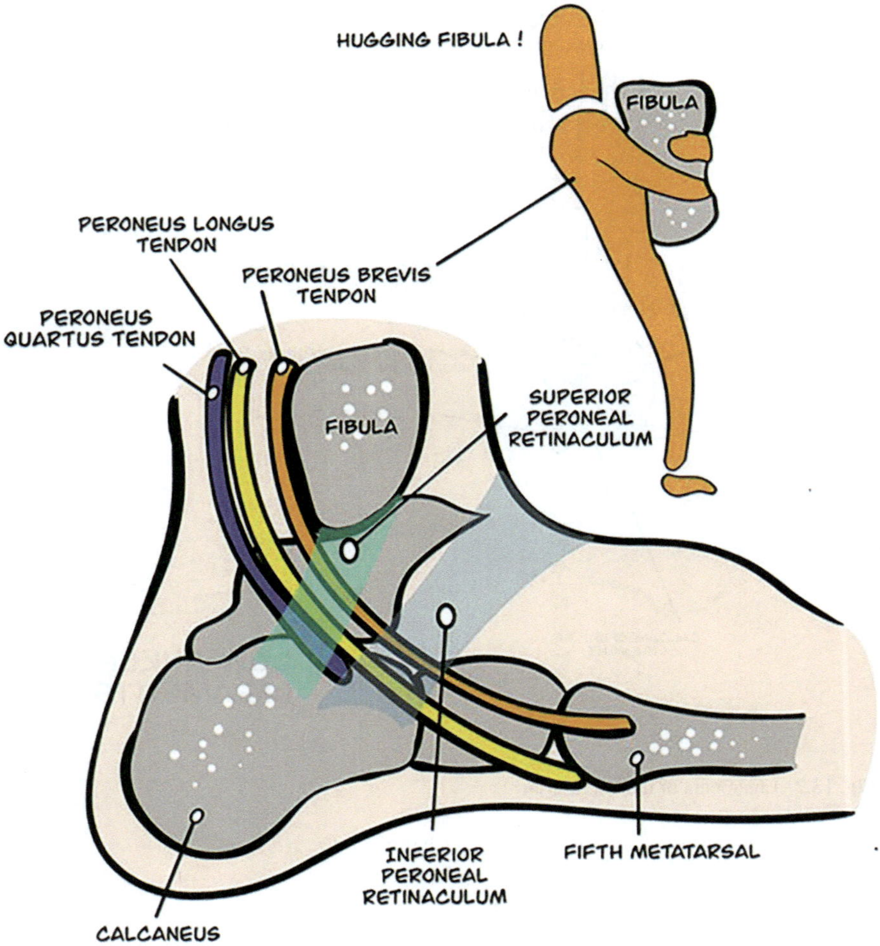

Fig. 13.3 Muscles/tendons of the lateral foot

their way to their respective insertions. The PBT is shorter ("briefer"), and the "B" in brevis reminds us that the tendon is a "bone hugger." In our protocol, we see it first hugging the distal posterior fibula and, a bit more distally, sitting on top of the PT on its way to its insertion. The PLT is longer than the PBT and goes lower ("L") and under the PBT on the way to its insertion.

Retinacula (Fig. 13.3)

The peroneal tendons are tethered by the superior and inferior peroneal retinacula [4].

1. The **superior peroneal retinaculum (SPR)** originates at the lateral malleolus and inserts on the lateral calcaneus.

2. The **inferior peroneal retinaculum (IPR)** originates from the inferior extensor retinaculum of the foot dorsum and inserts on the lateral calcaneus.

Clinical Comments

Lateral ankle injuries commonly occur during sports [5]. These injuries include damage to ligaments, peroneal tendons, and bones. Weight-bearing radiographs should be performed initially to evaluate for alignment and bone fracture. Ligaments and tendons of the lateral ankle may be assessed by magnetic resonance imaging (MRI) and ultrasound (US). MRI is the gold standard for evaluating intra-articular and osteochondral abnormalities [5]. However, US dynamically evaluates tendon motion, subluxation, and ligament integrity under stress.

Ligament Injury

About a fifth of sports-related injuries involve the lateral ankle; resultant sprains are the primary cause of posttraumatic ankle osteoarthritis [6–8]. Ankle sprains from inversion injuries cause ligament laxity and mechanical joint instability, which predisposes to osteoarthritis, osteophytes, secondary synovitis, and perhaps even anterior ankle impingement [6, 9, 10]. Chronic anterior ankle impingement occurs when bone spurs and soft tissue thickening impair normal ankle mechanical function, culminating in constant pain and reduced range of motion [7, 10].

The clinical context of the injury focuses the US exam on the specific ligament tear(s) that may have occurred. Locate the target ligament(s) to observe the overall appearance and for evidence of swelling, laxity, hypoechogenic areas, and partial or complete disruption. Apply stress to assess mechanical strength.

Ultrasound appearance of ligament sprain *grades* [5]:

Grade 1 sprain is a stretch injury without a tear: US is normal.
Grade 2 sprain is a partial tear; US reveals hypoechoic thickening with a lax or irregularly contoured ligament [11].
Grade 3 sprain is a complete tear. US demonstrates a lack of continuity (hypoechoic gap), an absent ligament, and an abnormal stress (dynamic) test [11].

Ultrasound appearance of ligament sprains [5, 8]:

1. **Acute partial thickness tears**: heterogeneous echotexture and thickening of the ligament, perhaps with power Doppler (PD) activity.
2. **Chronic partial thickness tears**: variable appearance with thickening or thinning, ligamentous elongation, or wavy contour.
3. **Chronic complete tears** will show the absence of the ligament.

Interestingly, the Latin derivation of fibula is "clasp, bolt, pin, peg, or brooch." All are fasteners. Without the distal fibula as an anchor for the lateral ankle ligaments, it is easy to see how quickly lateral ankle stability would unravel. Remember this role of the distal fibula if a lateral ligament tear is suspected.

Ligament Injury by Location

Low Ankle Ligament Injury

Low ankle ligament sprains typically arise with excessively forceful inversion, such as landing on the outside of the foot, as may occur when landing after jumping for a rebound in basketball. Low ankle ligament sprains are much more common than high ankle ligament sprains.

The following three ligaments listed comprise the Lateral Collateral Ligament (LCL) complex of the ankle [1]:

1. The ATaFL restricts excess ankle inversion and anterior displacement [5]. It is the weakest of the three ligaments and is almost always affected in an ankle sprain [5, 12, 13]. Test for an ATaFL tear with passive plantar flexion and internal rotation, known as the *anterior drawer test* [1]. The accuracy of US versus MRI detection of ATaFL injury is 91% versus 97%; however, sonography may yield false positives [14]. Be aware that the ATaFL is usually comprised of two bands, so do not misinterpret this normal anatomy for a longitudinal tear [14].
2. The CFL primarily limits excessive inversion and is second only to the ATaFL in injury frequency [5, 13]. Although isolated CFL sprains rarely occur, the CFL is usually torn sequentially after the ATaFL during an inversion-induced injury [13]. Since the CFL forms a sling for the peroneal tendons, CFL tears often cause a secondary, telltale peroneal tendon effusion due to proximity [11].
3. The PTaFL works with the ATaFL and the CFL to stabilize the ankle [15]. While sonographically accessible, the PTaFL is not routinely viewed since it is infrequently injured compared with its companion lateral ligaments.

High Ankle Ligament Injury

The ligamentous syndesmosis binding the distal tibia and fibula is located proximal to the lateral malleolus. This high ankle stabilizer is formed by the following ligaments [16, 17]:

1. The AITiFL is the least resistant to shearing forces produced by external rotation of the fibula [13]. An accessory AITiFL, also known as Bassett's ligament, is present in 94% of people and is located inferiorly to the AITiFL [17].
2. The PITiFL is the least often injured high ankle ligament and is best assessed by MRI.
3. The intraosseous tibiofibular ligament controversially contributes to ankle stability [17, 18].

High ankle ligament tears occur when an injury causes the foot and ankle to externally rotate together, exceeding the mechanical ability of the syndesmosis to withstand shearing forces acting to separate the fibula from the tibia [19]. Such force occurs with sudden twisting, cutting, or turning movements, such as when playing football, soccer, or basketball. High ankle ligament injuries occur in less than 11% of ankle sprains [20].

The index of suspicion for a high ankle injury should be increased since it may be challenging to diagnose and delayed treatment results in poor outcomes [11]. Based on the injury history, the presence of an AITiFL injury on US, radiographic widening between the tibia and fibula, or a lateral malleolar fracture, it is reasonable to order an MRI to confirm these findings and to examine the PITiFL and the ITL [5, 20]. Be aware that US diagnostic sensitivity to diagnose a high-grade injury of the AITiFL is only 66%. Hence, a negative US study warrants an MRI evaluation to avoid missing this critical diagnosis [21].

Tendon Injury

The MRI is the gold standard for imaging peroneal tendon pathology, including *fixed* subluxation or dislocation [5, 8, 22]. Ultrasound is 90% accurate for diagnosing peroneal tendon injuries [23]. An advantage of US over MR is that it delineates peroneal tendon subluxation/dislocation arising with movement [5].

The PBT and PLT are subject to acute tears and overuse injuries [5, 13]. Tendon damage (tendinosis), a common occurrence near the lateral malleolus, may be delineated by hypoechoic enlargement of the tendon with or without fiber disruption [5]. Tears appear as hypoechoic/anechoic fissures that may reach the tendon surface or travel longitudinally through the tendon, perhaps splitting the tendon. Longitudinal tears extend along the long axis and may split the tendon into two half tendons, typically occurring when retromalleolar peroneal tendon instability causes recurrent impingement between the tendon and the bone.

A common peroneal tendon sheath is present in the retromalleolar area and may exhibit tenosynovitis [24]. Tenosynovitis may occur with a tendon tear, repetitive injury, excessive tendon movement, inflammatory conditions, an enlarged PT, or other contiguous structural problems [25]. Tenosynovitis of the peroneal tendons is noted on US by anechoic or hypoechoic fluid in the tendon sheath with or without PD activity [5]. However, a small amount of tenosynovial fluid in this area is normal [26].

Peroneus Brevis Tendon

A frank longitudinal split tear affects the PBT more than the PLT [5]. In such a tear, the PLT may become interposed into the split PBT, and, in a transverse (TAX) view, the two tendons appear as three tendons and resemble a "hotdog in a bun," the bun

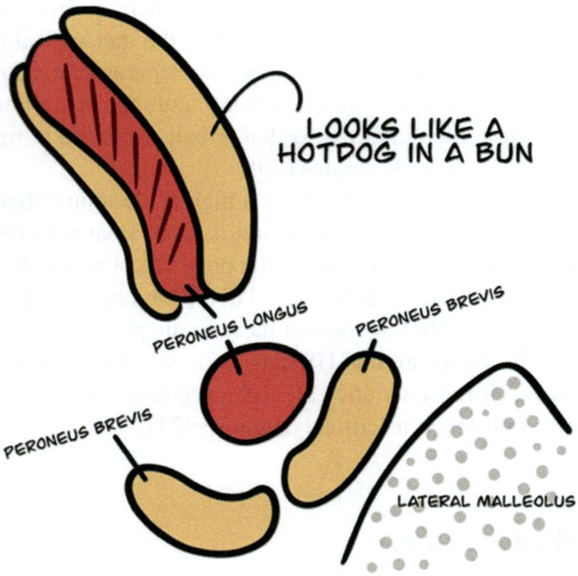

Fig. 13.4 Longitudinal split tear of the peroneus brevis tendon

being the split PBT (Fig. 13.4) [5, 27]. The PBT may also be affected by inflammatory arthropathies such as RA or SpA, producing GS or Doppler evidence of enthesitis at the insertion of the PBT on the fifth metatarsal [28].

Peroneus Longus Tendon

Isolated PLT tears are not typical and, if present, are usually observed near the cuboid bone [13]. Partial tears may sonographically present as thinning or thickening of the tendon with hypoechoic loss of normal fibular echotexture. Full-thickness tears are often associated with tenosynovitis and may be either longitudinal or complete tendon tears with retraction. An os peroneum (OP) is a normal accessory ossicle within the distal PLT near the cuboid bone; a comparable accessory bone on the medial foot is the os naviculare. An OP fracture may be associated with a distal PLT tear, instigating a search for PLT tendon stumps [24]. Likewise, a radiographic clue to a full-thickness PLT tear may be a more proximal than expected location of an OP [5].

Retinacula Injury

The SPR may be injured in athletes, thus predisposing them to tendon subluxation or dislocation [5, 22, 29]. This dorsiflexion and eversion injury typically occurs during skiing [30]. An acute SPR injury may be mistaken for an ankle sprain due to swelling, bruising, and pain at the lateral malleolus [13]. The SPR may detach with or without a bony fragment avulsion in such cases. Alternatively, the SPR may be rendered lax, allowing abnormal tendon movement, which may become symptomatic [13]. A radiographic clue to an SPR injury is an avulsion fracture anchor site on the lateral distal fibular cortex. Dynamic US maneuvers (ankle dorsiflexion, eversion, or circumduction) may show the migration of one or both peroneal tendons out of the normal position, indicating a loss of integrity of the SPR (Fig. 13.5) [5]. Pathologic peroneal tendon subluxation diagnosis is crucial since surgery may be required for successful treatment [5].

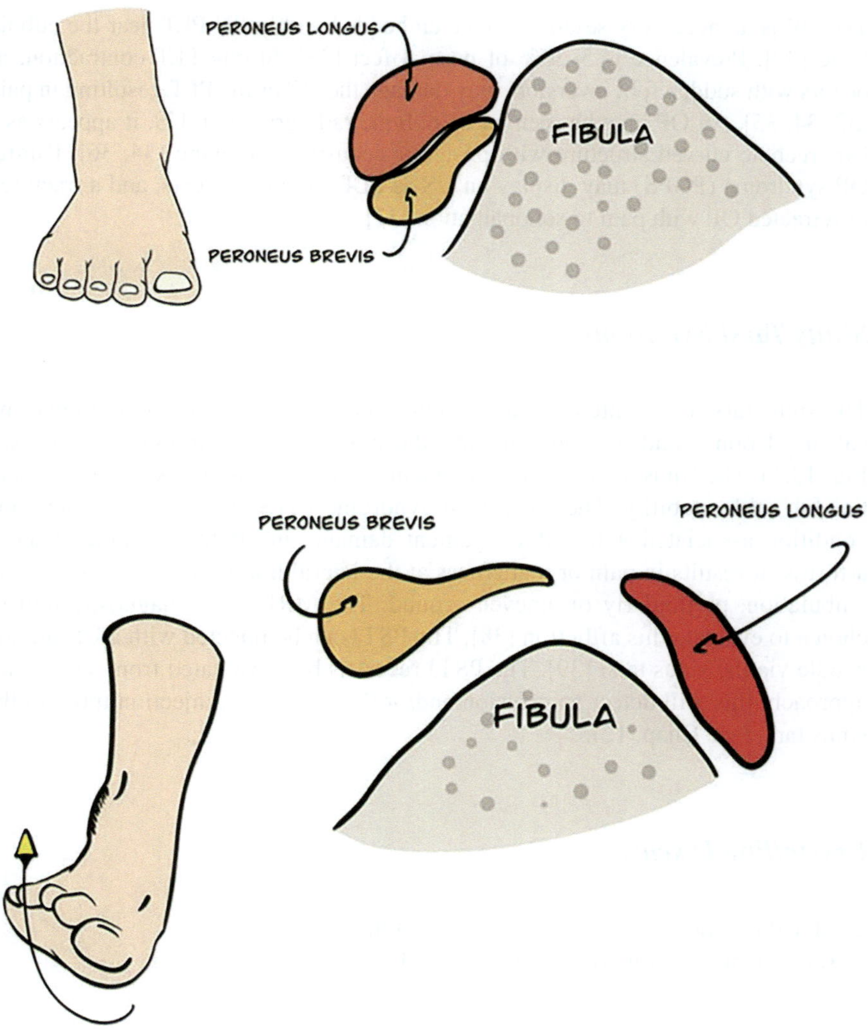

Fig. 13.5 Subluxation by dynamic exam

A subtype of peroneal tendon subluxation is intrasheath subluxation, in which the PLT and the PBT reverse regular positions within the sheath, but remain within the retromalleolar groove with an intact retinaculum [31]. Reproduction of symptoms with sonopalpation in a symptomatic patient with "popping" or "snapping" during ankle circumduction suggests that the intrasheath subluxation may be pathologic [5]. However, peroneal tendon instability is common in asymptomatic patients [31].

Painful Os Peroneum Syndrome

The OP is an accessory sesamoid bone embedded within the PLT near the cuboid bone [32]. Prevalence is 5–30% of normal feet [33]. Strong PLT contraction, as occurs with sudden foot inversion, may damage the OP or the PLT, resulting in pain [32, 34, 35]. An OP may be seen on an oblique radiograph; on US, it appears as a hyperechoic curved structure with posterior acoustic shadowing [34, 36]. Painful OP syndrome (POPS) may display on US as PLT tendinosis, tears, and a fractured or retracted OP with pain to sonopalpation [37].

Sinus Tarsi Syndrome

The sinus tarsi is an anterolateral funnel-shaped tunnel formed by the talar and calcaneal bones and is a conduit into the posterior subtalar joint (PSTJ) (see Fig. 13.1). The sinus tarsi contains ligaments and neurovascular structures essential for ankle stability. The sinus tarsi syndrome (STS) is a poorly understood condition associated with ankle ligament damage, instability, or inflammatory arthritis. It results in pain or tenderness at the lateral hindfoot, and worsens with ambulation, particularly on uneven ground. The MRI is the diagnostic tool of choice to evaluate this affliction [38]. The PSTJ can be injected with a US-guided needle via the sinus tarsi [39]. The PSTJ recess is best evaluated from a posterior approach; this will detect an effusion and/or the efficacy of injection through the sinus tarsi (see Chap. 12).

Crystalline Disease

Gout and pseudogout may present focally in the lateral ankle.

(See Chap. 24 for more information.) [5].

Variants

Variants are not uncommon and may lead to painful pathology.

Low-Lying Peroneus Brevis Muscle

A low-lying peroneus brevis muscle belly may cause overcrowding of the soft tissue in the retromalleolar groove, damaging the SPR; this may result in peroneal tendon subluxation [32, 40].

Peroneus Quartus Muscle and Tendon

The peroneus quartus muscle (PQM) is an accessory muscle present in 22% of the population [34]. The origin and insertion of the PQT vary greatly, but the most common insertion is at the lateral calcaneus [32]. The PQM may crowd the peroneal tendons in the retromalleolar area, causing tenosynovitis, swelling, pain, and even tendon tears [24, 36]. The PQM is best visualized on MRI [32]. The PQ is an imitator in the retromalleolar area; it may appear as a third tendon, mimic a low-lying PB muscle, or even simulate a longitudinal tear of the PBT [41].

Congenitally Shallow Retromalleolar Groove

This condition may facilitate peroneal tendon dislocation or subluxation [35, 42].

Peroneal Tubercle Prominence

Peroneal tubercle hypertrophy may irritate the PLT, causing tenosynovitis and a tear [32]. An adventitial bursa may also develop over the tubercle, which can become inflamed and symptomatic.

Pitfalls

1. The normal OP may be bipartite, so look for sonopalpation symptoms or retraction of the fractured bone segments. Fragment separation of >6 mm suggests an OP fracture and a full-thickness PLT tear [43].
2. If the AITiFL is torn, consider evaluating the IOM by US or MRI.

3. If there is suspicion of a high ankle tear based upon the mechanism of the injury, consider an MRI to evaluate further, *even if the symptoms improve rapidly*. These tears are difficult to detect on US and may have chronic residuals if not treated promptly.
4. If a lateral ligament tear is suspected after an inversion injury, focus on the ATaFL since that is the most injured, and then carefully inspect the CFL and PTaFL.
5. Ultrasound is not as sensitive for ATaFL tear detection as an MRI. Do not neglect to obtain an MRI if there is lingering suspicion of a lateral ligamentous sprain [44].
6. Do not forget to compare sonographically detected pathology to the contralateral side as a reference [45].
7. Since the ATaFL usually comprises two bands, do not misinterpret this normal anatomy for a longitudinal tear.
8. For a complete evaluation of the CFL, remember to dorsiflex the foot to enhance visualization.
9. When viewing the PBT and PLT in TAX along the lateral aspect of the ankle, remember that the two tendons are *not* parallel; thus, one may appear to be hypoechoic and imitate a ganglion cyst or other hypoechoic mass. Remember to tilt the probe to enhance the echogenicity of each circular structure to verify that you are looking at the tendon.

Method

The patient is supine with the knee flexed at 90° and the foot flat on the exam table. The foot is slightly inverted to reveal the lateral side of the ankle. An alternate position would be to have the supine patient extend the knee, placing the heel on a soft pad with an upright forefoot. Again, the foot is slightly inverted. A second alternative is to place the patient in a lateral decubitus position with the lateral ankle facing upward and a small pillow under the medial ankle. Use a stand-off pad if the contours of the ankle make contact difficult. Set the preset to superficial. Remember, the first seven protocol images focus on ligaments, and the remainder of the exam is concerned with tendons.

Protocol Image 1: Anterior Talofibular Ligament, Longitudinal + Anterior Drawer Test (Fig. 13.6)

Place the proximal portion of the probe on the lateral malleolus and aim the distal portion of the probe toward the large toe until you see the hyperechoic, linear ATaFL. You may need to sweep the distal end of the probe a bit more medially to better visualize this ligament. The initial view may show ligamentous anisotropy, which should aid in identification. Use the heel-to-toe maneuver to diminish

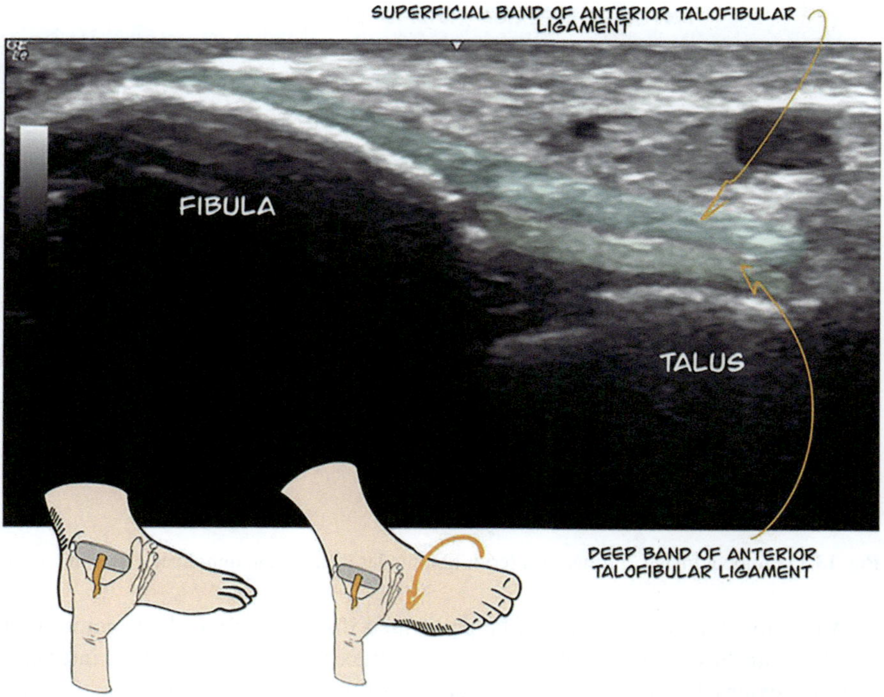

Fig. 13.6 Protocol Image 1: Anterior talofibular ligament, longitudinal + anterior drawer test

anisotropy. If the patient has increased body mass, you may need to increase the depth. Again, the ATaFL usually comprises two bands, so do not misinterpret this normal anatomy for a longitudinal tear.

To check for ligament integrity, perform the anterior drawer test. Hold the probe in place to watch the ATaFL. With your other hand, gently force the foot into simultaneous *plantar flexion* and *internal rotation* to see if the widening of the ATaFL occurs. If so, this indicates a full-thickness ATaFL tear. This maneuver improves ligament fibril visualization and thus helps to delineate partial-thickness tears as well.

Protocol Image 2: Anterior Inferior Tibiofibular Ligament, Longitudinal ± Dynamic View (Fig. 13.7)

Maintain the posterior edge of the probe on the lateral malleolus tip and cranially rotate the distal edge to see the distal aspect of the hyperechoic tibia. The ankle should be slightly inverted. Between these two bones, the hyperechoic fibrillar AITiFL is visualized. The angle of the probe should be about 45°. The AITiFL is short and thick. Elucidating the fibrillar echotexture may require slight probe rotation and tilting.

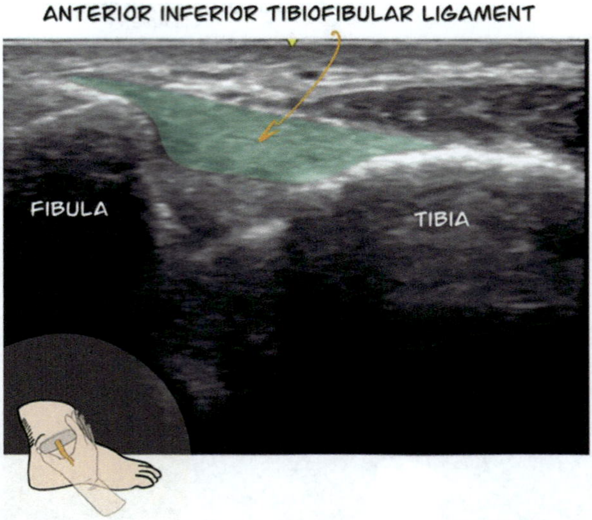

Fig. 13.7 Protocol Image 2: Anterior inferior tibiofibular ligament, longitudinal ± dynamic view

An optional dynamic view can be performed if there is clinical suspicion of an AITiFL tear. Passively dorsiflex the foot and maximally rotate the foot internally and then externally, observing for a widening of the distance between the tibia and fibula, which would indicate a complete tear [46]. This maneuver may likewise illuminate partial tears since it will tighten the ligament to better demonstrate discontinuity in the fibrillar echotexture.

Protocol Image 2a (Optional): Accessory Anterior Inferior Tibiofibular Ligament, Longitudinal

This accessory ligament is present in 94% of normal ankles and offers another view of a potentially injured ligament. Perform a distal TAX slide to visualize this ligament. Tilt and rock the transducer to maximally enhance the echogenicity of the ligament. Look for full and partial tears with the same dynamic maneuver as in Protocol Image 2.

Protocol Image 2b (Optional): Distal Interosseous Membrane

Evaluate the distal IOM if there is clinical suspicion of a high ankle sprain, mainly if there is evidence of injury to the AITiFL or the accessory AITiFL. Place the transducer in TAX between the distal tibia and fibula proximal to the AITiFL anteriorly. The IOM is thin and hyperechoic; nonvisualization may indicate injury.

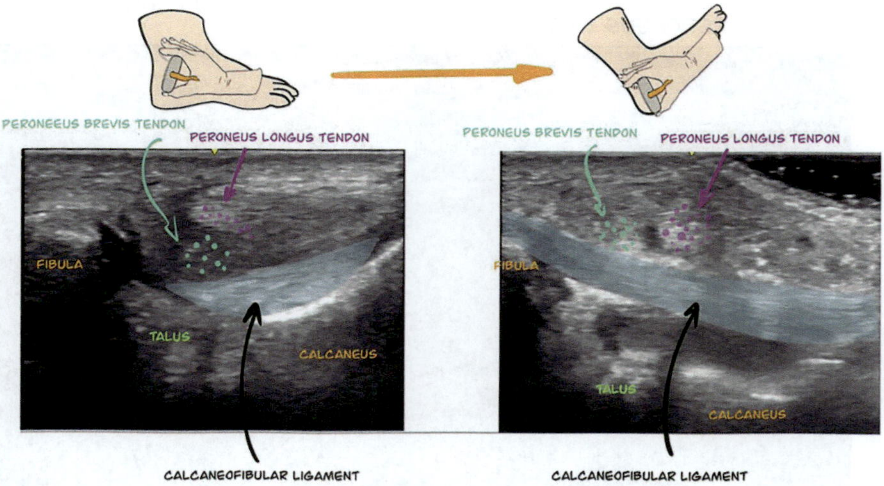

Fig. 13.8 Protocol Image 3: Calcaneofibular ligament, longitudinal + dynamic

Protocol Image 3: Calcaneofibular Ligament, Longitudinal + Dynamic (Fig. 13.8)

Keep the proximal edge of the probe on the inferior lateral malleolus, and aim the distal edge about 20° toward the heel tip. Look for the round peroneal tendons in TAX. The ligament is deep to the peroneal tendons and is curved, so its visibility will be *impaired* until dynamic maneuvers are performed.

For the dynamic view, the patient actively dorsiflexes the foot, or you passively invert the ankle. The now taut CFL is better delineated: it is now linear, fibrillar, and hyperechoic. If the ligament is intact, the peroneal tendons will be displaced superficially (in a lateral direction). No peroneal tendon displacement implies a complete ligament tear is present. Even with a dynamic view, there may still be anisotropy of at least one of the peroneal tendons, imitating a complex ganglion cyst.

Protocol Image 4: Sinus Tarsi (Fig. 13.9)

The foot is in slight plantar flexion and inversion, but relaxed. Rotate the transducer into a vertical position and perform a distal TAX slide so that the transducer slides off the lateral malleolus. Once the curved calcaneus is located, perform a small LAX cephalad slide so that the transducer is just distal (anterior) to the lateral malleolus. Between the cephalad talus and the caudal calcaneus lies the sinus tarsi.

Do not mistake anisotropy of the peroneal tendons and/or normal tenosynovial effusion for effusion or a cyst. The sinus tarsi is the gateway to injecting the PSTJ. The injection of the PSTJ via the sinus tarsi is described elsewhere. [39].

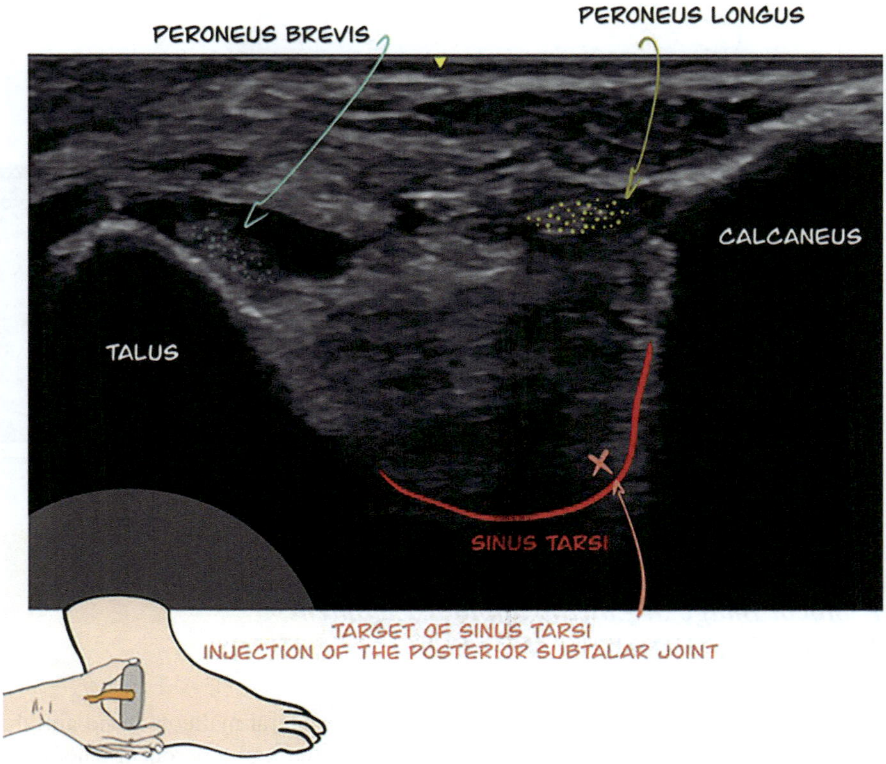

Fig. 13.9 Protocol Image 4: Sinus tarsi

Protocol Image 5: Peroneal Tendons at Supramalleolar Region, Transverse ± Power Doppler (Fig. 13.10)

Next, place one end of the probe against the posterior aspect of the lateral malleolus and aim the other end of the probe slightly downward toward the *posterior* portion of the talus; the probe should be transverse to the fibula. If both the PBT and PBM are "normal" (not torn, not displaced, or there is no low-lying PBM), the hyperechoic PBT is seen hugging the lateral malleolus contiguous to the larger, rounder PLT, which is more superficial. The even larger hypoechoic PBM is deep to the PBT. The SPR, attached to the lateral malleolus, is superficial to both the PBT and the PLT.

Protocol Image 6: Peroneal Tendons at Lateral Malleolus, Transverse ± Power Doppler + Dynamic (Fig. 13.11)

Angle the posterior portion of the probe down another 10–20° and perform a TAX slide distally. Note the hyperechoic bony cortex of the lateral malleolus. If there is a low-lying PBM, the PBT may not be fully present at this level. The hyperechoic

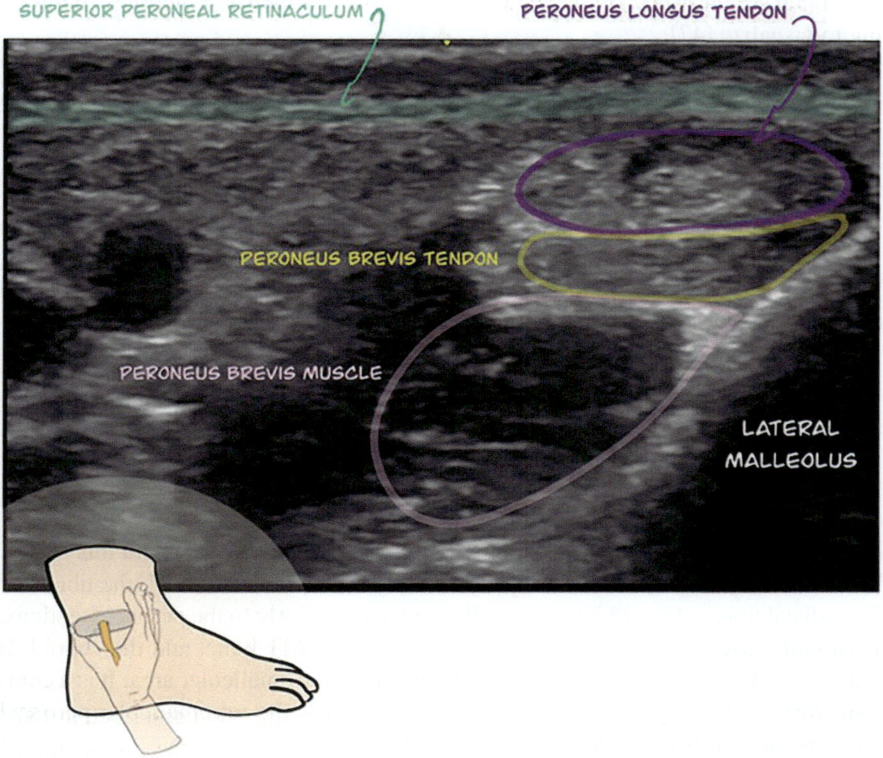

Fig. 13.10 Protocol Image 5: Peroneal tendons at supramalleolar region, transverse ± power Doppler

Fig. 13.11 Protocol Image 6: Peroneal tendons at lateral malleolus, transverse ± power Doppler + dynamic

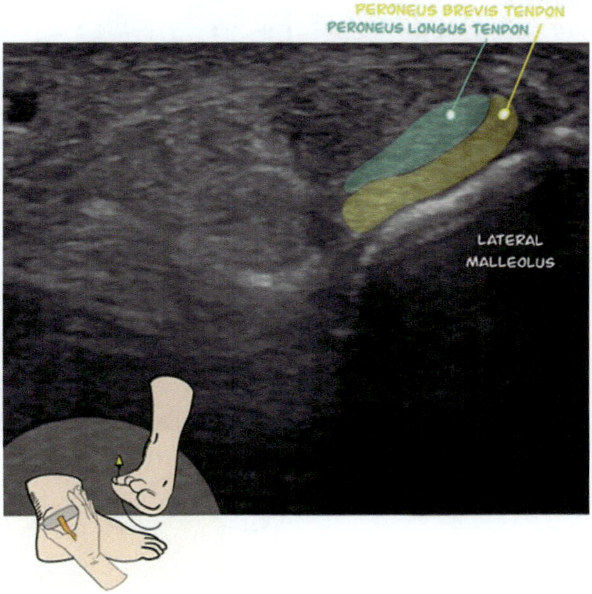

horizontal structure deep to the tendons is the PTaFL, which is sometimes challenging to visualize [47].

Next, dorsiflex and evert the ankle to test the integrity of the SPR. If the peroneal tendons transiently migrate (subluxate) out of their usual "stacked" position (with the PBT normally touching the bone), the SPR may not be intact. In such a case, the SPR may not be directly torn but may have avulsed from its periosteal insertion. Alternatively, ask the patient to circumduct the foot by drawing a large imaginary circle with the big toe. An alternative recommended patient position for dynamic US evaluation is to have the patient lie prone to allow for muscle relaxation [48]. Use copious gel and a light touch when performing dynamic maneuvers to avoid constraining abnormal tendon movement.

Protocol Image 6a: Peroneus Quartus at Lateral Malleolus, Transverse ± Power Doppler (Fig. 13.12)

Depending on the anatomy of the patient, you may not be able to obtain this image. This accessory muscle and tendon may variably arise (from the PBM, the fibula, or from distal fibers of the PLT) and usually course posteriorly to the peroneal tendons. It variably inserts on the calcaneal tubercle, the fifth MT bone, and the cuboid. It may be predominantly muscular or tendinous in the retromalleolar area. It is a common variant and may be a source of pain (if it crowds the retromalleolar peroneal tendons) or confusion (mimics a longitudinal split PBT).

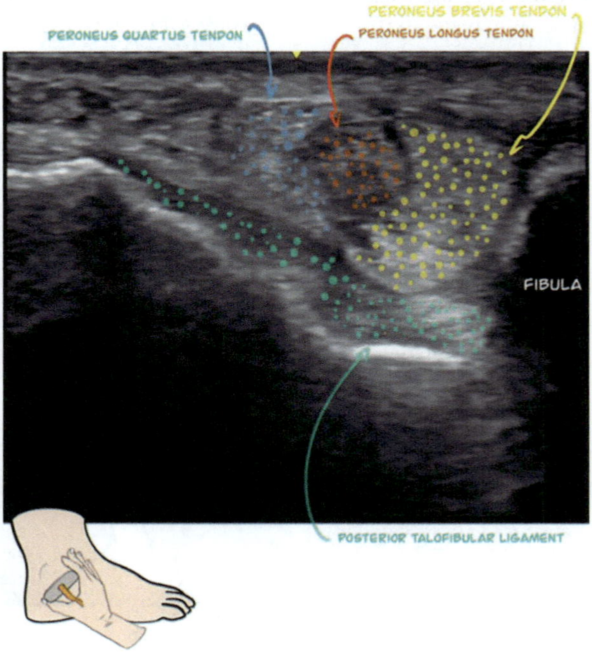

Fig. 13.12 Protocol Image 6a: Peroneus quartus at lateral malleolus, transverse

Protocol Image 7: Peroneal Tendons at Peroneal Tubercle, Transverse ± Power Doppler (Fig. 13.13)

Move the probe (TAX slide) more distally, with the inferior edge now posteriorly angled at about 80–90° to the horizontal. The probe is almost vertical. Slowly move the probe distally off the lateral malleolus. Note that up to 3 mm of fluid within the common peroneal tendon sheath is normal in this area [30, 49].

Continue to perform a TAX slide until you visualize a hyperechoic hill-like protuberance, the PT. The PBT and the PLT are on either side of the tubercle, with the PBT being cephalad and resting on the tubercle. The two tendons are *not* parallel, so one will exhibit anisotropy while the other does not, as seen when the probe is tilted in either direction. This image is reminiscent of a pair of aviator glasses, with the PT being the bridge and the peroneal tendons being the lenses. *Do not mistake either one of these two tendons for a ganglion cyst.*

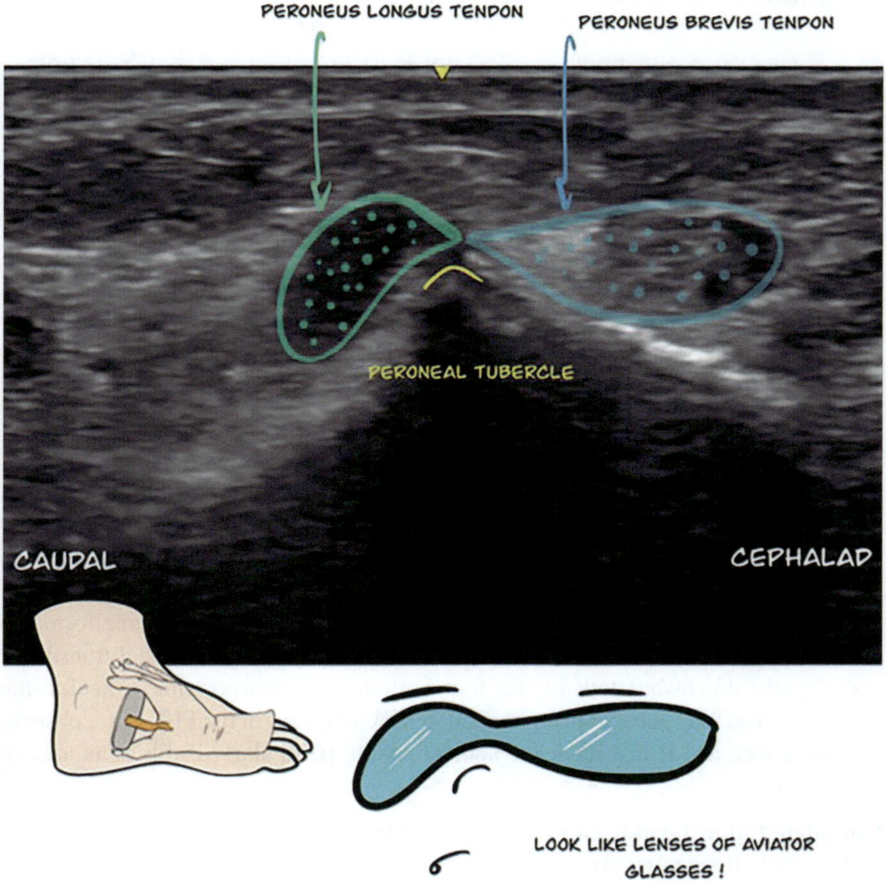

Fig. 13.13 Protocol Image 7: Peroneal tendons at peroneal tubercle, transverse ± power Doppler

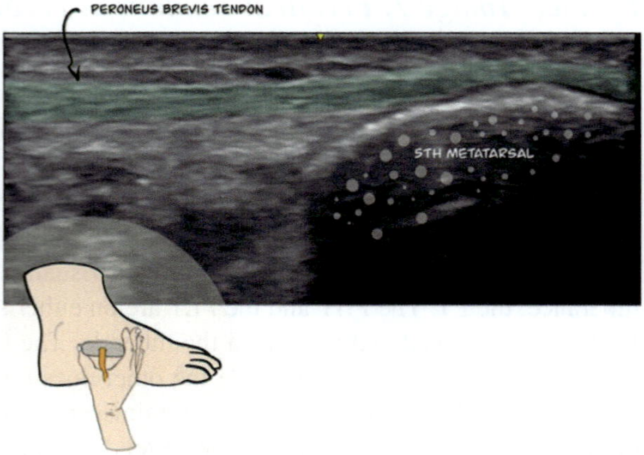

Fig. 13.14 Protocol Image 8: Peroneus brevis tendon at insertion, longitudinal ± power Doppler

Protocol Image 8: Peroneus Brevis Tendon Insertion, Longitudinal ± Power Doppler (Fig. 13.14)

Center the probe on the PBT and rotate 90° to see the fibrillar, hyperechoic PBT in LAX view. Follow the tendon distally to its insertion at the base of the fifth metatarsal; this insertion may be indistinct due to anisotropy, thus mimicking enthesitis [50]. However, Doppler activity at the insertion may indicate genuine enthesitis [28, 50]. As always, minimize anisotropy by keeping the transducer (and US beam) perpendicular to the insertion when evaluating for enthesopathy [28].

Protocol Image 9 (Optional): Peroneus Longus Tendon Distal to Peroneal Tubercle, Longitudinal ± Power Doppler

Go back to the view in Protocol Image 7 and center the probe in TAX on the PLT, which is located closer to the plantar surface than the PBT. Rotate the probe 90° to see the hyperechoic, fibrillar PLT in LAX. Perform a distal LAX slide, maintaining the tendon in the center and tilting the probe back and forth, sustaining the hyperechoic appearance of the tendon. The tendon will plunge toward the sole of the foot on its way to the insertion on the first metatarsal and the medial cuneiform. To maintain tendon visualization, counter anisotropy by angling the distal transducer end slightly downward toward the foot bottom and incorporating a heel-to-toe maneuver. You may see a hyperechoic OP bone embedded in the PLT. If so, observe its appearance and if it is tender to sonopalpation, particularly if this is an area of complaint.

Complete Lateral Ankle Ultrasound Checklist

☐ Protocol Image 1: Anterior talofibular ligament, longitudinal + anterior drawer test

☐ Protocol Image 2: Anterior inferior tibiofibular ligament, longitudinal ± dynamic view

- ☐ Protocol Image 2a (optional): Accessory anterior inferior tibiofibular ligament, longitudinal
- ☐ Protocol Image 2b (optional): Distal interosseous membrane
- ☐ Protocol Image 3: Calcaneofibular ligament, longitudinal + dynamic
- ☐ Protocol Image 4: Sinus tarsi
- ☐ Protocol Image 5: Peroneal tendons at supramalleolar region, transverse ± power Doppler
- ☐ Protocol Image 6: Peroneal tendons at lateral malleolus, transverse ± power Doppler + dynamic
- ☐ Protocol Image 6a: Peroneus quartus at lateral malleolus, transverse ± power Doppler
- ☐ Protocol Image 7: Peroneal tendons at peroneal tubercle, transverse ± power Doppler
- ☐ Protocol Image 8: Peroneus brevis tendon insertion, longitudinal ± power Doppler
- ☐ Protocol Image 9 (optional): Peroneus longus tendon distal to peroneal tubercle, longitudinal ± power Doppler

References

1. Sconfienza LM, Orlandi D, Lacelli F, Serafini G, Silvestri E. Dynamic high-resolution US of ankle and midfoot ligaments: normal anatomic structure and imaging technique. Radiographics. 2015;35(1):164–78.
2. Walt J, Massey P. Peroneal tendon syndromes. StatPearls. Treasure Island, FL: StatPearls Publishing; 2022.
3. Peroneal. The American Heritage Dictionary 2022. https://ahdictionary.com/word/search.html?q=peroneal.
4. Bianchi S, Becciolini M. Ultrasound features of ankle retinacula: normal appearance and pathologic findings. J Ultrasound Med. 2019;38(12):3321–34.
5. Kanal S, Saif M, Scher CE, Davis LC, editors. Ultrasound and MRI evaluation of the lateral ankle. American College of Osteopathic Radiology; 2020.
6. Battaglia PJ, Craig K, Kettner NW. Ultrasonography in the assessment of lateral ankle ligament injury, instability, and anterior ankle impingement: a diagnostic case report. J Chiropr Med. 2015;14(4):265–9.
7. Valderrabano V, Hintermann B, Horisberger M, Fung TS. Ligamentous posttraumatic ankle osteoarthritis. Am J Sports Med. 2006;34(4):612–20.
8. Rosenberg ZS, Beltran J, Bencardino JT. From the RSNA Refresher Courses. Radiological Society of North America. MR imaging of the ankle and foot. Radiographics. 2000;20 Spec No:S153–79.
9. Bonnel F, Toullec E, Mabit C, Tourné Y. Chronic ankle instability: biomechanics and pathomechanics of ligaments injury and associated lesions. Orthop Traumatol Surg Res. 2010;96(4):424–32.
10. Robinson P, White LM. Soft-tissue and osseous impingement syndromes of the ankle: role of imaging in diagnosis and management. Radiographics. 2002;22(6):1457–69; discussion 70–1.
11. McKiernan S, Fenech M, Fox D, Stewart I. Sonography of the ankle: the lateral ankle and ankle sprains. Sonography. 2017;4(4):146–55.
12. Martin RL, Davenport TE, Paulseth S, Wukich DK, Godges JJ. Ankle stability and movement coordination impairments: ankle ligament sprains. J Orthop Sports Phys Ther. 2013;43(9):A1–40.

13. Park JW, Lee SJ, Choo HJ, Kim SK, Gwak HC, Lee SM. Ultrasonography of the ankle joint. Ultrasonography. 2017;36(4):321–35.
14. Choo HJ, Lee SJ, Kim DW, Jeong HW, Gwak H. Multibanded anterior talofibular ligaments in normal ankles and sprained ankles using 3D isotropic proton density-weighted fast spin-echo MRI sequence. AJR Am J Roentgenol. 2014;202(1):W87–94.
15. Rasmussen O, Jensen IT, Hedeboe J. An analysis of the function of the posterior talofibular ligament. Int Orthop. 1983;7(1):41–8.
16. Golanó P, Vega J, de Leeuw PA, Malagelada F, Manzanares MC, Götzens V, et al. Anatomy of the ankle ligaments: a pictorial essay. Knee Surg Sports Traumatol Arthrosc. 2010;18(5):557–69.
17. Alves T, Dong Q, Jacobson J, Yablon C, Gandikota G. Normal and injured ankle ligaments on ultrasonography with magnetic resonance imaging correlation. J Ultrasound Med. 2019;38(2):513–28.
18. Hoefnagels EM, Waites MD, Wing ID, Belkoff SM, Swierstra BA. Biomechanical comparison of the interosseous tibiofibular ligament and the anterior tibiofibular ligament. Foot Ankle Int. 2007;28(5):602–4.
19. Lin CF, Gross ML, Weinhold P. Ankle syndesmosis injuries: anatomy, biomechanics, mechanism of injury, and clinical guidelines for diagnosis and intervention. J Orthop Sports Phys Ther. 2006;36(6):372–84.
20. Hermans JJ, Beumer A, de Jong TA, Kleinrensink GJ. Anatomy of the distal tibiofibular syndesmosis in adults: a pictorial essay with a multimodality approach. J Anat. 2010;217(6):633–45.
21. Milz P, Milz S, Steinborn M, Mittlmeier T, Putz R, Reiser M. Lateral ankle ligaments and tibiofibular syndesmosis. 13-MHz high-frequency sonography and MRI compared in 20 patients. Acta Orthop Scand. 1998;69(1):51–5.
22. Saragas NP, Ferrao PN, Mayet Z, Eshraghi H. Peroneal tendon dislocation/subluxation - case series and review of the literature. Foot Ankle Surg. 2016;22(2):125–30.
23. Grant TH, Kelikian AS, Jereb SE, McCarthy RJ. Ultrasound diagnosis of peroneal tendon tears. A surgical correlation. J Bone Joint Surg Am. 2005;87(8):1788–94.
24. Molini L, Bianchi S. US in peroneal tendon tear. J Ultrasound. 2014;17(2):125–34.
25. Al-Nuaimi A, Shalan K. Aging and tenosynovitis of the peroneal tendons: the cause and management. MOJ Anat Physiol. 2019;6:25–8.
26. Rawool NM, Nazarian LN. Ultrasound of the ankle and foot. Semin Ultrasound CT MR. 2000;21(3):275–84.
27. Philbin TM, Landis GS, Smith B. Peroneal tendon injuries. J Am Acad Orthop Surg. 2009;17(5):306–17.
28. Ward IM, Kissin E, Kaeley G, Scott JN, Newkirk M, Hildebrand BA, et al. Ultrasound features of the posterior tibialis tendon and peroneus brevis tendon entheses: comparison study between healthy adults and those with inflammatory arthritis. Arthritis Care Res (Hoboken). 2017;69(10):1519–25.
29. Mendicino RW, Orsini RC, Whitman SE, Catanzariti AR. Fibular groove deepening for recurrent peroneal subluxation. J Foot Ankle Surg. 2001;40(4):252–63.
30. Fessell DP, Vanderschueren GM, Jacobson JA, Ceulemans RY, Prasad A, Craig JG, et al. US of the ankle: technique, anatomy, and diagnosis of pathologic conditions. Radiographics. 1998;18(2):325–40.
31. Neustadter J, Raikin SM, Nazarian LN. Dynamic sonographic evaluation of peroneal tendon subluxation. AJR Am J Roentgenol. 2004;183(4):985–8.
32. Wang XT, Rosenberg ZS, Mechlin MB, Schweitzer ME. Normal variants and diseases of the peroneal tendons and superior peroneal retinaculum: MR imaging features. Radiographics. 2005;25(3):587–602.
33. Oh SJ, Kim YH, Kim SK, Kim MW. Painful os peroneum syndrome presenting as lateral plantar foot pain. Ann Rehabil Med. 2012;36(1):163–6.
34. Muir JJ, Curtiss HM, Hollman J, Smith J, Finnoff JT. The accuracy of ultrasound-guided and palpation-guided peroneal tendon sheath injections. Am J Phys Med Rehabil. 2011;90(7):564–71.

35. Niemi WJ, Savidakis J Jr, DeJesus JM. Peroneal subluxation: a comprehensive review of the literature with case presentations. J Foot Ankle Surg. 1997;36(2):141–5.
36. Martinelli B, Bernobi S. Peroneus quartus muscle and ankle pain. Foot Ankle Surg. 2002;8(3):223–5.
37. Smania L, Craig JG, von Holsbeeck M. Ultrasonographic findings in peroneus longus tendon rupture. J Ultrasound Med. 2007;26(2):243–6.
38. Kim TH, Moon SG, Jung HG, Kim NR. Subtalar instability: imaging features of subtalar ligaments on 3D isotropic ankle MRI. BMC Musculoskelet Disord. 2017;18(1):475.
39. Smith J, Maida E, Murthy NS, Kissin EY, Jacobson JA. Sonographically guided posterior subtalar joint injections via the sinus tarsi approach. J Ultrasound Med. 2015;34(1):83–93.
40. Mirmiran R, Squire C, Wassell D. Prevalence and role of a low-lying peroneus brevis muscle belly in patients with peroneal tendon pathologic features: a potential source of tendon subluxation. J Foot Ankle Surg. 2015;54(5):872–5.
41. Chepuri NB, Jacobson JA, Fessell DP, Hayes CW. Sonographic appearance of the peroneus quartus muscle: correlation with mr imaging appearance in seven patients. Radiology. 2001;218(2):415–9.
42. van Dijk PAD, Vopat BG, Guss D, Younger A, DiGiovanni CW. Retromalleolar groove deepening in recurrent peroneal tendon dislocation: technique tip. Orthop J Sports Med. 2017;5(5):2325967117706673.
43. Brigido MK, Fessell DP, Jacobson JA, Widman DS, Craig JG, Jamadar DA, et al. Radiography and US of os peroneum fractures and associated peroneal tendon injuries: initial experience. Radiology. 2005;237(1):235–41.
44. Oae K, Takao M, Uchio Y, Ochi M. Evaluation of anterior talofibular ligament injury with stress radiography, ultrasonography and MR imaging. Skelet Radiol. 2010;39(1):41–7.
45. Rossi F, Zaottini F, Picasso R, Martinoli C, Tagliafico AS. Ankle and foot ultrasound: reliability of side-to-side comparison of small anatomic structures. J Ultrasound Med. 2019;38(8):2143–53.
46. Baltes TPA, Arnáiz J, Geertsema L, Geertsema C, D'Hooghe P, Kerkhoffs G, et al. Diagnostic value of ultrasonography in acute lateral and syndesmotic ligamentous ankle injuries. Eur Radiol. 2021;31(4):2610–20.
47. Boonthathip M, Chen L, Trudell D, Resnick D. Lateral ankle ligaments: MR arthrography with anatomic correlation in cadavers. Clin Imaging. 2011;35(1):42–8.
48. Pesquer L, Guillo S, Poussange N, Pele E, Meyer P, Dallaudière B. Dynamic ultrasound of peroneal tendon instability. Br J Radiol. 2016;89(1063):20150958.
49. Nazarian LN, Rawool NM, Martin CE, Schweitzer ME. Synovial fluid in the hindfoot and ankle: detection of amount and distribution with US. Radiology. 1995;197(1):275–8.
50. Higgs JB. Ultrasound of the ankle and foot. In: Kohler MJ, editor. Musculoskeletal ultrasound in rheumatology review. 2nd ed. Springer; 2021.

Chapter 14
Medial Ankle

Reasons to Do the Study

1. Medial ankle pain
2. Medial ankle swelling
3. Evaluation for inflammatory polyarthropathy
4. Paresthesia or pain in the medial ankle or bottom of the foot

Questions We Want Answered

1. What is the cause of medial ankle pain?
2. Is there an underlying inflammatory condition such as spondyloarthropathy (SpA), rheumatoid arthritis (RA), or crystal disease?
3. Is a structural, functional, traumatic, or repetitive injury causing medial ankle pain, for example, a ligament, tendon, or retinaculum tear?
4. Is tenosynovitis present, and if so, what is the cause?
5. What is causing localized soft tissue swelling?
6. Does this patient have tarsal tunnel syndrome?

Basic Bone Anatomy (Fig. 14.1)

The bony anatomy is from superior to inferior to distal: **Ti**bia, **Tal**us, **Ca**lcaneus, **Na**vicular, **Cu**neiform, and **Meta**tarsal. The mnemonic "**Tim**'s **Tal**king **Ca**t **Na**te **Cu**ts **Meta**l" may be helpful to remember the bones in this location.

The contents do not represent the views of the U.S. Department of Veterans Affairs or the United States Government.

M. H. Greenberg et al., *Manual of Musculoskeletal Ultrasound*,
https://doi.org/10.1007/978-3-031-37416-6_14

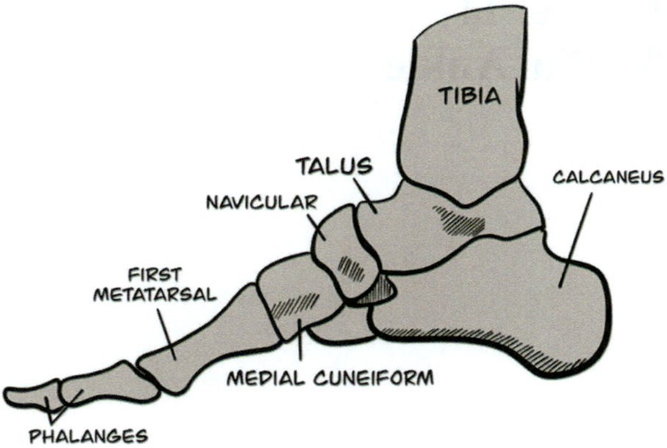

Fig. 14.1 Basic bony anatomy

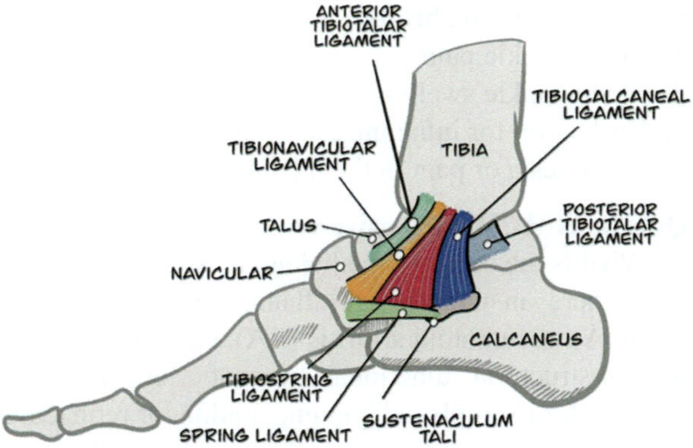

Fig. 14.2 Medial ankle ligaments

Ligament Anatomy (Fig. 14.2)

Deltoid Ligament

The deltoid ligament complex, shaped like a delta or triangle, stabilizes the medial ankle [1, 2]. These ligaments all start with "tibio." Again, think of the bones involved, proximal to distal: the tibia, the talus, the calcaneus, and the navicular. The *deeper* **tibiotalar ligaments** bind *adjacent* bones and have two parts, the **anterior** and **posterior**. The more *superficial* ligaments that bind *nonadjacent* bones are the **tibionavicular**, **tibiospring**, and **tibiocalcaneal ligaments**. One exception is the **posterior superficial tibiotalar ligament**, which binds adjacent bones but is superficial.

1. **Deep Deltoid Ligaments** [2]
   (a) **The anterior tibiotalar ligament (ATTL)** may be absent as a normal variant [1].
   (b) **The posterior tibiotalar ligament (PTTL)** is the thickest medial ankle ligament.

2. **Superficial Deltoid Ligaments**

   (a) **The tibionavicular ligament (TNL)** is present in 55% of the general population [1].
   (b) **The tibiospring ligament (TSL)** inserts into the spring ligament, which is unusual since most other ligaments tend to bind bone to bone [1, 3].
   (c) **The tibiocalcaneal ligament (TCL)** is present in 88% of the general population [1].
   (d) **The posterior superficial tibiotalar ligament (PSTL),** *not* labeled in Fig. 14.2, is the superficial layer of the PTTL.

Spring Ligament

Several ligaments constitute the spring ligament complex; however, the superomedial calcaneonavicular ligament is the main component, and we refer to it generally as the "spring ligament" [4]. The spring ligament stabilizes the foot's longitudinal arch and supports the talar head [2]. The spring ligament, although not officially part of the deltoid ligament complex, forms the distal horizontal base of the delta or triangle.

Tendon Anatomy (Fig. 14.3)

The ankle flexor tendons are the **tibialis posterior (TPT), the flexor digitorum longus (FDLT),** and the **flexor hallucis longus (FHLT)** [5]. All originate from muscles in the calf and enter the **tarsal tunnel (TT)** posterior to the medial **malleolus (MM)**. However, the FHLT may be more muscle than tendon within the TT. The three tendons continue posteriorly to the retromalleolar groove, which functions as a pulley. Each tendon makes a 90° curve and is encouraged to "stay in its lane" by the protruding medial calcaneal bony landmark, the sustentaculum tali (ST). Despite the name ending in "tali," the ST is part of the *calcaneus*. Above (cephalad to) the ST runs the TPT. The FDLT crosses the medial aspect of the ST. The FHLT courses beneath (caudal to) the ST in a groove. In cross-section, the three tendons and the ST are reminiscent of a hill with three suns. The ST is similar to the lateral calcaneus's peroneal tubercle (PT) in that the ST separates the tendons and serves as a sonographic landmark [5].

Tendon Comments [5]

1. **Tibialis Posterior Tendon**

   (a) The largest and most anterior of the three tendons
   (b) Held in place by the posterior MM sulcus and the flexor retinaculum

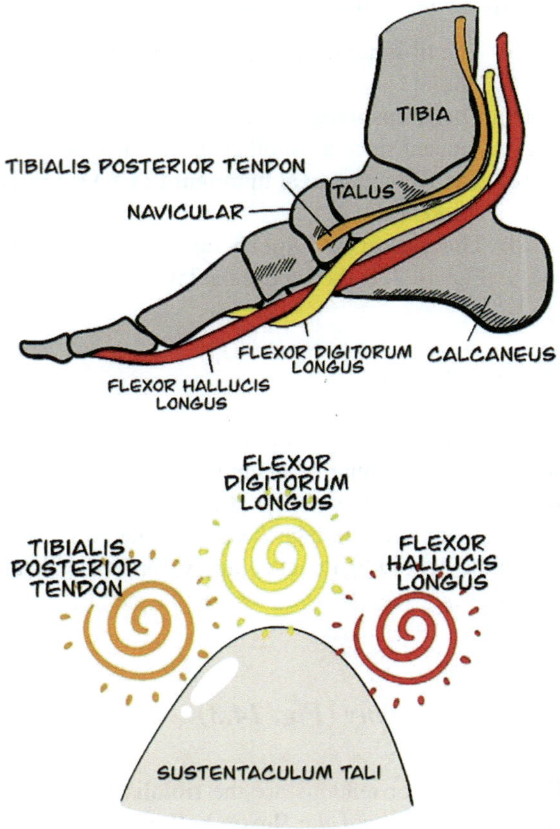

Fig. 14.3 Medial ankle tendons

(c) Inserts on the navicular, medial cuneiform bones, and metatarsals 2–4
(d) Facilitates ankle and foot plantar flexion and inversion
(e) An essential part of the medial longitudinal arch
(f) May contain a sesamoid bone that goes by three different names (os tibiale externum, os naviculare, or os navicularum), located proximal to the tendon's insertion onto the navicular, present in 23% [6]

2. **Flexor Digitorum Longus Tendon**

(a) Slimmer and posterior to the TPT
(b) Inserts on the plantar aspects of the distal phalanges of digits 2–5 to facilitate flexion of these digits

3. **Flexor Hallucis Longus Tendon**

(a) Inserts on the first toe base at the distal phalanx to enable first toe and ankle flexion
(b) Difficult to identify on ultrasound (US) in the TT (where it is more muscle than tendon)
(c) May communicate with the tibiotalar joint via the tendon sheath [7]

Be aware of the potential presence of an accessory FHLT or FDLT [5].

Tarsal Tunnel Anatomy (Fig. 14.4)

The TT is posterior to the MM. From anterior to posterior, the structures within the TT are memorialized by the mnemonic "Tom, Dick, and A Very Nervous Harry," which stands for **T**PT, F**D**LT, Posterior tibial **A**rtery, Posterior tibial **V**ein, Tibial **N**erve, and F**H**LT. We designate the tibial nerve (TN) in the medial ankle as just plain "tibial nerve," although others call it the posterior tibial nerve or tibialis posterior nerve. The **tibial nerve terminal branches are as follows** (Fig. 14.5) [8]:

1. **Medial calcaneal nerve** (MCN), which branches off the TN before entering the TT
2. **Lateral plantar nerve (LPN)**
3. **Medial plantar nerve** (MPN)
4. **Inferior calcaneal nerve (ICN)**, or **Baxter's nerve**, which branches off the LPN or the TN just before the LPN forms and has mixed motor and sensory components [9]

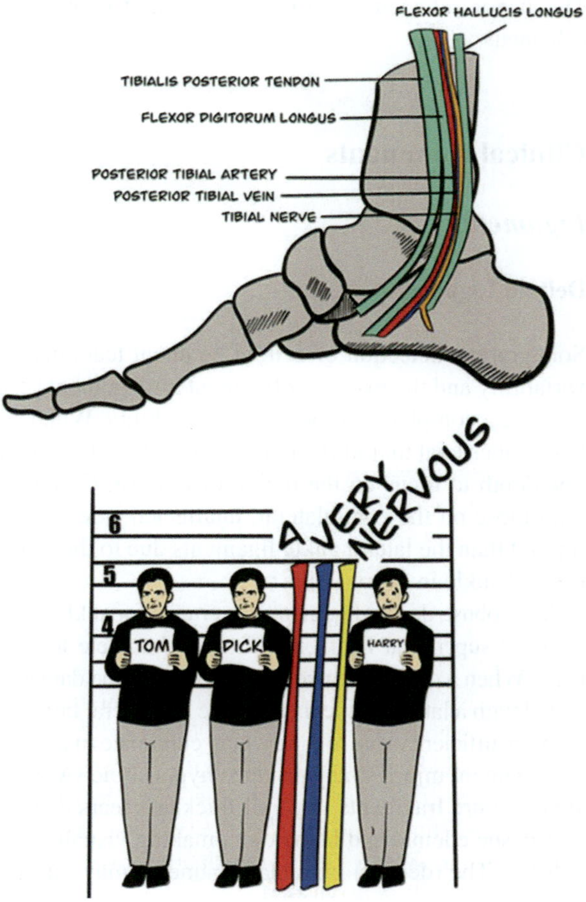

Fig. 14.4 Tibial nerve/tarsal tunnel anatomy

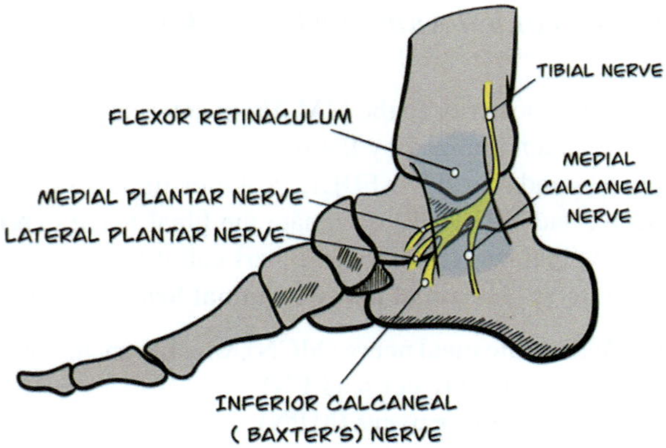

Fig. 14.5 Tibial nerve branches

Flexor Retinaculum

The flexor retinaculum (FR) covers the TT and stretches from the MM to the calcaneus.

Clinical Comments

Ligaments

Deltoid Ligament

Sonographic detection of deltoid ligament tears may be difficult due to individual variability and the two-layer ligament arrangement [2, 10]. An absent ligament may indicate a complete tear or normal variability. When assessing the ligaments, move from superficial to mid-depth to identify bony landmarks. Then, go back to the shallow depth to focus on the region of interest. Compare the affected foot ligaments with those on the contralateral, unaffected side. The deltoid ligament is less often injured than the lateral ankle ligaments due to the lower frequency of landing on an everted ankle in sports [3].

The robust, deeper ligaments are crucial for ankle stability but are more often injured than the superficial ligaments [2, 11]. Complete tears are less common than partial tears. When a deltoid ligament injury occurs, the damage may be severe, perhaps associated with a lateral malleolar fracture and lateral talar displacement [12]. Deltoid ligament insufficiency due to injury may culminate in ankle joint osteoarthritis [13].

Ligament injuries may produce hypoechoic swelling, fibril disruption, and avulsion fracture fragments [3]. Full-thickness tears demonstrate ligament interruption, soft-tissue edema, and hematoma, making visualization of specific ligaments difficult [2]. The (deeper) tibiotalar ligaments may falsely appear to have tears due to anisotropy. Sonographic sensitivity for partial tears is only 50%; therefore, magnetic resonance imaging remains the gold standard when evaluating ankle instability [2, 14].

Spring Ligament Complex

Injuries to the spring ligament (as well as to the TPT) promote pes planus (flatfoot deformity) [15]. A spring ligament injury on US is seen as a thickening, thinning, or frank discontinuity. The plantar components of the spring ligament complex are not consistently visible on US [2]. For that reason, what we term the "spring ligament" is technically the superomedial calcaneal ligament component, the more sonographically accessible component, and the one more likely to be injured [3].

Tendons/Tenosynovium

The tendons of the medial ankle may exhibit a range of pathologies, including tendinosis, partial or complete tears, tenosynovitis, and enthesopathy [16]. In addition, tendon subluxation, often associated with flexor retinaculum injury or dysfunction, may be demonstrated with dynamic maneuvers. Tendinosis, a degenerative change with hypoechoic enlargement, occurs in the tendon segments that abut bony structures, such as the posterior MM. Partial-thickness tears are seen as intrasubstance hypoechoic or anechoic areas, often in the setting of underlying tendinosis. Sometimes it isn't easy to separate the two, although a tear is more likely to be present if it is well-defined and anechoic. Complete tendon tears may also occur.

Each tendon is encased in a tenosynovial sheath. Infection, mechanical, or inflammatory conditions may instigate tenosynovitis, producing anechoic fluid and perhaps power Doppler (PD) activity. There may be associated tendon swelling, indicating tendinopathy. Comparison with the contralateral tendon may be helpful. The tibiotalar joint may normally communicate with medial tendon sheaths, especially the FHLT [17]. Thus, isolated tendon sheath distention *without* a tibiotalar effusion argues for the presence of tenosynovitis.

Tibialis Posterior Tendon

Evaluate for tendinosis of the TPT and the nearby FDLT by comparing their relative sizes. The TPT diameter is typically twice that of the FDLT, ranging from 4 to 6 mm [18, 19]. The TPT is the most frequently damaged of the ankle flexor tendons, with injury occurring at the MM and the distal insertions (entheses) [20]. Tendinosis and partial tears are more common than complete ruptures. Pathology starts as tenosynovitis with progression to a tear; however, early treatment may prevent severe disability [21].

Tears of the TPT most often occur just distal to the MM or at the navicular insertion. Complete TPT rupture often results in a flatfoot deformity [20, 22]. Rupture may be associated with RA and SpA. The normal variant accessory sesamoid bone, the os naviculare, may be embedded in the TPT proximal to its navicular bone insertion and may be associated with TPT pathology due to altered mechanical stresses [23]. The os naviculare can be mistaken for a bony avulsion or calcium deposition within the TPT.

TPT subluxation may be due to damage to the FR and is detected using dynamic maneuvers, namely ankle flexion and inversion [24]. Patients with RA and SpA may demonstrate TPT enthesopathy at the TPT insertion on the navicular, which is best appreciated when the transducer is tilted 30°–45° cephalad [25]. Tenosynovitis may occur near the malleolus, but not distally near the navicular insertion since this segment of the tendon lacks a tenosynovial sheath [26].

Flexor Digitorum Longus Tendon

Pathology of the FDLT is rare, but tendinosis may occur in ballet dancers [20, 22, 27, 28].

Flexor Hallucis Longus Tendon

Injury to the FHLT occurs in sports such as ballet, soccer, or basketball, resulting in tenosynovitis, stenosing tenosynovitis, and tendinosis [22, 29]. Nearby bony spurs and os trigonum (see below) may predispose to FHLT tendon pathology [22]. Visualization of the FHLT is impeded by its depth, but is improved by dynamic great toe flexion.

Specific Conditions

OS Trigonum Syndrome

The os trigonum variant, present in 7% of ordinary people, results from a lack of bone fusion at the posterior talus [30, 31]. The os trigonum may compress posterior ankle soft tissues, including the FHLT, causing tenosynovitis, chronic posterior ankle pain, and swelling [32–34]. However, similar symptoms may be caused by Achilles tendinosis or tear, Haglund's syndrome, retrocalcaneal bursitis, and osteochondral lesions [35]. Ultrasound has a role diagnostically and for injection guidance, but other imaging modalities are preferred [32, 36]. Dynamic evaluation for FHLT with plantar flexion may reveal bony impingement.

Intersection Syndrome

The FHLT and the FDLT cross the midfoot in the region known as the Master Knot of Henry [37]. This crossover may develop into an intersection syndrome with tendinosis, tenosynovitis, or tendon tears [38]. There may be fluid distention in both tendon sheaths and entrapment of the adjacent medial plantar nerve.

Rheumatoid Arthritis and Spondyloarthritis

Rheumatoid arthritis may present with TPT and FHLT tenosynovitis [16]. Rupture of the TPT distal to the MM or at the navicular insertion may be associated with RA and SpA [25]. Distal longitudinal TPT imaging may show enthesopathy features in both RA and SpA. Tilting the transducer in a 30°–45° cephalad direction improves tendon evaluation.

Flexor Retinaculum Dysfunction

The FR is a thickened fascia that restrains flexor tendon movement [39]. The FR is the TT roof, originating at the MM and extending to the medial calcaneus. On US, the FR appears as thin (1 mm) hyperechoic fibrillar bands overlying the tendons, best seen in the transverse (TAX) view [39]. When damaged, the FR has a thickened, hypoechoic appearance and may demonstrate PD activity. Damage to the FR may contribute to TPT dysfunction and, thus, to acquired pes planus.

Tarsal Tunnel Syndrome

Tarsal tunnel syndrome (TTS) occurs when the TN or one of its branches is compromised within the TT, resulting in numbness, tingling, burning, dysesthesia, cramping, or pain in the distribution of the TN or branches [40]. Symptoms may occur at rest or with different foot movements [41]. Compressive causes of TTS include myriad bone and soft tissue abnormalities. Foot deformities such as a varus or valgus ankle, pes planus, and tarsal coalition predispose to nerve branch compression [40]. Two other causes of TN, or branch compromise, are surgical injury and hardware compression.

The diagnosis of TTS is complicated by the large variety of etiologies and variance of nerve branches that may be affected [40]. Reproduction of symptoms may occur with transducer pressure or tapping over the involved nerve (Tinel's sign). Electrophysiologic studies are essential, but negative studies do not exclude TTS. Only imaging will establish the compression site, cause, and proper treatment.

Radiographs may reveal structures with compressive potential. Magnetic resonance imaging is the gold standard; however, high-resolution US is gaining support to delineate nerve compression and to rule out commonplace TTS mimics such as plantar fasciitis [40, 42–44]. Ultrasound evaluation while the patient stands may augment plantar venous distension and exacerbate compression caused by bony conditions, including compression of a TN branch by the talus with pes planus and valgus foot [40]. An alternative technique to augment venous pressure is to use a blood pressure cuff as a tourniquet in the lower extremity; this may reproduce symptoms [41]. In contrast to carpal tunnel syndrome, there are no well-defined US criteria for TTS. It is best to compare the cross-sectional area of the TN to that of the asymptomatic contralateral side.

Baxter's nerve impingement may cause pain at the medial calcaneus or the plantar fascia, mimicking TTS or plantar fasciitis [45]. Another mimic of TTS is TN neuropathy, which occurs with diabetes and other conditions. Treatment for TTS includes open decompression, US-guided hydrodissection, or minimally invasive TT release [46, 47].

Pitfalls

1. A positive US Tinel's sign occurs when transducer pressure over the TN or an affected branch reproduces symptoms. However, avoid overzealous compression of the nerve and branches during the examination since this may produce paresthesia even in ordinary people.
2. Always perform a dynamic examination to look for tendon subluxation, particularly for the TPT [24].
3. The TPT insertion on the navicular and the middle cuneiform may falsely appear to have tendinosis or a tendon tear due to anisotropy. Avoid mistaking this for pathology by performing sonopalpation to look for symptoms in this area [24, 48].
4. Do not mistake an accessory navicular bone (os naviculare) for a bony avulsion or tendon calcification. Sonopalpation will elicit symptoms if this finding is pathologic [24].
5. With a complete TPT rupture, avoid confusing an intact FDLT for the TPT [24].
6. Do not neglect to tilt the transducer to a 30°–45° cephalad orientation when evaluating the TPT in longitudinal (LAX) at the insertion on the navicular [25, 49].
7. When evaluating for compressive causes of TTS, consider performing US in the standing position to distend plantar veins or exacerbate bony compression [40].

Method

The patient is supine, the knee is flexed at 90°, and the foot is slightly externally rotated. Alternatively, the patient rolls onto the side to be examined with the foot off the table and supported by a towel under the lateral aspect of the distal foreleg. Protocol Images 1 through 6 evaluate the medial ankle ligaments. The deeper anterior and posterior tibiotalar ligaments are first assessed. Note that the posterior tibiotalar ligament has both deep and superficial layers.

The tendons are next studied in Protocol Images 7 through 11. Protocol Images 7 and 13 evaluate the TN and its branches. Use PD to locate the posterior tibial artery and Color Flow (CF) to find the posterior tibial veins. Avoid excessive probe pressure, which might inadvertently compress the tibial veins, rendering them sonographically invisible.

Protocol Image 1: Anterior Tibiotalar Ligament, Longitudinal (Fig. 14.6)

To visualize the (deep) ATTL, place one end of the probe on the MM and aim the distal end toward the base of the first metatarsal. Angle the probe downward about 30°. Look for the MM, the talus, and the joint space between the two. Toggle the probe until you see a hyperechoic band bridging the chasm between the two bones.

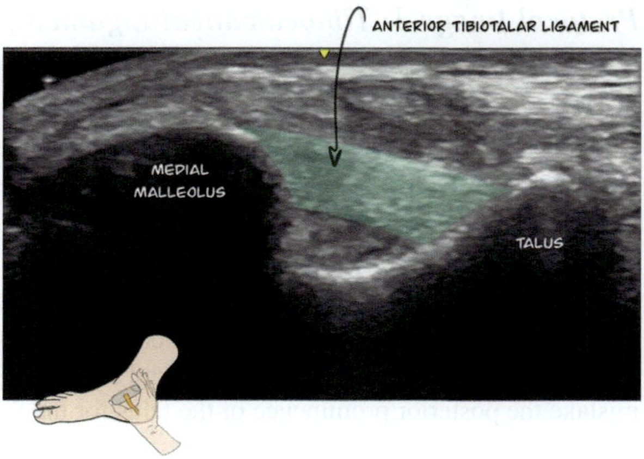

Fig. 14.6 Protocol Image 1: Anterior tibiotalar ligament, longitudinal

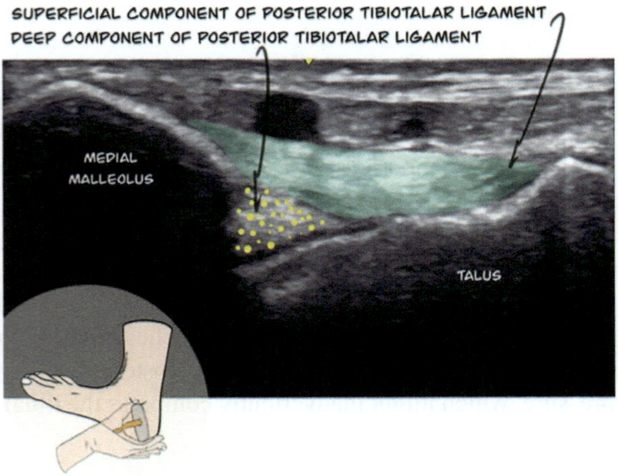

Fig. 14.7 Protocol Image 2: Posterior tibiotalar ligament, longitudinal

Protocol Image 2: Posterior Tibiotalar Ligament, Longitudinal (Fig. 14.7)

With the foot everted and dorsiflexed, lift the probe and place the proximal end on the inferior-posterior portion of the MM. Aim the distal probe end toward the tip of the heel to visualize the PTTL. Use the distal probe to look for the hyperechoic talus. This ligament's superficial and deep components bridge the valley between the MM and the talus.

Protocol Image 3: Tibiocalcaneal Ligament, Longitudinal (Fig. 14.8)

With the foot still everted and dorsiflexed, keep the proximal end of the probe on the inferior aspect of the MM and then rotate the distal end to a near-vertical position to display the TCL. You may need to rotate the distal end of the probe an additional 10° toward the tip of the heel to optimize the image.

Hint You should see the large TPT in TAX, just superficial to the ligament. Note the ligament's insertion on the ST of the calcaneus; the ST is about the same depth as the MM. The posterior prominence of the talus is still deeper than the ST. Do not mistake the posterior prominence of the talus for the ST.

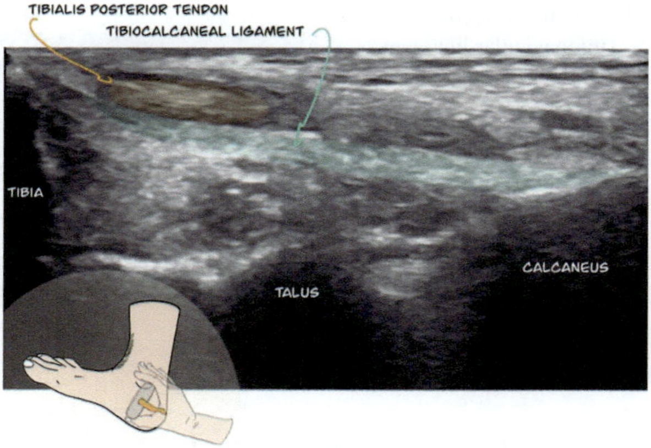

Fig. 14.8 Protocol Image 3: Tibiocalcaneal ligament, longitudinal

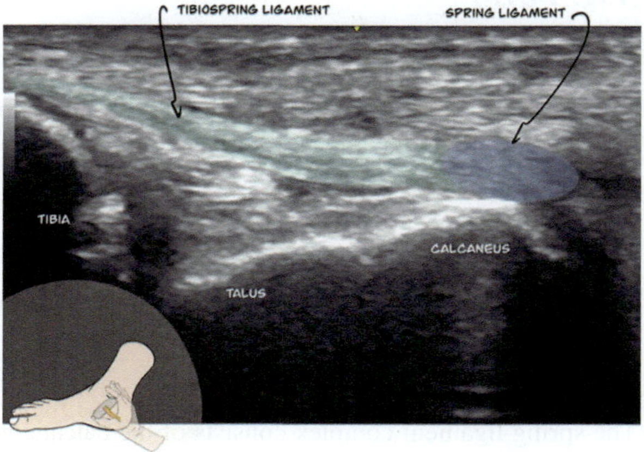

Fig. 14.9 Protocol Image 4: Tibiospring ligament, longitudinal

Protocol Image 4: Tibiospring Ligament, Longitudinal (Fig. 14.9)

Next, rotate the distal portion of the probe about 20° toward the toes. Perform a slight TAX slide anteriorly, moving the probe along the inferior MM. The TCL with its calcaneal insertion will disappear, leaving a view of the TSL that is proximally attached to the MM but without a distal bony attachment. Look for a subtle distal insertion of the TSL on the spring ligament, which, in TAX, appears as an oval hypoechoic structure. Once again, the TSL is unique in that it inserts into another ligament.

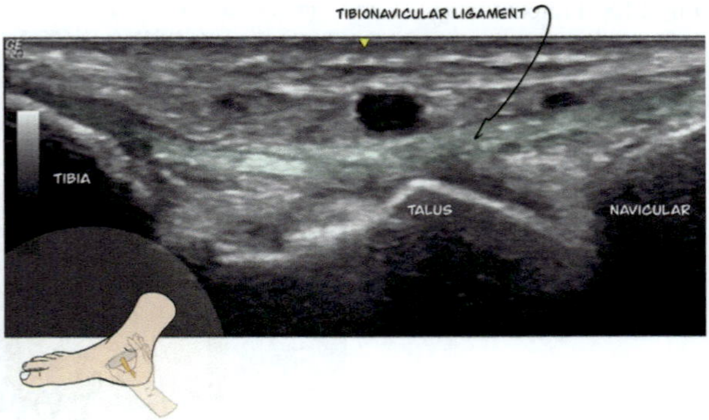

Fig. 14.10 Protocol Image 5: Tibionavicular ligament, longitudinal

Protocol Image 5: Tibionavicular Ligament, Longitudinal (Fig. 14.10)

Next, perform a LAX slide distally and just off the MM to see the tibiotalar joint and, more distally, the talonavicular joint. The superficial TNL connects the tibia to the navicular and is a curved ligament when the ankle is in a neutral position.

Protocol Image 6: Spring Ligament Complex, Longitudinal (Fig. 14.11)

The spring ligament complex consists of the calcaneonavicular ligaments, a group of ligaments that parallel the bottom of the foot. The ligament complex has dorsal and plantar components. We primarily focus on the most dorsal component, the superomedial calcaneonavicular ligament, and refer to this component as the "spring ligament."

Rotate the foot externally and place it in dorsiflexion. Next, palpate the ST just inferior to the MM. An alternative method to find the ST is to place the probe in a vertical position with one end on the MM. The more distal bony protuberance at about the same depth as the MM is the ST. Next, put one end of the probe on the ST and rotate the distal end anteriorly until it parallels the plantar surface of the foot. Move the probe in an LAX slide distally until the bony, curved navicular bone is seen. The ST and the navicular are at about the same depth and bracket the spring ligament, seen in LAX, connecting the ST and the navicular. The distal end of the probe may need to be angled up or down to visualize the spring ligament. This image may be challenging to obtain.

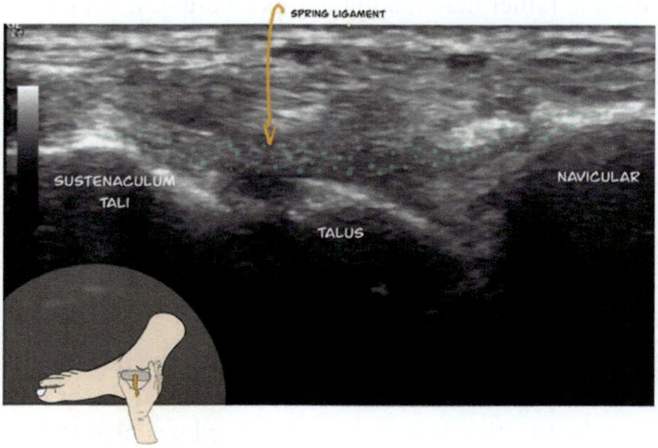

Fig. 14.11 Protocol Image 6: Spring ligament complex, longitudinal

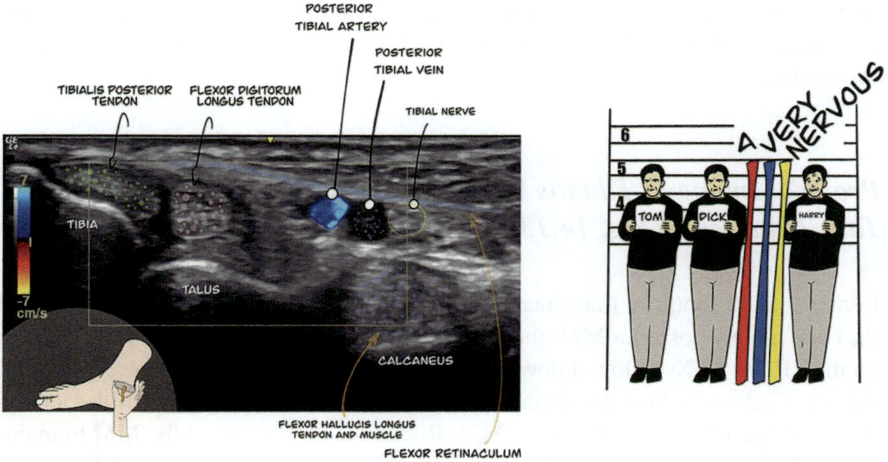

Fig. 14.12 Protocol Image 7: Tarsal tunnel tendons at medial malleolus, transverse ± Doppler

Protocol Image 7: Tarsal Tunnel Tendons at Medial Malleolus, Transverse ± Doppler (Fig. 14.12)

Next, examine the tendons within the TT. The patient's leg is in a frog-leg position with a pillow under the lateral malleolus. Alternatively, the patient may lie on their side. The flexor retinaculum runs from the MM to the calcaneus. It is the roof of the TT and, therefore, the most superficial part. Start in TAX with the probe posterior to the edge of the MM. Recognize from anterior to posterior: TPT, FDLT (half the size of the TPT), posterior tibial artery and veins, TN, and FHLT. The FHLT may be muscle rather than tendon in this location; if so, it is hypoechoic and poorly visualized compared with the other two tendons. The FHLT may also be deeper than the other two tendons. Again, the mnemonic "**T**om, **D**ick, and **A** **V**ery **N**ervous **H**arry" reminds us of the names. In this area, the tendon sheaths may collect synovial fluid.

Perform the following passive dynamic maneuvers to verify the identity of each tendon:

1. TPT: inversion and plantar flexion
2. FDLT: flexion and extension of digits 2 through 5
3. FHLT: flexion and extension of the great toe

Anterior subluxation of the TPT may be demonstrated with resisted dorsiflexion and inversion in the presence of a damaged flexor retinaculum [39]. Tibialis posterior dysfunction is a cause of acquired pes planus. Use PD or CF to detect the posterior tibial artery, usually located just deep to one of the posterior tibial veins. Excess probe pressure may collapse the veins, which is a way to identify the artery without Doppler. You may see "Mickey Mouse," with the ears being two veins and the artery representing Mickey's head. The TN is often just posterior to the vascular bundle.

Protocol Image 8: Tibialis Posterior Tendon at Sustentaculum Tali, Transverse (Fig. 14.13)

Center a small footprint transducer (still in TAX) on the ovoid TPT as it rests upon the curved bony posterior MM; the TPT is 4–6 mm in diameter [19]. While focusing on the TPT in TAX, follow it down as it curves around the retrocalcaneal groove to the ST. There may be a small amount of normal tenosynovial fluid in the tendon sheath of the TPT just distal to the MM. Rotate the probe around the MM to maintain TAX orientation to the TPT. Look for the ST, the bony protrusion of the medial calcaneus. The TPT runs superior (cephalad) to the ST; the FDLT is medial and superficial to the ST; and the FHLT is inferior (caudal) to the ST in a groove. These locations relative to the ST help verify the identity of each tendon when examined in TAX after exiting the TT. Again, this appears to be a "hill with three suns."

Protocol Image 9: Tibialis Posterior Tendon Navicular Insertion, Longitudinal (Fig. 14.14)

Center the TPT in the image and rotate the probe 90° to view the TPT in LAX. With a LAX slide, follow the TPT to its first insertion point, the navicular. Tilting the transducer upward (cephalad) at a 30°–45° angle improves the visualization of the TPT insertion [25]. Anisotropy near the navicular insertion may mimic a tendon tear. Near the navicular, you may find a curvilinear os naviculare within the TPT. Do not mistake this normal, variant accessory bone for an avulsion or tendon

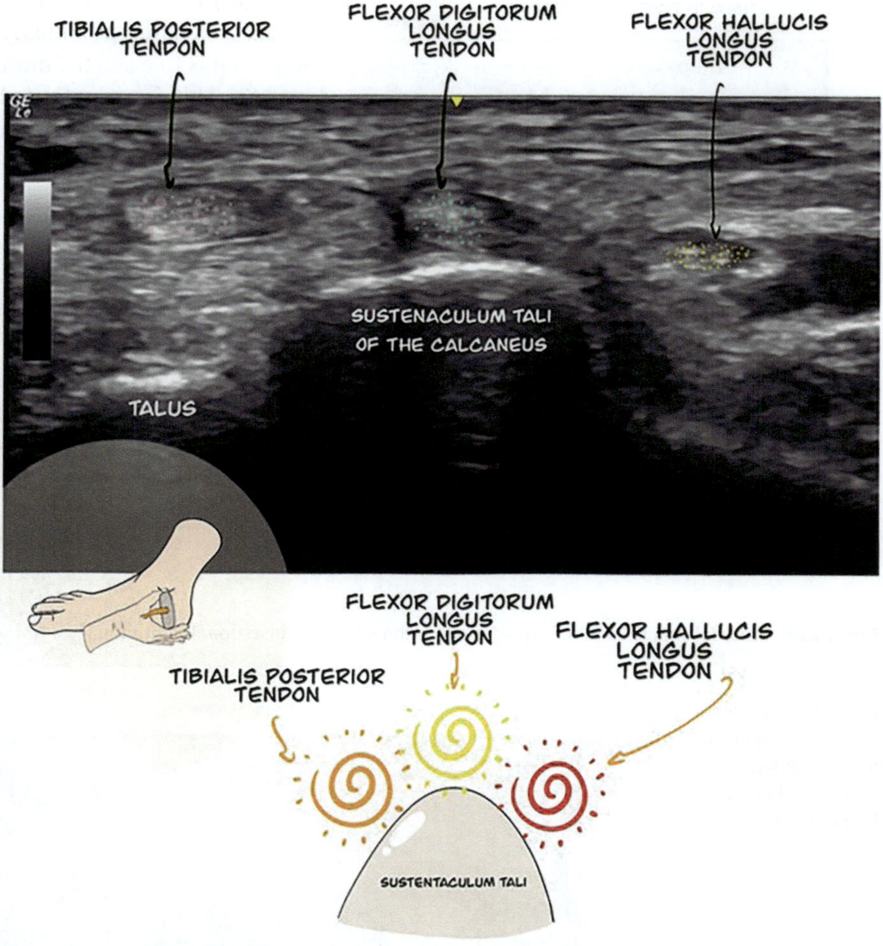

Fig. 14.13 Protocol Image 8: Tibialis posterior tendon at sustentaculum tali, transverse

calcification. Sonopalpation will elicit symptoms if this finding is pathologic. The rest of the TPT continues to insert on the plantar surface of the three cuneiforms and the second to fourth metatarsal bases.

Protocol Image 10: Flexor Digitorum Longus Tendon at Sustentaculum Tali, Longitudinal (Fig. 14.15)

Return to Protocol Image 7 (TT in TAX) and locate the FDLT, which is posterior to and half the size of the TPT. Manually dorsiflex toes 2 through 5 to verify that you are looking at the FDLT. Next, rotate the probe 90° into LAX and, in a LAX slide, follow the FDLT around the curve of the posterior MM. Remember to

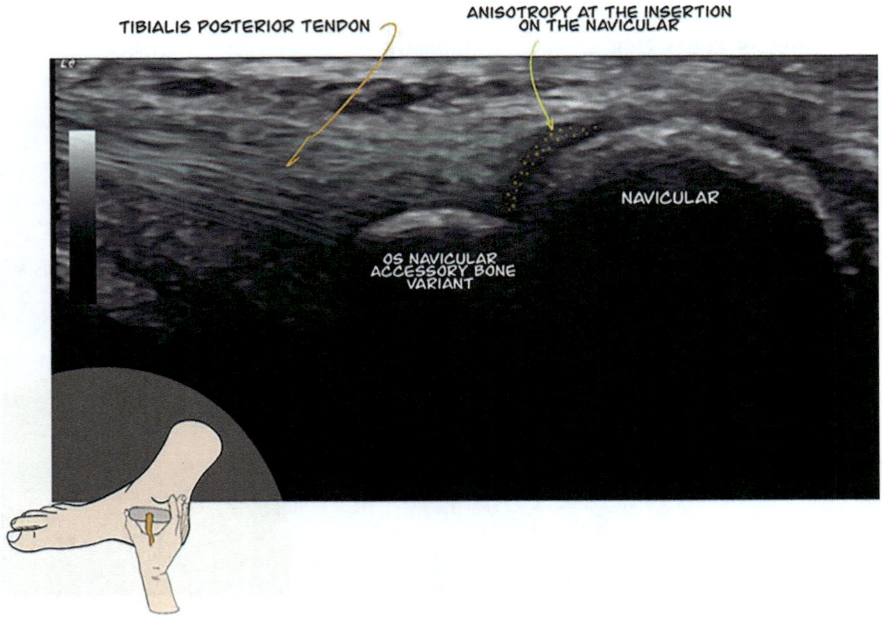

Fig. 14.14 Protocol Image 9: Tibialis posterior tendon navicular insertion, longitudinal

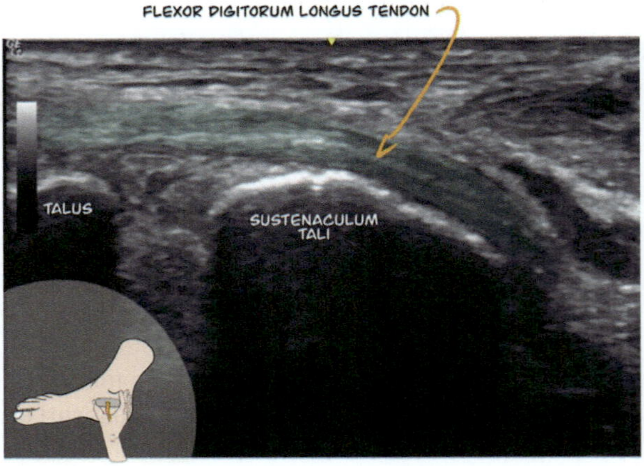

Fig. 14.15 Protocol Image 10: Flexor digitorum longus tendon at sustentaculum tali, longitudinal

look for tendinopathy near the retromalleolar groove. After clearing the rear MM, position the probe in a near-horizontal plane to see two hyperechoic rounded protrusions: the proximal one is the talus, and the more distal one is the protruding ST. The FDLT, in LAX orientation, lies superficial to these bony protrusions.

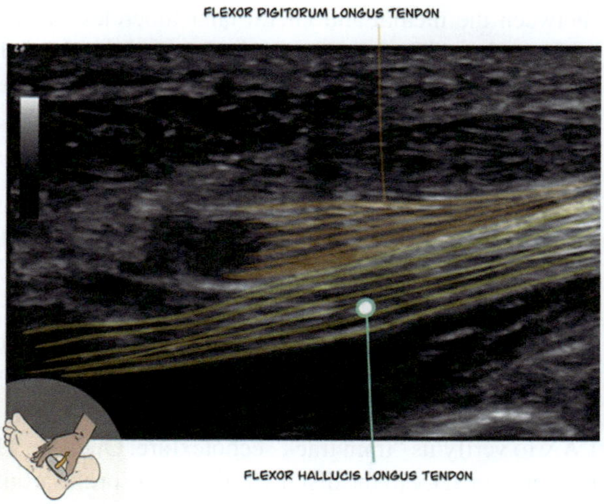

Fig. 14.16 Protocol Image 11: Knot of Henry, longitudinal

Protocol Image 11: Knot of Henry, Longitudinal (Fig. 14.16)

Continue the LAX slide and follow the FDLT in LAX distally to the plantar surface of the midfoot, where it crosses the FHLT to form the Knot of Henry (Image 11). The FDLT will angle slightly downward as it approaches the midfoot. The FDLT is cephalad to the FHLT, proximal to the intersection of the tendons. Performing tiny TAX slides (cephalad and plantar) will individually focus on each of these tendons since tendon crossing creates anisotropy. The FDLT eventually inserts on the base of the distal phalanges 2, 3, 4, and 5.

Protocol Image 12 (Optional): Flexor Hallucis Longus Tendon Starting at the Tarsal Tunnel, Longitudinal

Again, go back to the TT view from Protocol Image 7. The FHLT or its muscle may be seen deep to the TN; it may appear mostly hypoechoic if there is a preponderance of muscle. A hyperechoic structure deep to the neurovascular bundle is likely the emerging tendon near the myotendinous junction. Verify that this is, in fact, the FHLT by manually dorsiflexing the large toe.

Center the probe on the FHLT in TAX and rotate it 90° to visualize the FHLT in LAX. The FHLT may be challenging to visualize. Repeated passive flexion and dorsiflexion of the great toe during the examination will help with visualization. The FHLT curves around a shallow groove in the medial posterior aspect of the talus (between the medial and lateral talar tubercles). It then courses under the plantar surface of the ST of the calcaneus. It continues through the midfoot Knot of Henry (Protocol Image 10b), then between the medial and lateral sesamoids, finally inserting on the plantar surface of the distal phalanx of the large toe.

Protocol Image 13: Distal Tibial Nerve, Transverse (Fig. 14.17)

Go back to Protocol Image 7 to visualize the TN in TAX with a high-frequency probe. If there is a question as to the identity of the TN, tilt the probe to look for anisotropy and perform a proximal TAX slide to follow the TN proximal to the TT. Further, confirm the TN's identity by rotating the probe 90° to view the TN in LAX to verify its "tram track" echotexture. Once the TN has been identified, follow the nerve in TAX to the distal TT. Curve the probe around the retrocalcaneal groove. The proximal probe tip touches the inferior MM, and the distal end is aimed toward the direction of the heel. As the probe moves distally, the TN descends deep to the posterior tibial artery and veins.

Continuing a TAX slide distally will demonstrate the TN splitting into two plantar branches, the MPN and the LPN. The MPN is the most anterior of these two branches' structures; however, *both* the MPN and the LPN run anteriorly. The ICN, or Baxter's nerve, branches off either the distal TN or the LPN. The ICN is located posterior to the LPN and, importantly, lies posterior to the vascular bundle. The ICN is a mixed sensory and motor nerve that innervates the abductor digit minimi muscle and may mimic plantar fasciitis if entrapped [9, 45]. It is only 1–2 mm in diameter and may be challenging to visualize. To verify that the ICN has been identified in

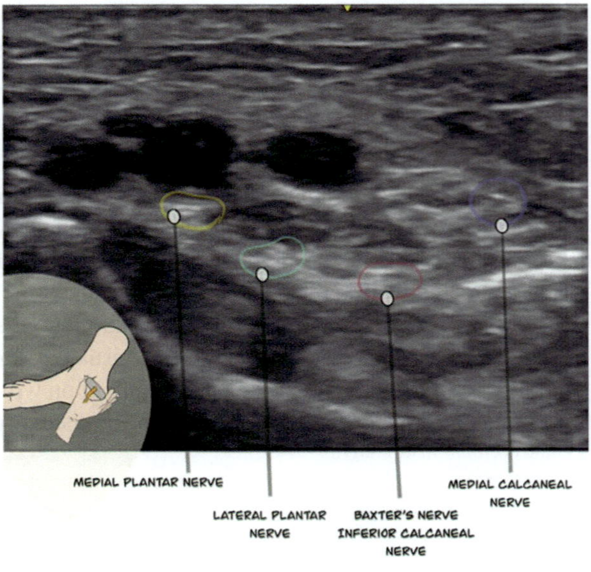

Fig. 14.17 Protocol Image 13: Distal Tibial Nerve, transverse

TAX, rotate the transducer 90° to the LAX view to visualize a tram-track pattern descending to the sole. Yet another branch, the MCN, stems from the TN, usually *proximal* to the TT, and runs posteriorly to the other nerve branches to innervate the medial calcaneus. To avoid confusing the MCN with the ICN, note again that the ICN tends to be posterior to the vascular bundle. In contrast, the MCN resides in a more posterior location.

Complete Medial Ankle Ultrasound Checklist

☐ Protocol Image 1: Anterior tibiotalar ligament, longitudinal
☐ Protocol Image 2: Posterior tibiotalar ligament, longitudinal
☐ Protocol Image 3: Tibiocalcaneal ligament, longitudinal
☐ Protocol Image 4: Tibiospring ligament, longitudinal
☐ Protocol Image 5: Tibionavicular ligament, longitudinal
☐ Protocol Image 6: Spring ligament complex, longitudinal
☐ Protocol Image 7: Tarsal tunnel tendons at medial malleolus, transverse ± Doppler
☐ Protocol Image 8: Tibialis posterior tendon at sustentaculum tali, transverse
☐ Protocol Image 9: Tibialis posterior tendon navicular insertion, longitudinal
☐ Protocol Image 10: Flexor digitorum longus tendon at sustentaculum tali, longitudinal
☐ Protocol Image 11: Knot of Henry, longitudinal
☐ Protocol Image 12 (Optional): Flexor hallucis longus tendon starting at the tarsal tunnel, longitudinal
☐ Protocol Image 13: Distal tibial nerve, transverse

References

1. Mengiardi B, Pfirrmann CW, Vienne P, Hodler J, Zanetti M. Medial collateral ligament complex of the ankle: MR appearance in asymptomatic subjects. Radiology. 2007;242(3):817–24.
2. Sconfienza LM, Orlandi D, Lacelli F, Serafini G, Silvestri E. Dynamic high-resolution US of ankle and midfoot ligaments: normal anatomic structure and imaging technique. Radiographics. 2015;35(1):164–78.
3. Alves T, Dong Q, Jacobson J, Yablon C, Gandikota G. Normal and injured ankle ligaments on ultrasonography with magnetic resonance imaging correlation. J Ultrasound Med. 2019;38(2):513–28.
4. Perrich KD, Goodwin DW, Hecht PJ, Cheung Y. Ankle ligaments on MRI: appearance of normal and injured ligaments. AJR Am J Roentgenol. 2009;193(3):687–95.
5. Precerutti M, Bonardi M, Ferrozzi G, Draghi F. Sonographic anatomy of the ankle. J Ultrasound. 2013;17(2):79–87.
6. Bareither DJ, Muehleman CM, Feldman NJ. Os tibiale externum or sesamoid in the tendon of tibialis posterior. J Foot Ankle Surg. 1995;34(5):429–34; discussion 509.
7. Na J-B, Bergman AG, Oloff LM, Beaulieu CF. The flexor hallucis longus: tenographic technique and correlation of imaging findings with surgery in 39 ankles. Radiology. 2005;236(3):974–82.
8. Desai SS, Cohen-Levy WB. Anatomy, bony pelvis and lower limb, tibial nerve. StatPearls. Treasure Island, FL: StatPearls Publishing; 2022.

9. Hung CY, Chang KV, Mezian K, Naňka O, Wu WT, Hsu PC, et al. Advanced ankle and foot sonoanatomy: imaging beyond the basics. Diagnostics (Basel). 2020;10(3):160.
10. Sarrafi SK, Kelikian AS. Functional anatomy of the foot and ankle. In: Sarrafian's anatomy of the foot and ankle. Wolters Kluwer; 2011.
11. Crim J, Longenecker LG. MRI and surgical findings in deltoid ligament tears. AJR Am J Roentgenol. 2015;204(1):W63–9.
12. McCollum GA, van den Bekerom MP, Kerkhoffs GM, Calder JD, van Dijk CN. Syndesmosis and deltoid ligament injuries in the athlete. Knee Surg Sports Traumatol Arthrosc. 2013;21(6):1328–37.
13. Mansour R, Teh J, Sharp RJ, Ostlere S. Ultrasound assessment of the spring ligament complex. Eur Radiol. 2008;18(11):2670–5.
14. Lechner R, Richter H, Friemert B, Palm HG, Gottschalk A. The value of ultrasonography compared with magnetic resonance imaging in the diagnosis of deltoid ligament injuries—is there a difference? Z Orthop Unfall. 2015;153(4):408–14.
15. Harish S, Kumbhare D, O'Neill J, Popowich T. Comparison of sonography and magnetic resonance imaging for spring ligament abnormalities: preliminary study. J Ultrasound Med. 2008;27(8):1145–52.
16. Micu MC, Nestorova R, Petranova T, Porta F, Radunovic G, Vlad V, et al. Ultrasound of the ankle and foot in rheumatology. Med Ultrason. 2012;14(1):34–41.
17. De Maeseneer M, Marcelis S, Jager T, Shahabpour M, Van Roy P, Weaver J, et al. Sonography of the normal ankle: a target approach using skeletal reference points. AJR Am J Roentgenol. 2009;192(2):487–95.
18. Bianchi S, Martinoli C, Gaignot C, De Gautard R, Meyer JM. Ultrasound of the ankle: anatomy of the tendons, bursae, and ligaments. Semin Musculoskelet Radiol. 2005;9(3):243–59.
19. Miller SD, Van Holsbeeck M, Boruta PM, Wu KK, Katcherian DA. Ultrasound in the diagnosis of posterior tibial tendon pathology. Foot Ankle Int. 1996;17(9):555–8.
20. Fessell DP, Vanderschueren GM, Jacobson JA, Ceulemans RY, Prasad A, Craig JG, et al. US of the ankle: technique, anatomy, and diagnosis of pathologic conditions. Radiographics. 1998;18(2):325–40.
21. Johnson KA, Strom DE. Tibialis posterior tendon dysfunction. Clin Orthop Relat Res. 1989;239:196–206.
22. Fessell DP, Jacobson JA. Ultrasound of the hindfoot and midfoot. Radiol Clin N Am. 2008;46(6):1027–43, vi.
23. Schweitzer ME, Caccese R, Karasick D, Wapner KL, Mitchell DG. Posterior tibial tendon tears: utility of secondary signs for MR imaging diagnosis. Radiology. 1993;188(3):655–9.
24. Patel S, Fessell DP, Jacobson JA, Hayes CW, van Holsbeeck MT. Artifacts, anatomic variants, and pitfalls in sonography of the foot and ankle. AJR Am J Roentgenol. 2002;178(5):1247–54.
25. Ward IM, Kissin E, Kaeley G, Scott JN, Newkirk M, Hildebrand BA, et al. Ultrasound features of the posterior tibialis tendon and peroneus brevis tendon entheses: comparison study between healthy adults and those with inflammatory arthritis. Arthritis Care Res (Hoboken). 2017;69(10):1519–25.
26. Higgs JB. Ultrasound of the ankle and foot. In: Kohler MJ, editor. Musculoskeletal ultrasound in rheumatology review. 2nd ed. Springer; 2021.
27. Cheung Y, Rosenberg ZS, Magee T, Chinitz L. Normal anatomy and pathologic conditions of ankle tendons: current imaging techniques. Radiographics. 1992;12(3):429–44.
28. Fornage BD, Rifkin MD. Ultrasound examination of tendons. Radiol Clin N Am. 1988;26(1):87–107.
29. Khoury V, Guillin R, Dhanju J, Cardinal E. Ultrasound of ankle and foot: overuse and sports injuries. Semin Musculoskelet Radiol. 2007;11(2):149–61.
30. Karasick D, Schweitzer ME. The os trigonum syndrome: imaging features. AJR Am J Roentgenol. 1996;166(1):125–9.
31. O'Rahilly R. A survey of carpal and tarsal anomalies. J Bone Joint Surg Am. 1953;35-a(3):626–42.

32. Berman Z, Tafur M, Ahmed SS, Huang BK, Chang EY. Ankle impingement syndromes: an imaging review. Br J Radiol. 2017;90(1070):20160735.
33. Pesquer L, Guillo S, Meyer P, Hauger O. US in ankle impingement syndrome. J Ultrasound. 2014;17(2):89–97.
34. Hedrick MR, McBryde AM. Posterior ankle impingement. Foot Ankle Int. 1994;15(1):2–8.
35. Jordan LK III, Helms CA, Cooperman AE, Speer KP. Magnetic resonance imaging findings in anterolateral impingement of the ankle. Skelet Radiol. 2000;29(1):34–9.
36. Robinson P, Bollen SR. Posterior ankle impingement in professional soccer players: effectiveness of sonographically guided therapy. AJR Am J Roentgenol. 2006;187(1):W53–8.
37. Donovan A, Rosenberg ZS, Bencardino JT, Velez ZR, Blonder DB, Ciavarra GA, et al. Plantar tendons of the foot: MR imaging and US. Radiographics. 2013;33(7):2065–85.
38. Rajakulasingam R, Murphy J, Panchal H, James SL, Botchu R. Master knot of Henry revisited: a radiologist's perspective on MRI. Clin Radiol. 2019;74(12):972.e1–8.
39. Bianchi S, Becciolini M. Ultrasound features of ankle retinacula: normal appearance and pathologic findings. J Ultrasound Med. 2019;38(12):3321–34.
40. Fantino O. Role of ultrasound in posteromedial tarsal tunnel syndrome: 81 cases. J Ultrasound. 2014;17(2):99–112.
41. Lau JT, Daniels TR. Tarsal tunnel syndrome: a review of the literature. Foot Ankle Int. 1999;20(3):201–9.
42. Erickson SJ, Quinn SF, Kneeland JB, Smith JW, Johnson JE, Carrera GF, et al. MR imaging of the tarsal tunnel and related spaces: normal and abnormal findings with anatomic correlation. AJR Am J Roentgenol. 1990;155(2):323–8.
43. Kerr R, Frey C. MR imaging in tarsal tunnel syndrome. J Comput Assist Tomogr. 1991;15(2):280–6.
44. Nagaoka M, Matsuzaki H. Ultrasonography in tarsal tunnel syndrome. J Ultrasound Med. 2005;24(8):1035–40.
45. Donovan A, Rosenberg ZS, Cavalcanti CF. MR imaging of entrapment neuropathies of the lower extremity. Part 2. The knee, leg, ankle, and foot. Radiographics. 2010;30(4):1001–19.
46. Moroni S, Zwierzina M, Starke V, Moriggl B, Montesi F, Konschake M. Clinical-anatomic mapping of the tarsal tunnel with regard to Baxter's neuropathy in recalcitrant heel pain syndrome: Part I. Surg Radiol Anat. 2019;41(1):29–41.
47. Fernández-Gibello A, Moroni S, Camuñas G, Montes R, Zwierzina M, Tasch C, et al. Ultrasound-guided decompression surgery of the tarsal tunnel: a novel technique for the proximal tarsal tunnel syndrome-Part II. Surg Radiol Anat. 2019;41(1):43–51.
48. Jamadar DA, Robertson BL, Jacobson JA, Girish G, Sabb BJ, Jiang Y, et al. Musculoskeletal sonography: important imaging pitfalls. AJR Am J Roentgenol. 2010;194(1):216–25.
49. Kissin E, DeMarco PJ, Malone DG, Bakewell C. Ultrasound training tips & pitfalls. Rheumatologist. 2018.

Chapter 15
Forefoot and Toes

Reasons to Do the Study

1. Pain in the forefoot
2. Soft tissue mass in the forefoot
3. Paresthesia in the forefoot

Questions We Want Answered

1. Could the patient have intermetatarsal bursitis or Morton's neuroma?
2. Is there evidence for rheumatoid arthritis (RA), gout, or osteoarthritis?
3. Is there a plantar plate tear?
4. Is there evidence of a foreign body or a metatarsal fracture?
5. If there is a mass, what is the cause?

Basic Anatomy

We include the metatarsals as part of the forefoot, although some authors define the forefoot as the anatomic region distal to the metatarsals. Please see Figs. 15.1a, b for an illustration of bony anatomy and the course and insertions of the flexor and extensor tendons. The fibular and tibial sesamoids are ovoid bones located plantar to the first metatarsal head. The flexor hallucis longus tendon (FHLT) runs between them to insert on the base of the distal phalanx [1]. The sesamoids are contained within the tendons of the flexor hallucis brevis (FHB) and form part of the plantar (volar) plate (PP). The sesamoids, the plantar plate, and the FHB tendon share a common connection to the proximal phalanx. The sesamoids elevate the metatarsal (MT) head off the ground and help bear weight.

The contents do not represent the views of the U.S. Department of Veterans Affairs or the United States Government.

M. H. Greenberg et al., *Manual of Musculoskeletal Ultrasound*,
https://doi.org/10.1007/978-3-031-37416-6_15

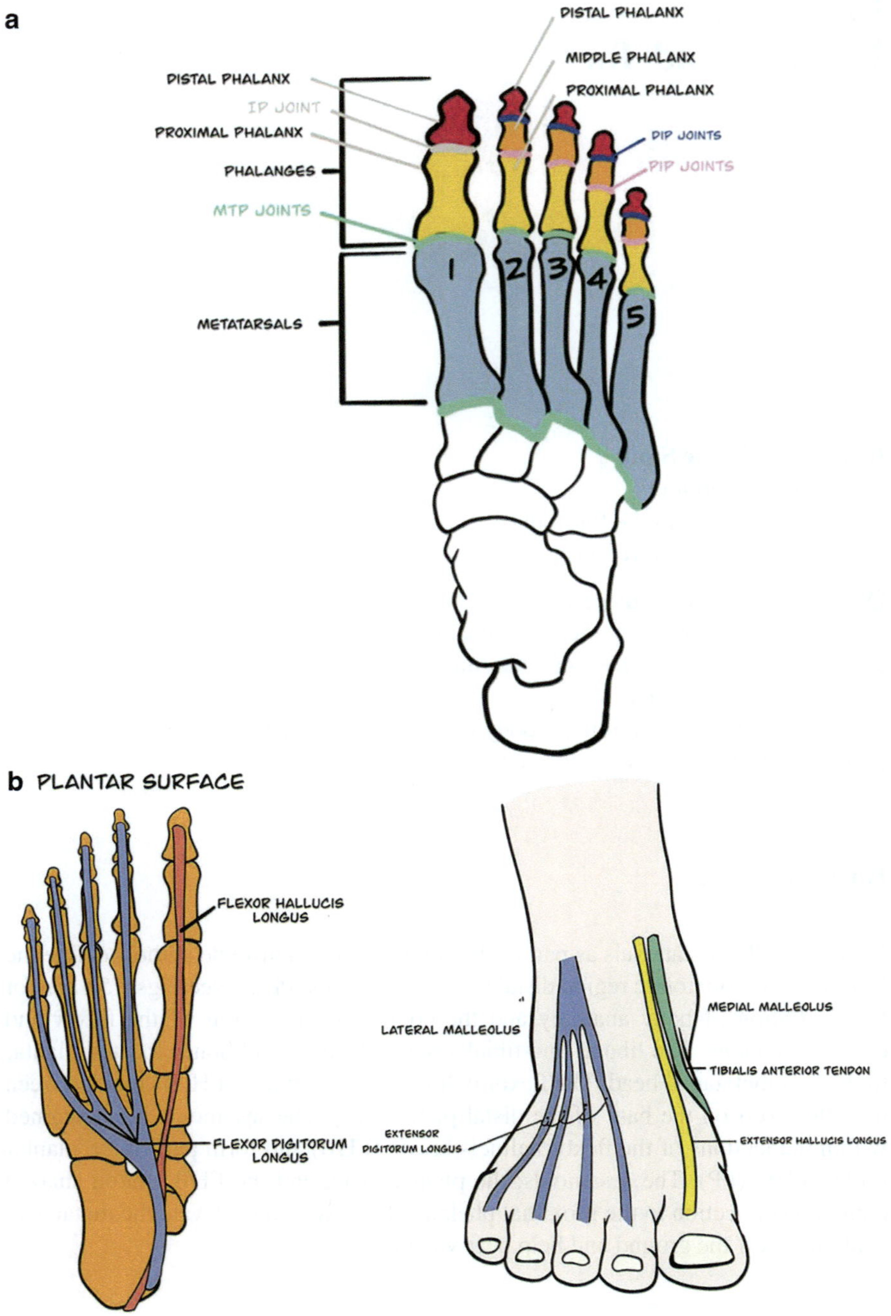

Fig. 15.1 (**a**) Bony anatomy of the forefoot (**b**) Tendons of the forefoot simplified

The PP is a flexible fibrocartilaginous structure that rests on the plantar surface of the cartilage of the MT head. It is attached to the collateral ligaments and the plantar fascia. It originates from the MT neck and inserts into the plantar aspect of the proximal phalanx (P1). The flexor tendon abuts the plantar surface of the PP. The PP limits the dorsiflexion of the P1, while the collateral ligaments constrain vertical and horizontal P1 movement.

Clinical Comments

The causes of forefoot pain are numerous and include plantar plate (PP) tears, joint disorders, stress fractures, osteonecrosis, Morton's neuroma, bursitis, tenosynovitis, and cyst formation [2]. Start the forefoot examination with a detailed clinical evaluation and *weight-bearing radiographs* to better display defects [3, 4]. If necessary, the painful area is assessed with ultrasound (US). Observe for pain exacerbated by transducer pressure. High-frequency transducers are necessary due to the superficial location of forefoot structures.

Plantar Plate Tear (Fig. 15.2)

On US, with the probe positioned on the plantar surface in longitudinal (LAX), the PP appears as a curved, mildly grainy hyperechoic structure hugging the anechoic or hypoechoic plantar cartilage of the MT head [2]. A normal triangular hyperechoic area at the distal PP may represent extra collagen fibers at the distal PP insertion. The bulk of PP tears occur at the second metatarsophalangeal joint (MTPJ) due to excess mechanical overload resulting from a preexisting hallux valgus deformity [5]. A hallux valgus deformity affects the first MTPJ, resulting in the first toe's lateral deviation and the first MT bone's medial deviation.

Although the second MTPJ is the site of most PP tears, such tears may occur in other digits [2, 6, 7]. Other causes of mechanical overload leading to PP tears include anatomic variance of the first and second toes (congenital or acquired), wearing narrow, high-heeled shoes that chronically hyperextend the lesser MTPJs, and PP trauma that may occur in sports [2].

There are typical features of a sizeable plantar plate tear [2, 4, 8, 9]:

1. A synovial effusion may be present.
2. There may be sonographic evidence of focal tears at the distal attachments to the proximal phalanx; these appear as ill-defined hypoechoic irregular defects.
3. Tears often occur at the second MTPJ, involve the lateral distal PP insertion, and then spread medially and proximally. Ultimately, the PP flattens and attenuates, becoming imperceptible on US.
4. If the PP tear progresses from a partial-thickness tear to a full-thickness tear, there may be a proximal retraction of the PP.

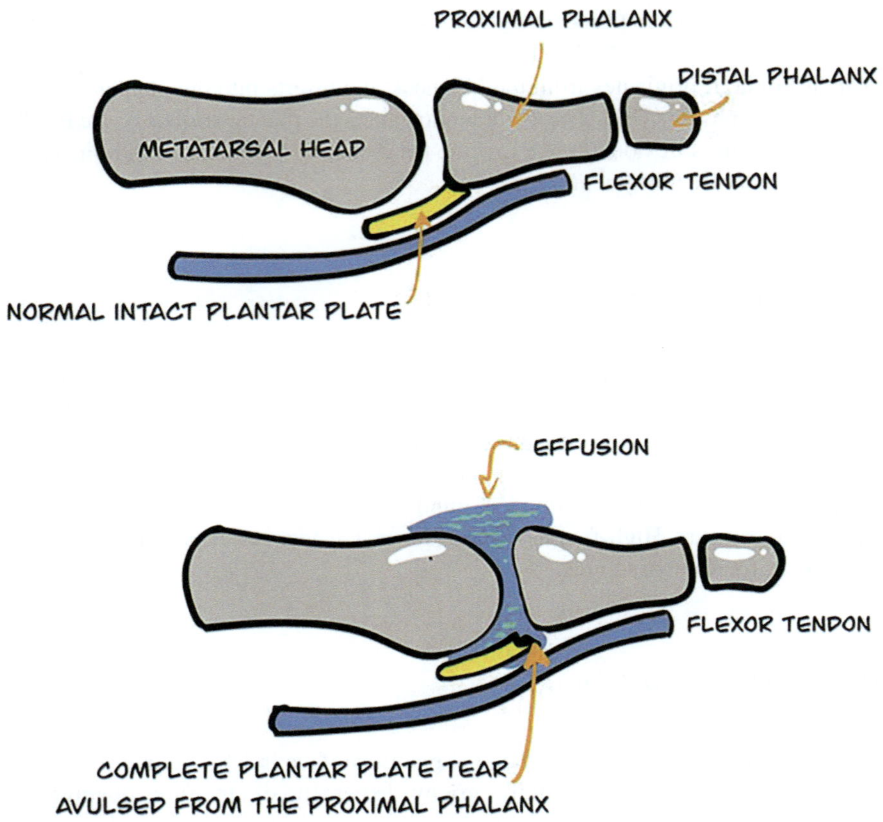

Fig. 15.2 Plantar plate normal and complete tear

5. The effusion becomes more evident as it communicates with the *dorsal* surface of the joint through the torn PP.
6. The deterioration of the PP leads to the *flexor tendon directly touching the plantar cartilage* that covers the MT head. A tenosynovial effusion surrounding the contiguous flexor tendon may appear on US as a hypoechoic halo in TAX.
7. The absence of a functional PP results in MTPJ instability, forefoot dysfunction, and deformity (second toe splaying, hyperextension, and perhaps even overlapping the first toe) with localized pain.
8. In addition to flexor tendon subluxation and/or tenosynovitis, other US findings may include a plantar cartilage interface sign, P1 enthesophytes, bony avulsion, MTPJ synovitis, and pericapsular fibrosis.

Additional dynamic views enhance tear visualization and exaggerate MTPJ subluxation if present [2]. Abnormalities found on static images should also be verified with transverse (TAX) views and Doppler, the latter being used to evaluate for hyperemia associated with a recent tear. Two static and two dynamic views are recommended [4].

Masses of the Forefoot

Forefoot masses may present as painless or painful lumps or be incidental findings on US or magnetic resonance imaging (MRI). Such masses may also occur in more proximal locations in the foot. The lesions can often be accurately characterized by their site, symptom association, and relationship to nearby structures such as nerves and blood vessels. Always perform a Doppler examination and measure the size of the lesions. In addition to US, further elucidation may require an MRI. Although a biopsy may be necessary, US can follow some lesions sequentially to quantitate enlargement over time.

Morton's Neuroma

The superficial and deep intermetatarsal ligaments run transversely, connecting the MT heads [4]. Along the bottom of the foot and between these layers runs the common plantar digital nerve, which divides into the interdigital nerves near the MT head. On the dorsal aspect of the intermetatarsal ligament lie the intermetatarsal bursae. The interdigital nerve in the third interspace (between the toe digits 3 and 4) is less mobile when compared to the other interdigital nerves [4]. This relatively fixed position may render this interdigital nerve less able to absorb mechanical stress and, hence, more prone to injury.

Due to chronic compression against the deep, transverse intermetatarsal ligament, this nerve may develop scarring (fibrosis) and swelling, resulting in a neuroma termed Morton's neuroma. This is the most frequent mass in the forefoot [4, 10, 11]. Causes of chronic entrapment or compression of this nerve include recurrent trauma, particularly from high-heeled shoes, and repeated dorsiflexion of the toes [4]. A Morton's neuroma may cause local excruciating pain, numbness, or paresthesia, perhaps radiating to adjacent toes. Patients liken the pain to "walking on a marble" [11]. Symptoms are aggravated by walking and improve with rest [12]. Morton's neuroma usually occurs between the third and fourth toes at the third web space [13].

Ultrasound can accurately detect Morton's neuroma and confirm a clinical diagnosis [12, 14, 15]. The ultrasonic appearance of Morton's neuroma is that of a hypoechoic, anechoic, or heterogeneous fusiform (ovoid) mass. In the sagittal (LAX) plane, US reveals a normal-sized fibrillar nerve to be continuous with the neuroma [12]. At the level of the MT heads, normal interdigital nerves measure 1–2 mm in diameter. Most symptomatic neuromas are greater than 5 mm in diameter; however, neuroma size does not necessarily correlate with symptoms [16]. If Morton's neuroma is accompanied by intermetatarsal bursitis, this is referred to as a "neuroma-bursal complex" [13]. Dynamic US may differentiate an IMB from Morton's neuroma (see section "Forefoot Bursae", below).

Ultrasound will also help guide local injections to treat Morton's neuroma via a dorsal approach [4, 17]. It is suggested that the area around the neuroma be injected along with any nearby bursa showing signs of inflammation. One study described a failure rate of only 15% with *direct* neuroma injection of corticosteroids [18]. The larger the neuroma, the less chance there is for injection success; however, there was no absolute size cutoff above which injection failure could be predicted. Since US detects the exact location and size of Morton's neuroma, it also helps guide surgery [4].

Forefoot Bursae

The two types of forefoot bursae, submetatarsal and intermetatarsal, are visible on US [19, 20]. Submetatarsal (adventitious) bursitis (SMB) is secondary to chronic overload of the MT heads, whereas intermetatarsal bursitis (IMB) may be associated with inflammatory arthropathy or Morton's neuroma [4]. Pain underneath the MT heads, particularly involving the first and fifth MT heads, is the typical history of a SMB [11]. On US, the SMB is hypoechoic and compressible within the plantar fat pad, superficial to the flexor tendon. In contrast, the IMB is located *between* the toes, on the dorsal aspect of the deep intermetatarsal ligament [21].

Clinically, an IMB presents with a painful forefoot lump-like sensation, intensified with walking, and possibly associated with paresthesia, mimicking Morton's neuroma [11]. On US, an IMB is centrally located between the MT heads and has thickened hyperechoic walls filled with an anechoic or hypoechoic fluid [4]. Local pressure may cause the internal swirling of fluid, confirming its liquid nature. There may occasionally be internal Doppler activity. Bursae are compressible, which helps differentiate them from peripheral solid masses such as Morton's neuroma and pericapsular fibrosis associated with plantar plate injury [11].

Dynamic US may differentiate an IMB from Morton's neuroma: place the probe in LAX on the dorsum of the interdigital webbing and exert upward thumb pressure from the plantar surface [4]. This maneuver shifts the neuroma in a *dorsal* direction and the IMB in a *posterior (proximal)* direction. However, as described above, a MN often has a concomitant IMB appearing on US as a large, complex, heterogeneous mass [12, 13]. It has been suggested that compression of the plantar nerve by an enlarged IMB may cause Morton's neuroma [11]. In practice, it may be challenging to sonographically distinguish Morton's neuroma from an IMB, particularly if the neuroma itself has an anechoic appearance.

Mucoid Cyst

Also known as digital mucous cysts, mucous pseudocysts, or myxoid cysts, these pearly colored benign cysts are associated with osteoarthritis. Mucoid cysts may sometimes be confused with ganglia and synovial cysts and occur near the proximal nail fold and distal interphalangeal joint. These periarticular cystic lesions have a fibrous wall, are filled with gelatinous material, and may communicate with the joint cavity or tendon [4]. Most often asymptomatic, these cysts compress sensitive structures, including nerves, producing pain when walking. On US, a mucoid cyst appears as a well-defined hypoechoic to anechoic mass adjacent to a joint; however, hypervascular changes may be noted if septae are present.

Ganglia

These are common benign, gelatinous-filled cystic masses that may communicate with nearby joints or tendon sheaths [22]. Ganglia appear on US as a uni- or multilocular anechoic or hypoechoic lesion with posterior acoustic enhancement and occasionally a stalk communicating with a tendon sheath or joint [11]. There may be internal septations, but no Doppler flow. If a ganglion is located between the metatarsals, it is challenging to differentiate this from an IMB on US. Although often called a "synovial cyst," the ganglion cyst *lacks* an actual synovial lining [23].

Synovial Cyst

These are synovial-lined periarticular fluid collections associated with joint arthropathy, which causes an effusion [24]. Synovial cysts communicate with the abnormal joint and may function as "drainage reservoirs" for excessive joint effusion in the setting of arthropathy [25]. Again, the synovial cyst contains synovial fluid, whereas the ganglion cyst is filled with gelatinous material. With its periarticular location and communication with a joint, the synovial cyst may easily be confused with a ganglion cyst.

Rheumatoid Nodules and Pannus

Rheumatoid nodules are granulomas with areas of central necrosis [24]. A history of RA is often elicited. The nodules, usually adjacent to bone, sonographically demonstrate slightly blurred edges with a mixed to hypoechoic echotexture [26]. There may be a central hypoechoic area, perhaps due to necrosis. The nodules can be mildly compressible, but lack posterior shadowing and Doppler activity. Rheumatoid pannus may also may present as a soft tissue mass in the foot [11]. On US, the pannus appears as hypoechoic synovial thickening within the joint capsule and may demonstrate Doppler activity.

Callus

This is a superficial thickening of the soft tissue secondary to mechanical pressure [24]. The typical location is the forefoot submetatarsal soft tissues. These are common and benign, but may ulcerate, particularly in diabetic patients.

Epidermoid Cyst or Implantation Dermoid

These occur due to the traumatic implantation of epidermal tissue into the dermis, forming a cyst containing viable epidermal cells that continue to produce keratin, filling the cyst [11]. Ultrasound reveals these lesions to be well-defined, solid, and hypoechoic, with variable internal echogenic debris and posterior acoustic enhancement. No Doppler signal is noted.

Inflammatory Granuloma

Retained foreign bodies provoke an immune response that may culminate in the formation of a granuloma. Pain, swelling, and a sinus tract to the skin may be present. Microorganisms introduced along with the foreign body may cause an abscess, which, on US, is well-defined and ranges from hypoechoic to hyperechoic with posterior acoustic enhancement [22]. Internal septations may be seen. Transducer compression may reveal Doppler activity in the surrounding area, but *not* within the abscess itself.

Ultrasound detects the foreign body and the secondary inflammatory response. Sonography may provide helpful information about the composition of the foreign body. If a granuloma forms, US may reveal a surrounding hypoechoic *halo* with posterior acoustic enhancement. There may or may not be a history of puncture wounds [4].

See section "Foreign Bodies", below.

Gouty Tophus

Extensive collections of uric acid may form tophi, which are periarticular solid masses [11]. Ultrasound reveals a central heterogeneous hyperechogenicity that may be surrounded by a thin hypoechoic halo. The first MTPJ is a common area to develop gouty tophi, but tophi may occur in other joints in the foot and, if in a plantar location, may result in the sensation of walking "on a lump."

Other evidence raising suspicion of a mass being a tophus includes a definitive history of gouty arthropathy, hyperuricemia, and other US findings such as synovitis, joint effusion, bony erosion (particularly with overhanging edges), or a double contour sign superficially coating hyaline cartilage. Multiple, hyperechoic uric acid foci may produce a "snowstorm" appearance on US. These foci are microtophi. Note that tophi may occur nearly *anywhere* in the foot and may result in multiple extensive soft tissue masses, perhaps associated with bony destruction from pressure erosion [24].

(See Chap. 24 for more information.)

Lipoma

Lipomas are benign aggregates of mature fat cells, often elliptical and well-defined, with the most prominent dimension parallel to the skin [22]. Echogenicity is variable, with no internal Doppler activity. The diagnosis of lipoma, strongly suggested by US, is confirmed by MRI.

Plantar Fibromatosis (Ledderhose's Disease)

Benign plantar fascia fibroblastic proliferation occurs in the mid-foot but may extend more distally to the plantar fascia of the forefoot and toes [22]. Patients usually present with either pain or a lump on the sole [27]. Ultrasound reveals nodular, fusiform, hypoechoic, or mixed echogenic masses in the middle or distal plantar fascia, sometimes demonstrating Doppler activity [22, 24]. Sonographic findings are often strongly suggestive or diagnostic of plantar fibromatosis, especially if the condition is bilateral. Plantar fibromatosis may occasionally become aggressive and surround skeletal muscle [11].

(See Chap. 12 for more information.)

Pigmented Villonodular Synovitis

Also known as a giant cell tumor, pigmented villonodular synovitis (PVNS) is a benign proliferation of synovial tissue that may affect the ankle joint or tendon sheaths [4, 24, 28, 29]. It presents in a diffuse or nodular form. The nodular form primarily affects distal tendon sheaths, appearing sonographically as a homogeneous, hypoechoic, well-defined solid mass bordering the tendon. There may also be Doppler activity [29]. Ultrasound may demonstrate the relationship to adjacent structures, pressure erosions of bone, or local recurrence after surgical excision. However, since the appearance of PVNS on US is *nonspecific*, an MRI is necessary for a more definitive diagnosis [4]. Although PVNS is described as benign, there is a 14% recurrence rate after excision [24].

Benign Nerve Sheath Tumors: Schwannoma and Neurofibroma

These benign tumors of peripheral nerves appear on US as hypoechoic to heteroechoic, fusiform, well-marginated masses found along the course of a peripheral nerve [4]. There may be internal Doppler activity. Neurofibromas are not encapsulated and may be infiltrative; schwannomas are encapsulated and may mimic Morton's neuromas [11]. Differentiating a schwannoma from a neurofibroma is often impossible on either US or MR imaging [22].

Glomus Tumor

These benign hamartomas of the neuromyoarterial glomus body may occur in the fingers or toes [22]. On US, these ovoid masses appear hypoechoic and are sharply delineated [30]. Color Doppler may delineate vasculature within the lesion, a helpful diagnostic sign [31]. Glomus tumors are well-defined on MRI [32].

Other Masses

In addition to the list of masses already considered, additional conditions described in other sections of this chapter may produce a mass-like structure. These include pericapsular fibrosis associated with a plantar plate fracture and osteoarthritis. Less common lesions presenting as a forefoot mass are beyond the scope of this chapter. These include synovial chondromatosis, soft tissue chondroma, synovial sarcoma, undifferentiated pleomorphic sarcoma, leiomyosarcoma, clear cell sarcoma, and subcutaneous granuloma annulare [4, 11, 22, 24].

Foreign Bodies

Plantar puncture wounds may result in foreign body (FB) retention, which predisposes to inflammation, infection, and damage to surrounding structures [33]. Radiographs may not detect certain materials, such as wood, plastic, fish bones, and aluminum. Other FBs may be too small to delineate on radiographs. However, most FBs are hyperechoic and therefore sonographically discernible.

Glass is radiopaque and therefore detectable on both plain radiographs and US scans [34]. In the case of a retained pencil tip, the linear graphite center is radiopaque, and thus detectable on radiographs; however, the surrounding wood is radiolucent. Wooden foreign bodies as small as 2.5 mm in length can be detected with 87% sensitivity and 97% specificity by US [35].

Accurate sonographic localization of FBs minimizes surgical exploration and may guide percutaneous removal [34]. In addition, US assesses secondary soft tissue reactions and neurovascular and tendon damage caused by the FB. Foreign bodies in soft tissue are *all initially hyperechoic* on sonography; however, wooden FBs may become less echogenic with time [35].

Sonographic artifacts help detect the presence and composition of the FB [34]:

1. **Acoustic Shadowing** (Figs. 15.3a, b): Objects with a small radius of curvature or a rough surface (e.g., a wooden toothpick) cause "clean" acoustic shadowing (crisp, well-defined sharp edges with complete darkening deep to the object). Objects with a large radius of curvature or a smooth surface (e.g., glass and metal) produce "dirty" (hazy) shadowing and reverberation artifacts.

2. **Reverberation Artifact** (Fig. 15.3c) occurs when the transducer is positioned parallel to the FB surface and thus depends on the orientation of the FB. Therefore, an expected reverberation artifact for a sewing needle may not be visualized on US if the probe cannot be placed perfectly parallel to the object.

Table 15.1 summarizes the sonographic findings of a study in which different types of foreign bodies were experimentally introduced into cadaver feet [34, 35]. Note that the probe should be aligned parallel to the object as much as possible for oblong objects to produce these telltale artifacts. In vivo, the FB may create edema, hemorrhage, and later granulation tissue [34]. This results in a hypoechoic halo which helps to identify the object as being a FB. However, the halo may *not* be present in the first 24 h since it represents an inflammatory response [36].

A hypervascular response may be demonstrated with Doppler positivity due to inflammation or infection [37]. Ultimately, a granuloma may be caused by a retained FB with clinical evidence of swelling, tenderness, or a sinus tract [11]. Ultrasound and MRI may both demonstrate a FB and surrounding reactive changes. In addition to US and MRI, other modalities that can detect the FBs include fluoroscopy and CT. See section "Inflammatory Granuloma", above.

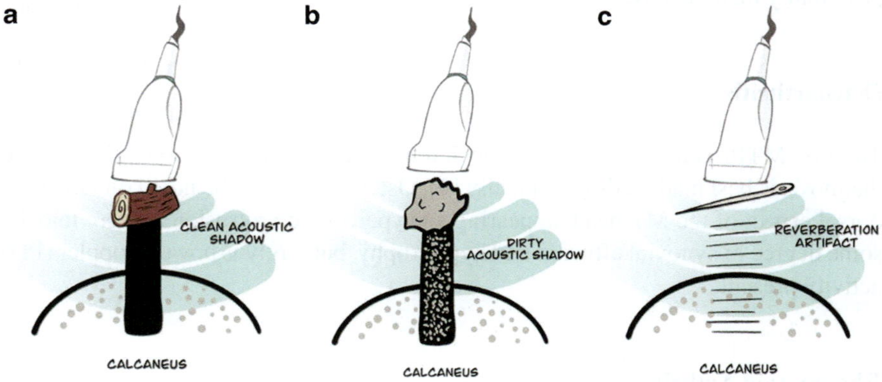

Fig. 15.3 (**a**) Clean acoustic shadowing with foreign bodies (**b**) Dirty acoustic shadowing with foreign bodies (**c**) Reverberation artifact with foreign bodies

Table 15.1 Summary of foreign body ultrasound findings in cadaver feet [33, 34]

	Acoustic shadowing	Reverberation
Steel	None	++
BB shot pellet	None	++
Glass	Dirty	+
Sewing needle	Dirty	+/++
Stone	Dirty	+
Plastic	Clean	None
Wooden toothpick	Clean	None
Pencil fragments	Clean	None

Tendon Pathology

The pathology of extensor and flexor tendons most often occurs proximal to the forefoot. Tendon injury in the forefoot may be a result of direct trauma, gout, or rheumatoid arthritis, although this is relatively rare [38]. Distal tendon tears, either partial or complete, are detectable by US [4]. A dynamic examination differentiates a partial tear from a complete tear. A complete tear will demonstrate separation of the tendon stumps if viewed in LAX by passively flexing or extending the toes or forefoot.

Arthritis

Forefoot arthropathy often involves the MTPJs, but the interphalangeal joints may also be involved. Do not mistake the normal first MTPJ physiologic fluid in the recess for an effusion. Recall that the average bone-joint capsule distance for the first MTPJ is 1.7 mm [39]. Also, bear in mind that the synovial recess reflects more proximally than distally.

Osteoarthritis

The first MTPJ is the most common location for degenerative forefoot change; the diagnosis is best made radiographically [4]. Osteophytes may be noted on US at the dorsal aspect of the MT head, appearing as hyperechoic protrusions. There may be some degree of synovial effusion and hypertrophy, but rarely is power Doppler (PD) activity present.

Rheumatoid Arthritis

The foot is often overlooked when evaluating RA. Synovial hypertrophy (pannus) and cartilage and bone erosion of the MTPJs may be evident [4]. Synovitis appears as a hypoechoic thickening of the synovium within the recess, which may or may not contain Doppler activity [40]. The fifth MT head may be the earliest area where rheumatoid erosion is noted; fifth MT head erosions are 85% specific for RA [41, 42]. Bone erosion appears as an interruption in the hyperechoic cortex, with a small hyperechoic area appearing as a step off beneath the main level of the cortex surface [4]. In addition, thinning or interruption of the hypoechoic/anechoic cartilage covering the bony cortex may be seen. Ultrasound has proven to be a reliable technique for detecting early RA erosions [43]. Ultrasound-guided injections into RA-afflicted joints of the foot improve accuracy [4].

Gout

Gouty arthropathy has a predilection for the first metatarsophalangeal joint and the foot dorsum. Please see Chap. 24 and Gouty Tophus in Clinical Comments above for more information.

Psoriatic Arthritis

Psoriatic arthritis may involve the toes, including the MTPJs and interphalangeal joints. Ultrasound reveals synovitis, bone erosion, and enthesitis at tendon insertions [44]. Gout may mimic psoriatic arthritis and vice versa [45]. Ultrasound is beneficial in differentiating the two.

Bone Disorders

Metatarsal Fracture (Fig. 15.4)

Metatarsal stress fractures often occur in regular military recruits, runners, dancers, and gymnasts [11]. In addition, insufficiency fractures may occur with everyday stress on pathologically weak bones [46]. It is essential to diagnose MT fractures early to avoid progression to a complete fracture, delayed or poor healing, and non-union [47, 48]. However, plain radiographs have low sensitivity for the diagnosis of early MT fractures; they may be normal initially and only *weeks later* demonstrate changes indicating that a fracture had previously occurred [4, 11, 47, 49].

It is noteworthy that even follow-up radiographs may remain normal [50]. Ultrasound may diagnose early stress fractures of the MT bones even when conventional radiographs are unremarkable [4, 11, 49, 51]. Ultrasound is 83% sensitive and 76% specific for the early diagnosis of MT stress fractures [47].

The US signs of a MT fracture [4, 11, 49–51]:

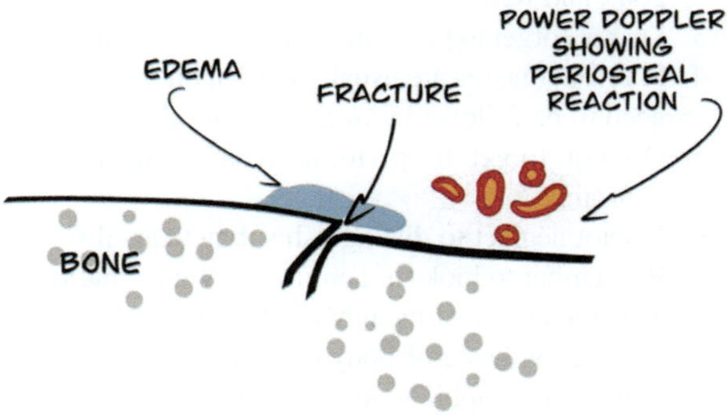

Fig. 15.4 Metatarsal fracture

1. Periosteal reaction: periosteal thickening or elevation, Doppler activity indicating hypervascularity, a minute fluid collection along the bony cortex, focal soft tissue edema
2. Surrounding soft tissue edema
3. Hematoma (hypoechoic fluid collection)
4. Discontinuities (breaks) in the hyperechoic cortex
5. Bony callus formation later: thickened hyperechoic cortex

MRI is the gold standard for diagnosing MT fractures [47]. A bone scan (scintigraphy) may identify occult stress fractures, but requires 24–36 h to become positive [11, 47, 48].

A proposed algorithm for evaluating suspected MT fractures [47]:

1. Radiographs (oblique view) improve diagnostic accuracy [52]
2. Ultrasound
3. MRI, if available
4. Bone scan, if there is still a question of a fracture

An abnormal bone scan requires a confirmatory imaging procedure such as a CT or MRI targeting the abnormality; this is important to exclude a bone condition different from a fracture, such as a bony neoplasm. When performing US for MT fracture evaluation, apply gentle transducer pressure over the painful area. Start in LAX orientation, scanning the entire width of the MT. Use TAX views to complete the exam and confirm suspected pathology.

Pitfalls

1. Always complete a detailed history and physical examination, followed by radiographs (if necessary), when evaluating foot pain.
2. Despite myriad causes of forefoot discomfort, common conditions are common and should be considered first, as directed by the clinical picture. This includes osteoarthritis, Morton's neuroma, plantar plate tear, bursitis, stress fracture, gout, and RA.
3. Do not forget to perform dynamic views when evaluating for plantar plate tears.
4. Forefoot masses are usually benign but need delineation. At the very least, these need to be followed sequentially for changes in size or shape.
5. Do not forget to perform dynamic maneuvers when evaluating Morton's neuroma.
6. Do not neglect to distinguish submetatarsal from intermetatarsal bursitis.
7. Remember to look for confirmatory information when considering gouty tophi. This includes a clinical history consistent with gout, hyperuricemia, and radiographic evidence of bony erosion with overhanging bone. There is often additional sonographic evidence of uric acid deposition, namely a double contour sign, which may be present in the forefoot or other body locations.

8. Regarding FBs, obtain a history of the FB composition, examine scout radiographs, and look for a puncture wound. Realize that the FB may be remote from the puncture wound and may have broken up within the soft tissue. In addition, try to align the transducer parallel to the FB based on its radiographic location. Use a sterile field for your US examination to avoid introducing bacteria into an open puncture wound. Artifacts associated with the FB help verify the composition. Look for sonographic signs of an inflammatory response.
9. Remember that RA may initially present in the foot. The earliest detectable erosion of bone may be in the fifth MT head, an area sonographically accessible. Do not depend on plain radiographs alone to rule out bony erosions in this area since US is more sensitive at detecting bony erosion.
10. Oblique radiographs increase the detection rate of MT fractures. However, do not depend on radiographs to rule out a MT fracture. Ultrasound may diagnose early MT stress fractures with normal radiographs. If the US is negative and a stress fracture is still suspected, a bone scan or possibly an MRI should be performed.

Method

Examine the forefoot dorsum with a supine patient and the foot flat on the table [4]. Plantar evaluation is performed with the leg stretched out, exposing the plantar surface of the foot. The foot may dangle off the table to facilitate dynamic maneuvers. The lateral and medial aspects of the forefoot may also be evaluated. Remember to do Doppler studies to look for hyperemia. Extensor tendons are thin, hyperechoic fibrillar structures that can be identified at the anterior ankle and followed to their distal insertions. Flexion-extension of the toes facilitates tendon identification and evaluates the integrity of the tendons and plantar plates. Likewise, US can follow flexor tendons to the distal insertion into the distal phalanx.

Evaluate the cortical aspect of the metatarsals, which is usually hyperechoic, continuous, and regular [53]. Perform transverse and LAX imaging of the dorsal and plantar aspects of the MTPJs [4]. For the first and fifth MTPJs, the medial and lateral views are accessible for inspection as well. The synovial recess of the MTPJs does not typically contain fluid; however, sometimes, a normal first MTP may demonstrate a small effusion. The cartilage of the distal MT head may be assessed on the dorsal and plantar surfaces.

The interphalangeal joints are challenging to evaluate due to their small size. Intermetatarsal spaces can be examined with the probe in both dorsal and plantar positions and in both TAX and LAX planes. The fifth MT head may demonstrate RA bony erosion imperceptible on plain radiographs. Inspect the dorsal, lateral, and plantar aspects of the fifth MT head if RA is suspected. The plantar forefoot often requires a lower frequency due to the presence of dense soft tissue through which the US beam needs to penetrate.

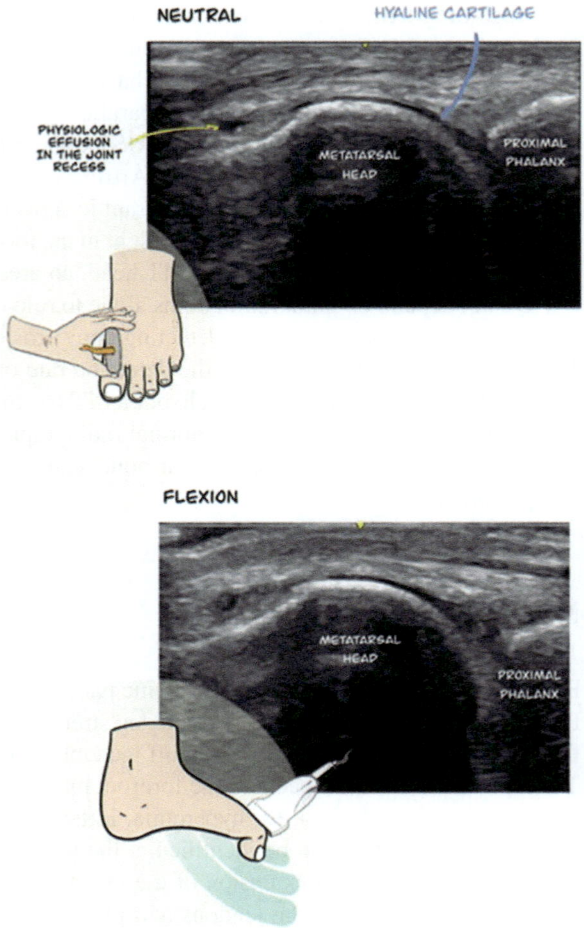

Fig. 15.5 Protocol Image 1: Dorsal first metatarsophalangeal joint, longitudinal ± power Doppler

Protocol Image 1: Dorsal First Metatarsophalangeal Joint, Longitudinal ± Power Doppler (Fig. 15.5)

With the foot flat on the exam table, place the transducer in LAX across the dorsum of the first MTPJ, looking for the hyperechoic MT head and P1; the joint space is defined between these two bones. Use a liberal amount of transmission gel. Evaluate the cartilage and the joint recess to look for effusion, cartilage/bone erosion, calcium deposition, gouty tophi, and double contour sign of uric acid deposition on the cartilage surface.

Next, passively flex the toe to examine the distal cartilage surface and the extensor tendon. An interface reflex on the dorsum of the cartilage implies a smooth, intact cartilage surface. Synovial effusion will distend the joint recess distally and, to a greater extent, proximally. For the first MTPJ, the mean bone-to-joint capsule distance is 1.7 mm, with an additional 0.9 mm representing one standard deviation [39]. Thus, values larger than 2.6 mm imply abnormal capsular distension.

If bony erosion is suspected, then perform a TAX slide around the first MT head from dorsal to plantar, confirming a putative erosion in TAX (orthogonal) views. Turn on the PD to look for hyperemia. Look at the insertion of the extensor tendon along the distal bony phalanx to evaluate for active enthesitis. This requires a high-frequency transducer.

Protocol Image 2: Dorsal First Metatarsophalangeal Joint, Transverse ± Power Doppler (Fig. 15.6)

Rotate the probe 90° and perform a distal TAX slide starting at the MT head and moving over the joint space to the P1. Reverse the movement if necessary. Look for erosion, crystal deposits, and hyperemia, confirming prior LAX view findings.

Protocol Image 3: Plantar First Metatarsophalangeal Joint, Longitudinal ± Power Doppler + Dynamic (Fig. 15.7)

Stretch the leg out and place the probe in LAX orientation over the plantar first MTP. Increasing the depth as needed and lowering the frequency will help identify the joint space and PP. The latter appears typically as an echogenic, homogeneous

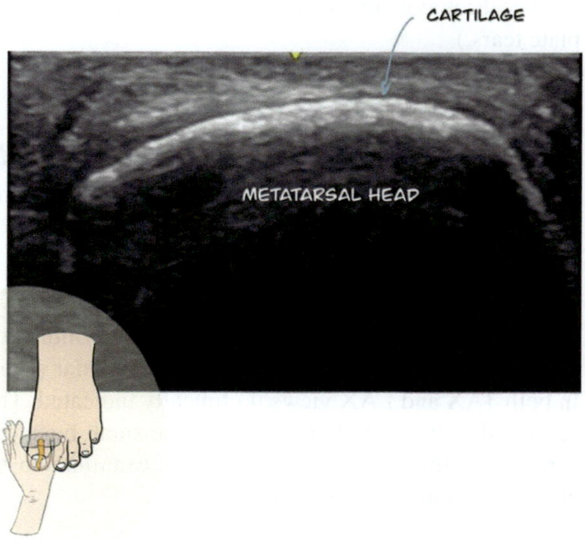

Fig. 15.6 Protocol Image 2: Dorsal first metatarsophalangeal joint, transverse ± power Doppler

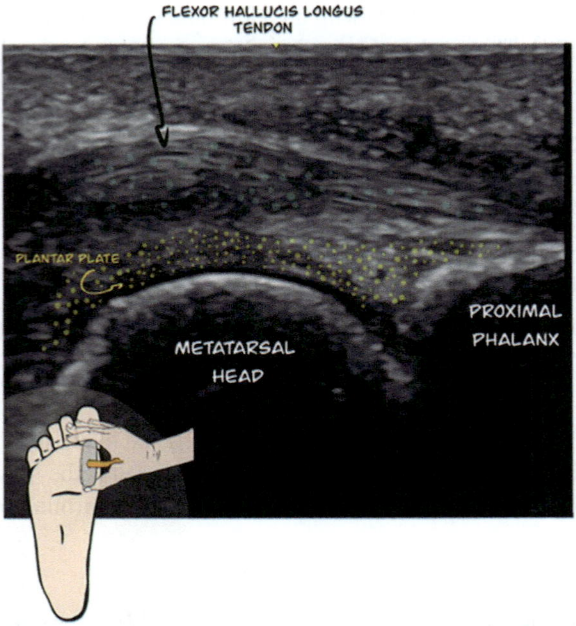

Fig. 15.7 Protocol Image 3: Plantar first metatarsophalangeal joint, longitudinal ± power Doppler + dynamic

curvilinear structure hugging the anechoic or hypoechoic cartilage of the metatarsal head [54]. A PP tear will have a hypoechoic or heterogeneous defect and sometimes be detectable only with passive toe dorsiflexion, so passively flex and extend the toe to verify the integrity of the PP and the flexor tendon. Turn on the PD to evaluate for hyperemia. Do not mistake an interface reflex superficial to the MT head cartilage for a DCS.

(See Protocol Images 6a–c for more details on imaging of suspected plantar plate tears.)

Protocol Image 4: Plantar First Metatarsophalangeal Joint, Transverse ± Power Doppler ± Dynamic Exam (Fig. 15.8)

Rotate the probe 90° and perform a distal TAX slide starting at the MT head and moving over the joint space to the P1. Reverse the movement if necessary. Look for erosion, crystal deposits, and hyperemia, confirming prior LAX view findings. Note the two sesamoids, the tibial (medial) and fibular (lateral). Examine the sesamoids in both TAX and LAX views if clinically indicated. The oval structure between the sesamoids is the FHLT in TAX, and the small bone deep to the sesamoids located centrally is the first MT. The dynamic examination may be repeated if there are concerning findings in Protocol Image 3.

Fig. 15.8 Protocol Image 4: Plantar first metatarsophalangeal joint, transverse ± power Doppler ± dynamic exam

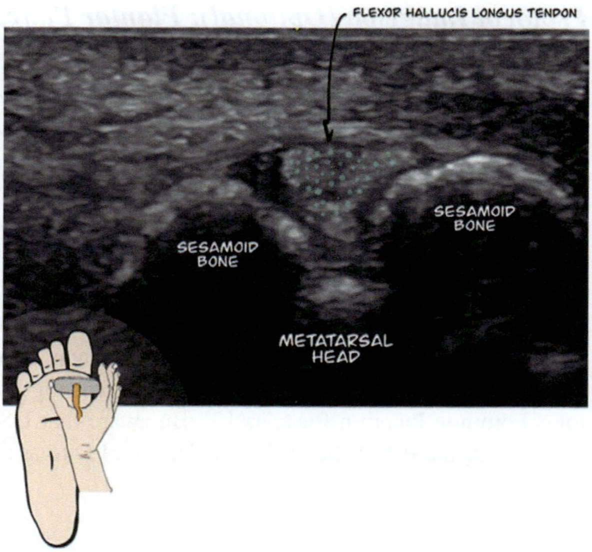

Protocol Image 5: Dorsal Fifth Metatarsophalangeal Joint, Longitudinal ± Power Doppler + Dynamic

Place the small footprint probe in an LAX configuration on the dorsal surface of the fifth MTPJ. Evaluate the fifth MT head for erosion and examine the joint for calcium and uric acid deposits. Passively flex and extend the toe to examine the distal extensor tendon. Slowly move the probe in a TAX slide to reach the lateral aspect of the joint, focusing on the MT head and the joint space. Continue to curve the probe in a TAX slide around the fifth MT head, looking for bone erosion. Passively flex and extend the toe to examine the distal flexor tendon. Confirm any putative erosions in the TAX view. Remember that fifth-MT head erosions are 85% specific for RA [42].

Protocol Images 6a–c (Optional): Evaluation of a Suspected Metatarsophalangeal Plantar Plate Tear

These images may be applied to any toe suspected of having a plantar plate tear.

Protocol Image 6a (Optional): Plantar Plate, Longitudinal ± Power Doppler

Place the probe in LAX over the plantar surface of the MTPJ of concern to visualize the MT head with covering cartilage, the proximal phalanx, and the curved, slightly hyperechoic PP superficial to the anechoic or hypoechoic cartilage. A lower frequency might be necessary. Also, note the fibrillar flexor tendon location and if it directly abuts the MT head cartilage, as would occur with severe PP damage (the PP normally separates the flexor tendon from the MT head cartilage). Manually dorsiflexing the toe may increase tension on the PP and improve visualization. Look for discontinuity in the PP, which might indicate a tear. Start at the lateral portion of the PP and perform a slow medial TAX slide, focusing on the distal PP insertion, the most common location for tears [2]. Be aware that US cannot assess the proximal insertion of the PP on the MT neck. Turn on PD to assess for hyperemia.

Protocol Image 6b (Optional): Plantar Plate, Transverse ± Power Doppler (Fig. 15.9)

Rotate the probe 90° to look at the MTPJ in TAX. Use this view to *confirm* PP tears by looking for a *halo* surrounding the flexor tendon (tenosynovial fluid) or an *abnormal migration of the flexor tendon* to touch the MT head cartilage if the PP is

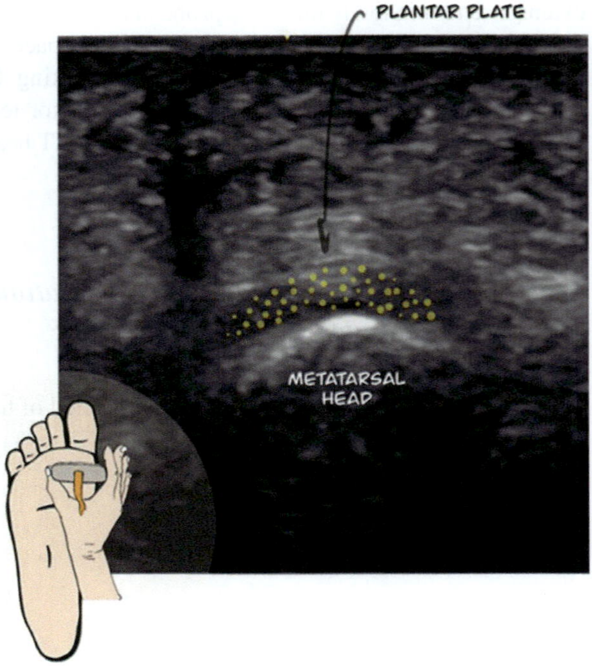

Fig. 15.9 Protocol Image 6b (Optional): Plantar plate, transverse ± power Doppler

substantially damaged. This view will also portray the eccentric location of pericapsular fibrosis, which may be associated with a PP tear [2]. Perform a slow TAX slide from distal to proximal, focusing on the PP.

Protocol Image 6c (Optional): Dorsal Metatarsophalangeal Joint, Longitudinal + Dynamic

Next, move the probe to the dorsal aspect of the MTPJ, place it in LAX (sagittal plane), and observe for dorsal subluxation of the proximal phalanx, which might indicate a full-thickness PP tear. Use the free hand to dorsiflex the toe to better delineate a PP tear by making the PP tauter. In the case of a full-thickness PP tear, this maneuver may exaggerate dorsal subluxation of the proximal phalanx, as noted by increased space between the proximal phalanx and the MT head. Next, place the free hand thumb on the plantar surface of P1 and exert upward pressure, which might further displace P1 dorsally should a full-thickness PP tear exist. Observe for a joint effusion, which may be associated with a PP tear.

Protocol Image 7: Morton's Neuroma and Intermetatarsal Bursa Evaluation, Dorsal, Longitudinal + Dynamic (Fig. 15.10)

A small footprint, high-frequency probe is placed in TAX over the dorsum of the interdigital webbing, typically interspace number three, when evaluating for Morton's neuroma. Once the honeycomb echotexture of the interdigital nerve is located, slowly rotate the transducer 90° to visualize the nerve in its longitudinal orientation.

At the level of the MT heads, normal interdigital nerves measure 1–2 mm in diameter [16]. Thus, identifying a normal interdigital nerve may be challenging.

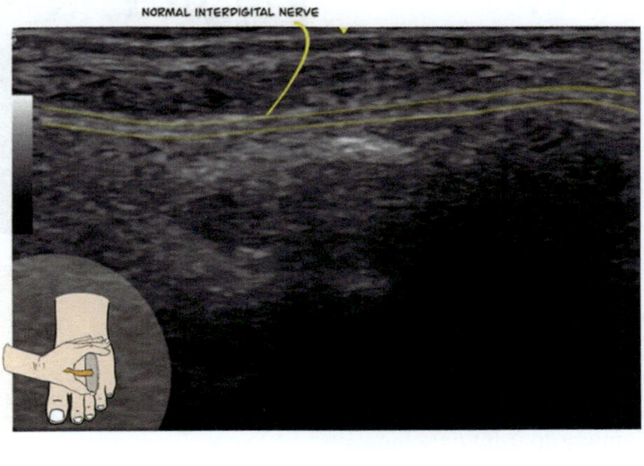

Fig. 15.10 Protocol Image 7: Morton's neuroma and intermetatarsal bursa evaluation, dorsal, longitudinal + dynamic

Using a high-frequency transducer (at least 16 MHz) along with the zoom function (if available) on the US machine will enhance the visualization of this small nerve. You may see an enlarged interdigital nerve, consistent with Morton's neuroma. If detected, a normal-sized interdigital nerve leading into the neuroma helps corroborate the diagnosis.

Morton's neuroma may have an accompanying IMB. The free thumb of the examiner presses upward from the plantar surface. This maneuver decreases the thickness of the soft tissue and shifts the neuroma *dorsally*, enhancing visualization. Furthermore, an IMB will be shifted *posteriorly* with this maneuver.

See the section "Forefoot Bursae and Morton's Neuroma", above for more information.

Protocol Image 8: Morton's Neuroma and Intermetatarsal Bursa Evaluation, Plantar, Transverse + Dynamic Exam (Fig. 15.11)

The high-frequency probe is placed in TAX along the plantar aspect of the painful interdigital space [12]. The examiner's free hand compresses the MT heads together. A Morton's neuroma will be displaced in a plantar direction, perhaps associated with a "click" and the reproduction of pain. This has been described as the sonographic Mulder sign [16].

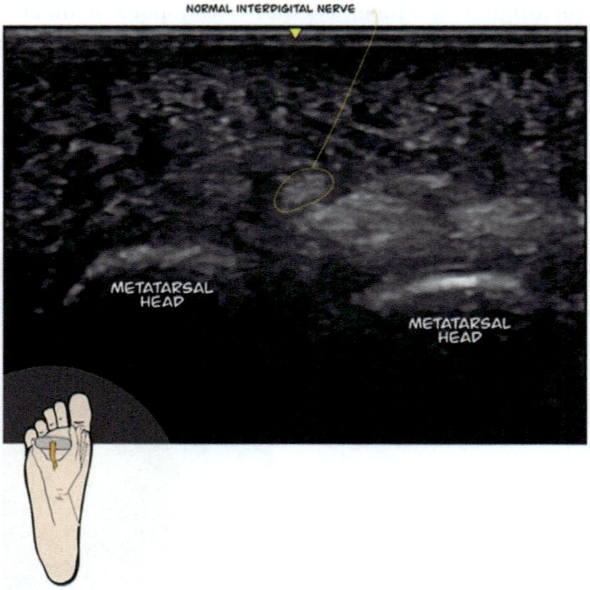

Fig. 15.11 Protocol Image 8: Morton's neuroma and intermetatarsal bursa evaluation, plantar, transverse + dynamic exam

Protocol Image 9 (Optional): Evaluation for Metatarsal Fracture

When performing a US examination for an MT fracture, use minimal transducer pressure over the painful area. Start in LAX orientation, slowly scanning the entire width of the MT with gentle TAX slides. Use TAX views to complete the exam and confirm suspected pathology. Use PD to look for further evidence of a periosteal reaction. Record additional pertinent images as Protocol Images 9a, 9b, etc. See the section "Metatarsal Fracture", above for more information.

Protocol Image 10 (Optional): Evaluation for a Foreign Body

After taking a pertinent history, review the radiographs. Plastic, aluminum, and wooden objects may not be appreciated on plain films. Locate the maximal point of pain or swelling and the point of entrance. Sterilize the area and perform the US exam with sterile gel and a probe cover to avoid introducing bacteria into a puncture wound. Center the probe on this area and slowly rotate and search in 360° since the object and attendant artifacts, if linear, will be best visualized parallel to the probe.

Perform the exam methodically using both TAX and LAX views as well as views in between due to orientation issues with the foreign body. Verify all findings with orthogonal views. Look for the echogenicity of the foreign body, specifically posterior acoustic shadowing and reverberation artifacts. Secondary swelling, granulation tissue, or pus may cause a hypoechoic rim or halo. Do not forget to use Color or power Doppler to evaluate for hypervascularity. Refer to Table 15.1 and the "Clinical Comments" section on foreign body artifacts to determine if a visualized structure has sonographic characteristics compatible with the FB composition expected from the patient history. The FB may also be somewhat remote from the entrance wound. Be aware that multiple foreign bodies may be present, and a single foreign body may have broken into two or more fragments.

Complete Forefoot/Toes Ultrasonic Examination Checklist

☐ Protocol Image 1: Dorsal first metatarsophalangeal joint, longitudinal ± power Doppler

☐ Protocol Image 2: Dorsal first metatarsophalangeal joint, transverse ± power Doppler

☐ Protocol Image 3: Plantar first metatarsophalangeal joint, longitudinal ± power Doppler + dynamic

☐ Protocol Image 4: Plantar first metatarsophalangeal joint, transverse ± power Doppler ± dynamic exam

☐ Protocol Image 5: Dorsal fifth metatarsophalangeal joint, longitudinal ± power Doppler + dynamic
☐ Protocol Image 6a (optional): Plantar plate, longitudinal ± power Doppler
☐ Protocol Image 6b (optional): Plantar plate, transverse ± power Doppler
☐ Protocol Image 6c (optional): Dorsal metatarsophalangeal joint, longitudinal + dynamic
☐ Protocol Image 7: Morton's neuroma and intermetatarsal bursa evaluation, dorsal, longitudinal + dynamic
☐ Protocol Image 8: Morton's neuroma and intermetatarsal bursa evaluation, plantar, transverse + dynamic exam
☐ Protocol Image 9 (optional): Evaluation for metatarsal fracture
☐ Protocol Image 10 (optional): Evaluation for a foreign body

References

1. Sims AL, Kurup HV. Painful sesamoid of the great toe. World J Orthop. 2014;5(2):146–50.
2. McCarthy CL, Thompson GV. Ultrasound findings of plantar plate tears of the lesser metatarsophalangeal joints. Skelet Radiol. 2021;50(8):1513–25.
3. Morvan G, Wybier M, Mathieu P, Vuillemin V, Guérini H, Stérin P, et al. Foot imaging. Rev Prat. 2010;60(3):335–41.
4. Bianchi S. Practical US of the forefoot. J Ultrasound. 2014;17(2):151–64.
5. Doty JF, Coughlin MJ. Metatarsophalangeal joint instability of the lesser toes. J Foot Ankle Surg. 2014;53(4):440–5.
6. Umans H, Srinivasan R, Elsinger E, Wilde GE. MRI of lesser metatarsophalangeal joint plantar plate tears and associated adjacent interspace lesions. Skelet Radiol. 2014;43(10):1361–8.
7. Nery C, Coughlin MJ, Baumfeld D, Mann TS. Lesser metatarsophalangeal joint instability: prospective evaluation and repair of plantar plate and capsular insufficiency. Foot Ankle Int. 2012;33(4):301–11.
8. Carlson RM, Dux K, Stuck RM. Ultrasound imaging for diagnosis of plantar plate ruptures of the lesser metatarsophalangeal joints: a retrospective case series. J Foot Ankle Surg. 2013;52(6):786–8.
9. Gregg JM, Schneider T, Marks P. MR imaging and ultrasound of metatarsalgia—the lesser metatarsals. Radiol Clin N Am. 2008;46(6):1061–78, vi–vii.
10. Abreu E, Aubert S, Wavreille G, Gheno R, Canella C, Cotten A. Peripheral tumor and tumor-like neurogenic lesions. Eur J Radiol. 2013;82(1):38–50.
11. Ganguly A, Warner J, Aniq H. Central metatarsalgia and walking on pebbles: beyond Morton neuroma. AJR Am J Roentgenol. 2018;210(4):821–33.
12. Quinn TJ, Jacobson JA, Craig JG, van Holsbeeck MT. Sonography of Morton's neuromas. AJR Am J Roentgenol. 2000;174(6):1723–8.
13. Cohen SL, Miller TT, Ellis SJ, Roberts MM, DiCarlo EF. Sonography of Morton neuromas: what are we really looking at? J Ultrasound Med. 2016;35(10):2191–5.
14. Redd RA, Peters VJ, Emery SF, Branch HM, Rifkin MD. Morton neuroma: sonographic evaluation. Radiology. 1989;171(2):415–7.
15. Mahadevan D, Venkatesan M, Bhatt R, Bhatia M. Diagnostic accuracy of clinical tests for Morton's neuroma compared with ultrasonography. J Foot Ankle Surg. 2015;54(4):549–53.
16. Torriani M, Kattapuram SV. Technical innovation. Dynamic sonography of the forefoot: the sonographic Mulder sign. AJR Am J Roentgenol. 2003;180(4):1121–3.

17. Klontzas ME, Koltsakis E, Kakkos GA, Karantanas AH. Ultrasound-guided treatment of Morton's neuroma. J Ultrason. 2021;21(85):e134–8.
18. Park YH, Kim TJ, Choi GW, Kim HJ. Prediction of clinical prognosis according to intermetatarsal distance and neuroma size on ultrasonography in Morton neuroma: a prospective observational study. J Ultrasound Med. 2019;38(4):1009–14.
19. Koski JM. Ultrasound detection of plantar bursitis of the forefoot in patients with early rheumatoid arthritis. J Rheumatol. 1998;25(2):229–30.
20. Bowen CJ, Culliford D, Dewbury K, Sampson M, Burridge J, Hooper L, et al. The clinical importance of ultrasound detectable forefoot bursae in rheumatoid arthritis. Rheumatology (Oxford). 2010;49(1):191–2.
21. Hammer HB, Kvien TK, Terslev L. Intermetatarsal bursitis is frequent in patients with established rheumatoid arthritis and is associated with anti-cyclic citrullinated peptide and rheumatoid factor. RMD Open. 2019;5(2):e001076.
22. Pham H, Fessell DP, Femino JE, Sharp S, Jacobson JA, Hayes CW. Sonography and MR imaging of selected benign masses in the ankle and foot. AJR Am J Roentgenol. 2003;180(1):99–107.
23. Wu JS, Hochman MG. Soft-tissue tumors and tumorlike lesions: a systematic imaging approach. Radiology. 2009;253(2):297–316.
24. Bancroft LW, Peterson JJ, Kransdorf MJ. Imaging of soft tissue lesions of the foot and ankle. Radiol Clin N Am. 2008;46(6):1093–103, vii.
25. Neto N, Nunnes P. Spectrum of MRI features of ganglion and synovial cysts. Insights Imaging. 2016;7(2):179–86.
26. Nalbant S, Corominas H, Hsu B, Chen LX, Schumacher HR, Kitumnuaypong T. Ultrasonography for assessment of subcutaneous nodules. J Rheumatol. 2003;30(6):1191–5.
27. Griffith JF, Wong TY, Wong SM, Wong MW, Metreweli C. Sonography of plantar fibromatosis. AJR Am J Roentgenol. 2002;179(5):1167–72.
28. Masih S, Antebi A. Imaging of pigmented villonodular synovitis. Semin Musculoskelet Radiol. 2003;7(3):205–16.
29. Wan JM, Magarelli N, Peh WC, Guglielmi G, Shek TW. Imaging of giant cell tumour of the tendon sheath. Radiol Med. 2010;115(1):141–51.
30. Fornage BD. Glomus tumors in the fingers: diagnosis with US. Radiology. 1988;167(1):183–5.
31. Aytekin K, Esenyel CZ, Coskun ZU. Rare localisation of glomus tumor under greater toenail bed, presenting with pain for a decade-a case report. J Orthop Sports Med. 2019;1:060–5.
32. Drapé JL, Idy-Peretti I, Goettmann S, Wolfram-Gabel R, Dion E, Grossin M, et al. Subungual glomus tumors: evaluation with MR imaging. Radiology. 1995;195(2):507–15.
33. Nwawka OK, Kabutey NK, Locke CM, Castro-Aragon I, Kim D. Ultrasound-guided needle localization to aid foreign body removal in pediatric patients. J Foot Ankle Surg. 2014;53(1):67–70.
34. Horton LK, Jacobson JA, Powell A, Fessell DP, Hayes CW. Sonography and radiography of soft-tissue foreign bodies. AJR Am J Roentgenol. 2001;176(5):1155–9.
35. Jacobson JA, Powell A, Craig JG, Bouffard JA, van Holsbeeck MT. Wooden foreign bodies in soft tissue: detection at US. Radiology. 1998;206(1):45–8.
36. Crankson S, Oratis P, Al Maziad G. Ultrasound in the diagnosis and treatment of wooden foreign bodies in the foot. Ann Saudi Med. 2004;24(6):480–1.
37. Boyse TD, Fessell DP, Jacobson JA, Lin J, van Holsbeeck MT, Hayes CW. US of soft-tissue foreign bodies and associated complications with surgical correlation. Radiographics. 2001;21(5):1251–6.
38. Burge AJ, Gold SL, Potter HG. Imaging of sports-related midfoot and forefoot injuries. Sports Health. 2012;4(6):518–34.
39. Schmidt WA, Schmidt H, Schicke B, Gromnica-Ihle E. Standard reference values for musculoskeletal ultrasonography. Ann Rheum Dis. 2004;63(8):988–94.
40. Szkudlarek M, Narvestad E, Klarlund M, Court-Payen M, Thomsen HS, Østergaard M. Ultrasonography of the metatarsophalangeal joints in rheumatoid arthritis: compari-

son with magnetic resonance imaging, conventional radiography, and clinical examination. Arthritis Rheum. 2004;50(7):2103–12.
41. Klocke R, Glew D, Cox N, Blake DR. Sonographic erosions of the rheumatoid little toe. Ann Rheum Dis. 2001;60(9):896–7.
42. Zayat AS, Ellegaard K, Conaghan PG, Terslev L, Hensor EM, Freeston JE, et al. The specificity of ultrasound-detected bone erosions for rheumatoid arthritis. Ann Rheum Dis. 2015;74(5):897–903.
43. Hassan R, Hussain S, Bacha R, Gillani SA, Malik SS. Reliability of ultrasound for the detection of rheumatoid arthritis. J Med Ultrasound. 2019;27(1):3–12.
44. Patience A, Helliwell PS, Siddle HJ. Focussing on the foot in psoriatic arthritis: pathology and management options. Expert Rev Clin Immunol. 2018;14(1):21–8.
45. López-Reyes A, Hernández-Díaz C, Hofmann F, Pineda C. Gout mimicking psoriatic arthritis flare. J Clin Rheumatol. 2012;18(4):220.
46. Soubrier M, Dubost JJ, Boisgard S, Sauvezie B, Gaillard P, Michel JL, et al. Insufficiency fracture. A survey of 60 cases and review of the literature. Joint Bone Spine. 2003;70(3):209–18.
47. Banal F, Gandjbakhch F, Foltz V, Goldcher A, Etchepare F, Rozenberg S, et al. Sensitivity and specificity of ultrasonography in early diagnosis of metatarsal bone stress fractures: a pilot study of 37 patients. J Rheumatol. 2009;36(8):1715–9.
48. Peris P. Stress fractures. Best Pract Res Clin Rheumatol. 2003;17(6):1043–61.
49. Banal F, Etchepare F, Rouhier B, Rosenberg C, Foltz V, Rozenberg S, et al. Ultrasound ability in early diagnosis of stress fracture of metatarsal bone. Ann Rheum Dis. 2006;65(7):977–8.
50. Drakonaki EE, Garbi A. Metatarsal stress fracture diagnosed with high-resolution sonography. J Ultrasound Med. 2010;29(3):473–6.
51. Bodner G, Stöckl B, Fierlinger A, Schocke M, Bernathova M. Sonographic findings in stress fractures of the lower limb: preliminary findings. Eur Radiol. 2005;15(2):356–9.
52. De Smet AA, Doherty MP, Norris MA, Hollister MC, Smith DL. Are oblique views needed for trauma radiography of the distal extremities? AJR Am J Roentgenol. 1999;172(6):1561–5.
53. Morvan G, Brasseur J, Sans N. Superficial US of superficial bones. J Radiol. 2005;86(12 Pt 2):1892–903.
54. Stone M, Eyler W, Rhodenizer J, van Holsbeeck M. Accuracy of sonography in plantar plate tears in cadavers. J Ultrasound Med. 2017;36(7):1355–61.

Chapter 16
Anterior Knee

Reasons to Do the Study
Evaluation or assessment of:

1. Possible suprapatellar recess knee effusion
2. Quadriceps tendon architecture
3. Patellar tendon architecture
4. Hoffa's fat pad
5. Femoral trochlear cartilage
6. Medial and lateral patellar retinacula
7. Prepatellar bursa, infrapatellar bursae, or pes anserine bursa
8. Inflammatory arthritis
9. Enthesitis
10. Presence of plicae
11. Patella

Questions We Want Answered
1. Is there evidence of a suprapatellar knee effusion?
2. If there are structural changes, does the pathology align with the symptoms?
3. If there is evidence of inflammatory arthritis, are there specific findings that point to the etiology (e.g., double contour sign or tophus)?
4. If there is evidence of bursitis, is the bursitis inflammatory?
5. What is the next step for treatment?
6. Is tendon pain caused by an inflammatory or degenerative process?

The contents do not represent the views of the U.S. Department of Veterans Affairs or the United States Government.

M. H. Greenberg et al., *Manual of Musculoskeletal Ultrasound*,
https://doi.org/10.1007/978-3-031-37416-6_16

Basic Anatomy

Bony Anatomy

The bones of the anterior knee are illustrated in Fig. 16.1.

Soft Tissues

Tendons and Muscles (Fig. 16.2)

Quadriceps Tendons

The *three* layers of the quadriceps tendon, from superficial to deep, are the rectus femoris, the vastus medialis/lateralis, and the vastus intermedius [1]. Look for signs of tendinosis at the insertion of the quadriceps tendon onto the superior patella, but this may be mimicked by insertional anisotropy.

Patellar Tendon

Called by some the patellar ligament, the **patellar tendon** connects the distal anterior patella to a bony prominence on the anterior tibia, the tibial tubercle (or tuberosity).

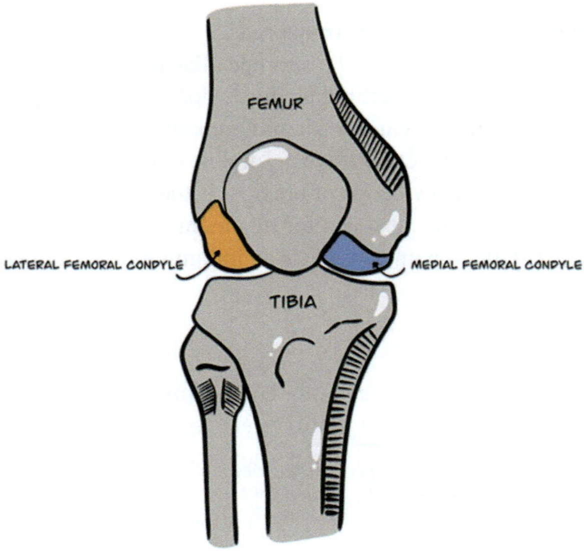

Fig. 16.1 Anterior knee bony anatomy

Fig. 16.2 Simplified anterior knee soft tissue anatomy

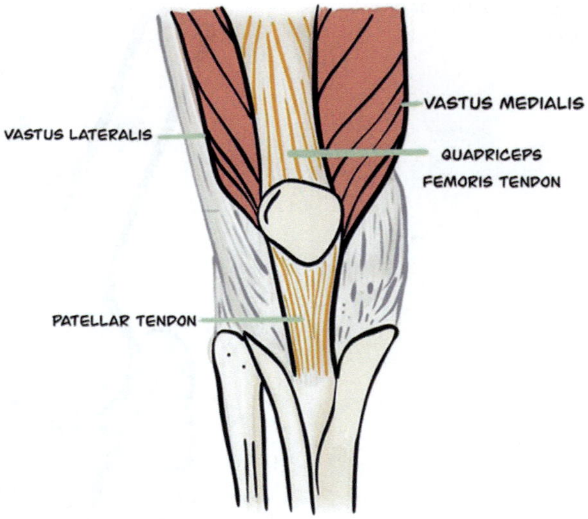

Bursae (Fig. 16.3)

Suprapatellar Recess

The suprapatellar recess (bursa or pouch), or joint recess, is a channel that tracks proximally from beneath the patella. The recess may contain synovial fluid, synovial hypertrophy, synovitis, loose bodies, plicae, or a combination. The recess communicates with and is proximal to the knee joint.

Prepatellar Bursa

This superficial bursa lies anterior to the bony patella and is only evident when pathologically distended.

Infrapatellar Bursae

At the quadriceps tendon insertion, there are two infrapatellar bursae, the **superficial** and the **deep**. A small amount of fluid in the deep infrapatellar bursa is common.

Pes Anserine Bursa

The **pes anserine bursa (PAB)** is located near the attachment of the gracilis, sartorius, and semitendinosus tendons to the anteromedial border of the tibia [2].

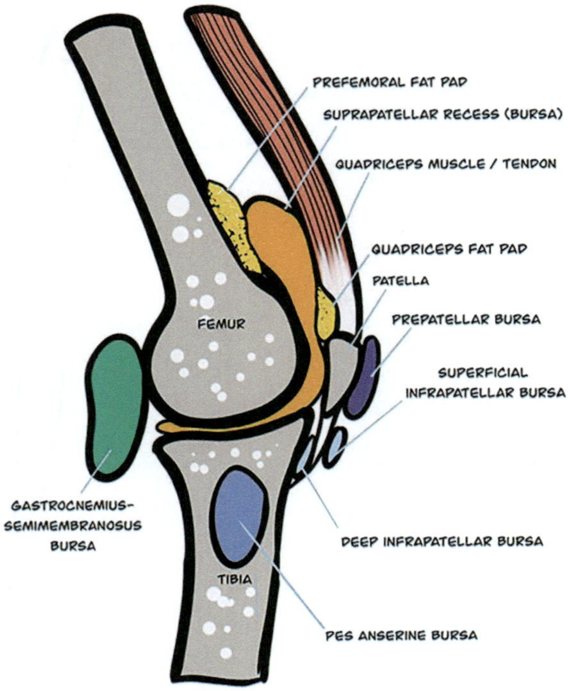

Fig. 16.3 Anterior knee bursae

Fat Pads

Just posterior (deep) to the quadriceps tendon are two hyperechoic structures, the **quadriceps** (or suprapatellar) and **prefemoral fat pads** (Fig. 16.3). The suprapatellar recess courses between the two fat pads; fat pad separation is enhanced with suprapatellar recess distension. The **infrapatellar** (**Hoffa's**) **fat pad** lies deep in the patellar tendon; it serves as a shock absorber for the patellar tendon and helps to nourish and repair contiguous structures [3].

Plica

Occasionally, a suprapatellar synovial plica may divide the recess into two distinct compartments, connecting the two fat pads [4]. Such a plica is typically not pathologic or relevant. However, medication injected into the suprapatellar recess may not reach the knee joint if an obstructing plica is present. Plicae may evade sonographic detection, obscured by the bony patella.

Ligaments and Retinacula (Fig. 16.4)

The **medial and lateral retinacula** are fibrous tissues that tether the patella to underlying soft tissue and bone. The **medial patellofemoral ligament** (MPFL), located within the medial retinaculum, attaches the patella to the femur, thus inhibiting lateral patella displacement [5].

Clinical Comments

Effusion

A suprapatellar recess effusion will be seen just deep to the quadriceps tendon as a fluid-filled distention separating the prefemoral and quadriceps fat pads. The joint effusion may appear anechoic or heterogeneously hypoechoic depending on the nature of the fluid [6]. In addition to the suprapatellar recess, effusions may be identified laterally or medially to the patella in the parapatellar recesses. A simple synovial fluid effusion is displaceable with probe compression but lacks a Doppler signal [7].

Synovitis

In contrast to synovial fluid, synovial hypertrophy may be appreciated as nondisplaceable and poorly compressible [7]. Synovial hypertrophy appears as hyperechoic tissue or has a mixed hyperechoic and hypoechoic signal occasionally with synovial fronds. Synovial hypertrophy may or may not reveal a Doppler signal [7, 8].

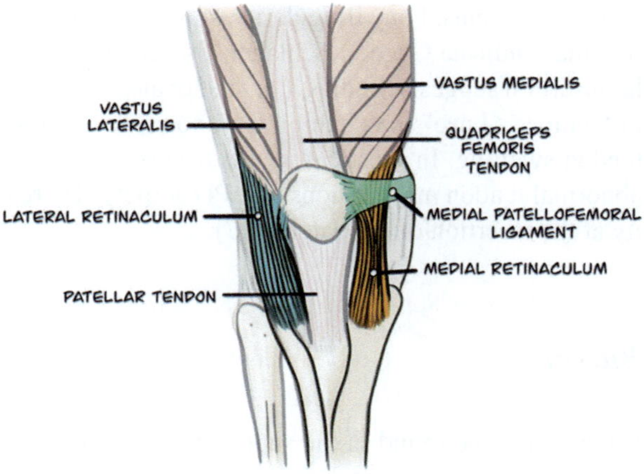

Fig. 16.4 Ligaments and retinacula of the anterior knee

Loose Bodies

Loose bodies may appear as hyperechoic, rounded forms. Additionally, in the same joint or bursa, there may be simultaneous evidence of a mixture of simple effusion, synovial hypertrophy, and synovitis.

Tendon Damage

Quadriceps and Patellar Tendons

The quadriceps and patellar tendons can plainly exhibit tendinosis and partial- or full-thickness tears [6, 9]. Again, the classic sonographic signs of tendinosis are loss of fibrillar pattern, thickening, hypoechoic areas, and small anechoic clefts due to intrasubstance tearing (and perhaps calcium deposits) [10]. Note that power Doppler (PD) activity may be present, but realize that this may be neovascularization as a response to tendon injury rather than a sign of inflammation [11]. Tophaceous gout deposits can also be seen intratendinously in the distal quadriceps tendon, with a predilection for the patellar tendon [12, 13]. These deposits occasionally exhibit a hypoechoic rim.

Enthesopathy

The enthesis, or insertion of the tendon or ligament into the bone, can be a site of inflammatory processes known as enthesitis [14]. This is often due to spondyloarthritis, such as psoriatic arthritis or ankylosing spondylitis. Enthesopathy, a more general term to encompass all pathological abnormalities of an enthesis, can be seen due to overuse, microtrauma, advanced age, and even the metabolic syndrome [15, 16]. Findings suggesting enthesopathy include neovascularization, bursitis, cortical abnormalities, enthesophytes, and calcium deposition.

In adolescents, bony irregularity of the patellar tendon distally at the tibial insertion may indicate Osgood-Schlatter disease, whereas bony irregularity at the patellar insertion suggests Sinding-Larsen-Johansson syndrome [17]. The patellar tendon in "jumper's knee" may exhibit tendinosis, severe loss of fibrillar echotexture, and tendon swelling. In most cases, this occurs primarily just distal to the patella. The abnormal tendon may demonstrate PD activity. There may also be a bony irregularity at the insertion of the patella [17].

Bursitis

Bursitis may be found in the suprapatellar, prepatellar, infrapatellar, or pes anserine bursae [6]. For bursal evaluation, including the suprapatellar recess, a light touch of the probe avoids artificial compression. Bursitis may also affect the superficial or deep infrapatellar bursae. However, a small amount of fluid in the deep infrapatellar bursa is often normal. Similar to the prepatellar bursa, the PAB is only detected on ultrasound with bursal distention, typically caused by inflammation.

(Please see Chap. 19 for more information.)

Cartilage Damage

The femoral trochlear cartilage can be assessed for thinning, as seen in osteoarthritis [18]. Loss of thickness and irregularity of the superficial chondral surface indicate osteoarthritis [19]. Hyperechoic enhancement at the superficial margin of the hyaline cartilage can be seen with monosodium urate deposition in gout (a double contour sign). In contrast, hyperechoic enhancement, or spots within the hyaline cartilage, can be seen with calcium pyrophosphate dihydrate deposition [20]. However, patients with advanced osteoarthritis may have difficulty maximally flexing their knees to obtain this view.

Ligament Damage

The MPFL prevents lateral patellar displacement and may be damaged in acute lateral patella dislocation. Sonographic evaluation of MPFL tears after acute lateral patellar dislocation showed diagnostic accuracy comparable to that of magnetic resonance imaging [21].

Fat Pad Damage

In response to repeated excessive stress, the infrapatellar (Hoffa's) fat pad may become inflamed and eventually fibrotic. Ultrasound may reveal hypoechoic thickening, fatty architecture changes, and ultimately hyperechoic fibrous tissue. There may be PD activity as well. Hoffa's fat pad may also become impinged between the patellar tendon and the patella, termed infrapatellar fat pad impingement or Hoffa's syndrome [3].

Pitfalls

1. Use a rolled towel under the knee, since knee flexion at 30° is the most sensitive position to detect effusion [22].
2. Be mindful of anisotropy mimicking tendon pathology.

3. Maintain light probe pressure or float the transducer to avoid inadvertent compression of effusions or bursae.
4. Don't mistake a physiologic amount of fluid in the suprapatellar recess or the deep infrapatellar bursa for a pathologic effusion.
5. Positive PD activity does not equate to inflammation. It means neovascularization, which may or may not be due to inflammation. Neovascularization may also occur with a chronic injury, such as tendinosis.
6. Do not confuse a normal variant, a bipartite patella, with a patellar fracture. Obtain a radiograph to confirm this.
7. Complete tears of the quadriceps or patellar tendon are apparent clinically. However, always do dynamic exams to expose partial tears if suspected.

Method

The patient is positioned supine with the leg resting comfortably at approximately 30° of flexion. A small pillow or rolled towel beneath the knee is recommended for comfort. A knee or medium-depth preset is suggested to visualize the femur as the bottommost structure.

Protocol Image 1: Quadriceps Tendon, Longitudinal (Fig. 16.5)

Place the probe on the long axis (LAX) to the midline of the knee with the probe marker, or fin, in a proximal position. The distal end of the probe should touch the proximal patella. The quadriceps tendon is seen in LAX, with the tendon fibrils

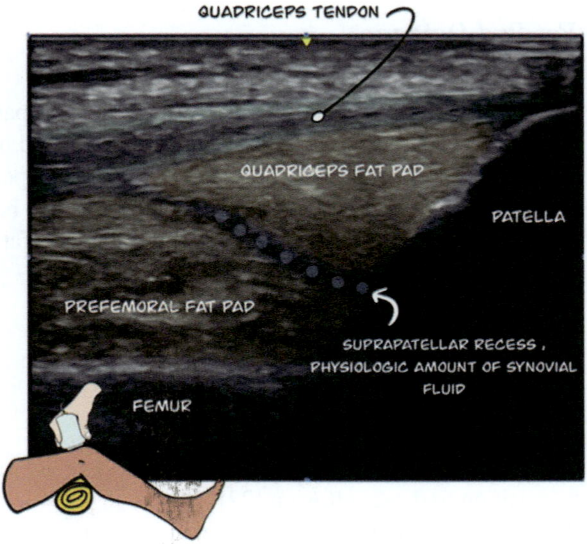

Fig. 16.5 Protocol Image 1: Quadriceps tendon, longitudinal

creating a paintbrush-like pattern. Move the transducer medially and laterally to evaluate the entire quadriceps tendon. If an anechoic cleft in the quadriceps tendon is detected, add a dynamic exam with knee flexion and extension to see if the fissure enlarges upon stretching.

Protocol Image 2: Quadriceps Tendon, Synovial Fluid Expression, Longitudinal ± Power Doppler (Fig. 16.6)

While keeping the suprapatellar recess in sonographic view, milk the joint upward by taking a free hand and placing the distal patella tip between the webbing of your first two fingers. Exert firm pressure superiorly and downward to push synovial fluid from the parapatellar recesses into the suprapatellar recess. If an obstructing plica is present, the proximal displacement of fluid can cause bulging of the plica, known as a "sail sign" [23]. Be sure to evaluate the quadriceps insertion into the patella and the suprapatellar recess with PD.

Protocol Image 3: Medial and Lateral Parapatellar Recesses, Longitudinal + Dynamic (Fig. 16.7)

Straighten the knee to enhance fluid in the parapatellar recesses. Move the probe medially and laterally and look at the recesses with and without manual squeezing of the infrapatellar area.

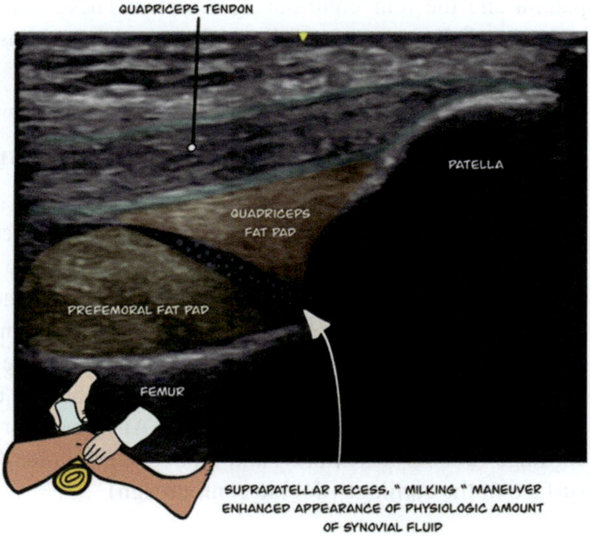

Fig. 16.6 Protocol Image 2: Quadriceps tendon, synovial fluid expression, longitudinal ± power Doppler

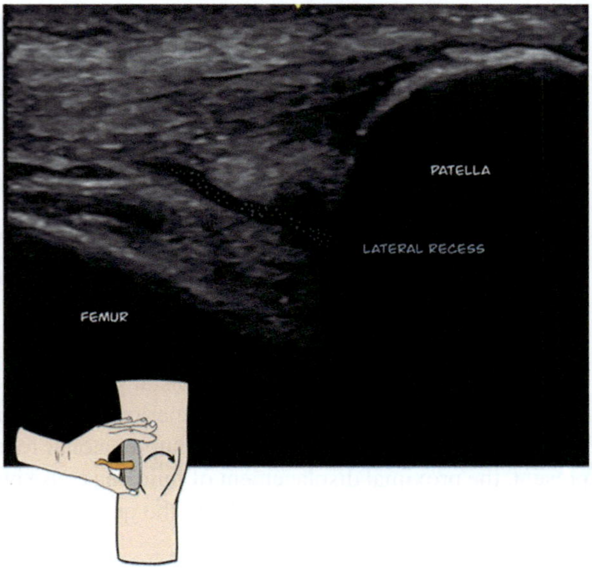

Fig. 16.7 Protocol Image 3: Medial and lateral parapatellar recesses, longitudinal + dynamic

Protocol Image 4: Quadriceps Tendon, Transverse ± Power Doppler (Fig. 16.8)

Return to the initial position in **Protocol Image 1**. Identify the quadriceps tendon in LAX and rotate the probe 90° to see the oval-shaped quadriceps tendon in the transverse axis (TAX). Tilting the probe to create anisotropy can help with tendon identification. Perform distal and proximal TAX slides to examine the insertion on the patella and the four contributing muscles. These muscles are the rectus femoris, vastus medialis, vastus lateralis, and vastus intermedius.

Protocol Image 5: Femoral Cartilage, Transverse (Fig. 16.9)

Have the patient bend the knee to at least 100° and place the probe perpendicular to the examination table, just proximal to the tip of the patella. Move the probe in medial and lateral directions to see the hyperechoic bone of the distal anterior femur covered with anechoic or hypoechoic cartilage. Remember, you are *not* looking at a joint but at cartilage residing within the joint. Observe cartilage thickness and any irregularities at the superficial surface. In the absence of pathological changes, the cartilage is thicker in the central portion. Also, look for calcium deposits within the cartilage (chondrocalcinosis) and monosodium urate deposition on the superficial surface of the cartilage (double contour sign).

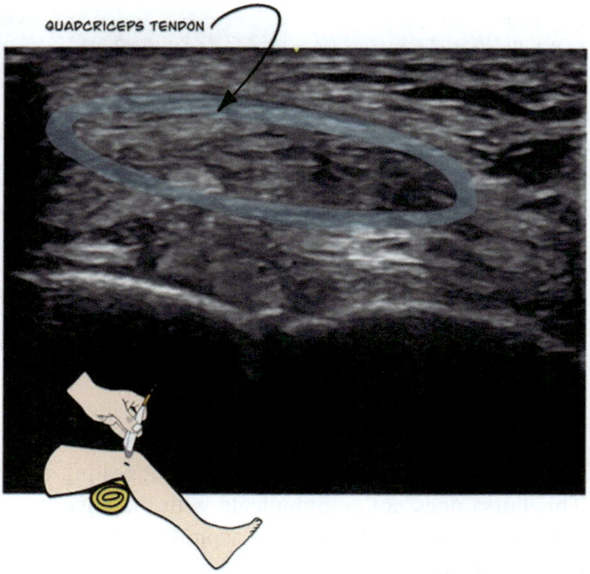

Fig. 16.8 Protocol Image 4: Quadriceps tendon, transverse ± power Doppler

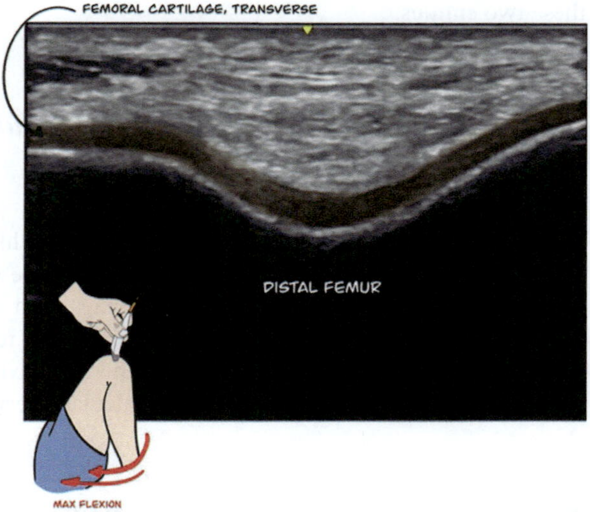

Fig. 16.9 Protocol Image 5: Femoral cartilage, transverse

Protocol Image 6: Patella, Transverse (Fig. 16.10)

Now, extend the knee and apply copious transmission gel to the patella, sufficient to float the transducer. You should see an utterly anechoic region above the skin to indicate that no probe pressure disrupts the view. Gently perform TAX slides over the entire patella to examine the bony contour and look for the presence of a prepatellar bursa. This bursa becomes visible only when pathologically distended.

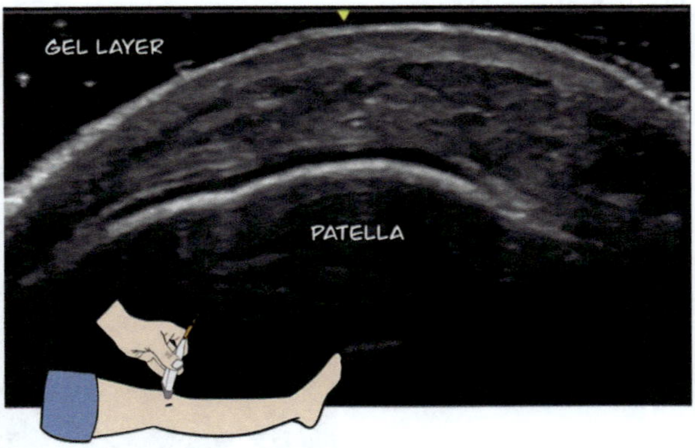

Fig. 16.10 Protocol Image 6: Patella, transverse

Evaluate for distension of fluid on probe compression (simple effusion), PD activity, and hyperechoic areas, the latter suggesting a hematoma with clotted blood. This bursa does not communicate with the knee joint; however, if it is distended, consider infection and perform aspiration. A complete superior to inferior scan of the patella may reveal bony "valleys," which could be from a fracture but more commonly indicate a normal variant, a bipartite patella. A radiograph will differentiate these two entities.

Protocol Image 7: Lateral Patellar Retinaculum, Transverse + Dynamic ± Power Doppler (Fig. 16.11)

Now move the probe laterally off of the mid-patella while maintaining TAX orientation. The lateral patellar retinaculum (LPR) should be visible as a horizontal, elongated, hyperechoic, fibrillar structure arising from the patella. Perform small proximal and distal TAX slides to fully examine the retinaculum. If a retinaculum tear is suspected, evaluate dynamically with knee flexion and extension. PD imaging in this region may reveal the lateral genicular artery.

Protocol Image 8: Medial Patellar Retinaculum, Medial Patellofemoral Ligament, Transverse + Dynamic ± Power Doppler (Fig. 16.12)

Now move the probe in a medial LAX slide and center it on the medial patellar retinaculum (MPR), a structure with a similar echotexture and horizontal orientation as the LPR. The probe should bridge the medial patella and the distal femur. From the lateral aspect of the patella, use two fingers of your free hand to push the lateral edge of the patella medially and posteriorly. This dynamic maneuver tilts the medial

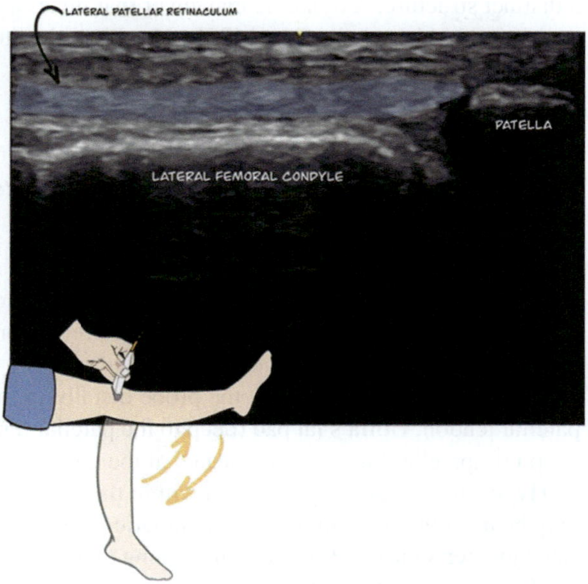

Fig. 16.11 Protocol Image 7: Lateral patellar retinaculum, transverse + dynamic ± power Doppler

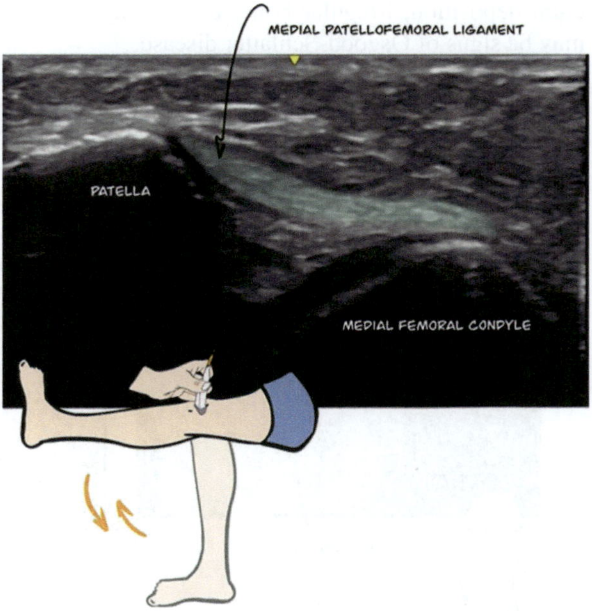

Fig. 16.12 Protocol Image 8: Medial patellar retinaculum, medial patellofemoral ligament, transverse + dynamic ± power Doppler

patella anteriorly to expose more of the cartilage on the underside of the patella and may better demonstrate the MPFL. The MPFL originates at the adductor tubercle of the femur. It is a superficial thickening of the MPR and may or may not be seen as a distinct structure. A dynamic maneuver that flexes and extends the knee assists in evaluating hidden tears of the MPFL. Use PD to locate the medial genicular artery as well.

Protocol Image 9: Patellar Tendon, Longitudinal ± Power Doppler (Fig. 16.13)

Next, bend the knee to 90°, place the probe just distal to the patella, and rotate it 90° to LAX to look at the patellar tendon. The proximal portion of the transducer should rest on the distal end of the patella. Continually sweep the probe (TAX slides) laterally and medially as you move the probe distally to thoroughly examine the distal patellar tendon, Hoffa's fat pad (deep to the patellar tendon), and the superficial and deep infrapatellar bursae, both located at the distal patellar tendon.

Hypoechoic areas may be noted within the tendon, indicating tendinosis, which may be asymptomatic in athletes. Dramatic proximal patellar tendinosis may indicate "jumper's knee." However, this finding may suggest Sinding-Larsen-Johansson syndrome or pediatric patellar tendinosis in a child or adolescent. At the more distal attachment of the patella tendon to the tibial tubercle, evidence of tendinosis, calcium deposition, irregular bone, PD activity, and an enlarged infrapatellar bursa may be signs of Osgood-Schlatter disease.

This tendon's length often exceeds the size of the transducer, necessitating an extended field of view or simply capturing images of the proximal and distal insertions. Straighten the knee and use PD to examine the proximal and distal attachments. If neovascularization is present, determine if it is within the tendon (tendinosis) or at the tendon insertion (enthesitis), as seen in inflammatory arthritis.

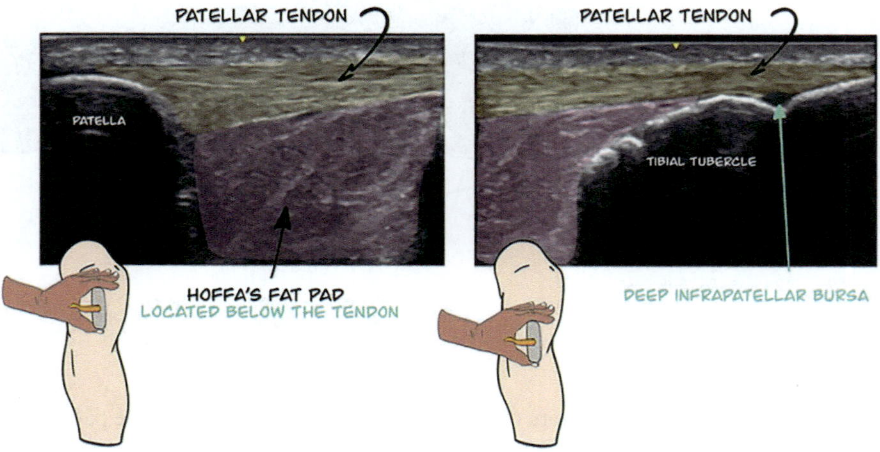

Fig. 16.13 Protocol Image 9: Patellar tendon, longitudinal ± power Doppler

Protocol Image 10: Patellar Tendon, Transverse (Fig. 16.14)

Next, rotate the probe 90° to examine the patellar tendon in transverse view and perform proximal and distal TAX slides. The tendon should be thin and oval-shaped.

Complete Anterior Knee Ultrasonic Examination Checklist

□ Protocol Image 1: Quadriceps tendon, longitudinal
□ Protocol Image 2: Quadriceps tendon, synovial fluid expression, longitudinal ± power Doppler
□ Protocol Image 3: Medial and lateral suprapatellar recesses, longitudinal + dynamic
□ Protocol Image 4: Quadriceps tendon, transverse ± power Doppler
□ Protocol Image 5: Femoral cartilage, transverse
□ Protocol Image 6: Patella, transverse
□ Protocol Image 7: Lateral patellar retinaculum, transverse + dynamic ± power Doppler
□ Protocol Image 8: Medial patellar retinaculum, medial patellofemoral ligament, transverse + dynamic ± power Doppler
□ Protocol Image 9: Patellar tendon, longitudinal ± power Doppler
□ Protocol Image 10: Patellar tendon, transverse.

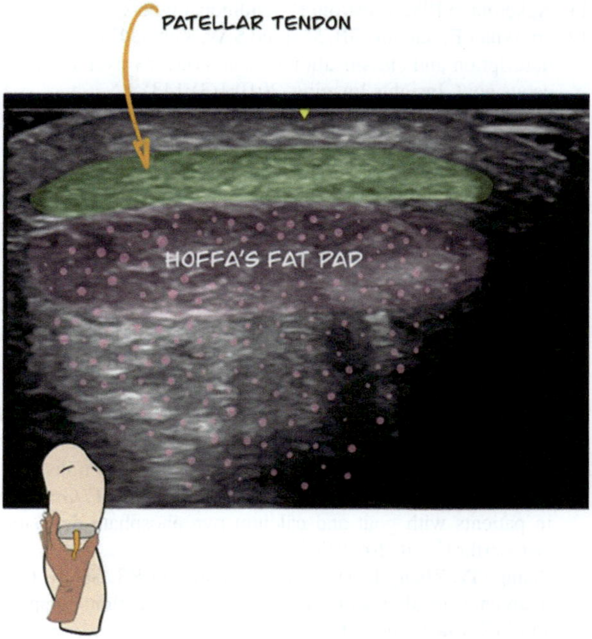

Fig. 16.14 Protocol Image 10: Patellar tendon, transverse

References

1. De Maeseneer M, Marcelis S, Boulet C, Kichouh M, Shahabpour M, de Mey J, et al. Ultrasound of the knee with emphasis on the detailed anatomy of anterior, medial, and lateral structures. Skelet Radiol. 2014;43(8):1025–39.
2. Toktas H, Dundar U, Adar S, Solak O, Ulasli AM. Ultrasonographic assessment of pes anserinus tendon and pes anserinus tendinitis bursitis syndrome in patients with knee osteoarthritis. Mod Rheumatol. 2015;25(1):128–33.
3. Lapègue F, Sans N, Brun C, Bakouche S, Brucher N, Cambon Z, et al. Imaging of traumatic injury and impingement of anterior knee fat. Diagn Interv Imaging. 2016;97(7–8):789–807.
4. Liu YW, Skalski MR, Patel DB, White EA, Tomasian A, Matcuk GR Jr. The anterior knee: normal variants, common pathologies, and diagnostic pitfalls on MRI. Skelet Radiol. 2018;47(8):1069–86.
5. Starok M, Lenchik L, Trudell D, Resnick D. Normal patellar retinaculum: MR and sonographic imaging with cadaveric correlation. AJR Am J Roentgenol. 1997;168(6):1493–9.
6. Alves TI, Girish G, Kalume Brigido M, Jacobson JA. US of the knee: scanning techniques, pitfalls, and pathologic conditions. Radiographics. 2016;36(6):1759–75.
7. Wakefield RJ, Balint PV, Szkudlarek M, Filippucci E, Backhaus M, D'Agostino MA, et al. Musculoskeletal ultrasound including definitions for ultrasonographic pathology. J Rheumatol. 2005;32(12):2485–7.
8. Bruyn GA, Iagnocco A, Naredo E, Balint PV, Gutierrez M, Hammer HB, et al. OMERACT definitions for ultrasonographic pathologies and elementary lesions of rheumatic disorders 15 years on. J Rheumatol. 2019;46(10):1388.
9. La S, Fessell DP, Femino JE, Jacobson JA, Jamadar D, Hayes C. Sonography of partial-thickness quadriceps tendon tears with surgical correlation. J Ultrasound Med. 2003;22(12):1323–9. quiz 30-1
10. Rasmussen OS. Sonography of tendons. Scand J Med Sci Sports. 2000;10(6):360–4.
11. Ackermann PW, Renström P. Tendinopathy in sport. Sports Health. 2012;4(3):193–201.
12. de Ávila FE, Sandim GB, Mitraud SAV, Kubota ES, Ferrari AJL, Fernandes ARC. Sonographic description and classification of tendinous involvement in relation to tophi in chronic tophaceous gout. Insights Imaging. 2010;1(3):143–8.
13. Girish G, Glazebrook KN, Jacobson JA. Advanced imaging in gout. Am J Roentgenol. 2013;201(3):515–25.
14. D'Agostino MA, Terslev L. Imaging evaluation of the entheses: ultrasonography, MRI, and scoring of evaluation. Rheum Dis Clin. 2016;42(4):679–93.
15. Kaeley GS, Eder L, Aydin SZ, Gutierrez M, Bakewell C. Enthesitis: a hallmark of psoriatic arthritis. Semin Arthritis Rheum. 2018;48(1):35–43.
16. Abate M, Di Carlo L, Salini V, Schiavone C. Metabolic syndrome associated to non-inflammatory Achilles enthesopathy. Clin Rheumatol. 2014;33(10):1517–22.
17. Carr JC, Hanly S, Griffin J, Gibney R. Sonography of the patellar tendon and adjacent structures in pediatric and adult patients. AJR Am J Roentgenol. 2001;176(6):1535–9.
18. Naredo E, Acebes C, Moller I, Canillas F, de Agustin JJ, de Miguel E, et al. Ultrasound validity in the measurement of knee cartilage thickness. Ann Rheum Dis. 2009;68(8):1322–7.
19. Minagawa H, Wong KH. Musculoskeletal ultrasound: echo anatomy & scan technique: Amazon Digital Services. Tokyo, Japan, Ohmsha Publishing; 2017.
20. Filippucci E, Gutierrez M, Georgescu D, Salaffi F, Grassi W. Hyaline cartilage involvement in patients with gout and calcium pyrophosphate deposition disease. An ultrasound study. Osteoarthr Cartil. 2009;17(2):178–81.
21. Zhang GY, Zheng L, Ding HY, Li EM, Sun BS, Shi H. Evaluation of medial patellofemoral ligament tears after acute lateral patellar dislocation: comparison of high-frequency ultrasound and MR. Eur Radiol. 2015;25(1):274–81.

22. Mandl P, Brossard M, Aegerter P, Backhaus M, Bruyn GA, Chary-Valckenaere I, et al. Ultrasound evaluation of fluid in knee recesses at varying degrees of flexion. Arthritis Care Res (Hoboken). 2012;64(5):773–9.
23. Bianchi S, Martinoli C. Ultrasound of the musculoskeletal system. New York, NY: Springer, Berlin, Heidelberg; 2007.

Chapter 17
Posterior Knee

Reasons to Do the Study
Evaluation or assessment of:

1. Posterior knee pain
2. Baker's or popliteal cyst (PC)
3. Popliteal vasculature
4. Tibial, sciatic, and common peroneal nerves
5. Posterior cruciate ligament

Questions We Want Answered
1. Is there evidence of a PC?
   (a) Does the PC communicate with the knee joint?
   (b) Is the PC intact or ruptured?
   (c) Is the PC complex or simple?
   (d) Is synovitis or synovial hypertrophy present?
2. Are the tendons or other structures normal in appearance?
   (a) If abnormal, does the pathology align with the symptoms?
3. Does the popliteal vasculature appear normal?
   (a) Is a popliteal aneurysm or pseudoaneurysm present?
   (b) Is a popliteal deep venous thrombosis (DVT) present?
4. Is there evidence of injury, entrapment, or an intraneural ganglion involving the nerves?
5. What is the cause of the posterior knee pain?

The contents do not represent the views of the U.S. Department of Veterans Affairs or the United States Government.

M. H. Greenberg et al., *Manual of Musculoskeletal Ultrasound*,
https://doi.org/10.1007/978-3-031-37416-6_17

Necessary Basic Anatomy: Posterior Knee

See Fig. 17.1 for the location of the bones, posterior distal femoral cartilage, ligaments, and menisci. Some people have a calcified accessory sesamoid bone, the fabella. The soft tissues of the posterior knee (Fig. 17.2) comprise medial, central, and lateral zones. From medial to lateral, the structures of interest include the **sar**torius, **grac**ilis, semi**t**endinosus (ST), semi**m**embranosus (SM), gastrocnemius-semimembranosus **bu**rsa (GSB), **m**edial head gastrocnemius (MHG), tibial **ne**rve (TN), popliteal **ve**in (PV), and **a**rtery (PA), and **la**teral head of the gastrocnemius (LHG). The mnemonic "**S**ergeant **Grac**e **St**opped **Sm**iling, **But** **M**ajor **Ne**lson, **Ve**ry **A**mused, **La**ughed" may help to remember the order of these structures.

Muscles and Tendons

The **pes anserine (or anserinus) tendons** are the sartorius, gracilis, and ST, memorialized by "Say Grace before Tea." Before these tendons insert into the anteromedial knee, they pass through the posteromedial corner of the knee. At this level, the sartorius is still a muscle rather than a tendon. Likewise, the bulk of the MHG is muscular, with only the medial tip being tendinous. The ST is superficial to, and almost appears to be a roof or tent over, the GSB. A mnemonic is that the **ST** is a *tent* over the bursa. The **lateral and medial gastrocnemius muscles** ascend to insert upon the lateral femoral condyle (LFC) and medial femoral condyle (MFC), respectively.

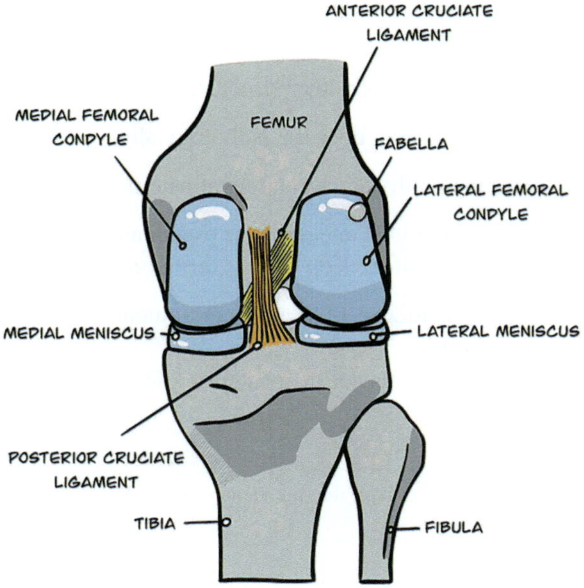

Fig. 17.1 Bones and ligaments of the posterior knee simplified

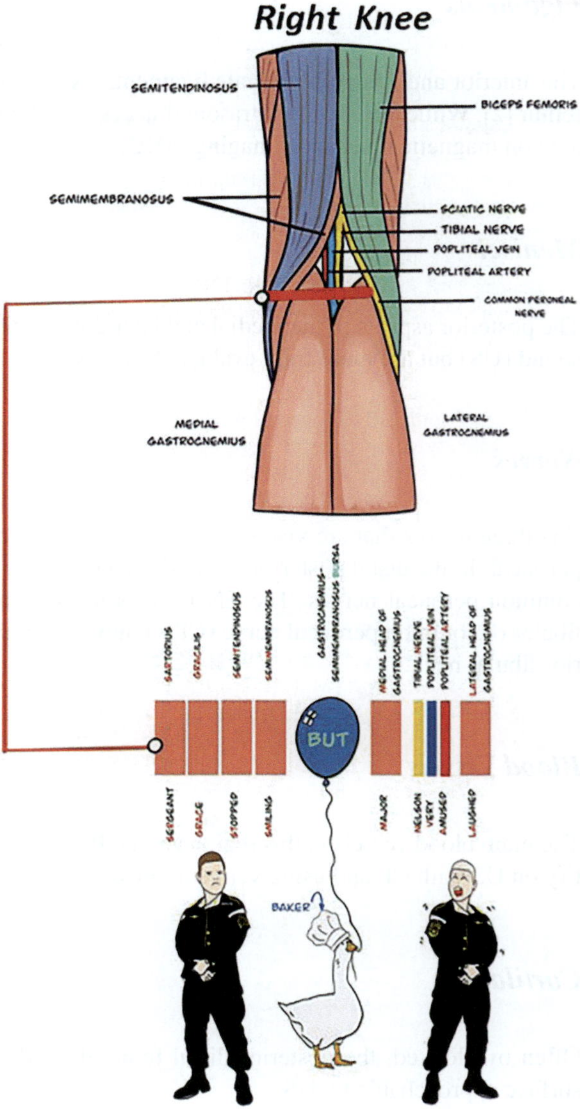

Fig. 17.2 Soft tissues of the posterior knee simplified

Bursae

The **GSB,** located between the tendons of the SM and MHG, communicates with the healthy native knee joint in 40–54% of people [1]. The GSB is named for its location between the MHG and the SM tendon. The **pes anserine bursa** lies deep to the pes anserine tendons at the anteromedial proximal tibia. It is only visible when distended.

Ligaments

The anterior and posterior cruciate ligaments (ACL and PCL) attach the tibia to the femur [2]. While technically ultrasound-accessible, these ligaments are better evaluated on magnetic resonance imaging (MRI).

Menisci

The posterior aspects of the medial and lateral menisci may also be visible on ultrasound (US) but are more fully evaluated by MRI.

Nerves

The three nerves that are visualized posteriorly are the tibial, sciatic, and common peroneal. In the distal posterior thigh, the sciatic nerve bifurcates into the tibial and common peroneal nerves. The TN follows the PA and PV posteriorly, while the fibular or common peroneal nerve (CPN) moves laterally to wrap around the posterior fibular neck.

Blood Vessels

The main blood vessels in this region are the PA and the PV, which are easy to identify on US with CF and using vein compression maneuvers.

Cartilage

Often overlooked, the posterior distal femoral condyles have a sizeable cartilage surface approachable by US.

Clinical Comments

Bursitis

Popliteal Cyst

The GSB may enlarge into a PC, a common pathologic entity first described in 1877 by W. Morant Baker [3]. Most PCs are secondary cysts due to the production of excess synovial fluid in the knee in response to underlying intraarticular pathologies such as meniscal tears, articular cartilage lesions, or inflammatory or noninflammatory joint diseases [4]. Synovial fluid may flow to the cyst via a communicating channel [1, 5].

Popliteal cysts often present as asymptomatic swelling in the popliteal fossa with or without posterior knee pain; however, continued enlargement of the PC may result in rupture [6]. The clinical signs of a ruptured cyst may mimic deep-vein thrombosis or cellulitis [7]. A PC should be viewed in two planes to determine if a rupture has occurred [8]. The PC may or may not communicate with the knee joint. When the bursa enlarges to become a PC, US can determine if communication with the joint is present since a connection to the joint may be closed in extension but is open with knee flexion [9]. First, look for a connecting stalk, and then passively flex and extend the knee to see if the fluid-filled channel size increases in flexion, thereby indicating a conduit between the GSB and the knee joint.

Ultrasound-guided aspiration and steroid injection may be considered if the PC is thought to be a pain generator, but the cyst may return unless the principal cause is addressed. The PC, being a synovial-lined structure, can also become inflamed and distended due to an underlying inflammatory arthropathy such as rheumatoid arthritis (RA) [6]. This is defined as primary GSB enlargement and may be seen in the setting of a noncommunicating GSB. A PC may also arise after total knee arthroplasty and may be primary or secondary, the latter with a communicating channel to the prosthesis.

Other cyst-like masses that may be confused with a PC are PA aneurysms or pseudoaneurysms. Popliteal artery aneurysms are uncommon but account for up to 70% of peripheral artery aneurysms [10]. These often occur in older men with significant cardiovascular disease [11]. There is some controversy regarding the size of a popliteal artery aneurysm that requires intervention, but a diameter of 2 cm is considered a possible threshold [12]. Unlike a true aneurysm, where the weakened vessel wall dilates abnormally, a pseudoaneurysm is caused by damage to the vessel wall, producing a hematoma surrounding the damaged vessel [13]. Swirling flow can be appreciated on the color Doppler as a "yin-yang" configuration [14]. Remember to turn on the color Doppler when evaluating any structure that looks like a PC.

Pes Anserine Bursitis

(See Chap. 19 for more information.)

Ligament Damage

The ACL and PCL can be visualized on US, albeit incompletely. MRI remains the gold standard for assessing these structures [8]. A uniformly hypoechoic PCL is considered normal, whereas a torn PCL shows heterogeneous hypoechogenicity and is often thicker than normal [15]. Avulsion fractures at the tibial insertion may also be visible [16].

Nerve Damage

Entrapment of the CPN due to compression or stretching against the fibular neck represents the most common cause of lower extremity neuropathy [17]. Triggers include repetitive trauma (runners), habitual leg crossing, proximal fibula fracture, and rapid weight loss [18]. After knee replacement and perhaps associated with prosthetic overhang, the CPN may enlarge due to repetitive knee bending, causing compression or trauma. This CPN dysfunction may cause localized knee pain with or without radiating paresthesia [19]. The diagnosis is made by measuring the cross-sectional area (CSA) of the involved CPN and comparing it with the contralateral CPN. The diagnosis can be verified with electrophysiologic studies. Using a CPN cutoff of a CSA of 10.9 mm^2, US diagnostic sensitivity was 90%, with a specificity of 69% for CPN neuropathy at the fibular head (FH) [20].

Intraneural ganglia can also cause lower extremity neuropathy [21]. In patients with knee osteoarthritis, synovial fluid may track down from the tibiofibular joint along the epineural sheath of a branch of the CPN, potentially compressing the CPN. This can result in knee pain, paresthesia, and even foot drop [18]. This fluid collection may also track proximally to the sciatic nerve and down the TN. This is an unusual ganglion since it involves a nerve sheath, whereas most other ganglia originate from joints and tendon sheaths.

Another potential source of compression of the CPN at the posterior knee is the peroneus longus muscle. After winding around the posterolateral aspect of the fibular neck, the CPN dives between the peroneus longus muscle and the FH, described as the peroneal or fibular tunnel. CPN compression may occur here [22].

Blood Vessel Pathology

Although not a pulsatile mass, DVT of the PV may also be appreciated [8]. A visible thrombus (an echogenic mass within the lumen of the vein) or lack of compressibility is considered diagnostic of a DVT [23]. However, it is essential to note that a complete diagnostic point-of-care DVT examination requires a comprehensive evaluation starting at the common femoral vein [24]. A PC can also be mimicked by a soft tissue injury [9]. However, in contrast to a PC, soft tissue swelling does not originate between the MHG and SM tendons or connect to the joint cavity. Again, a popliteal artery aneurysm and a pseudoaneurysm may each mimic a PC. See Popliteal Cyst, above.

Fabella Pathology

Another cause of posterolateral knee pain is the *fabella*. The fabella, a small "bean-like" structure in the proximal LHG tendon, is a sesamoid bone that articulates with the LFC cartilage [25]. The fabella may normally be absent or present and may or may not be calcified. It is calcified in 10–30% of the general population [26]. If present but noncalcified, it may mimic a tear of the LHG. If present and calcified, it is a hyperechoic bony semicircle and may mimic a bony avulsion. Since the anterior face of the fabella communicates with the cartilage, it may develop osteoarthritis [9]. The fabella may also cause mechanical compression neuropathy of the CPN [25].

Cartilage Damage

Ultrasound enables visualization of the posterior femoral cartilage. This may demonstrate signs of degeneration (loss of cartilage surface sharpness or cartilage thinning), chondrocalcinosis, or the double contour sign of uric acid deposition [27].

Meniscal Damage

Ultrasound enables inspection of the posterolateral and posteromedial menisci. For details on meniscal pathology, please see the Lateral and Medial Knee chapters.

Pitfalls

1. Don't attempt an examination with an externally rotated leg. Instead, have the patient lie prone.
2. Anisotropy of the semimembranosus or MHG tendons may mimic a PC.
3. A noncalcified fabella may be hypoechoic, mimicking an LHG muscle tear on US.
4. Always attempt to trace the origin of a PC to the GSB and perhaps to the knee joint itself. A dynamic exam with knee flexion and extension may help determine if the GSB truly communicates with the joint.
5. Determine if a PC is primary or secondary, since this will influence treatment.
6. If you see a possible PC, turn on the color flow Doppler since a popliteal aneurysm or pseudoaneurysm may mimic a PC.
7. The PCL and the menisci are best evaluated with MRI rather than US.
8. Evaluate the PV for a thrombus in the popliteal fossa, but a dedicated venous duplex study best evaluates venous thrombosis.
9. The insertion of the distal semimembranosus tendon on the tibia mimics a parameniscal cyst of the medial meniscus (MM) due to anisotropy.
10. A hypoechoic cleft in the lateral meniscus (LM) may be due to the popliteus tendon sheath mimicking a meniscal tear.

Method

Begin with the patient in the prone position. The FH may be marked for reference. Use mid-frequency settings.

Protocol Image 1: Medial Femoral Condyle, Transverse (Fig. 17.3)

Place the probe in a medial position on the posterior knee in the transverse axis (TAX) (near the knee crease) and visualize the hyperechoic distal MFC. Directly superficial to the MFC is the hyperechoic SM tendon. Just lateral to the SM tendon is the hypoechoic MHG muscle and its hyperechoic tendon. Next, move the probe to center the image on the triangular-shaped MHG muscle and tendon. The tendon is the medial and superior apex of the muscle and may be hypo- or hyperechoic. Again, medial to the MHG tendon is the SM tendon.

Between the two tendons is the GSB, where the neck of a PC occurs; the bursa hugs the MHG tendon. Normally, this bursa contains minimal fluid and is challenging to detect if not distended. Knee flexion opens the channel to the joint cavity and may help demonstrate the stalk. A trick to locating the GSB is to perform a dynamic longitudinal axis (LAX) slide to see the crossing of the MHG muscle and tendon with the SM tendon. Recall that the MHG tendon inserts on the femur and the SM tendon inserts on the posterior tibia.

If a PC is identified, separate images (labeled 1a, 1b, etc.) should be obtained in multiple planes to look for rupture and whether or not the PC tracks to the posterior knee joint. If present, the connection to the joint closes in extension and opens with

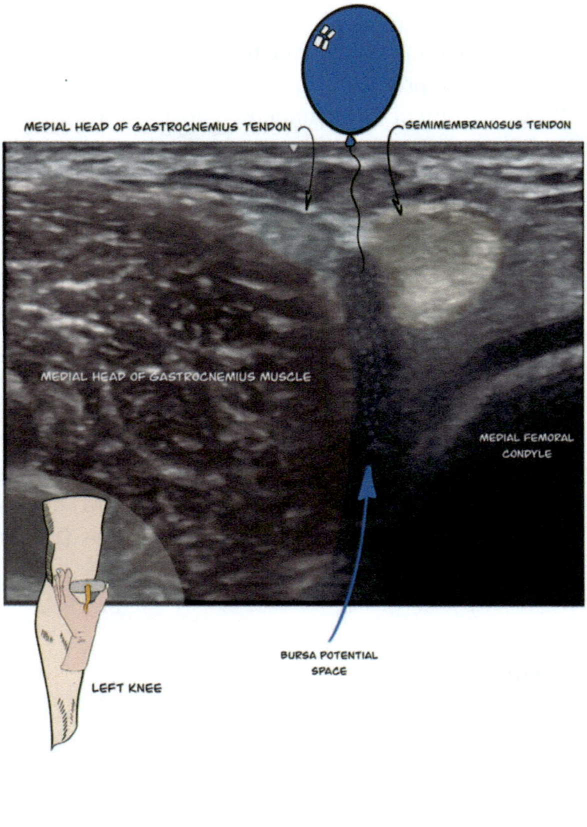

Fig. 17.3 Protocol Image 1: Medial femoral condyle, transverse

knee flexion [9]. Passively flex and extend the knee to see if the PC *channel* to the posterior knee joint is more visible in flexion, thus establishing a conduit to the knee joint. Observe the presence of synovial fluid, debris, synovial septations, synovitis, and PD signals within the PC. Measure the cyst's size in two dimensions. A ruptured PC may be painful, and hypoechoic fluid will move into the surrounding soft tissue.

Protocol Image 2: Medial Femoral Condylar Cartilage, Transverse (Fig. 17.4)

Increase the depth if needed to visualize the posterior one-third of the articular cartilage of the MFC in TAX. Look for cortical defects. With the probe in TAX, have the patient very slowly flex the knee to 100° and scan the cartilage superiorly and inferiorly to look for evidence of cartilage degeneration or crystal disease.

Protocol Image 3: Medial Femoral Condylar Cartilage, Longitudinal (Fig. 17.5)

Extend the knee and rotate the probe 90° into LAX to view the articular cartilage. Do a TAX slide from medial to lateral to fully evaluate the posteromedial articular cartilage.

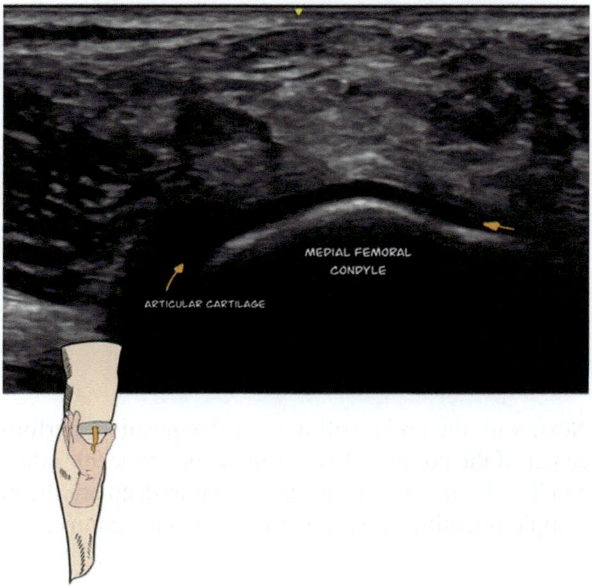

Fig. 17.4 Protocol Image 2: Medial femoral condylar cartilage, transverse

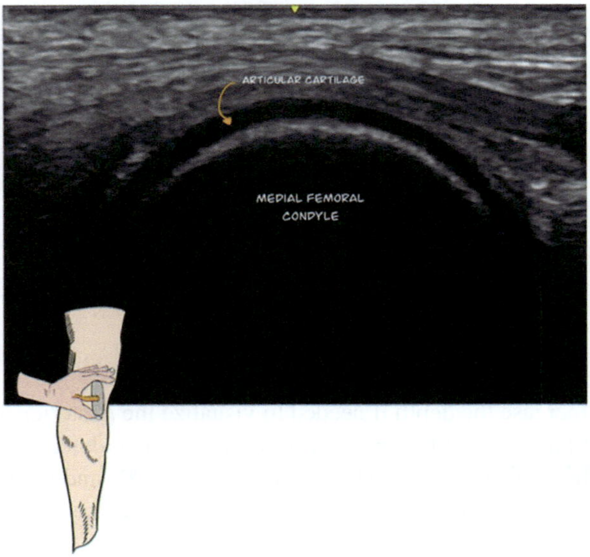

Fig. 17.5 Protocol Image 3: Medial femoral condylar cartilage, longitudinal

Protocol Image 4: Lateral Femoral Condylar Cartilage, Transverse

Now, rotate the probe 90° back to TAX. A lateral LAX slide will bring the LFC into view. Now perform the same dynamic scan as in Protocol Image 3 to examine the posterior femoral cartilage more fully.

Protocol Image 5: Lateral Femoral Condylar Cartilage, Longitudinal

Extend the knee and rotate the probe 90° for a LAX view of the articular cartilage. Do a slight TAX slide from medial to lateral to thoroughly scan the cartilage.

Protocol Image 6: Neurovascular Bundle, Transverse + Color Doppler (Fig. 17.6)

Next, with the probe still in the LAX position, perform a medial TAX slide to the center of the popliteal fossa. Rotate the probe 90° into TAX to visualize the PA, PV, and TN. Remember, from superficial to deep are the nerve, vein, and artery. Color Doppler identifies vascular structures and examines vessel pathology.

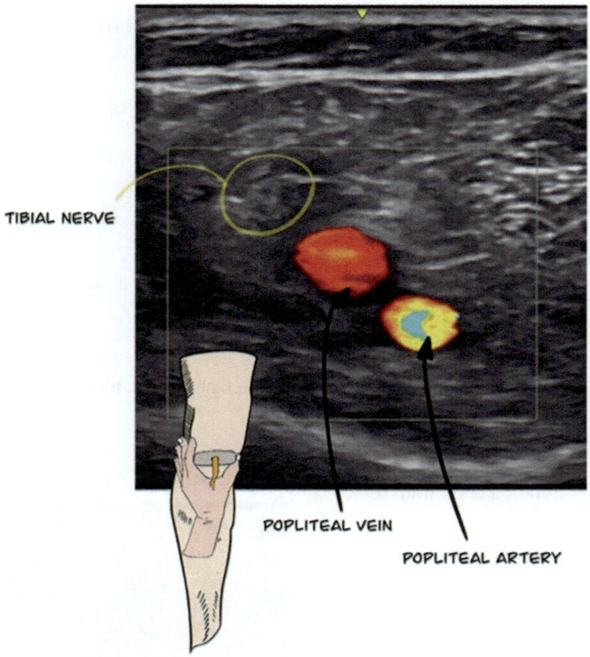

Fig. 17.6 Protocol Image 6: Neurovascular bundle, transverse + color Doppler

Protocol Image 7: Neurovascular Bundle, Transverse + Color Doppler + Sonocompression (Fig. 17.7)

Superficial to the PA is the PV. The patient can flex the knee approximately 30° to fill the vein. Probe pressure will collapse a normal PV vein and verify a lack of thrombosis. Again, color Doppler can help identify the vessels.

Protocol Image 8: Sciatic Nerve, Transverse (Fig. 17.8)

The honeycombed TN in TAX is located superficially and laterally to the PV. With the probe in TAX, follow the TN with a proximal TAX slide to the large sciatic nerve before bifurcation into the TN and CPN.

Protocol Image 9: Common Peroneal Nerve, Transverse (Fig. 17.9)

Next, still in TAX, center the probe on the CPN in a cross-sectional view and follow the CPN distally in a TAX slide as it courses more laterally down to the FH. The FH is seen as a hyperechoic "hill." The CPN in TAX is best seen medially

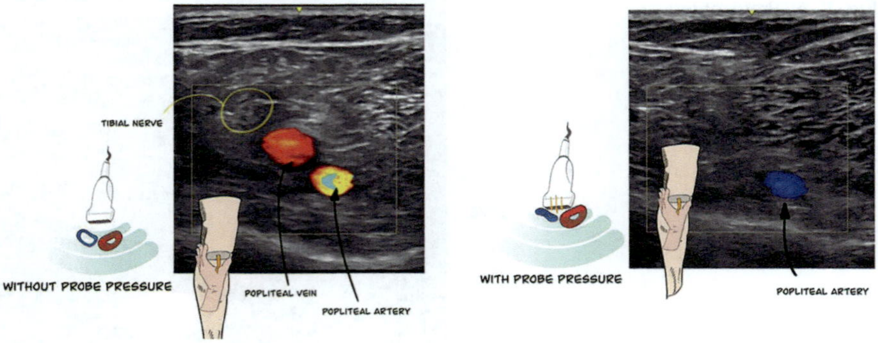

Fig. 17.7 Protocol Image 7: Neurovascular bundle, transverse + color Doppler + sonocompression

Fig. 17.8 Protocol Image 8: Sciatic nerve, transverse

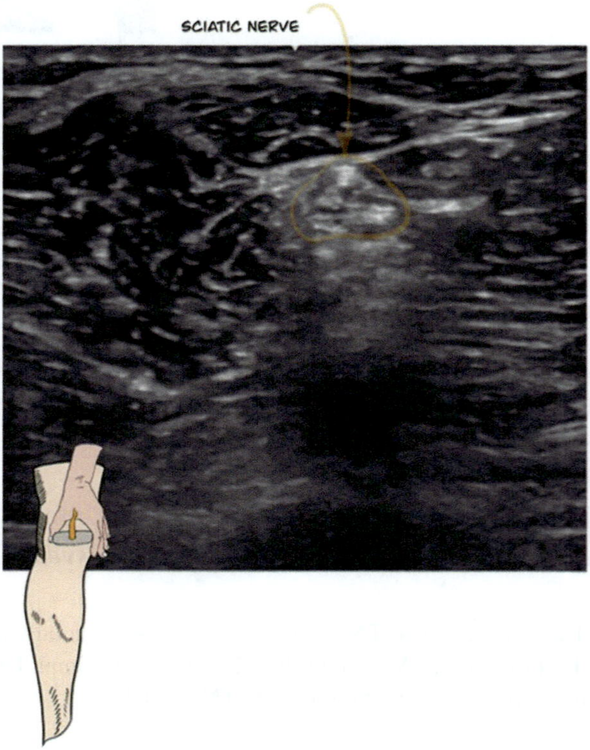

to the FH and has a rounded or ovoid honeycomb structure. Tilt the probe back and forth to utilize anisotropy to help identify the CPN, which is often isoechoic with adjacent soft tissue. If there is a question about CPN dysfunction, measure the CSA and compare it with the contralateral side. Please see the Clinical Comments for further discussion.

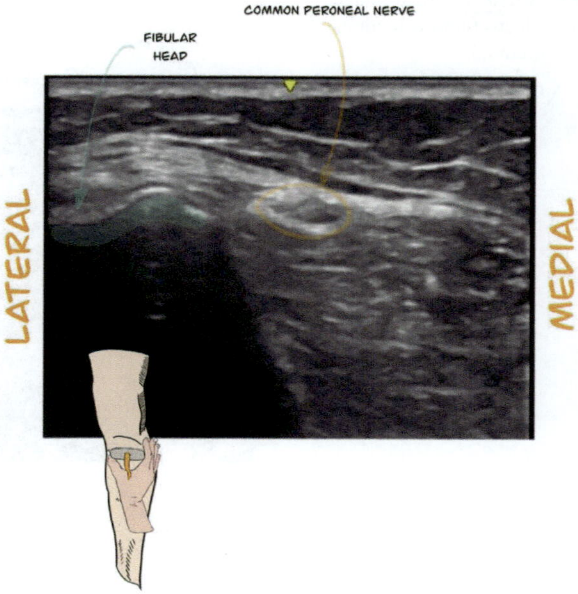

Fig. 17.9 Protocol Image 9: Common peroneal nerve, transverse

Alternatively, take the foreleg, bend the knee 15°, and place the ankle on its contralateral counterpart. This externally rotates the hip and the entire knee. Take the probe in TAX and scan the proximal FH, looking for the bony linear diagonal cortex. The CPN candidate is located just medially to this bony cortex. Move the probe distally and watch the CPN move laterally in an oblique path, superficial to and across the fibular neck, thus identifying the CPN. The CPN then dives beneath the FH of the peroneus longus muscle, also seen in TAX. This maneuver is akin to what we use when identifying the median nerve in the wrist by exploiting the distal anatomic course of the CPN.

To obtain a longitudinal view of the CPN, rotate the probe 90°. This view is challenging. You may see anechoic fluid tracking along the CPN epineural sheath, consistent with an intraneural ganglion. This view may demonstrate the peroneal longus muscle compressing the CPN as the nerve dives deep into this tendon; the CPN may swell proximal to the point of compression.

Protocol Image 10: Biceps Femoris Tendon, Longitudinal (Fig. 17.10)

Maintain the probe in LAX with respect to the femur, placing the distal end on the FH. The FH has a diagonal, hyperechoic cortex. The fibrillar biceps femoris tendon (BFT) is the product of the long and short heads of the biceps tendon and inserts into the FH. Verify that this is the BFT by moving the probe proximally to see the transition to muscle.

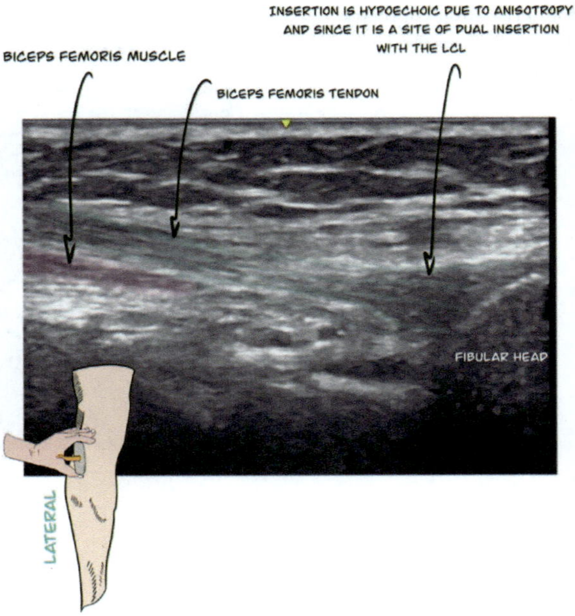

Fig. 17.10 Protocol Image 10: Biceps femoris tendon, longitudinal

Protocol Image 11: Posterior Horn of the Lateral Meniscus, Longitudinal (Fig. 17.11)

Still in LAX, deep to the BFT, locate the curved bony LFC and, more distally, the proximal tibia. The triangular posterior horn of the LM is between these two bones. Perform a lateral TAX slide to evaluate the posterior horn. Remember that US can only penetrate the more superficial portion of the meniscus, so deeper meniscal pathology will not be appreciated. MRI is the gold standard for meniscal evaluation.

A potential pitfall is to conclude that a hypoechoic cleft in the LM is a tear since this may be mimicked by the popliteus tendon sheath [28]. Another pitfall is to assume that a vertical cleft through the LM is a tear since this may be fibrous tissue [29]. On the femur in the popliteus groove, the popliteus tendon may appear as a hypoechoic structure due to anisotropy and can be mistaken for a parameniscal cyst. A trick is passively flexing and extending the knee, a maneuver to bring the meniscus into focus and enlarge a structure that you might think is a parameniscal cyst. This positioning lends itself to performing varus stress views of the knee to look for possible meniscal extrusion. The patient's leg may be prone or in external hip rotation, better exposing the posterolateral meniscus. When applying this technique, keeping the probe LAX to the tibia, not the femur, is essential.

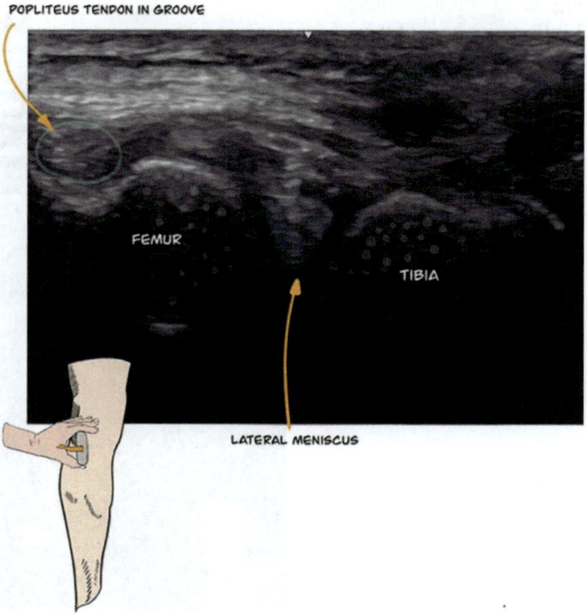

Fig. 17.11 Protocol Image 11: Posterior horn of the lateral meniscus, longitudinal

Protocol Image 12: Posterior Horn of the Medial Meniscus, Longitudinal (Fig. 17.12)

A medial TAX transverse slide to the posteromedial knee will visualize the posterior horn of the MM. You may see a hypoechoic round area distal to the MM, just superficially at the proximal tibia. This anisotropic insertion of the distal semimembranosus tendon on the tibia may mimic a parameniscal cyst. Tilting the probe will cause the hyperechoic tendon insertion to appear.

Protocol Image 13 (Optional): Fabella (Fig. 17.13)

Still in LAX, perform a lateral TAX slide toward the FH to visualize the LFC and the LHG. You may see a variant, a small bony hyperechoic "bean-like" structure embedded in the proximal LHG tendon. If the radiograph reveals a calcified fabella, this will be detectable on US. If there is a hypoechoic area within the LHG tendon, this may be a noncalcified fabella. Please see *Fabella Pathology* in the clinical comments above.

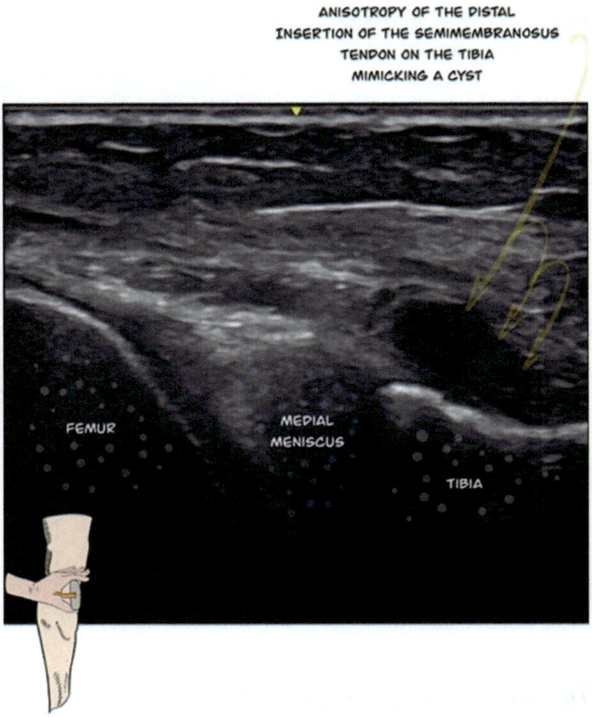

Fig. 17.12 Protocol Image 12: Posterior horn of the medial meniscus, longitudinal

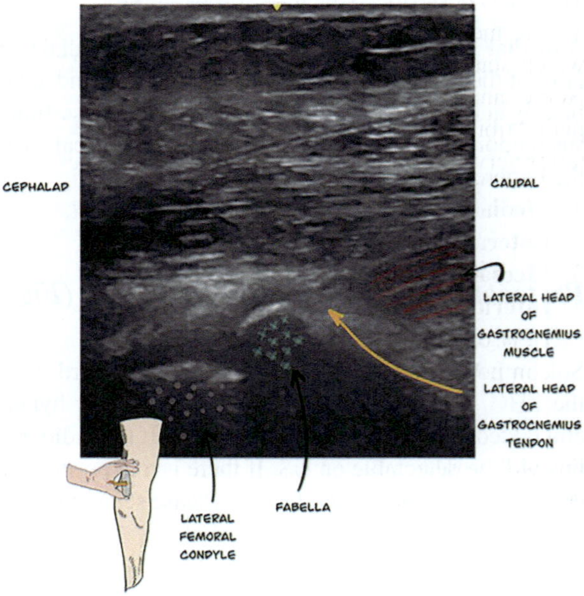

Fig. 17.13 Protocol Image 13 (optional): Fabella

Complete Posterior Knee Ultrasonic Examination Checklist

☐ Protocol Image 1: Medial femoral condyle, transverse
☐ Protocol Image 2: Medial femoral condylar cartilage, transverse
☐ Protocol Image 3: Medial femoral condylar cartilage, longitudinal
☐ Protocol Image 4: Lateral femoral condylar cartilage, transverse
☐ Protocol Image 5: Lateral femoral condylar cartilage, longitudinal
☐ Protocol Image 6: Neurovascular bundle, transverse + color Doppler
☐ Protocol Image 7: Neurovascular bundle, transverse + color Doppler + sonocompression
☐ Protocol Image 8: Sciatic nerve, transverse
☐ Protocol Image 9: Common peroneal nerve, transverse
☐ Protocol Image 10: Biceps femoris tendon, longitudinal
☐ Protocol Image 11: Posterior horn of the lateral meniscus, longitudinal
☐ Protocol Image 12: Posterior horn of the medial meniscus, longitudinal
☐ Protocol Image 13 (optional): Fabella.

References

1. Frush TJ, Noyes FR. Baker's cyst: diagnostic and surgical considerations. Sports Health. 2015;7(4):359–65.
2. Gupton M, Imonugo O, Terreberry RR. Anatomy, bony pelvis and lower limb, knee. StatPearls. Treasure Island, FL: StatPearls Publishing. Copyright © 2022, StatPearls Publishing LLC.; 2022.
3. Baker W. On the formation of synovial cysts in the leg in connection with disease of the knee-joint. St Bart Hosp Rep. 1877;15:245–61.
4. Rupp S, Seil R, Jochum P, Kohn D. Popliteal cysts in adults: prevalence, associated intraarticular lesions, and results after arthroscopic treatment. Am J Sports Med. 2002;30(1):112–5.
5. Fritschy D, Fasel J, Imbert JC, Bianchi S, Verdonk R, Wirth CJ. The popliteal cyst. Knee Surg Sports Traumatol Arthrosc. 2006;14(7):623–8.
6. Greenberg MH, Patel P, Mitcham E, Fant JW, Voss FR. Ultrasound aids diagnosis of a man with knee pain and swelling. Rheumatologist. 2019.
7. Kim JS, Lim SH, Hong BY, Park SY. Ruptured popliteal cyst diagnosed by ultrasound before evaluation for deep vein thrombosis. Ann Rehabil Med. 2014;38(6):843–6.
8. Alves TI, Girish G, Kalume Brigido M, Jacobson JA. US of the knee: scanning techniques, pitfalls, and pathologic conditions. Radiographics. 2016;36(6):1759–75.
9. Minagawa H, Wong KH. Musculoskeletal ultrasound: Echo anatomy & scan technique: Amazon Digital Services; 2017.
10. Duffy ST, Colgan MP, Sultan S, Moore DJ, Shanik GD. Popliteal aneurysms: a 10-year experience. Eur J Vasc Endovasc Surg. 1998;16(3):218–22.
11. Galland RB, Magee TR. Management of popliteal aneurysm. Br J Surg. 2002;89(11):1382–5.
12. Wolf YG, Kobzantsev Z, Zelmanovich L. Size of normal and aneurysmal popliteal arteries: a duplex ultrasound study. J Vasc Surg. 2006;43(3):488–92.
13. Rivera P, Dattilo J. Pseudoaneurysm. [Updated 2019 Nov 14]. In: StatPearls [Internet]. Treasure Island, FL: StatPearls Publishing; 2020. https://www.ncb.inlm.nih.gov/books/NBK542244/.
14. Kapoor BS, Haddad HL, Saddekni S, Lockhart ME. Diagnosis and management of pseudoaneurysms: an update. Curr Probl Diagn Radiol. 2009;38(4):170–88.

15. Cho K-H, Lee D-C, Chhem RK, Kim S-D, Bouffard JA, Cardinal E, et al. Normal and acutely torn posterior cruciate ligament of the knee at US evaluation: preliminary experience. Radiology. 2001;219(2):375–80.
16. Hsu C-C, Tsai W-C, Chen CP-C, Yeh W-L, Tang SF-T, Kuo J-K. Ultrasonographic examination of the normal and injured posterior cruciate ligament. J Clin Ultrasound. 2005;33(6):277–82.
17. Bowley MP, Doughty CT. Entrapment neuropathies of the lower extremity. Med Clin North Am. 2019;103(2):371–82.
18. Grant TH, Omar IM, Dumanian GA, Pomeranz CB, Lewis VA. Sonographic evaluation of common peroneal neuropathy in patients with foot drop. J Ultrasound Med. 2015;34(4):705–11.
19. Greenberg MH, Mitcham E, Patel P, Fant JW, Voss FR. Case report: ultrasound reveals cause of post-arthroplasty knee pain. Rheumatologist. 2020.
20. Visser LH, Hens V, Soethout M, De Deugd-Maria V, Pijnenburg J, Brekelmans GJ. Diagnostic value of high-resolution sonography in common fibular neuropathy at the fibular head. Muscle Nerve. 2013;48(2):171–8.
21. Spinner RJ, Desy NM, Amrami KK. Sequential tibial and peroneal intraneural ganglia arising from the superior tibiofibular joint. Skelet Radiol. 2008;37(1):79–84.
22. Baima J, Krivickas L. Evaluation and treatment of peroneal neuropathy. Curr Rev Musculoskelet Med. 2008;1(2):147–53.
23. Narasimhan M, Koenig SJ, Mayo PH. A whole-body approach to point of care ultrasound. Chest. 2016;150(4):772–6.
24. Kory PD, Pellecchia CM, Shiloh AL, Mayo PH, DiBello C, Koenig S. Accuracy of ultrasonography performed by critical care physicians for the diagnosis of DVT. Chest. 2011;139(3):538–42.
25. Driessen A, Balke M, Offerhaus C, White WJ, Shafizadeh S, Becher C, et al. The fabella syndrome - a rare cause of posterolateral knee pain: a review of the literature and two case reports. BMC Musculoskelet Disord. 2014;15:100.
26. Ehara S. Potentially symptomatic fabella: MR imaging review. Jpn J Radiol. 2014;32(1):1–5.
27. Khalil NFW, El-sherif S, El Hamid MMA, Elnemr R, Taleb RSZ. Role of global femoral cartilage in assessing severity of primary knee osteoarthritis. Egypt Rheumatol Rehabil. 2022;49(1):16.
28. Jacobson JA. Fundamentals of musculoskeletal ultrasound. 3rd ed. Philadelphia, PA: Elsevier; 2018.
29. Lee SY, Lee CA, Lee JH, Yim YJ, Kim CS. 0531: central hypoechogenic line on the ultrasonographic image of the meniscus: does it represent a meniscal tear? Ultrasound Med Biol. 2009;35(8, Supplement):S76.

Chapter 18
Lateral Knee

Reasons to Do the Study

Evaluation or assessment of soft tissue structures, that may be responsible for knee pain or dysfunction.

Questions We Want Answered

1. Is the iliotibial band (ITB) normal in appearance?
   (a) Is there evidence of bursitis deep to the ITB?
2. Are the tendons and ligaments normal in appearance?
   (a) If abnormal, does the pathology align with the symptoms?
   (b) If abnormal, is the pathology caused by an inflammatory or degenerative process?
3. Is there evidence of a meniscal tear?
   (a) Is a parameniscal cyst present?
   (b) Is meniscal extrusion present?
4. Is the common peroneal nerve normal in appearance?
   (a) If abnormal, does the pathology align with the symptoms?

The contents do not represent the views of the U.S. Department of Veterans Affairs or the United States Government.

M. H. Greenberg et al., *Manual of Musculoskeletal Ultrasound*,
https://doi.org/10.1007/978-3-031-37416-6_18

Basic Anatomy

The **iliotibial band (ITB),** sometimes called the iliotibial tract, is a complex fibrous connective tissue structure that originates on the iliac crest and has contributions from the gluteus maximus and tensor fascia lata muscles [1]. The ITB inserts onto **Gerdy's tubercle (GT)** on the anterolateral aspect of the tibia (Fig. 18.1) but has other attachment sites more proximally, including the intermuscular septum and supracondylar femoral tubercle [2, 3]. The ITB transfers forces from the hip to the knee while stabilizing the lateral knee [4].

The **anterolateral ligament (ALL)** is a thin ligament that inserts on the tibia halfway between the GT and the fibular head (FH). It attaches to the lateral femoral epicondyle (LFE). At knee flexion of 90°, the ALL resists internal rotation (Fig. 18.2). Deep to the ALL is the lateral inferior geniculate artery (LIGA), which is an important landmark to verify the identity of the ALL as well as test its integrity. Just posterior to the ITB and the ALL is the **lateral collateral ligament (LCL).** The LCL originates at the distal femur and attaches to the lateral femoral head. It is one of several lateral knee stabilizers.

Deep to the LCL runs the **popliteus tendon,** which inserts just proximal to the LFE in the popliteus groove. This is an unusual tendon since it runs *deep* to a ligament. The popliteus muscle is flat and lies deep to the plantar and gastrocnemius

Fig. 18.1 Distal Iliotibial band

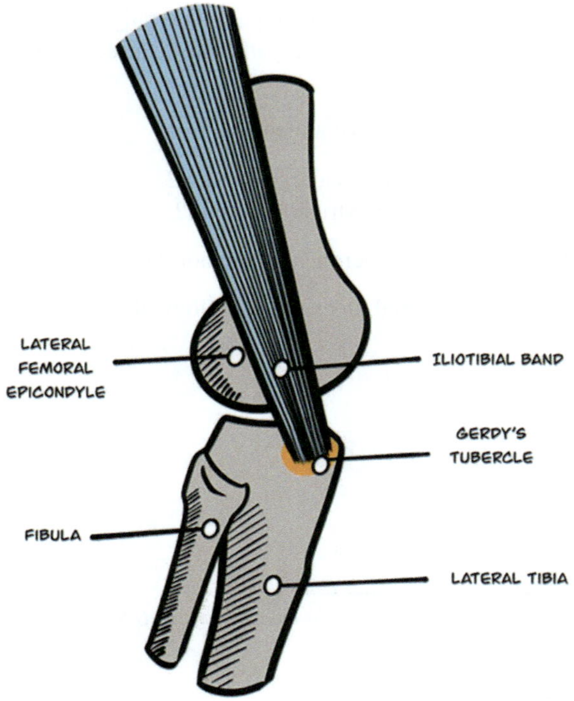

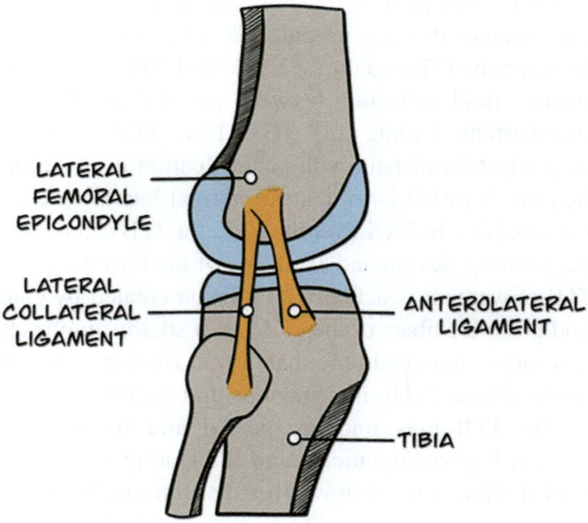

Fig. 18.2 Ligamentous anatomy of the lateral knee simplified

muscles. It connects the posterior tibia to the external aspect of the lateral femoral condyle (LFC).

The **lateral meniscus (LM)** is a semicircular fibrocartilaginous “shock absorber” located between the lateral articulation of the tibia and femur. The outer aspect can be sonographically inspected and appears as a hyperechoic triangular structure deep to the LCL, separating the femur from the tibia. The **biceps femoris tendon (BFT)** and LCL share a common distal attachment at the lateral FH, with the BFT located posterior to the LCL.

The **common peroneal (fibular) nerve (CPN)** is the most posterior structure in the lateral knee and is found just posterior to the BFT. Sonographically, the nerve is best visualized transverse (TAX) to the FH [5]. The nerve curves anteriorly around the lateral fibular neck and then splits into the superficial and peroneal nerves.

Clinical Comments

Iliotibial Band Syndrome

Repetitive knee flexion and extension, such as running and cycling, may cause the ITB to scrape against the LFE [4]. This may produce iliotibial band syndrome (ITBS) (“runner’s knee”), a common overuse injury. Pain on the lateral aspect of the knee occurs during physical activity and worsens during the initial 25–30° of knee flexion [4, 6]. Predisposing factors include knee varus deformity, leg length discrepancy, and weak hip abductors [4]. Generally considered a friction syndrome due to overuse, alternative etiologies include compression **of fat and connective tissue** deep to the ITT or chronic inflammation of the IT bursa [7–9].

On ultrasound (US), the normal ITB in longitudinal (LAX) reveals a linear fibrillar structure that inserts onto GT [10]. The adventitial iliotibial bursa is located between the ITB and the LFE [4]. With ITBS, US may reveal soft tissue edema or a distinct fluid collection between the ITB and LFE. Thickening of the ITB is an inconsistent finding [4]. The LFE may display cortical irregularities [9]. Reproduction of pain with sonopalpation over the involved area is diagnostically helpful. A pitfall is to mistake normal lateral recess synovial fluid for adventitial bursitis [11]. In TAX over the LFE, the ITB can be dynamically evaluated for snapping during flexion and extension of the knee [12]. Some authors maintain that the ITB movement is actually an illusion created by changing tension in the anterior and posterior fibers of the ITB [7]. If so, this implies that the ITBS may be more of a compression syndrome than a true friction syndrome [13]. However, a different study affirmed ITB movement using US [14].

The ITB may undergo partial and full-thickness tears [15]. Sonography reveals hypoechogenicity and thickening with fluid surrounding a partial tear. Full-thickness tears show fiber disruption [15]. Finally, distal ITB tendinopathy, an entity separate from ITBS, may occur at or near the GT insertion and appears on US as thickening and/or hypoechogenicity compared with the contralateral knee [16]. Such tendinopathy may be due to knee prosthesis impingement or abnormal varus or valgus stress on the ITB due to knee osteoarthritis. Pain may be localized to the insertion or be perceived along the lateral thigh. Using US guidance, a fluid collection deep to the ITB can be aspirated and injected for treatment, and a positive therapeutic response favors the diagnosis of ITBS [4, 17, 18].

Anterolateral Ligament

This thin ligament, whose existence was first suspected in 1879, has recently been rediscovered and found to be present in 97% of the population in cadaver studies [19, 20]. In one study, US was 100% sensitive in detecting the ALL [21]. The anterior cruciate ligament (ACL) is the primary stabilizer of the internal rotation of the knee. The ALL also stabilizes internal knee rotation. Since tears of the ALL may occur with an ACL tear, it is essential to evaluate for a concomitant ALL tear if an ACL tear is suspected [22]. Ultrasound detection of ALL tears is at least as accurate as magnetic resonance imaging (MRI).

Lateral Collateral Ligament

The LCL of the knee is the primary stabilizer to counter varus instability [23]. On US, the normal LCL appears hyperechoic and fibrillar. Injuries to the LCL include partial or complete tears as well as soft tissue or bony avulsions, the latter most commonly occurring at the FH attachment [24]. On US, pathology ranges from ligament thickening, heterogeneity, or a wavy pattern, to complete ligament disruption or even avulsion of the LCL from the fibula [9].

There are several pitfalls in the sonographic evaluation of the LCL. Valgus knee angulation slackens the LCL, producing a wavy, anisotropic appearance that mimics injury. Another trap is to mistake the anisotropy of the dual insertion of the LCL and the BFT on the FH for LCL enthesopathy or BFT tendinosis [9]. A third pitfall is to mistake decreased echogenicity of the proximal LCL for pathology in the absence of lateral knee symptoms [25]. This normal sonographic finding most likely relates to connective tissue mucin. It is unusual to have an isolated LCL injury since it is but one of the several lateral knee stabilizers [9]. Therefore, further evaluation for concomitant soft tissue damage should be considered if LCL injury is detected.

Lateral Meniscus

To accurately evaluate the entire LM, MRI is the best choice; however, specific findings on US may indicate pathology [9]. On US, the ordinary LM is hyperechoic and triangular-shaped. Degeneration of the LM appears as meniscal heterogeneity with possible extrusion and fragmentation [9]. The knee menisci, positioned between the femoral and tibial components, protect knee cartilage from excessive wear and tear [26]. Meniscal tears may accelerate knee osteoarthritis and are associated with meniscal extrusion [9]. Tears may appear as focal anechoic or hypoechoic defects extending to the meniscal surfaces.

One study *comparing US to MRI* to detect meniscal tears revealed only 64% specificity for US [27]. The medial and lateral menisci *posterior horns* yielded the best specificity of 100% and 94%, respectively. However, sensitivity levels were only 63 and 75%, respectively. Furthermore, US cannot detect bucket-handle tears due to the beam's inability to penetrate the meniscus [27].

Meniscal (parameniscal) cysts may present as palpable, sometimes painful, masses along the knee joint [28]. The cyst on US is anechoic or hypoechoic and may contain septa or debris. Ultrasound is 94% accurate for detecting parameniscal cysts [28]. An underlying meniscal tear is associated with 90% of LM cysts, making the latter a bellwether finding. If a meniscal cyst is detected, knee flexion may enhance its prominence [29]. A complex meniscal cyst may mimic a solid mass; however, continuity with the meniscus argues in favor of a cyst [15]. Transducer compression may demonstrate a connection in which the cyst partially empties its contents via a connecting tract with the meniscus.

Meniscal extrusion is described as the displacement of meniscal tissue beyond the *tibial* margin [30]. In addition to meniscal tears, the degree of meniscal extrusion is affected by age, body weight, knee alignment, cartilage damage, and the presence of osteoarthritis [26, 31]. The MRI is the gold standard for assessing meniscal extrusion [32]. Meniscal extrusion on MRI is defined as >3 mm [33]. Ultrasound reliably measures LM extrusion but tends to overestimate extrusion by about 1.1 mm compared to MRI [26]. However, the same study demonstrated an overlap of values of LM extrusion in normal and abnormal subjects.

Biceps Femoris Tendon

The BFT may develop tendinosis, noted as thickening and hypoechogenicity at the conjoined tendon with the LCL, where both insert on the FH [9]. However, to reiterate the pitfall previously described for the LCL, the dual insertion of the LCL and the BFT on the FH may give a false impression of heteroechogenicity and swelling, thus mimicking enthesopathy of the LCL and tendinosis of the BFT. Tears of the BFT have been reported to occur more often at the distal myotendinous junction [9].

Common Peroneal Nerve

Peroneal neuropathy may cause lateral foreleg paresthesia and foot weakness. The CPN is found just posterior to the BFT. Due to its proximity to the FH, the CPN is prone to direct injury [9]. It is also subject to entrapment, neoplasms, and intraneuronal ganglia. The nerve may develop hypoechoic swelling proximal to the point of compression from the entrapment [9]. Repetitive injury from an overhanging prosthetic knee component may also cause neuropathy [14].

Pitfalls

1. Use a 12 MHz or higher transducer since the region of interest is superficial.
2. Accurately mark three bony landmarks: the LFE, the GT, and the FH. The FH is more posterior than most people think.
3. Remember, when evaluating the LCL, valgus knee angulation slackens this ligament, producing a wavy appearance with anisotropy that may mimic injury. A pillow between the knees moderates valgus knee angulation, thus reducing artifacts.
4. Do not mistake the lateral recess synovial fluid or synovial hypertrophy for adventitial bursitis associated with ITBS.
5. Do not mistake the dual insertion of the LCL and the BFT on the FH for enthesopathy or tendinosis since this may give a false impression of hypoechogenicity and swelling.
6. Do not mistake decreased echogenicity of the proximal LCL for pathology without lateral knee symptoms; this is a normal sonographic finding.
7. Since the LCL is only one of several lateral knee stabilizers, further evaluation for concomitant soft tissue damage should be undertaken if an LCL injury is detected.
8. While some meniscal tears may be visible sonographically, MRI remains the gold standard to confirm and evaluate for bucket-handle tears since the latter remain invisible to US.

9. Look for telltale signs of meniscal damage, such as a meniscal cyst or extrusion.
10. If a meniscal cyst is detected, knee flexion may enhance its prominence, and transducer compression may partly empty the cyst, thus demonstrating a tract leading to the meniscus.
11. Minimize the anisotropy of the lateral knee stabilizing ligaments by examining them when taut. The LCL is tight when the knee is extended, whereas the ALL is taut when the knee is flexed 90° and the foot is internally rotated.

Method

The patient is in the lateral decubitus position with a bolster under the knee; the knee is bent 10–20°. An alternative position is to have the supine patient internally rotate the hip and flex the knee [9]. Mark off GT, the bony prominence on the anterolateral tibia just distal to the patella. This should not be confused with the FH at the same level, which is posterolateral. Mark off the FH and LFE as well (Fig. 18.3).

Protocol Image 1: Iliotibial Band, Longitudinal + Dynamic (Fig. 18.4)

With the transducer in LAX with respect to the femur, place the distal end on GT, a brightly hyperechoic bone with a gentle proximal slope. Rotate the proximal end posterolaterally until you see hyperechoic ITB fibers inserted onto GT. The ITB courses *over* the underlying LFE; look for sonographic signs related to ITBS and distal tendinopathy. Deep to the ITB, just proximal to the knee joint, you may see the ITB bursa, which is more apparent when ITB bursitis is present. See Clinical Comments for more information.

An alternative method to visualize the ITB is to view the patellar tendon in LAX and then perform a lateral TAX slide to visualize the ITB extending from GT [9].

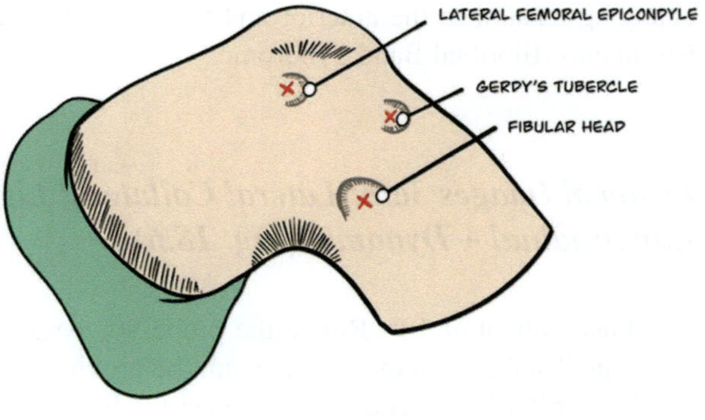

Fig. 18.3 Lateral knee topographic landmarks

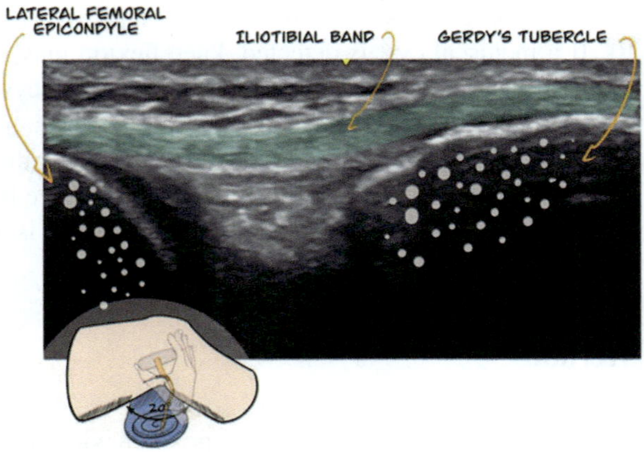

Fig. 18.4 Protocol Image 1: Iliotibial band, longitudinal + dynamic

Note that with this method, the transducer is initially positioned parallel to the tibial axis, so you will need to rotate the proximal end of the probe in a posterolateral direction.

Protocol Image 2 (Optional): Iliotibial Band, Transverse + Dynamic (Fig. 18.5)

At GT, rotate the transducer 90° to examine the ITB in TAX view to confirm any pathology found on the LAX or if you suspect ITBS with or without snapping. Slowly move the probe in a proximal TAX slide to thoroughly explore the ITB in a cross-section. If considering ITBS, try to identify an adventitial iliotibial bursa or simply soft tissue swelling between the ITB and the LFE. There may or may not be ITB thickening or LFE cortical irregularity. Sonopalpitory pain is diagnostically useful. Do not mistake the normal lateral recess synovial fluid for adventitial bursitis [11]. At the LFE, have the patient slowly bend the knee to see if the ITB snaps over the LFE. At 30° of flexion, the ITB is said to move from anterior to posterior over the LFE. This can be evaluated in a supine or standing position [34]. Be aware that some authors maintain that the ITB movement is only an illusion created by changing tension in the anterior and posterior fibers of the ITB [7]. See Clinical Comments, Iliotibial Band Syndrome.

Protocol Images 3a/b: Lateral Collateral Ligament, Longitudinal + Dynamic (Fig. 18.6)

The knee is bent at 10°. Rotate the probe 90° back to a longitudinal orientation, place the distal end on the FH, and aim the proximal end toward the LFE to visualize the fibrillar LCL. You may visualize the hyperechoic bone between the femur

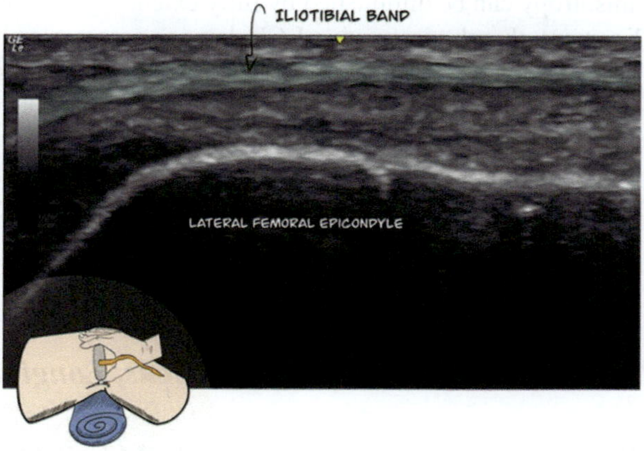

Fig. 18.5 Protocol Image 2 (optional): Iliotibial band, transverse + dynamic

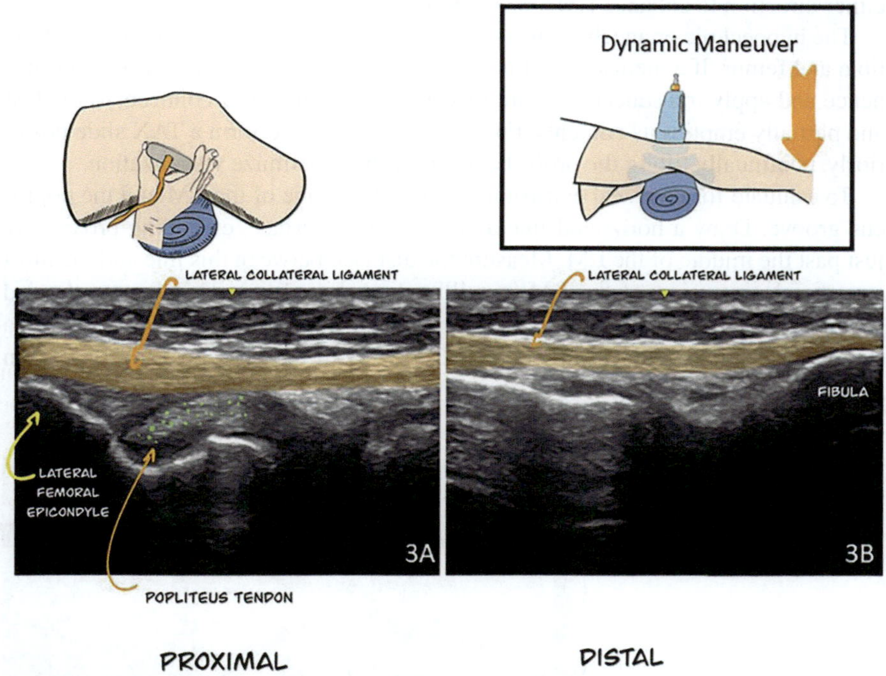

Fig. 18.6 Protocol Images 3a/b: Lateral collateral ligament, longitudinal + dynamic

and the FH; this is the tibia. Perform a proximal LAX slide and scrutinize the entire LCL. Separate images can be saved for the proximal (**Protocol Image 3a**) and distal attachments (**Protocol Image 3b**).

A valgus knee deformity may slacken the LCL, creating anisotropy with a wavy appearance of the LCL, thus mimicking injury [9]. If this is present, a pillow between the knees will somewhat correct the valgus angulation [35]. Normal anisotropy can be diminished by fully extending the knee to increase tension on the ligament, thereby improving LCL damage assessment [29]. Deep to the LCL is the popliteus tendon origin at the popliteus groove or sulcus of the femur. This sulcus is at the distal LFC. Tilt the probe to counteract anisotropy since this may mimic a cystic mass otherwise. To place dynamic varus stress on the LCL in the lateral decubitus position, place a block under the knee so that the foreleg is suspended. Press down on the foreleg to look at the integrity of the LCL and note any joint space widening.

Protocol Image 4: Lateral Meniscus, Longitudinal (Fig. 18.7)

Next, focus on the structures deep to the LCL. Between the distal femur and the tibia lie the femorotibial joint and the triangular, hyperechoic LM. Tilt the probe in either direction to enhance LM visualization.

The hypoechoic or anechoic areas on each side of the LM are the cartilage of the tibia and femur. If a meniscal cyst is suspected, flex the knee to enhance its prominence and apply transducer pressure to see if the putative cyst connects to the LM and partially empties its contents. Focusing on the LM, perform a TAX slide posteriorly, continually tilting the probe back and forth to optimize visualization.

To evaluate for meniscal extrusion, capture the image of the LM and the popliteus groove. Draw a horizontal line along the tibial surface, extending proximally just past the middle of the LM. Measure the distance between this line and the most superficial protrusion of the LM in millimeters (Fig. 18.8). This is an unloaded (non-weight-bearing) view. One study protocol suggests examining the patient in a standing bipedal, weight-bearing (loaded) view with the knee bent in 10° flexion

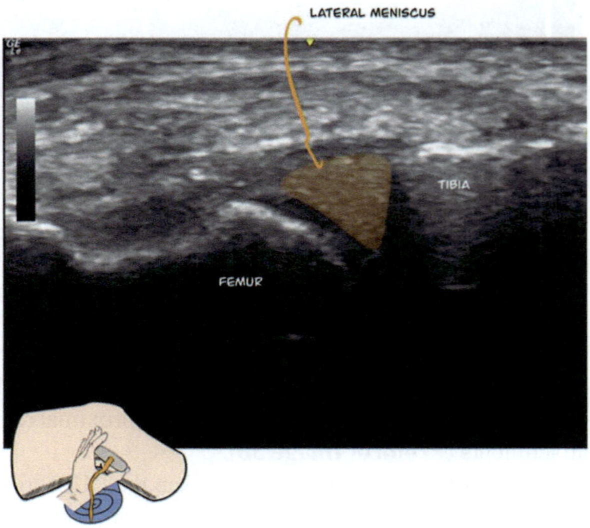

Fig. 18.7 Protocol Image 4: Lateral meniscus, longitudinal

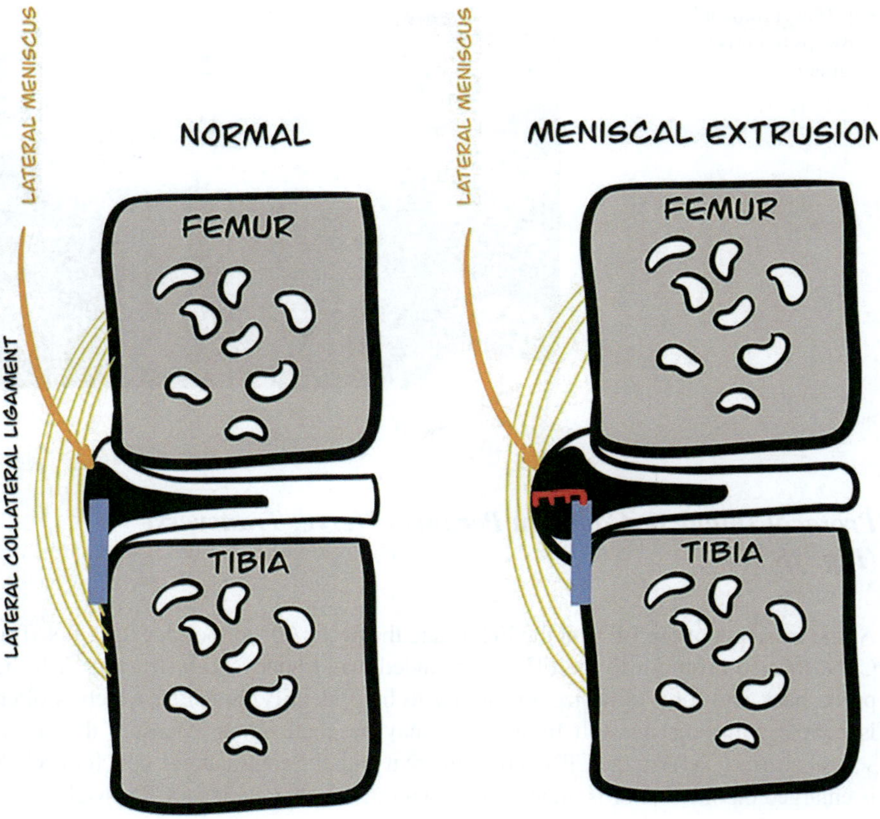

Fig. 18.8 Lateral meniscus extrusion measurement

[26]. This requires marking the transducer position with a ballpoint pen when supine so that the identical probe position can be duplicated in the standing position. Make sure there is no tibial rotation in the supine or standing examination.

Protocol Image 5: Biceps Femoris Tendon, Longitudinal (Fig. 18.9)

With the patient supine or in lateral decubitus, the knee is bent to 10° and the hip is internally rotated. Affix the distal end of the transducer to the *posterolateral edge* of the FH, which appears as a hyperechoic diagonal line. Rotate the proximal transducer end posteriorly to parallel the femoral axis to visualize the BFT in a LAX view. The myotendinous junction is seen proximally and is deep to the tendon. Note that the LCL and the BFT both insert into the FH; this dual insertion produces hypoechogenicity, mimicking enthesopathy.

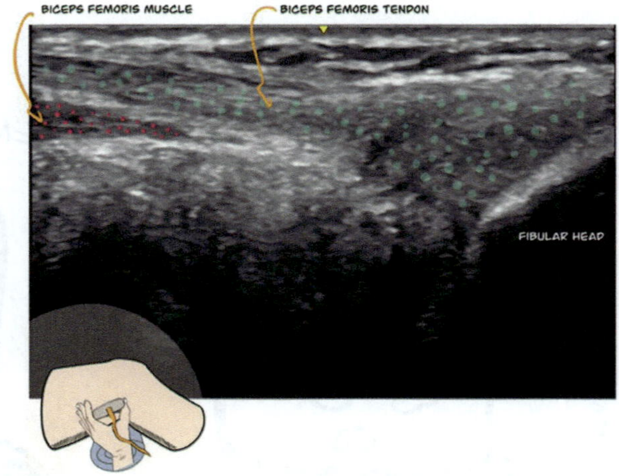

Fig. 18.9 Protocol Image 5: Biceps femoris tendon, longitudinal

Protocol Image 6: Common Peroneal Nerve, Transverse (Fig. 18.10)

At the insertion of the BFT on the FH, rotate the probe 90° to look for the CPN. The CPN, located posterior to the FH, is a rounded/ovoid honeycomb structure. Tilt the probe back and forth to utilize anisotropy to help identify the CPN, which is often isoechoic with adjacent soft tissue. This may be challenging. Measure the cross-sectional area (CSA) of the CPN and compare it with the contralateral side. If the CSA is enlarged on one side or is greater than 10 mm^2, this may indicate CPN swelling.

An alternative method of finding the CPN:

With the patient in a prone position, bend the knee 15° and place the ankle on the contralateral ankle, thus externally rotating the hip. With the probe in TAX, identify the proximal FH by its hyperechoic linear diagonal cortex. In cross-section, the CPN is just medial to the straight bony cortex of the FH. Move the probe distally and watch the CPN move in an oblique path, superficial to and then lateral to the fibular neck, thus identifying the CPN. The CPN then dives beneath one of the heads of the peroneus longus muscle, also seen in TAX. This maneuver exploits the anatomic course of the CPN, akin to the method of identifying the median nerve in the wrist.

Protocol Image 7: Common Peroneal Nerve, Longitudinal (Fig. 18.11)

Once the CPN is identified in TAX, rotate the probe 90° to evaluate the nerve in LAX, remembering its oblique path as described above. In this view, there may be anechoic fluid tracking along the CPN epineural sheath, consistent with an intraneural ganglion. This view may also demonstrate the peroneus longus tendon compressing the CPN as the nerve dives deep to this tendon to hug the fibular neck. Again, the CPN may swell proximal to compression.

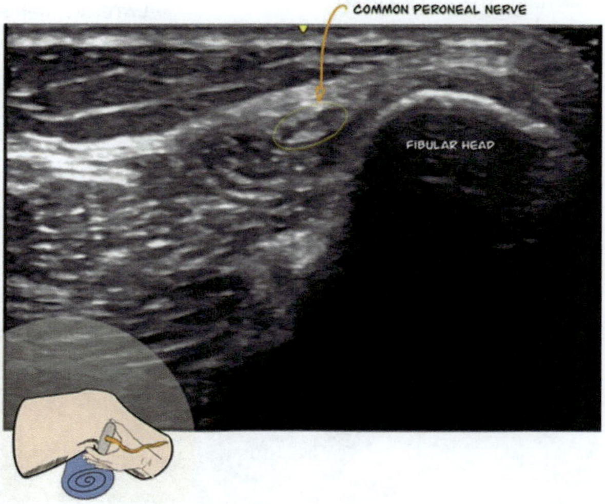

Fig. 18.10 Protocol Image 6: Common peroneal nerve, transverse

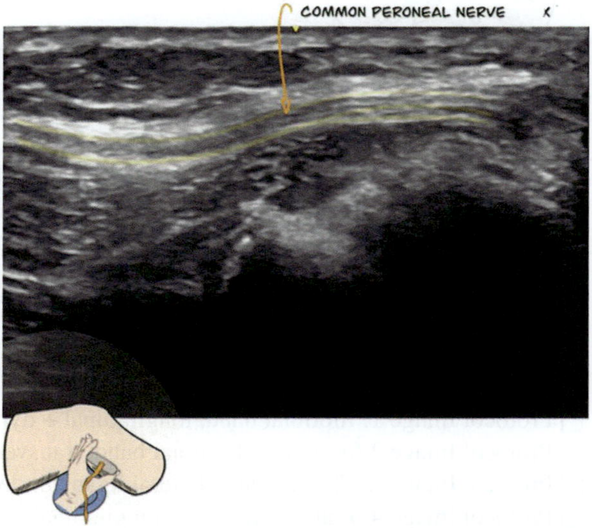

Fig. 18.11 Protocol Image 7: Common peroneal nerve, longitudinal

Protocol Image 8 (Optional): Anterolateral Ligament, Longitudinal ± Power Doppler (Fig. 18.12)

Return the transducer to the position in Protocol Image 1 of the ITB. Next, flex the knee to 90° and internally rotate the knee by slightly inverting the foot [19]. From GT, perform a posterolateral TAX slide to a point halfway between GT and the FH to visualize the distal ALL insertion point onto the tibia. If you reach the FH, you have gone too far. Once the ALL insertion point is seen, rotate the proximal end of the probe slightly posteriorly to visualize the full ligament in LAX and its insertion onto the LFE. Deep to the ligament is the popliteus groove and the LM [19]. Turn

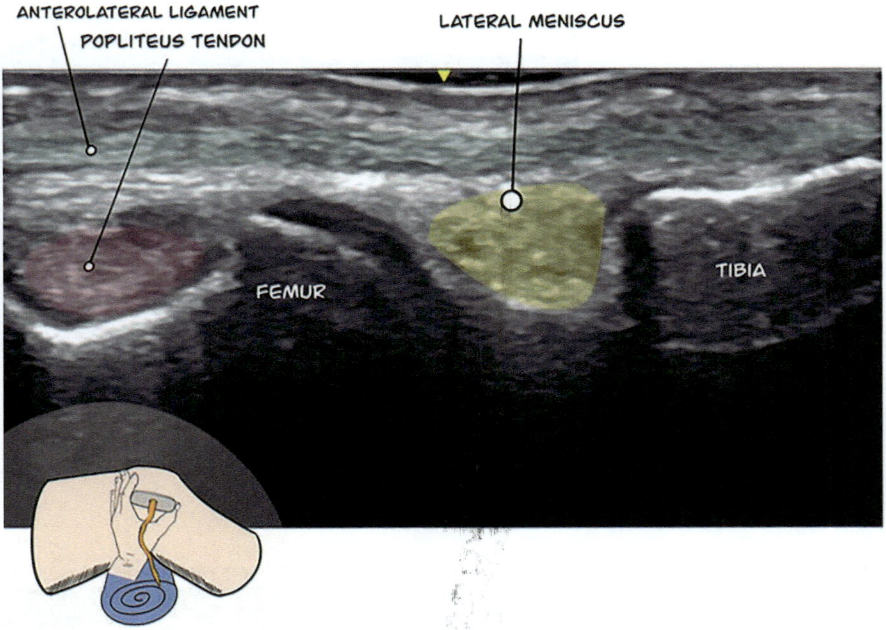

Fig. 18.12 Protocol Image 8 (Optional): Anterolateral ligament, longitudinal + dynamic

the PD on to locate the LIGA in cross-section deep to the ALL [21]. Since the ALL and the LIGA are closely related, the presence of the LIGA helps confirm that the nearby thin linear structure is the ALL. To dynamically demonstrate the integrity of the ALL, turn on PD and further internally rotate the foot to increase tension on this ligament [21]. This rotation should impede the blood flow to the LIGA if the ALL is intact.

Complete Lateral Knee Ultrasonic Examination Checklist

☐ Protocol Image 1: Iliotibial band, longitudinal + dynamic
☐ Protocol Image 2 (optional): Iliotibial band, transverse, + dynamic
☐ Protocol Images 3a/b: Lateral collateral ligament, longitudinal + dynamic
☐ Protocol Image 4: Lateral meniscus, longitudinal
☐ Protocol Image 5: Biceps femoris tendon, longitudinal
☐ Protocol Image 6: Common peroneal nerve, transverse
☐ Protocol Image 7: Common peroneal nerve, longitudinal
☐ Protocol Image 8 (optional): Anterolateral ligament, longitudinal ± power Doppler

References

1. Vieira EL, Vieira EA, da Silva RT, Berlfein PA, Abdalla RJ, Cohen M. An anatomic study of the iliotibial tract. Arthroscopy. 2007;23(3):269–274.
2. Flato R, Passanante GJ, Skalski MR, Patel DB, White EA, Matcuk GR Jr. The iliotibial tract: imaging, anatomy, injuries, and other pathology. Skelet Radiol. 2017;46(5):605–22.

3. Strauss EJ, Kim S, Calcei JG, Park D. Iliotibial band syndrome: evaluation and management. J Am Acad Orthop Surg. 2011;19(12):728–36.
4. Jiménez Díaz F, Gitto S, Sconfienza LM, Draghi F. Ultrasound of iliotibial band syndrome. J Ultrasound. 2020;23(3):379–85.
5. Jacobson JA. Fundamentals of musculoskeletal ultrasound. 3rd ed. Philadelphia, PA: Elsevier; 2018.
6. Ellis R, Hing W, Reid D. Iliotibial band friction syndrome—a systematic review. Man Ther. 2007;12(3):200–8.
7. Fairclough J, Hayashi K, Toumi H, Lyons K, Bydder G, Phillips N, et al. The functional anatomy of the iliotibial band during flexion and extension of the knee: implications for understanding iliotibial band syndrome. J Anat. 2006;208(3):309–16.
8. Muhle C, Ahn JM, Yeh L, Bergman GA, Boutin RD, Schweitzer M, et al. Iliotibial band friction syndrome: MR imaging findings in 16 patients and MR arthrographic study of six cadaveric knees. Radiology. 1999;212(1):103–10.
9. Alves TI, Girish G, Kalume Brigido M, Jacobson JA. US of the knee: scanning techniques, pitfalls, and pathologic conditions. Radiographics. 2016;36(6):1759–75.
10. De Maeseneer M, Marcelis S, Boulet C, Kichouh M, Shahabpour M, de Mey J, et al. Ultrasound of the knee with emphasis on the detailed anatomy of anterior, medial, and lateral structures. Skelet Radiol. 2014;43(8):1025–39.
11. Jelsing EJ, Finnoff J, Levy B, Smith J. The prevalence of fluid associated with the iliotibial band in asymptomatic recreational runners: an ultrasonographic study. PM R. 2013;5(7):563–7.
12. Marchand AJ, Proisy M, Ropars M, Cohen M, Duvauferrier R, Guillin R. Snapping knee: imaging findings with an emphasis on dynamic sonography. AJR Am J Roentgenol. 2012;199(1):142–50.
13. Lavine R. Iliotibial band friction syndrome. Curr Rev Musculoskelet Med. 2010;3(1–4):18–22.
14. Greenberg MH, Mitcham E, Patel P, Fant JW, Voss FR. Case report: ultrasound reveals cause of post-arthroplasty knee pain. Rheumatologist. 2020.
15. Lee MJ, Chow K. Ultrasound of the knee. Semin Musculoskelet Radiol. 2007;11(2):137–48.
16. Bianchi S, Martinoli C. Ultrasound of the musculoskeletal system. New York, NY: Springer, Berlin, Heidelberg; 2007. p. 458–67.
17. Hong JH, Kim JS. Diagnosis of iliotibial band friction syndrome and ultrasound guided steroid injection. Korean J Pain. 2013;26(4):387–91.
18. Gunter P, Schwellnus MP. Local corticosteroid injection in iliotibial band friction syndrome in runners: a randomised controlled trial. Br J Sports Med. 2004;38(3):269–72; discussion 72
19. Cianca J, John J, Pandit S, Chiou-Tan FY. Musculoskeletal ultrasound imaging of the recently described anterolateral ligament of the knee. Am J Phys Med Rehabil. 2014;93(2):186.
20. Claes S, Vereecke E, Maes M, Victor J, Verdonk P, Bellemans J. Anatomy of the anterolateral ligament of the knee. J Anat. 2013;223(4):321–8.
21. Cavaignac E, Wytrykowski K, Reina N, Pailhé R, Murgier J, Faruch M, et al. Ultrasonographic identification of the anterolateral ligament of the knee. Arthroscopy. 2016;32(1):120–6.
22. Faruch Bilfeld M, Cavaignac E, Wytrykowski K, Constans O, Lapègue F, Chiavassa Gandois H, et al. Anterolateral ligament injuries in knees with an anterior cruciate ligament tear: contribution of ultrasonography and MRI. Eur Radiol. 2018;28(1):58–65.
23. Grawe B, Schroeder AJ, Kakazu R, Messer MS. Lateral collateral ligament injury about the knee: anatomy, evaluation, and management. J Am Acad Orthop Surg. 2018;26(6):e120–e7.
24. Porrino J, Sharp JW, Ashimolowo T, Dunham G. An update and comprehensive review of the posterolateral corner of the knee. Radiol Clin N Am. 2018;56(6):935–51.
25. Falkowski AL, Jacobson JA, Gandikota G, Lucas DR, Magerkurth O, Zaottini F. Imaging characteristics of the proximal lateral collateral ligament of the knee: findings on ultrasound and mri with histologic correlation. J Ultrasound Med. 2022;41(4):827–34.
26. Winkler PW, Csapo R, Wierer G, Hepperger C, Heinzle B, Imhoff AB, et al. Sonographic evaluation of lateral meniscal extrusion: implementation and validation. Arch Orthop Trauma Surg. 2021;141(2):271–81.
27. Unlu EN, Ustuner E, Saylisoy S, Yilmaz O, Ozcan H, Erden I. The role of ultrasound in the diagnosis of meniscal tears and degeneration compared to MRI and arthroscopy. Acta Medica Anatolia. 2014;2(3):80–7.

28. Rutten MJ, Collins JM, van Kampen A, Jager GJ. Meniscal cysts: detection with high-resolution sonography. AJR Am J Roentgenol. 1998;171(2):491–6.
29. Bianchi E, Martinoli C. Knee. Ultrasound of the musculoskeletal system. Springer; 2007. p. 637–744.
30. Masouros SD, McDermott ID, Amis AA, Bull AM. Biomechanics of the meniscus-meniscal ligament construct of the knee. Knee Surg Sports Traumatol Arthrosc. 2008;16(12):1121–32.
31. Crema MD, Roemer FW, Felson DT, Englund M, Wang K, Jarraya M, et al. Factors associated with meniscal extrusion in knees with or at risk for osteoarthritis: the Multicenter Osteoarthritis study. Radiology. 2012;264(2):494–503.
32. Podlipská J, Koski JM, Kaukinen P, Haapea M, Tervonen O, Arokoski JP, et al. Structure-symptom relationship with wide-area ultrasound scanning of knee osteoarthritis. Sci Rep. 2017;7:44470.
33. Gajjar SM, Solanki KP, Shanmugasundaram S, Kambhampati SBS. Meniscal extrusion: a narrative review. Orthop J Sports Med. 2021;9(11):23259671211043797.
34. Jelsing EJ, Finnoff JT, Cheville AL, Levy BA, Smith J. Sonographic evaluation of the iliotibial band at the lateral femoral epicondyle: does the iliotibial band move? J Ultrasound Med. 2013;32(7):1199–206.
35. Jacobson J. Knee ultrasound. Fundamentals of musculoskeletal ultrasound. Philadelphia, PA: Elsevier; 2018. p. 284–327.

Chapter 19
Medial Knee

Reasons to Do the Study
Evaluation or assessment of the following:

1. Medial (tibial) collateral ligament
2. Medial meniscus
3. Pes anserine tendon insertion
4. Pes anserine bursa
5. Snapping pes anserine tendons

Questions We Want Answered
1. Are the tendons and ligaments normal in appearance?
    (a) If abnormal, does the pathology align with the symptoms?
    (b) If abnormal, is the pathology caused by an inflammatory or degenerative process?
2. Is there evidence of a meniscal tear?
    (a) Is a parameniscal cyst present?
    (b) Is meniscal extrusion present?
3. Is there evidence for pes anserine bursitis?
4. Is there evidence for snapping pes anserine tendons?

The contents do not represent the views of the U.S. Department of Veterans Affairs or the United States Government.

M. H. Greenberg et al., *Manual of Musculoskeletal Ultrasound*,
https://doi.org/10.1007/978-3-031-37416-6_19

Basic Anatomy (Fig. 19.1)

The **medial collateral ligament (MCL),** also known as the tibial collateral ligament, traverses from the **medial femoral epicondyle (MFE)** to the proximal tibia [1]. It resists valgus stress on the knee. The MCL has superficial and deep fibers, also known as anterior and posterior bands. A third, even deeper layer consists of the **meniscofemoral** and **meniscotibial ligaments,** the latter known as the coronary ligament. On ultrasound (US), the superficial layer is hyperechoic, and the deeper layer is hypoechoic. The even deeper meniscofemoral and meniscotibial ligaments are more hyperechoic. This trilaminate structure evokes a sandwich cookie, but with inverse coloration. The deeper layers may be increasingly difficult to visualize on US. Between the superficial and deep bands of the MCL lies the **MCL bursa**.

The **medial meniscus (MM)**, a crescent-shaped fibrocartilage between the femoral condyle and tibial plateau, functions as a shock absorber. The meniscotibial ligaments attach the MM to the tibia. In contrast to the lateral meniscus, the MM is fused to the collateral ligament via meniscofemoral and via the meniscotibial ligaments [1].

Pes Anserine Tendons and Bursa

The **pes anserine tendons (sartorius, gracilis,** and **semitendinosus)** cross superficially at an oblique angle to the distal MCL along their way to the proximal anteromedial tibia. The three **pes anserine tendons (PATs)** from anterior to posterior [2]:

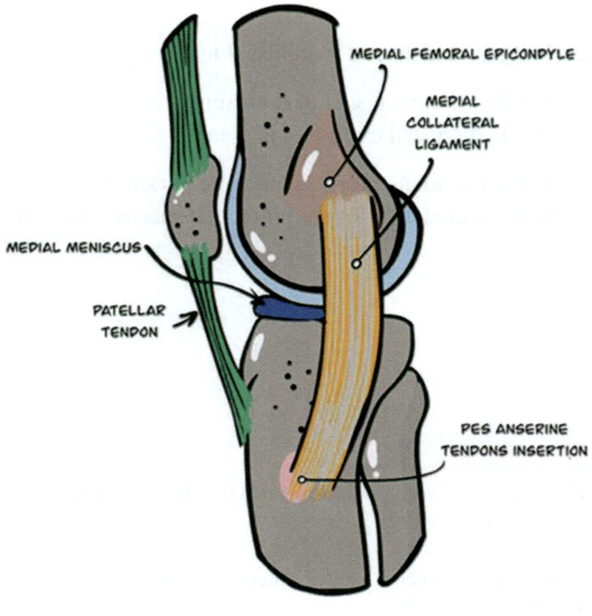

Fig. 19.1 Simplified medial knee anatomy

1. **Sartorius**

   This superficial, flat muscle runs obliquely from the anterior superior iliac spine across the anterior thigh to the medial tibial tubercle. "Sartor" is Latin for tailor; the sartorius muscle is the longest in the body and evokes a tailor's tape measure.
2. **Gracilis**

   This thin muscle runs along the medial thigh, deep to the adductor longus and magnus, from the pubic symphysis to the medial tibial tubercle. "Gracilis" is Latin for thin, slender, and svelte.
3. **Semitendinosus**

   This superficial muscle in the posterior medial thigh runs from the ischial tuberosity to insert on the medial tibial tubercle. The semitendinosus name reminds us that the distal free tendon comprises about 25% of the length of the entire muscle-tendon unit.

The tendons are mnemonically recalled by "**S**ay **G**race before **T**ea" [3]. An alternative mnemonic is "**SGT**" or "sergeant." The pes anserinus (Latin for "goose foot") is the joining of the three tendons that are *superficial* to the MCL and inserted into the anteromedial proximal tibia just *distal* to the insertion of the MCL [3]. The tendons splay out at their *insertion*, suggesting the shape of a goosefoot [4]. Our favorite mnemonic is "**S**illy **G**oose **T**endons," which evokes both the tendon names and insertional configuration (Fig. 19.2).

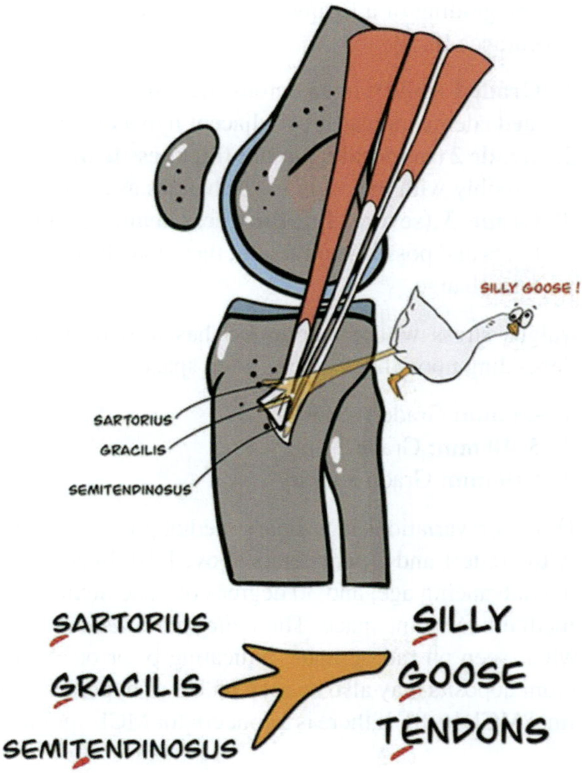

Fig. 19.2 Pes anserine tendons

The **pes anserine (anserinus) bursa (PAB)** is sandwiched between the deeper MCL and the more superficial PATs at the anteromedial border of the tibia [5]. This bursa is undetectable unless it is pathologically distended due to inflammation [4].

Clinical Comments

Medial Collateral Ligament

Valgus stress or twisting knee injuries are the leading cause of MCL injuries, most often occurring in sports participants such as soccer players and skiers [3, 6]. Ultrasound had 94% accuracy in one small series for MCL injury detection [7].

The most common locations for MCL tears are the proximal portion of the superficial MCL and the meniscofemoral ligaments; bony avulsion of the proximal insertion may occur for either layer [6]. Sonographic findings of MCL injury are ligamentous thickening and heterogeneous hypoechogenicity [3, 7]. Ligamentous thickening implies damage and is defined as greater than 6 mm at the femoral attachment or greater than 3.6 mm at the tibial attachment. Injury may also be inferred from loss of the fibular echotexture when the ligament is taut (knee extended) [3].

The grading of a ligamentous sprain may also be suggested by the sonographic appearance [8, 9]:

1. **Grade 1 (mild):** ligamentous stretching without fiber disruption but with associated edema indicated by adjacent hypoechoic or anechoic fluid.
2. **Grade 2 (moderate, partial thickness tear):** abnormal focal hypoechogenicity, possibly with the edematous changes as in Grade 1.
3. **Grade 3 (severe, full-thickness tear):** complete disruption of the ligament fibers and possibly retraction; there may be hemorrhage and surrounding edematous change.

Valgus stress with dynamic US has also been proposed to define injury grade depending upon the degree of joint space widening [3, 7]:

1. **<5 mm:** Grade 1 injury
2. **5–10 mm:** Grade 2 injury
3. **>10 mm:** Grade 3 injury.

However, variations in ordinary medial joint spaces under valgus stress may overlap with grade 1 and 2 parameters above [10]. In addition, be aware that the male gender, advancing age, and 30 degrees of knee flexion are all associated with increased medial knee joint space. The *Pellegrini-Stieda sign* is calcification of the proximal MCL seen on radiographs, indicating prior or chronic MCL injury [3]. Such calcium deposits may also be seen on US as hyperechoic echogenicity within the proximal MCL [6, 8]. If there is a concern for MCL sprain, compare the MCL echotexture and joint space width of the affected knee with the normal contralateral knee. The MCL echotexture should be examined in full knee extension while the joint space is evaluated in 30-degree flexion under valgus stress.

Medial Collateral Ligament Bursitis

The bursa between the superficial and deep MCL layers may become inflamed and produce medial joint line pain on palpation or valgus stress [11]. Lidocaine injections may be both diagnostic and therapeutic. Ultrasound may be used to guide the injection of the bursa [12]. Causes of MCL bursitis include rheumatoid arthritis, genu valgus, direct trauma, pes planus, osteophytes, and activities such as riding a horse or motorcycle [13]. The latter two pursuits generate friction in the medial aspect of the knee.

Medial Meniscus

Ultrasound may detect meniscal abnormalities but, by itself, is insufficient to diagnose meniscal tears [14]. Magnetic resonance imaging (MRI) is the gold standard imaging modality for detecting meniscal pathology [15]. Evaluation of the meniscus by US is limited by a lack of visualization of the entire meniscus, particularly the inner margins [14]. The knee menisci, positioned between the femoral and tibial components, protect knee cartilage from excessive wear and tear [16]. Consequently, meniscal damage may encourage the development of osteoarthritis.

On US, the normal MM is hyperechoic and triangular-shaped [3]. **Meniscal degeneration** appears on US as a heterogeneous echotexture with possible fragmentation and extrusion. **Medial meniscal tears** may arise as focal anechoic or hypoechoic defects extending to the meniscal surface [3]. Using high-frequency transducers (12 or 14 MHz), sensitivity and specificity for sonographic detection of meniscal tears have been reported as 85 and 86% in one study and 88 and 85% in another [15, 17]. Varus stress may enhance visualization of a meniscal tear [18].

A **meniscal (parameniscal) cyst** may present as a palpable, sometimes painful, mass along the knee joint [19]. The cyst on US is anechoic or hypoechoic and may contain septa or debris. Ultrasound is 94% accurate for detecting parameniscal cysts [19]. An underlying meniscal tear is often, but not always, associated with a meniscal cyst. Thus, sonographic identification of a meniscal cyst should heighten suspicion of an underlying meniscal tear. If a meniscal cyst is detected, knee flexion may enhance its prominence [6]. A complex meniscal cyst may mimic a solid mass; however, continuity with the meniscus favors a cyst [18]. Linkage is demonstrated when transducer compression causes the cyst to empty its contents via a tract connected to the meniscus.

Meniscal extrusion (**ME**) is described as the displacement of meniscal tissue beyond the *tibial* margin [20]. Meniscal extrusion, also known as subluxation, is associated with meniscal tears and osteoarthritis [21, 22]. MRI is considered the gold standard for the assessment of ME [21–23]. Meniscal extrusion is detectable by US; the degree of ME is affected by age, body weight, knee alignment (varus deformity for medial ME), cartilage damage, and the presence of osteoarthritis [21, 24].

When compared with MRI results, US had excellent sensitivity (95–96%) and reasonable specificity (70–82%) for the detection of ME [22]. A study of primarily *medial* menisci suggests that MRI-delineated ME greater than *2 mm* often indicates an underlying meniscal tear confirmed by arthroscopy [25]. Medial meniscal extrusion (MME) may displace the MCL. One interesting study describes that 61% of patients with knee pain and radiographic osteoarthritis had MME coupled with MCL displacement [24]. This held true for patients with milder degrees of radiographic joint space narrowing. This suggests but does not prove that MME and MCL displacement may be pain generators in some patients with knee osteoarthritis. It is unclear if the MME contributes to radiographic medial joint space narrowing or vice-versa.

Pes Anserine Tendons and Bursitis

Pes anserine bursitis is associated with obesity, osteoarthritis, MM tear, direct knee trauma, overuse, obesity, valgus knee deformity, pes planus, and type 2 diabetes [4, 26–28]. Symptoms of PAB include local pain, swelling, and tenderness at the anteromedial knee, approximately 5–7 cm below the joint line [4, 29]. The pain is exacerbated by arising from a seated position or using stairs.

Bursitis appears as anechoic fluid deep to the pes anserine tendons with thickening and decreased echogenicity of the tendons [3]. The distended oblong bursa is sandwiched between the PAT and the deeper MCL [4]. Ultrasound-guided injection of the PAB is more accurate than blind injection [30, 31]. It is important to remember that PAB is only one reason for proximal tibial anterior and anteromedial knee pain. Other causes, depending on the clinical setting, include MM tear, proximal medial tibial fracture or osteonecrosis, saphenous nerve compression or damage, and snapping pes anserine tendons [32–36].

Snapping Pes Anserine Tendons

Snapping pes anserine tendons (gracilis or semitendinosus) over the medial femur or tibia during active knee flexion or extension may cause extra-articular knee snapping and pain [35, 36]. The snapping may be audible and visible in the medial distal hamstring area [35]. This condition has been described in athletes or associated with trauma, overuse, tumors, and postoperative changes. Structural etiologies of snapping pes anserine tendons include bony protuberance, unstable pes tendons due to deficient accessory bands or ligamentous laxity following trauma, and chronic overload of the anterior knee [37]. Dynamic US of the involved area of the posterior medial knee is diagnostic [35, 37]. Ultrasound-guided corticosteroid and anesthetic injections may assist in diagnosis [35].

Infrapatellar Branch of the Saphenous Nerve Damage

The infrapatellar branch of the saphenous nerve (IPBSN) is a terminal sensory branch that innervates the anteromedial knee. Injury may occur from surgical incision or retraction, sports injury, direct trauma such as in a motor vehicle accident, or entrapment neuropathy between the sartorius tendon and the MFE [32]. Symptoms include acute or chronic anterior knee pain, stiffness, dysesthesia, numbness, abnormal gait, swelling, and warmth; the latter two symptoms mimic complex regional pain syndrome [32, 34]. Pain may be poorly localized and exacerbated by knee flexion [34]. A positive Tinel's sign over the involved nerve may be diagnostically helpful [32].

High-resolution US (15–22 MHz) correctly identifies the IPBSN in 86–100% and demonstrates hypoechoic neural thickening described as a neuroma [33]. Ultrasound may visualize the causative condition, such as fibrous scar tissue or surgical fixation [32]. Diagnostic sensory nerve conduction studies have also been suggested [34]. Treatment options include ultrasound-guided injection, radiofrequency ablation, or neuroma excision [32, 34].

Pitfalls

1. To avoid mistaking a meniscal cyst for a solid mass, apply probe pressure to establish a connection with the meniscus since the cyst may empty some of its contents via a tract to the meniscus.
2. The Pellegrini-Stieda sign is radiographically or sonographically detected MCL calcification indicating prior or chronic MCL injury.
3. If there is a concern for **MCL sprain**, sonographically compare the MCL echotexture and joint space width of the affected and the normal contralateral knee. Due to variations in *normal* medial joint spaces under valgus stress, there may be an overlap of grade 1 and 2 parameters. In addition, the male gender and advanced age are associated with increased medial knee joint space width.
4. The MCL echotexture should be examined in full knee extension, while the joint space is best evaluated in 30-degree flexion and under valgus stress.
5. Ultrasound may detect abnormalities in the MM but is insufficient as a standalone imaging technique to diagnose meniscal tears. MRI is the gold standard.

6. When evaluating sources of knee pain, consider that MME or MCL displacement may be pain generators in some patients with knee osteoarthritis.
7. It is unclear if the MME contributes to radiographic medial joint space narrowing or vice versa.
8. Ultrasound-guided injection of the PAB is more accurate and effective.
9. Do not misinterpret the insertion of the semimembranosus tendon in its bony groove (sulcus) on the tibia for a parameniscal cyst. Tilting the probe to counteract anisotropy will reveal the hyperechoic tendon.
10. Pes anserine bursitis is only one cause of proximal anteromedial knee pain. Depending on the clinical setting, other reasons include MM tear, proximal medial tibial fracture or osteonecrosis, saphenous nerve compression or damage, and snapping pes anserine tendons.

Method

The patient is recumbent with the knee flexed to about 20–30 degrees using a bolster or rolled-up towel. The hip should be comfortably externally rotated, allowing full access to the medial knee. With a ballpoint pen, mark off the palpable protruding MFE, which is about 3 cm superior to the joint line. A trick to verify the location of the MFE is to slowly scan the medial femur in transverse axis (TAX) with a TAX slide until you see a hyperechoic triangular protrusion, which is the MFE [38]. Also, mark off the joint line by palpation. The evaluation of the medial patellar retinaculum and the medial patellofemoral ligament is described in Chap. 16.

Protocol Image 1: Medial Collateral Ligament, Longitudinal + dynamic (Fig. 19.3)

The MCL stretches from the MFE to the medial tibia in a slightly oblique path. Align the probe with the femur, placing the proximal end on the MFE. The probe is in the coronal plane of the femur. Rotate the distal transducer end posteriorly toward the examination table about 20–30 degrees until you see the MCL, a hyperechoic band overlying the V-shaped joint space. The long axis of the probe should lie somewhere between the femoral and tibial long axes. Note that the bony tibial portion of the knee joint is slightly more superficial than the femoral section.

To best delineate the MCL, tilt the probe back and forth, performing slight TAX slides anteriorly and posteriorly to look at the margins of the MCL; anisotropy will demarcate it from other soft tissue structures. Evaluate the layers of the MCL. Please see Basic Anatomy above for more details. The deeper layers are challenging to visualize. Between the superficial and deep bands of the MCL, the MCL bursa may be visible. Perform longitudinal axis (LAX) slides to examine the proximal

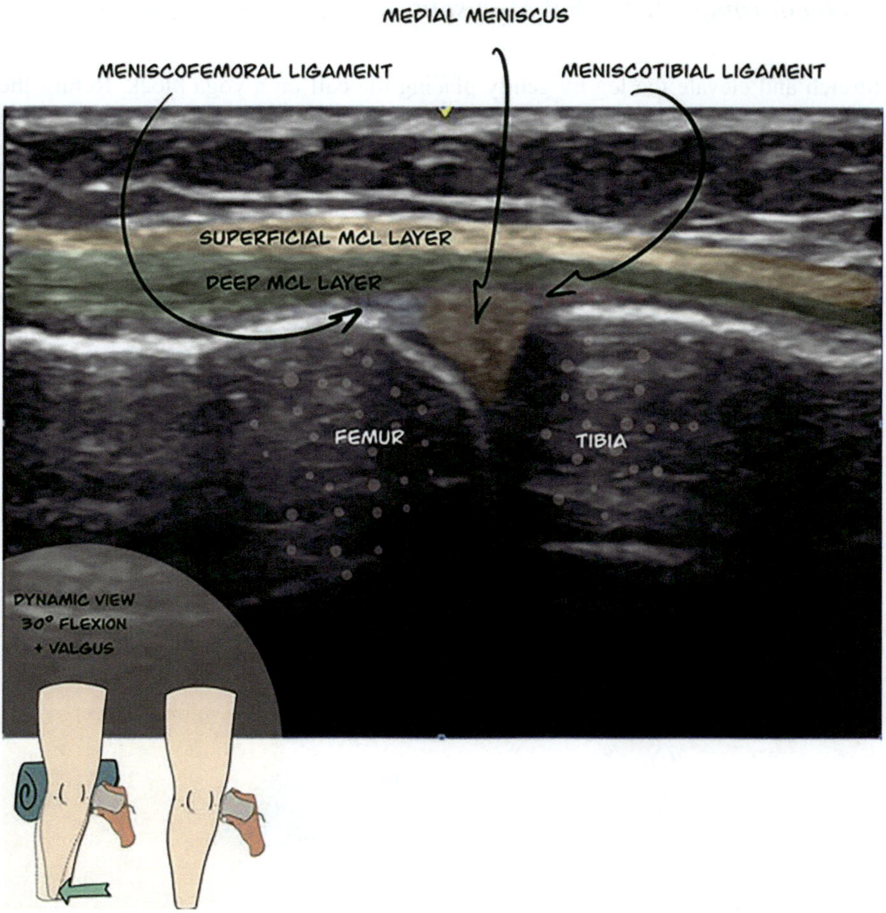

Fig. 19.3 Protocol Image 1: Medial collateral ligament, longitudinal + dynamic

insertion of the MCL on the MFE and the distal insertion on the tibia about 5 cm distal to the joint line. If anisotropy is significant, the knee can be placed into full extension to reduce artifacts by increasing MCL tautness.

To examine the MCL's integrity, apply valgus stress to a knee at 20–30-degree flexion by pushing the foreleg in a lateral direction; a damaged MCL may permit joint space widening [3, 7, 24]. In a fully extended knee, look for MME against the MCL, possibly displacing the MCL. If necessary, rotate the transducer 90 degrees to visualize the MCL in TAX view. Identify the MCL in cross-section by tilting the transducer since the anisotropy of the ligament will help define the margins of this band-like oblong structure. The TAX view may verify any damage suspected from the LAX images. The entire length of the MCL, from the femur to the tibia, should be carefully scanned to look for pathology that may have been overlooked on the LAX scan.

Protocol Image 2: Medial Meniscus, longitudinal (Fig. 19.4)

Stretch and elevate the leg by gently placing the calf on a yoga block. Return the probe to the coronal position in Protocol Image 1, that is, the LAX view of the MCL. Focus on the triangular-shaped hyperechoic MM. Tilting the probe one way

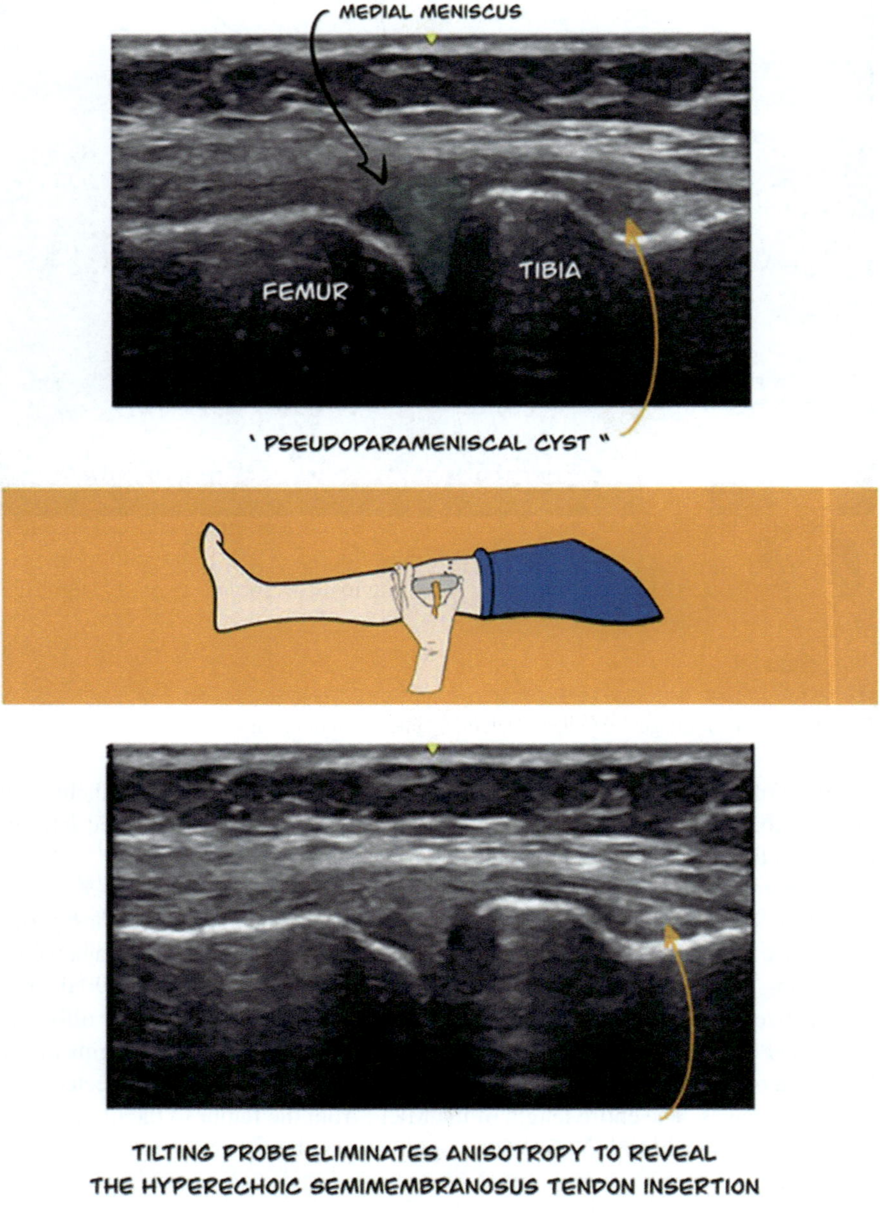

Fig. 19.4 Protocol Image 2: Medial meniscus, longitudinal

and then the other will aid visualization. Perform a slow TAX slide posteriorly to see the posterior horn of the MM, tilting the probe back and forth to sharpen the view at each stage. Look for meniscal degeneration, tears, cysts, or extrusion. Do not misinterpret the insertion of the semimembranosus tendon in its groove at the tibia as a parameniscal cyst. Tilting the probe to counteract anisotropy will reveal the hyperechoic tendon and avoid this pitfall.

Protocol Image 3 (Optional): Medial Meniscal Extrusion Measurement, Longitudinal

Look for MME, which is the protrusion of the MM beyond the tibial bone. If present, also note the displacement of the MCL. To quantitate MM extrusion, draw a line across the external edge of the medial tibial plateau. Measure the height of any protrusion of the meniscus superficial to this line [22]. See Clinical Comments above for more information.

Protocol Image 4: Pes Anserine Tendons, transverse (Fig. 19.5)

Perform a distal LAX slide to center the probe on the distal MCL insertion on the medial tibia. This is about 5–6 cm distal to the joint line. Rotate the distal probe 10–20 degrees medially toward the anterior tibia to look for the PATs in TAX. Just superficial to the MCL insertion, look for three subtle hyperechoic "hills": the pes anserine tendons (sartorius, gracilis, and semitendinosus) in transverse view, on their way to their insertion points slightly distal and anterior to the MCL insertion, the tibia. This may require tilting the probe in both directions to use anisotropy to delineate these tendons in cross-section.

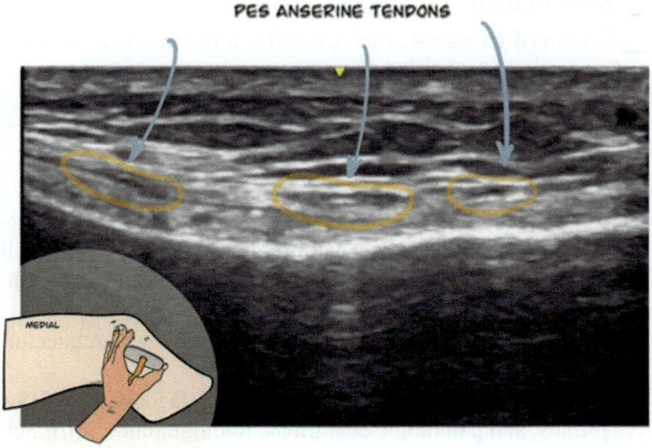

Fig. 19.5 Protocol Image 4: Pes anserine tendons, transverse

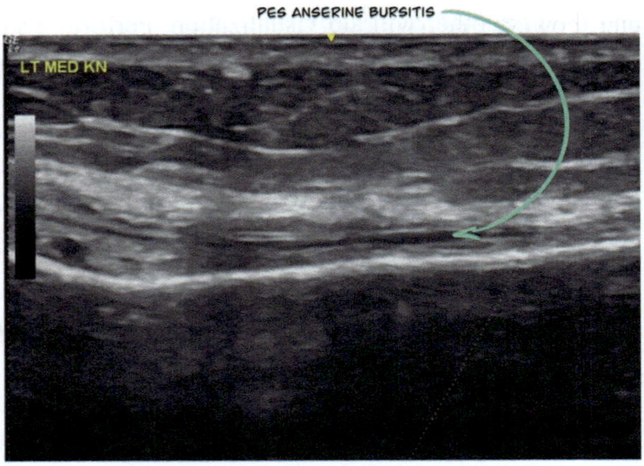

Fig. 19.6 Pes anserine bursitis

The PAB, sandwiched between the MCL and the PATs, is typically detected only when filled with anechoic fluid, such as when bursitis is present (Fig. 19.6). Slight TAX slides in either direction may help to detect the PAB, which will appear as an oblong anechoic fluid collection. Verify the PAB in an orthogonal view. The inferior medial genicular artery may be seen by using the PD; it is deep to the MCL insertion, resting on the tibial cortex. If an injection is contemplated, rotate the transducer to best define the PAB and survey the area with PD to avoid damaging the genicular artery.

Turn the probe 90-degrees if needed for an orthogonal view of the pes anserine tendons. The tendon complex may appear as a single entity in this position; moving the probe in a TAX slide posteriorly may help to demarcate individual tendons as they are followed proximally [3].

Complete Medial Knee Ultrasonic Examination Checklist

☐ Protocol Image 1: Medial collateral ligament, longitudinal, + dynamic
☐ Protocol Image 2: Medial meniscus, longitudinal
☐ Protocol Image 3 (optional): Medial meniscal extrusion measurement, longitudinal
☐ Protocol Image 4: Pes anserine tendons, transverse

References

1. Juneja P, Hubbard JB. Anatomy, bony pelvis and lower limb, knee medial collateral ligament. StatPearls. Treasure Island, FL: StatPearls Publishing. Copyright © 2022, StatPearls Publishing LLC.; 2022.
2. Woodley SJ, Mercer SR. Hamstring muscles: architecture and innervation. Cells Tissues Organs. 2005;179(3):125–41.
3. Alves TI, Girish G, Kalume Brigido M, Jacobson JA. US of the knee: scanning techniques, pitfalls, and pathologic conditions. Radiographics. 2016;36(6):1759–75.

4. Gupta A, Saraf A, Yadav C. High-resolution ultrasonography in pesanserinus bursitis: case report and literature review. Scholars J Appl Med Sci. 2013;1:753–7.
5. Toktas H, Dundar U, Adar S, Solak O, Ulasli AM. Ultrasonographic assessment of pes anserinus tendon and pes anserinus tendinitis bursitis syndrome in patients with knee osteoarthritis. Mod Rheumatol. 2015;25(1):128–33.
6. Bianchi E, Martinoli C. Knee. Ultrasound of the musculoskeletal system. New York: Springer; 2007. p. 637–744.
7. Lee JI, Song IS, Jung YB, Kim YG, Wang CH, Yu H, et al. Medial collateral ligament injuries of the knee: ultrasonographic findings. J Ultrasound Med. 1996;15(9):621–5.
8. Jacobson J. Knee ultrasound fundamentals of musculoskeletal ultrasound. Philadelphia, PA: Elsevier; 2018. p. 284–327.
9. Ghosh N, Kruse D, Subeh M, Lahham S, Fox JC. Comparing point-of-care-ultrasound (POCUS) to MRI for the diagnosis of medial compartment knee injuries. J Med Ultrasound. 2017;25(3):167–72.
10. Lutz PM, Feucht MJ, Wechselberger J, Rasper M, Petersen W, Wörtler K, et al. Ultrasound-based examination of the medial ligament complex shows gender- and age-related differences in laxity. Knee Surg Sports Traumatol Arthrosc. 2021;29(6):1960–7.
11. Kerlan RK, Glousman RE. Tibial collateral ligament bursitis. Am J Sports Med. 1988;16(4):344–6.
12. Jose J, Schallert E, Lesniak B. Sonographically guided therapeutic injection for primary medial (tibial) collateral bursitis. J Ultrasound Med. 2011;30(2):257–61.
13. De Maeseneer M, Shahabpour M, Van Roy F, Goossens A, De Ridder F, Clarijs J, et al. MR imaging of the medial collateral ligament bursa: findings in patients and anatomic data derived from cadavers. AJR Am J Roentgenol. 2001;177(4):911–7.
14. Azzoni R, Cabitza P. Is there a role for sonography in the diagnosis of tears of the knee menisci? J Clin Ultrasound. 2002;30(8):472–6.
15. Wareluk P, Szopinski KT. Value of modern sonography in the assessment of meniscal lesions. Eur J Radiol. 2012;81(9):2366–9.
16. Winkler PW, Csapo R, Wierer G, Hepperger C, Heinzle B, Imhoff AB, et al. Sonographic evaluation of lateral meniscal extrusion: implementation and validation. Arch Orthop Trauma Surg. 2021;141(2):271–81.
17. Akatsu Y, Yamaguchi S, Mukoyama S, Morikawa T, Yamaguchi T, Tsuchiya K, et al. Accuracy of high-resolution ultrasound in the detection of meniscal tears and determination of the visible area of menisci. J Bone Joint Surg Am. 2015;97(10):799–806.
18. Lee MJ, Chow K. Ultrasound of the knee. Semin Musculoskelet Radiol. 2007;11(2):137–48.
19. Rutten MJ, Collins JM, van Kampen A, Jager GJ. Meniscal cysts: detection with high-resolution sonography. AJR Am J Roentgenol. 1998;171(2):491–6.
20. Masouros SD, McDermott ID, Amis AA, Bull AM. Biomechanics of the meniscus-meniscal ligament construct of the knee. Knee Surg Sports Traumatol Arthrosc. 2008;16(12):1121–32.
21. Crema MD, Roemer FW, Felson DT, Englund M, Wang K, Jarraya M, et al. Factors associated with meniscal extrusion in knees with or at risk for osteoarthritis: the multicenter osteoarthritis study. Radiology. 2012;264(2):494–503.
22. Nogueira-Barbosa MH, Gregio-Junior E, Lorenzato MM, Guermazi A, Roemer FW, Chagas-Neto FA, et al. Ultrasound assessment of medial meniscal extrusion: a validation study using MRI as reference standard. AJR Am J Roentgenol. 2015;204(3):584–8.
23. Podlipská J, Koski JM, Kaukinen P, Haapea M, Tervonen O, Arokoski JP, et al. Structure-symptom relationship with wide-area ultrasound scanning of knee osteoarthritis. Sci Rep. 2017;7:44470.
24. Naredo E, Cabero F, Palop MJ, Collado P, Cruz A, Crespo M. Ultrasonographic findings in knee osteoarthritis: a comparative study with clinical and radiographic assessment. Osteoarthr Cartil. 2005;13(7):568–74.
25. Muzaffar N, Kirmani O, Ahsan M, Ahmad S. Meniscal extrusion in the knee: should only 3 mm extrusion be considered significant? An assessment by MRI and arthroscopy. Malays Orthop J. 2015;9(2):17–20.

26. Kang I, Han SW. Anserine bursitis in patients with osteoarthritis of the knee. South Med J. 2000;93(2):207–9.
27. Alvarez-Nemegyei J, Canoso JJ. Evidence-based soft tissue rheumatology IV: anserine bursitis. J Clin Rheumatol. 2004;10(4):205–6.
28. Cohen SE, Mahul O, Meir R, Rubinow A. Anserine bursitis and non-insulin dependent diabetes mellitus. J Rheumatol. 1997;24(11):2162–5.
29. Handy JR. Anserine bursitis: a brief review. South Med J. 1997;90(4):376–7.
30. Lee JH, Lee JU, Yoo SW. Accuracy and efficacy of ultrasound-guided pes anserinus bursa injection. J Clin Ultrasound. 2019;47(2):77–82.
31. Finnoff JT, Nutz DJ, Henning PT, Hollman JH, Smith J. Accuracy of ultrasound-guided versus unguided pes anserinus bursa injections. PM R. 2010;2(8):732–9.
32. Boyle J, Eason A, Hartnett N, Marks P. Infrapatellar branch of the saphenous nerve: a review. J Med Imaging Radiat Oncol. 2021;65(2):195–200.
33. Riegler G, Jengojan S, Mayer JA, Pivec C, Platzgummer H, Brugger PC, et al. Ultrasound anatomic demonstration of the infrapatellar nerve branches. Arthroscopy. 2018;34(10):2874–83.
34. Trescot AM, Brown MN, Karl HW. Infrapatellar saphenous neuralgia - diagnosis and treatment. Pain Physician. 2013;16(3):E315–24.
35. Shapiro SA, Hernandez LO, Montero DP. Snapping pes Anserinus and the diagnostic utility of dynamic ultrasound. J Clin Imaging Sci. 2017;7:39.
36. Akagawa M, Kimura Y, Saito H, Kijima H, Saito K, Segawa T, et al. Snapping pes syndrome caused by the gracilis tendon: successful selective surgery with specific diagnosis by ultrasonography. Case Rep Orthop. 2020;2020:1783813.
37. Oh HS, Kim TE. Snapping pes Anserinus caused by gracilis tendon: a new mechanism proposed by dynamic knee ultrasonography. J Korean Soc Radiol. 2019;80(5):969–74.
38. De Maeseneer M, Vanderdood K, Marcelis S, Shabana W, Osteaux M. Sonography of the medial and lateral tendons and ligaments of the knee: the use of bony landmarks as an easy method for identification. AJR Am J Roentgenol. 2002;178(6):1437–44.

Chapter 20
Anterior Hip

Reasons to Do the Study
1. Anterior hip and groin pain
2. Anterior thigh pain
3. Paresthesia of the lateral thigh

Questions We Want Answered
1. Is the anterior hip pain referred from the lumbosacral spine or pelvis?
2. Is the anterior hip pain coming from the femur, the hip joint, or the labrum?
3. If the hip pain is from the joint, is it due to an inflammatory or structural problem?
4. Is the cause of the pain greater trochanteric pain syndrome? If so, what is the specific underlying pathology?
5. Is anterolateral thigh paresthesia due to meralgia paresthetica?

Basic Anatomy

Bones (Fig. 20.1)

Bony landmarks and protuberances pertinent to US examination are depicted. Note that the femoral neck angles inward about 30–45° from the axis of the femur.

Labrum (Fig. 20.2)

The fibrocartilaginous labrum along the acetabulum's circumference augments the contact area with the femoral head (FH). On ultrasound (US), the labrum appears as a hypo- or hyperechoic triangular-shaped structure between the acetabulum and FH.

The contents do not represent the views of the U.S. Department of Veterans Affairs or the United States Government.

M. H. Greenberg et al., *Manual of Musculoskeletal Ultrasound*,
https://doi.org/10.1007/978-3-031-37416-6_20

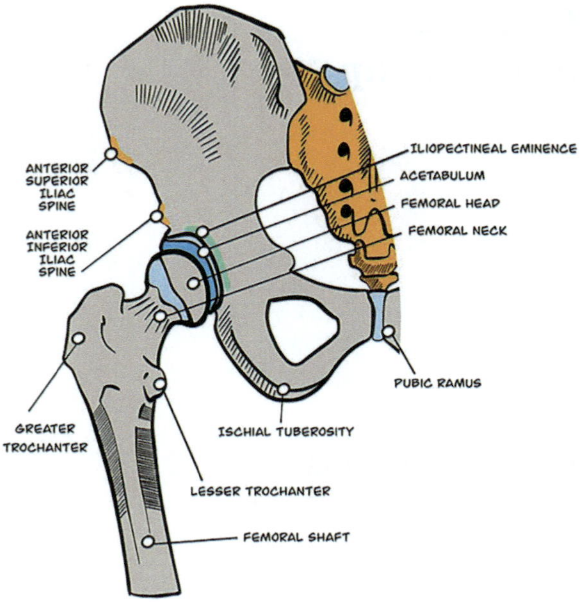

Fig. 20.1 Bony anatomy of the hip

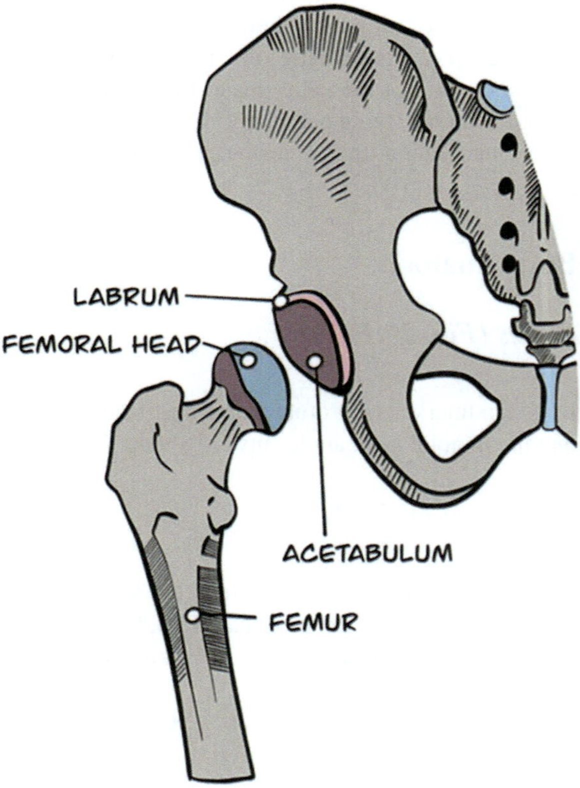

Fig. 20.2 Hip labrum

Ligaments

The three ligaments reinforcing the hip joint all end with "femoral "and start with the pelvic origin: *ischiofemoral, pubofemoral, and iliofemoral*. Another ligament, the *inguinal ligament* (IL), runs diagonally from the anterior superior iliac spine (ASIS) to the pubic tubercle (Fig. 20.3). Structures deep to the IL, from lateral to

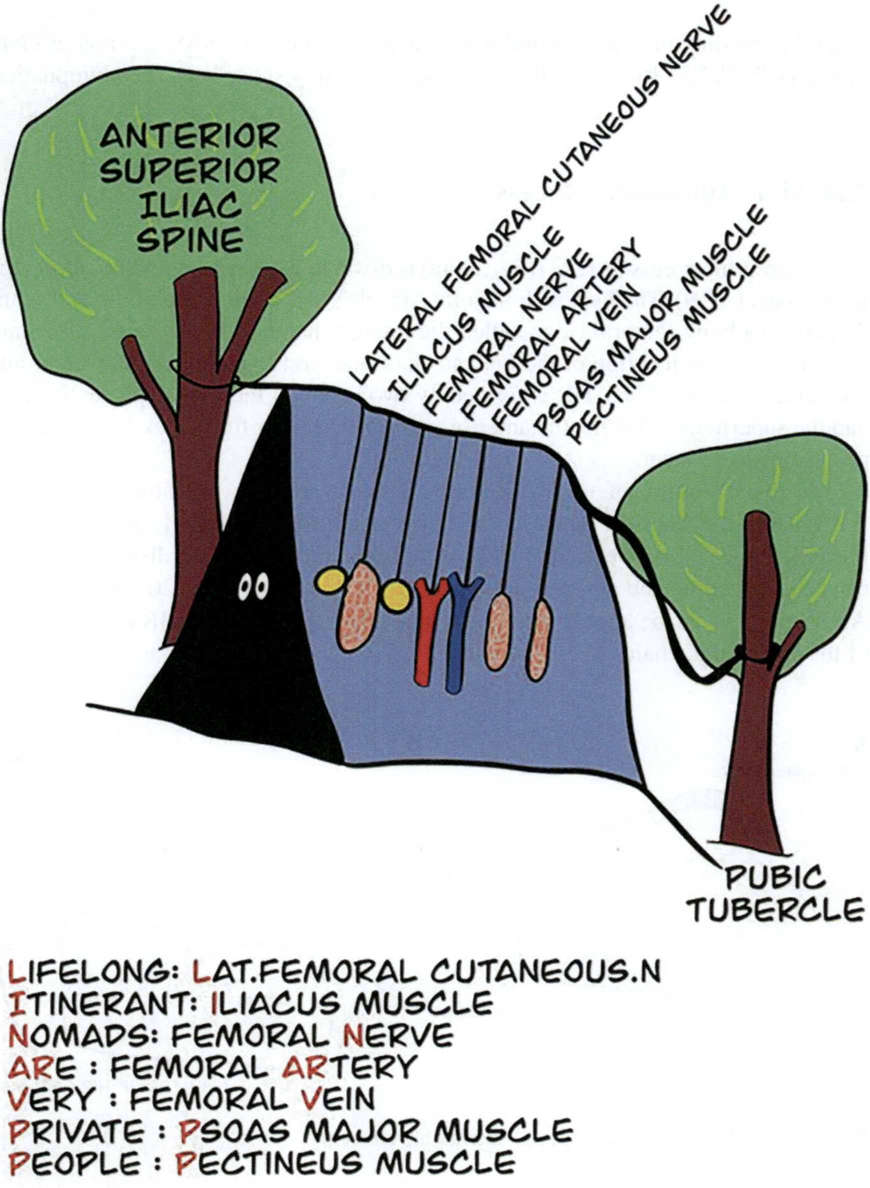

Fig. 20.3 Inguinal ligament + neurovascular bundle

medial, are the lateral femoral cutaneous nerve (LFCN), the iliacus muscle, the femoral nerve/artery/vein, the psoas major, and the pectineus muscle. A mnemonic for these structures deep to the IL is "Lifelong Itinerant Nomads Are Very Private People."

Neurovascular Bundle

Lateral to medial, find the femoral nerve, artery, or vein (or "NAV"). Another mnemonic is "NAVEL" by adding the more medial "**e**mpty space" and then **l**ymphatics.

Anterior Hip (Joint) Recess

The anterior hip recess (AHR) (Fig. 20.4a) is distal to the hip joint and lies along the femoral neck (FN). The hyperechoic joint capsule starts at the acetabulum and spans distally to a bony femoral ridge called the intertrochanteric line. The intertrochanteric line is at the junction of the FN and shaft and stretches between the lesser and greater trochanters. The AHR is formed by two layers of the joint capsule: the deep and the superficial. Distally, the anterior (superficial) layer folds back to become the posterior (deep) layer.

The distal portion of the iliofemoral ligament (IFL) is just superficial to the AHR. Thus, the AHR is "sandwiched" between the FN and the IFL. An effusion, if present, is located between the two layers of the joint recess and distends the joint capsule, spreading the anterior and posterior layers apart (Fig. 20.4b). A normal AHR on US gives the appearance of a snake, while a distended AHR is reminiscent of the shape of a whale.

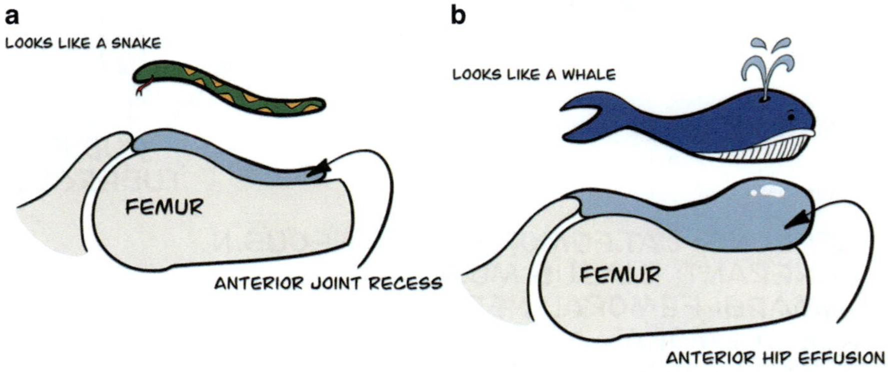

Fig. 20.4 (**a**) Anterior joint recess. (**b**) Anterior hip effusion

Hip Joint

The hip joint, or femoroacetabular joint (FAJ), is comprised of the femoral head (FH) (the ball) and the acetabulum (socket). The labrum lines the edge of the acetabulum. Superficial to the FH are the iliopsoas tendon (IPT) and muscle (IPM).

Iliopsoas Muscle, Tendon, and Bursa

The IPM-IPT unit, formed from the merger of the psoas major muscle and iliacus, crosses the FAJ and the FH to insert medially on the lesser trochanter (LT) to function as a hip flexor. The iliopsoas bursa (IPB) is deep to the IPM and medial to the IBT. The IPB is usually invisible when not distended and may typically contain some synovial fluid in about 15% of normals due to communication with the FAJ [1].

Tensor Fascia Lata and Sartorius

The ASIS is the origin of the tensor fascia lata (TFL) and the sartorius. The sartorius, the longest muscle in the body, crosses the anterior thigh to insert posteromedially at the knee. The TFL contributes anterior fibers to the iliotibial band, thus stabilizing the knee with walking and running.

(Please see Chap. 22 for more information.)

Rectus Femoris

The rectus femoris (RF) is one of the four quadriceps muscles. It has two heads: the *direct head* originates from the anterior inferior iliac spine (AIIS); the *indirect head* stems from the iliac bone at the lateral edge of the acetabulum. The RF flexes the hip and extends the knee. At the juncture of the indirect and direct heads, the interdigitation of the tendon fibers creates anisotropy, which can mimic an RF tendon tear.

Clinical Comments

A dynamic US examination of the hip may be extraordinarily helpful [2].

Joint Effusion and Synovitis

The AHR at the FN is the best location to visualize an effusion and synovitis [2]. When a hip effusion is present, synovial fluid often pools in the AHR (Fig. 20.4b). The AHR allows access to the FAJ and is thus commonly aspirated and injected. In the longitudinal axis (LAX) view on US, measure the distance from the joint capsule to the FN; an effusion is suspected if the space is >7 mm or if there is a >1 mm difference between the two hips [3]. Often, no fluid is normally present [4]. However, internal rotation of the leg may cause bulging of the normal joint capsule, falsely mimicking an effusion [2]. The transverse axis (TAX) view should confirm an effusion detected in LAX.

The AHR may appear hypoechoic or hyperechoic, depending on anisotropy and patient body habitus. Recess widening may be due to a joint effusion or synovial hypertrophy [2]. Distention of the AHR by an effusion may be anechoic, isoechoic, or hypoechoic, depending on the composition of synovial fluid. Doppler activity might indicate synovial hypertrophy but is not always diagnostic. However, Doppler activity may also occur with an infection. If an infection is a consideration, aspiration is *always* required to confirm an effusion and establish its nature [2]. If initial aspiration yields no fluid, lavage of the AHR and repeat aspiration may increase the yield for infected synovial fluid by 25% [5]. Synovial hypertrophy in the AHR may be due to inflammatory polyarthropathy, pigmented villonodular synovitis, synovial chondromatosis, and amyloid deposition [2, 6]. The sonographic finding of AHR distension by itself is not diagnostic of hip effusion in native and prosthetic hips; aspiration is necessary for confirmation [7].

Ultrasound can also detect a noncommunicating soft-tissue fluid collection that might be an adjacent abscess. Never aspirate the anterior recess through such a fluid collection since this might seed the joint/prosthesis with bacteria [7].

Iliopsoas Bursitis

The IPB is usually located medially to the iliopsoas tendon and deep to the muscle. Together, this iliopsoas complex passes over the ilium [2]. The nondistended IPB is usually undetectable [8]. Fifteen percent of normal people will have communication between the IPB and the FAJ, although this percentage increases after arthroplasty and with inflammatory arthritis [9]. With IPB distension, attempt to verify communication between the IPB and the FAJ since IPB distention may be due to a chronic hip problem causing excess synovial fluid. This communication may be seen in the TAX view, medial to the FH near the psoas major tendon [2].

Distention of the IPB may be due to synovial fluid or synovitis, possibly with positive power Doppler (PD). Absent direct hip joint pathology, IPB distension from synovitis may be due to an underlying inflammatory arthropathy such as gout or rheumatoid arthritis, infection, trauma, or overuse [10–12]. However, the IPB may be distended asymptomatically. Concomitant iliopsoas bursitis and tendinitis often occur due to proximity; this is described as *iliopsoas syndrome* [10].

Be aware that IPB enlargement may typically occur after hip arthroplasty; however, aspiration is necessary if an infection is suspected [13]. Infection is suggested by transducer-induced pain or PD activity. A distended IPB may extend anteriorly, anterolaterally, and even proximally into the abdomen, mimicking a psoas abscess [2].

Labrum Abnormalities

The anterior labrum is the most common site for labral tears and is the portion of the hip labrum best visualized by US [14]. Labral hypo- or heteroechogenicity suggests degeneration, but a well-defined anechoic cleft indicates a labral tear [15]. Ultrasound is 82% sensitive but only 60% specific for labral tear detection, making the MR arthrogram the premier modality for delineating labral pathology [15].

Femoral acetabular impingement (FAI) with hip flexion and internal rotation may cause a labrum tear [16]. This is typically a cam-type impingement whose presence may be inferred from plain radiographs. A bony protuberance or irregularity of the anterior superior femoral head-neck junction may impinge on the hyaline cartilage and labrum, causing a tear [16]. Impingement on the labrum may be dynamically visualized by US; however, US is unreliable as a screening tool for FAI [17]. A paralabral cyst, contiguous to the labrum, also implies a labral tear [18].

Postsurgical Hip

To delineate pain after total hip replacement, an US examination is recommended as the next step after radiographs [19]. Causes of hip pain after arthroplasty include infection, iliopsoas bursitis or impingement, prosthesis loosening, arthroplasty particle disease, trochanteric bursitis (possibly septic), abductor tendon insufficiency, incision site infection, hematoma, seroma, and heterotopic ossification [19]. The prosthetic cup, FH, and femoral shaft may all exhibit reverberation artifacts, whereas normal bone demonstrates acoustic shadowing [20]. A hypoechogenic area that is superficial to the FN is a common finding after hip replacement and should not be misinterpreted as pathologic [21].

Since the native joint capsule may have been resected with surgery, a poorly demarcated pseudocapsule may form over the prosthesis neck. The pseudocapsule may distend, but the degree of distention does not accurately predict the presence or absence of an effusion [7]. The suggested threshold for pseudocapsule enlargement indicative of a septic prosthesis varies from 3.2 mm to 10 mm [13]. Detection of effusion after hip arthroplasty may be highly challenging, and a lack of AHR or pseudocapsule distension on US does not rule out an effusion [7, 19]. An effusion may be loculated, so when you suspect a fluid collection, also evaluate the soft tissue superficial to the lateral and posterior FN [2].

In addition to possible infection, joint effusion may be aseptic, such as in particle disease and joint loosening [19]. Particle disease is a granulomatous reaction to polyethylene or polymethylmethacrylate debris and may cause periprosthetic osteolysis and a soft tissue mass called a pseudotumor [19]. Also, consider the iliopsoas tendon and bursal impingement directly upon the anterior femoral component or acetabular cup as a sources of pain.

Additionally, evaluate the gluteal tendons for abnormalities, particularly if the arthroplasty was inserted via the anterolateral approach since the abductor tendons are manipulated with this approach [19]. Iliotibial band abnormalities may also be due to the lateral surgical approach or friction over the GT [19]. Ultrasound may detect tendinosis or tears. Heterotopic bone, occurring in approximately 6% of total hip arthroplasties, may cause pain or decreased range of motion in the hip replacement [19, 22]. Ultrasound will detect heterotopic bone as hyperechoic ossifications with posterior acoustic shadowing [19].

Tendon and Muscle Abnormalities

The RF may experience a tear at the origin of the direct or indirect heads [23]. However, tears of the RF most commonly occur in the muscle belly, or myotendinous junction, and less often at the tendon origin itself. The RF crosses two joints and is the most frequently injured quadriceps muscle or tendon [24]. The direct and indirect heads of the RF, the IPT, and the sartorius origin may exhibit sonographic signs of tendinosis or a frank tendon tear. Tears of any of these muscles may also be sonographically detectable, appearing as a loss of normal muscle echotexture, perhaps associated with a hypoechoic hematoma [24]. Later, an organized hematoma may have a hyperechoic appearance [24].

Snapping Hip Syndrome

Extra-Articular Medial

Just medial to the AIIS are the lateral and medial iliacus muscles. The medial iliacus muscle normally interposes between the psoas major tendon (PMT) and the iliopectineal eminence when the leg is in a frog-leg or FABER position (hip flexion, abduction, and external rotation). In some individuals, this muscle becomes transiently entrapped so that straightening the leg abruptly releases the ensnared muscle laterally, causing the PMT to forcibly snap against the iliopectineal eminence [2, 25, 26].

This is the frog-leg test for medial extra-articular snapping hip syndrome. However, such entrapment and release may be asymptomatic.

Remember that even though some sources describe this as the IPT snapping, the *PMT* does the snapping since the iliopsoas muscle and tendon have not yet formed in this location. However, the *IPT* may impinge and snap upon a protruding acetabular component of a total hip arthroplasty [27].

Extra-Articular Lateral

The ITB or the gluteus maximus snaps anteriorly over the GT with hip flexion.

(Please see Chap. 22 for more information.)

Intra-Articular: Femoroacetabular Impingement

This, the least common form of snapping hip, is due to hip capsule injuries such as loose bodies in the synovial folds, a torn acetabular labrum, synovial chondromatosis, or recurrent hip dislocation/subluxation [28].

See Labrum Abnormalities, above.

Calcific Tendinosis

Calcific tendinosis, primarily calcium hydroxyapatite deposition, may also occur in several hip tendons, including the RF, gluteus maximus, and gluteus medius tendons. Ultrasound reveals hyperechoic foci, perhaps with posterior acoustic shadowing and increased Doppler flow [29]. This may be a source of pain.

Diabetic Muscle Infarction

Diabetic muscle infarction may occur in the calf and, less commonly, in the thigh. The cause is suspected to be vascular occlusive disease in the setting of longstanding diabetes [30]. A differential diagnosis includes soft tissue infection. Muscle infarcts are well-marginated, hypoechoic intramuscular lesions with internal linear structures felt to be muscle fibers. In contrast to an abscess, a muscle infarct lacks fluid swirling with transducer pressure [30].

Meralgia Paresthetica

Meralgia paresthetica (MP) may occur when the LFCN is compressed as it courses deep to the IL, just medial to the ASIS (Fig. 20.5) [31]. Symptoms of LFCN impingement include paresthesia, numbness, burning, dysesthesia, and pain in an oval shape along the long axis of the anterolateral thigh [31]. Symptoms may worsen with prolonged walking and standing. Known causes of MP are leg length discrepancy, weight gain, tight belts or clothing, prolonged standing or hyperextension of the leg or trunk, trauma, and pelvic tumors [31]. The diagnosis is often made on clinical grounds since the electrophysiologic study of this nerve is challenging. Ultrasound may demonstrate a hypoechoic, swollen LFCN near the lateral aspect of the inguinal ligament, particularly when compared to the contralateral, unaffected side [31]. Sonopalpation over the affected nerve may reproduce symptoms. Treatment is conservative but may require injection under the ligament near the nerve; US guidance markedly improves injection efficacy [32].

Inguinal Lymph Nodes

Inguinal lymph nodes are a common finding. Normal lymph nodes sonographically have an oval appearance with a hypoechoic cortex and a hyperechoic hilum [2]. On short-axis, asymptomatic inguinal lymph node measurements range from 2.1 to

Fig. 20.5 Meralgia paresthetica mnemonic

13.6 mm [33]. Be aware that average-size inguinal lymph nodes may still be malignant, and enlarged reactive lymph nodes may be benign [2]. Sonographic features that imply possible malignancy include enlargement with a round or asymmetric shape, absence of normal echogenic hilum, nonuniform cortical thickness, and a peripheral blood flow pattern [2, 34]. A biopsy is often required if uncertainty exists, depending on the clinical setting.

Pitfalls

1. When imaging the AHR, position the transducer perpendicular to the recess since anisotropy may cause the hyperechoic recess layers to appear hypoechoic, thus mimicking an effusion [2].
2. Be aware that internal hip rotation may produce AHR bulging, simulating distention [2].
3. The sonographic delineation of a hip effusion is *inaccurate* for both native and prosthetic hips. Aspiration is mandatory to confirm an effusion and determine its nature [7].
4. Never aspirate the AHR through a fluid collection since the latter might be infected and seed the joint or prosthesis with bacteria [7].
5. Sonographic evaluation of hip arthroplasty has several potential pitfalls:
   (a) An effusion may be loculated; thus, US evaluation for an effusion over the lateral and posterior FN should also be performed [2].
   (b) Hip arthroplasty makes it difficult to detect a small joint effusion [7].
   (c) Consider the complications of hip arthroplasty, such as particle disease and pseudotumor formation [35].
6. For medial extra-articular hip snapping, the psoas major tendon (not the iliopsoas tendon) snaps against the iliopectineal eminence.
7. A small size of an inguinal lymph node is not assurance that it is benign; likewise, a large nodal size does not mean malignancy.
8. A hypoechoic area in the direct head of the RF may be due to anisotropy from the interdigitation of the indirect head tendon fibers, thereby mimicking a tendon tear.

Method

The patient is recumbent with the foot in a slight external rotation of about 15°. Palpate and mark the bony prominences of the AIIS and the ASIS. Most patients require a curvilinear probe with the lower frequency; however, a linear probe may be used in thin patients. When a linear probe is used, lower the frequency and use the virtual convex setting to widen the field of view.

Protocol Image 1: Femoral Head and Neck, Anterior Hip (Joint) Recess, Longitudinal (Fig. 20.6)

Place the probe in LAX orientation, aligned with the long axis of the *femoral shaft*, *proximal* to the *FAJ*. Perform a proximal LAX slide to visualize the *femoral neck* and *head*. Toggle and rock the probe to sharpen the hyperechoic FN and FH image, which resembles a staff with a hooked end. The goal is to image the AHR crisply; this requires aiming the cephalad end of the probe toward the umbilicus and varying the transducer angle anywhere between 10 and 45°. Perform a further proximal LAX slide to see part of the FAJ. Performing tiny TAX slides in either direction will help to delineate the AHR better. Look at the curved FH, the smaller curved hyperechoic acetabulum, and the slightly hyperechoic/hypoechoic anterior acetabular labrum between these two bones.

Just superficial to the labrum is the IFL, and just superficial to that is the IPT, which extends distally. Tendinosis of the IPT may be represented by hypoechogenicity, swelling, and loss of fibrillar echotexture. The IPB lies medially to the IPT and remains invisible when not distended.

Look for the hyperechoic joint capsule, which starts at the acetabulum and spans the FN distally to the bony femoral ridge, the intertrochanteric line. The hypoechoic AHR is "sandwiched" between the FN bony cortex and the IFL. The AHR is comprised of two layers within the joint capsule: the deep and the superficial. An effusion, if present, is located between the two layers of the AHR and may distend the capsule. Measure the distance from the joint capsule to the FN; an effusion is

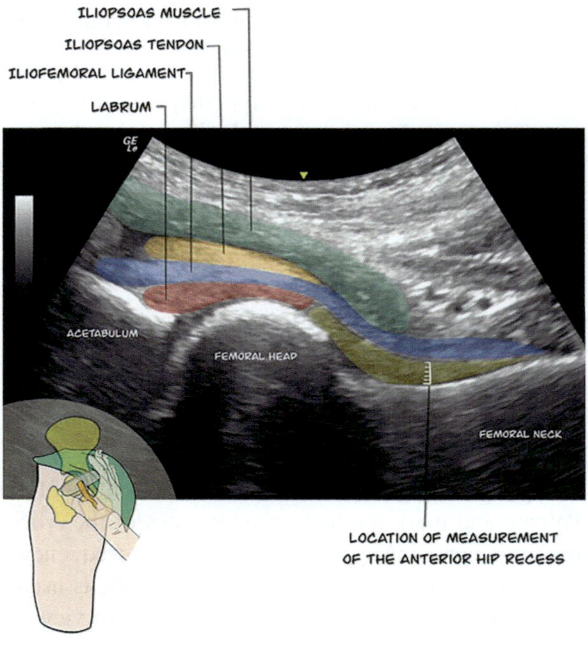

Fig. 20.6 Protocol Image 1: Femoral head and neck, anterior hip (joint) recess, longitudinal

diagnosed if the space is >7 mm or if there is a >1 mm difference in this measurement between the two hips [3].

Perform a slight TAX slide in each direction to evaluate the full extent of the AHR. A small amount of joint fluid may be physiologic. A pitfall is that if the foot is internally rotated, the AHR may bulge, mimicking an effusion. Another pitfall is misinterpreting an effusion as normal postoperative changes after hip arthroplasty.

Protocol Image 2: Femoral Head, Iliopsoas, Transverse (Fig. 20.7)

Next, center the image over the FH, and then rotate the probe to a horizontal position. In the example image of the right anterior hip, the finned end of the transducer is on the left side (lateral). This view evaluates the iliopsoas muscle and tendon in cross-section as it crosses the medial aspect of the FAJ. Focus on the hyperechoic FH, which should have a thin covering of anechoic hyaline cartilage. The IPT is a hyperechoic structure that is just deep to the IPM. Superficial to the IPT, you *may* see the cross-section of the iliopsoas bursa, *if distended*, containing anechoic or hypoechoic synovial fluid or synovitis. In addition to the iliopsoas muscle, the sartorius and the rectus femoris muscles can also be viewed in cross-section. This image verifies the putative pathology seen in the LAX view.

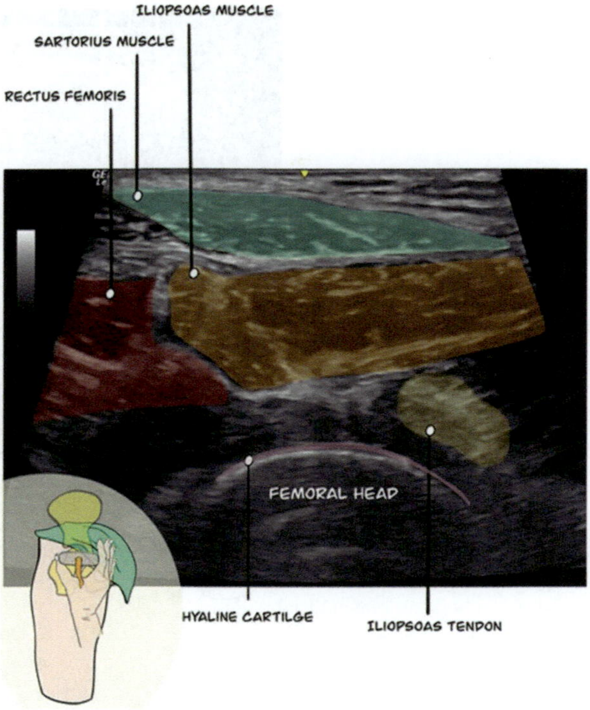

Fig. 20.7 Protocol Image 2: Femoral head, iliopsoas, transverse

Protocol Image 3: Anterior Labrum, Longitudinal (Fig. 20.8)

Rotate the transducer approximately 90° to a near-vertical position, with the finned end being more cephalad (proximal). Angle the finned end medially by about 10°. If the probe is over the FAJ, the bony hyperechoic acetabulum will be on the left side of the screen and the FH on the right side. Just superficial to the FH is the anechoic joint cartilage, and just superficial to that is the hypoechoic labrum, which attaches to the acetabulum. From deep to shallow, the IFL, the IPT, and the iliopsoas muscle are three concentric, longitudinal layers.

Focus on the labrum, the slightly hyper- or hypoechoic tissue between the acetabulum and the humeral head, and perform TAX slides medially and laterally to evaluate the anterior labrum for tears, cysts, and calcium deposition. Also, examine the hyaline cartilage covering the FH. The acetabular labrum may be challenging to visualize in many normal patients. Switching to a linear probe for a thin patient may enhance visualization.

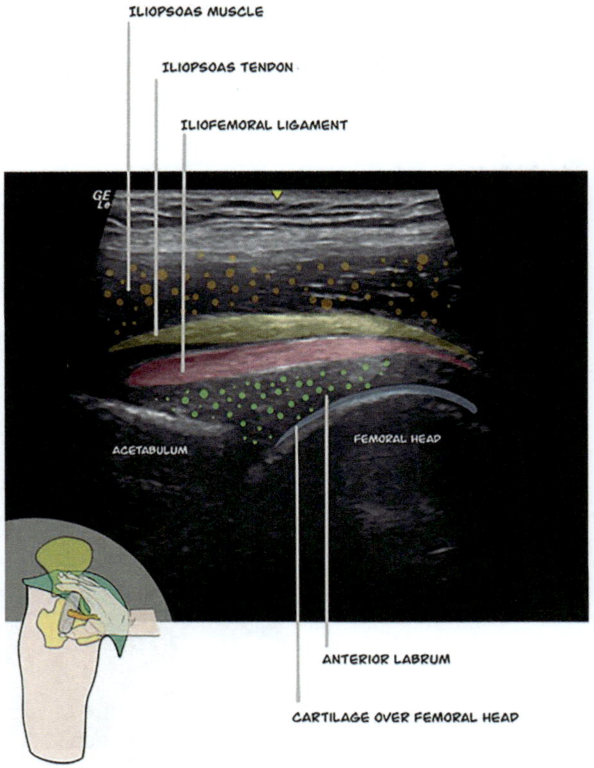

Fig. 20.8 Protocol Image 3: Anterior labrum, longitudinal

Protocol Image 4: Neurovascular Structures, Transverse ± Doppler (Fig. 20.9)

Next, go back to the transducer position in **Protocol Image 2** and perform a medial LAX slide to see the neurovascular structures. Turn on Color Flow Doppler to aid identification of the femoral artery and vein. Alternatively, transducer pressure will collapse the vein, thus distinguishing it from the artery. Angling the medial end of the transducer approximately 20° in a caudal direction may enhance visualization. Lateral to medial, find the femoral nerve, artery, or vein or "NAV."

Protocol Image 5: Anterior Inferior Iliac Spine, Transverse (Fig. 20.10)

The AIIS is the "home base" bony protuberance that will lead to the next few images. Again, go back to the transducer position for Protocol Image 2. Find the AIIS by performing a proximal (cephalad) TAX slide until you see the hyperechoic

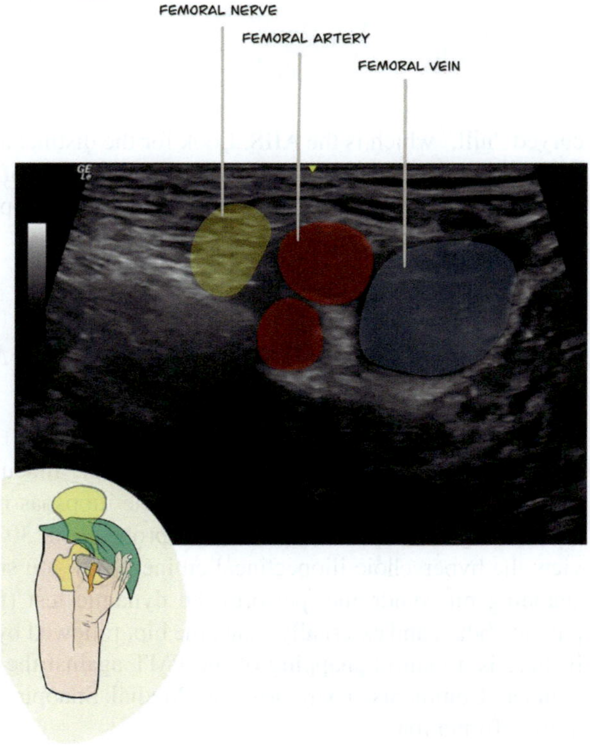

Fig. 20.9 Protocol Image 4: Femoral Neurovascular structures, transverse ± Doppler

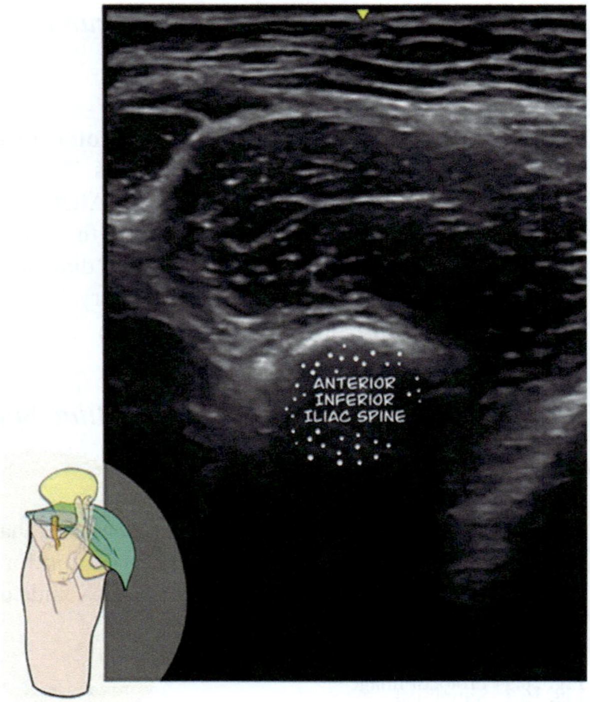

Fig. 20.10 Protocol Image 5: Anterior inferior iliac spine, transverse

curved "hill," which is the AIIS. Look for the distinct anechoic acoustic shadowing deep to the AIIS. The AIIS shape is also reminiscent of a comet with a dark tail. Do not confuse the AIIs with the ASIS, which is also a hyperechoic curved "hill" but is more cephalad and lateral to the AIIS.

Protocol Image 6: Iliacus and Psoas Major Muscles, Transverse ± Dynamic (Fig. 20.11)

Next, perform a medial LAX slide to view the iliacus and psoas major muscles. These muscles will more distally form the iliopsoas muscle and tendon. You will need to angle the medial aspect of the probe about 30–45° in a caudal direction to view the hyperechoic iliopectineal eminence. If you suspect medial extra-articular snapping hip syndrome, perform the dynamic test (frog-leg test) by having the patient abduct and externally rotate the hip, followed by straightening the leg, to see if there is a painful snapping of the PMT against the iliopectineal eminence. See Clinical Comments, Extra-articular Medial Snapping Hip Syndrome, above for more information.

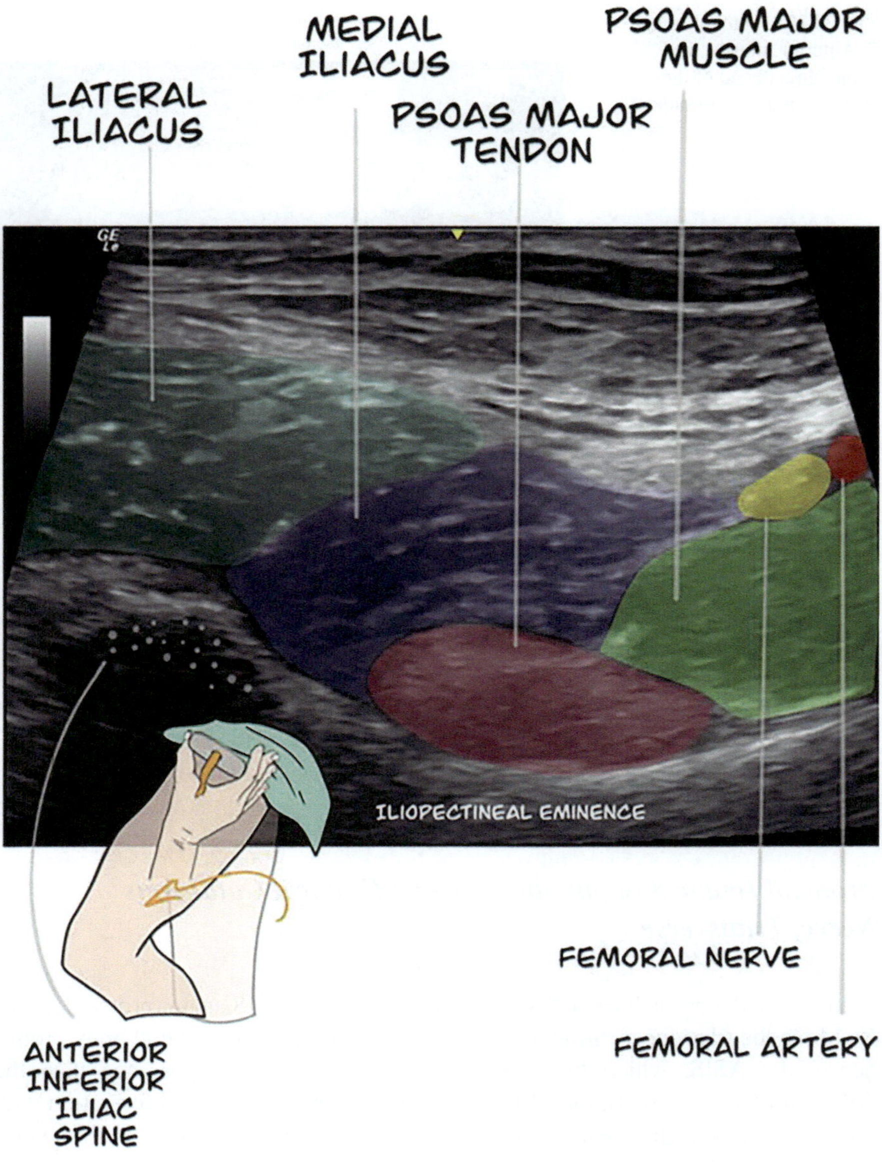

Fig. 20.11 Protocol Image 6: Iliacus and psoas major muscles, transverse ± dynamic

Protocol Image 7: Anterior Inferior Iliac Spine, Direct Head of the Rectus Femoris, Longitudinal (Fig. 20.12)

Next, return to the view from Protocol Image 5 (AIIS TAX). Center the transducer on the AIIS and slowly rotate it 90° to see the origin of the direct head of the RF in

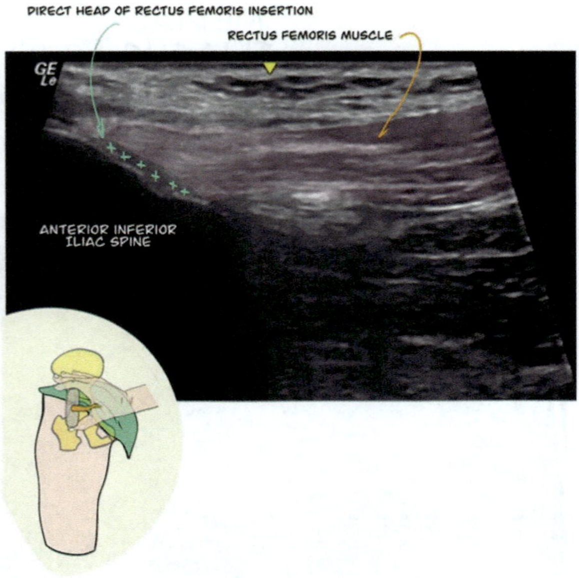

Fig. 20.12 Protocol Image 7: Anterior inferior iliac spine, direct head of the rectus femoris, longitudinal

LAX at the AIIS. Within the hyperechoic tendon, there may be a hypoechoic area, which is the union of the direct and indirect heads of the RF. This does *not* represent tendinopathy or a tendon tear. The origin of the indirect head is the lateral aspect of the superior acetabular ridge. Look for tears of the RF tendon, although most RF tears occur in the muscle belly, myotendinous junction, and tendon origin [23].

Protocol Image 8 (Optional): Lateral Femoral Cutaneous Nerve, Transverse

Next, rotate the probe back to TAX and place it over the AIIS again (protocol image 5). Move the transducer further cephalad to see a slightly more lateral bony prominence, the ASIS, which has a similar but more robust appearance than the AIIS. Similar to the AIIS, the ASIS is a broad hill or comet-shaped formation with posterior acoustic shadowing causing an anechoic appearance deep to the hyperechoic curved bony cortex. The ASIS is the origin of the TFL (laterally) and the sartorius (medially).

The ASIS is also the origin of the IL and is a landmark for the nearby LFCN. From the TAX position, rotate the medial probe end downward about 30–45°, aiming for the bony pubic tubercle. This highlights the lateral aspect of the hyperechoic linear IL and the LFCN. To locate the LFCN sonographically, look for ovoid hypo- or hyperechoic fascicles of the LFCN, which in most people are just deep to the junction of the IL and next to the ASIS. A linear transducer is beneficial in this more superficial location. Entrapment of the LFCN at the attachment of the IL to the ASIS causes proximal nerve swelling and the symptoms of meralgia paresthetica.

The normal LFCN may be challenging to visualize. The image example compares a normal with an abnormal LFCN in a person who suffers from meralgia paresthetica (Fig. 20.13a, b).

Helpful hints [31]:

(a) Use a high-frequency linear probe since the nerve is relatively small.
(b) Perform small TAX slides in cephalad and caudal directions to visualize the nerve, which sometimes passes superficially to, deep to, or even through the IL. Sometimes it is visualized as two fascicles that bifurcate distally.
(c) Compare the LFCN cross-sectional area to the contralateral, presumably unaffected side.
(d) Look for swelling and hypoechogenicity proximal to the compression area, similar to other nerves in tight fibro-osseous canals, such as the median nerve in the carpal tunnel.
(e) Sonopalpation may reproduce the symptoms of meralgia paresthetica.

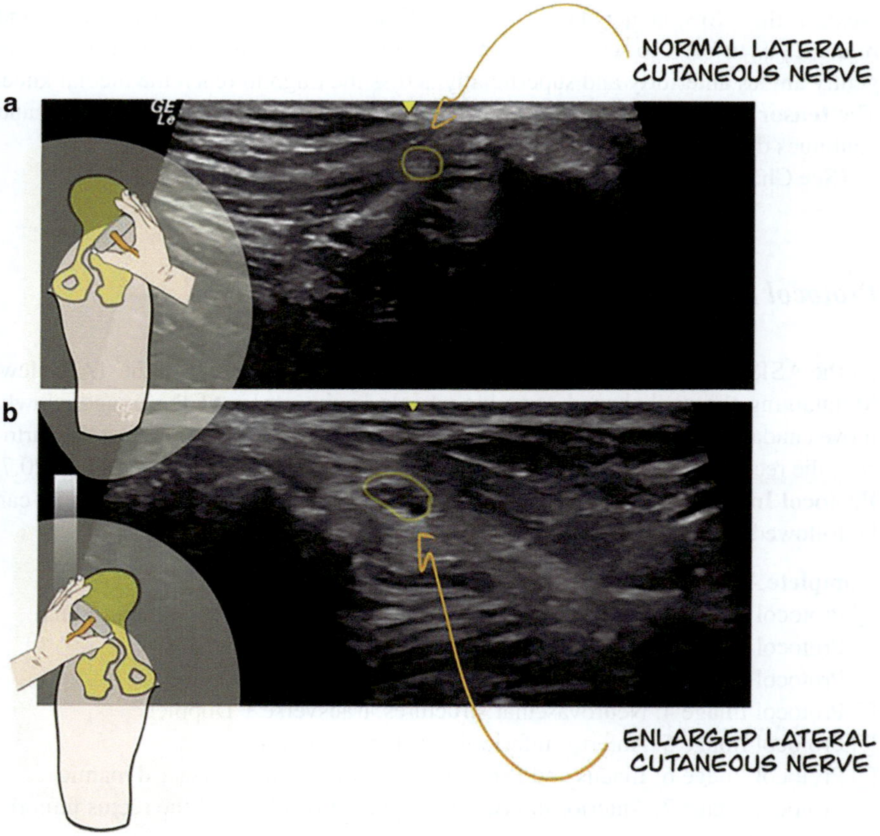

Fig. 20.13 Lateral femoral cutaneous nerve, transverse (**a**) Normal nerve on the left side (**b**) Enlarged, abnormal nerve on the right side

Protocol Image 9 (Optional): Rectus Femoris, Vastus Muscles, Transverse

Return to the AIIS in TAX (Image 5), perform a TAX slide distally toward the knee, and see the vastus intermedius deep to the RF. Perform medial and lateral LAX slides to view these muscles fully. More distally, the vastus medialis (VM) and lateralis (VL) muscles join the RF and the deeper vastus intermedius (VI) to form the quadriceps tendon. These are the four muscles, or quadriceps, of the anterior thigh. Look for tears, contusions, edema, hematomas, and myositis ossificans. Turn the probe 90° for an orthogonal view (LAX) to verify any pathologic findings.

Protocol Image 10 (Optional): Sartorius and Tensor Fascia Latae Origins, Longitudinal

Next, at the ASIS, center the probe in TAX at the *medial* slope of the ASIS and rotate it 90° to LAX to see the tendon origin of the **sartorius.** The sartorius muscle gently angles anteriorly and superficially across the thigh to reach the medial knee. The **tensor fascia latae** (TFL) originates from the *lateral* border of the ASIS and continues distally to form the anterior fibers of the iliotibial band (ITB).

(See Chap. 22 for further details and images.)

Protocol Image 11 (Optional): Sartorius, Transverse

At the ASIS, center the probe on the sartorius and then rotate it to the TAX view. Maintaining the oval-shaped sartorius muscle in the center of the image, slowly move caudally in a TAX slide to identify these three contiguous muscles: the sartorius, the rectus femoris, and the iliopsoas. Distal to the AIIS, the image in Fig. 20.7, **Protocol Image 2,** emerges. From this view, if necessary, the sartorius muscle can be followed distally in TAX for evaluation.

Complete Anterior Hip Ultrasonic Examination Checklist

☐ Protocol Image 1: Femoral head and neck, anterior joint recess, longitudinal
☐ Protocol Image 2: Femoral head, iliopsoas, transverse
☐ Protocol Image 3: Anterior labrum, longitudinal
☐ Protocol Image 4: Neurovascular structures, transverse ± Doppler
☐ Protocol Image 5: Anterior inferior iliac spine, transverse
☐ Protocol Image 6: Iliacus and Psoas Major muscles, transverse ± dynamic
☐ Protocol Image 7: Anterior inferior iliac spine, direct head of the rectus femoris, longitudinal
☐ Protocol Image 8 (optional): Lateral femoral cutaneous nerve, transverse

☐ Protocol Image 9 (optional): Rectus femoris, vastus muscles, transverse
☐ Protocol Image 10 (optional): Sartorius and tensor fascia latae origins, longitudinal
☐ Protocol Image 11 (optional): Sartorius, transverse

References

1. Nguyen MS, Kheyfits V, Giordano BD, Dieudonne G, Monu JU. Hip anatomic variants that may mimic pathologic entities on MRI: nonlabral variants. AJR Am J Roentgenol. 2013;201(3):W401–8.
2. Jacobson JA, Khoury V, Brandon CJ. Ultrasound of the groin: techniques, pathology, and pitfalls. AJR Am J Roentgenol. 2015;205(3):513–23.
3. Koski JM, Anttila PJ, Isomäki HA. Ultrasonography of the adult hip joint. Scand J Rheumatol. 1989;18(2):113–7.
4. Koski JM. Ultrasound detection of plantar bursitis of the forefoot in patients with early rheumatoid arthritis. J Rheumatol. 1998;25(2):229–30.
5. Kung JW, Yablon C, Huang ES, Hennessey H, Wu JS. Clinical and radiologic predictive factors of septic hip arthritis. AJR Am J Roentgenol. 2012;199(4):868–72.
6. Pai VR, van Holsbeeck M. Synovial osteochondromatosis of the hip: role of sonography. J Clin Ultrasound. 1995;23(3):199–203.
7. Weybright PN, Jacobson JA, Murry KH, Lin J, Fessell DP, Jamadar DA, et al. Limited effectiveness of sonography in revealing hip joint effusion: preliminary results in 21 adult patients with native and postoperative hips. AJR Am J Roentgenol. 2003;181(1):215–8.
8. Lin Y-T, Wang T-G. Ultrasonographic examination of the adult hip. J Med Ultrasound. 2012;20:201–9.
9. Wunderbaldinger P, Bremer C, Schellenberger E, Cejna M, Turetschek K, Kainberger F. Imaging features of iliopsoas bursitis. Eur Radiol. 2002;12(2):409–15.
10. Corvino A, Venetucci P, Caruso M, Tarulli FR, Carpiniello M, Pane F, et al. Iliopsoas bursitis: the role of diagnostic imaging in detection, differential diagnosis and treatment. Radiol Case Rep. 2020;15(11):2149–52.
11. Toohey AK, LaSalle TL, Martinez S, Polisson RP. Iliopsoas bursitis: clinical features, radiographic findings, and disease associations. Semin Arthritis Rheum. 1990;20(1):41–7.
12. Johnston CA, Wiley JP, Lindsay DM, Wiseman DA. Iliopsoas bursitis and tendinitis. A review. Sports Med. 1998;25(4):271–83.
13. Douis H, Dunlop DJ, Pearson AM, O'Hara JN, James SL. The role of ultrasound in the assessment of post-operative complications following hip arthroplasty. Skelet Radiol. 2012;41(9):1035–46.
14. Blankenbaker DG, De Smet AA, Keene JS, Fine JP. Classification and localization of acetabular labral tears. Skelet Radiol. 2007;36(5):391–7.
15. Jin W, Kim KI, Rhyu KH, Park SY, Kim HC, Yang DM, et al. Sonographic evaluation of anterosuperior hip labral tears with magnetic resonance arthrographic and surgical correlation. J Ultrasound Med. 2012;31(3):439–47.
16. Bedi A, Kelly BT. Femoroacetabular impingement. J Bone Joint Surg Am. 2013;95(1):82–92.
17. Buck FM, Hodler J, Zanetti M, Dora C, Pfirrmann CW. Ultrasound for the evaluation of femoroacetabular impingement of the cam type. Diagnostic performance of qualitative criteria and alpha angle measurements. Eur Radiol. 2011;21(1):167–75.
18. Mervak BM, Morag Y, Marcantonio D, Jacobson J, Brandon C, Fessell D. Paralabral cysts of the hip: sonographic evaluation with magnetic resonance arthrographic correlation. J Ultrasound Med. 2012;31(3):495–500.

19. Long SS, Surrey D, Nazarian LN. Common sonographic findings in the painful hip after hip arthroplasty. J Ultrasound Med. 2012;31(2):301–12.
20. van Holsbeeck MT, Eyler WR, Sherman LS, Lombardi TJ, Mezger E, Verner JJ, et al. Detection of infection in loosened hip prostheses: efficacy of sonography. AJR Am J Roentgenol. 1994;163(2):381–4.
21. Hoefnagels EM, Obradov M, Reijnierse M, Anderson PG, Swierstra BA. Sonography after total hip replacement: reproducibility and normal values in 47 clinically uncomplicated cases. Acta Orthop. 2007;78(1):81–5.
22. Board TN, Karva A, Board RE, Gambhir AK, Porter ML. The prophylaxis and treatment of heterotopic ossification following lower limb arthroplasty. J Bone Joint Surg Br. 2007;89(4):434–40.
23. Martinoli C, Garello I, Marchetti A, Palmieri F, Altafini L, Valle M, et al. Hip ultrasound. Eur J Radiol. 2012;81(12):3824–31.
24. Minagawa H, Wong KH. Musculoskeletal ultrasound: Echo anatomy & scan technique: Amazon digital services; 2017.
25. Guillin R, Cardinal E, Bureau NJ. Sonographic anatomy and dynamic study of the normal iliopsoas musculotendinous junction. Eur Radiol. 2009;19(4):995–1001.
26. Deslandes M, Guillin R, Cardinal E, Hobden R, Bureau NJ. The snapping iliopsoas tendon: new mechanisms using dynamic sonography. AJR Am J Roentgenol. 2008;190(3):576–81.
27. Bureau NJ. Sonographic evaluation of snapping hip syndrome. J Ultrasound Med. 2013;32(6):895–900.
28. Idjadi J, Meislin R. Symptomatic snapping hip: targeted treatment for maximum pain relief. Phys Sportsmed. 2004;32(1):25–31.
29. Lee HS, Lee YH, Sung NK, Jung KJ, Park YC, Kim HK, et al. Sonographic findings of calcific tendinitis around the hip. Ultrasonography. 2005;24(3):139–44.
30. Delaney-Sathy LO, Fessell DP, Jacobson JA, Hayes CW. Sonography of diabetic muscle infarction with MR imaging, CT, and pathologic correlation. AJR Am J Roentgenol. 2000;174(1):165–9.
31. Onat SS, Ata AM, Ozcakar L. Ultrasound-guided diagnosis and treatment of meralgia paresthetica. Pain Physician. 2016;19(4):E667–9.
32. Tagliafico A, Serafini G, Lacelli F, Perrone N, Valsania V, Martinoli C. Ultrasound-guided treatment of meralgia paresthetica (lateral femoral cutaneous neuropathy): technical description and results of treatment in 20 consecutive patients. J Ultrasound Med. 2011;30(10):1341–6.
33. Bontumasi N, Jacobson JA, Caoili E, Brandon C, Kim SM, Jamadar D. Inguinal lymph nodes: size, number, and other characteristics in asymptomatic patients by CT. Surg Radiol Anat. 2014;36(10):1051–5.
34. Esen G. Ultrasound of superficial lymph nodes. Eur J Radiol. 2006;58(3):345–59.
35. Ostlere S. How to image metal-on-metal prostheses and their complications. AJR Am J Roentgenol. 2011;197(3):558–67.

Chapter 21
Posterior Hip

Reasons to Do the Study

1. Posterior hip/thigh pain
2. Hip dysfunction
3. Evaluation for piriformis syndrome
4. Guidance for sacroiliac injections

Questions We Want Answered

1. Is the cause of the pain ischial bursitis or a torn hamstring muscle?
2. Is sciatica-type pain due to piriformis syndrome?

Anatomy

Posterior Pelvic Bones (Fig. 21.1)

Deep Posterior Pelvic Muscles (Fig. 21.2a)

The external (lateral) rotator muscles, from superior to inferior, are the **P**iriformis, **S**uperior gemellus, **O**bturator internus, **I**nferior gemellus, and **Q**uadratus femoris [1].

A mnemonic for the external rotator muscles is "**P**ear **S**its **O**n **I**ndigo **Q**uilt" (Fig. 21.2b).

The contents do not represent the views of the U.S. Department of Veterans Affairs or the United States Government.

M. H. Greenberg et al., *Manual of Musculoskeletal Ultrasound*,
https://doi.org/10.1007/978-3-031-37416-6_21

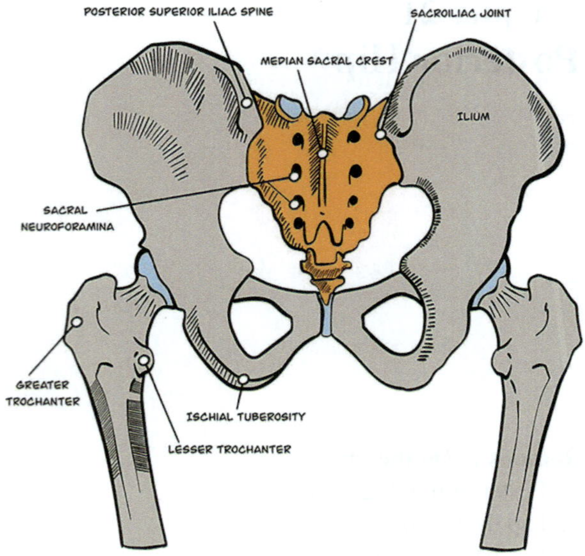

Fig. 21.1 Posterior hip bony anatomy

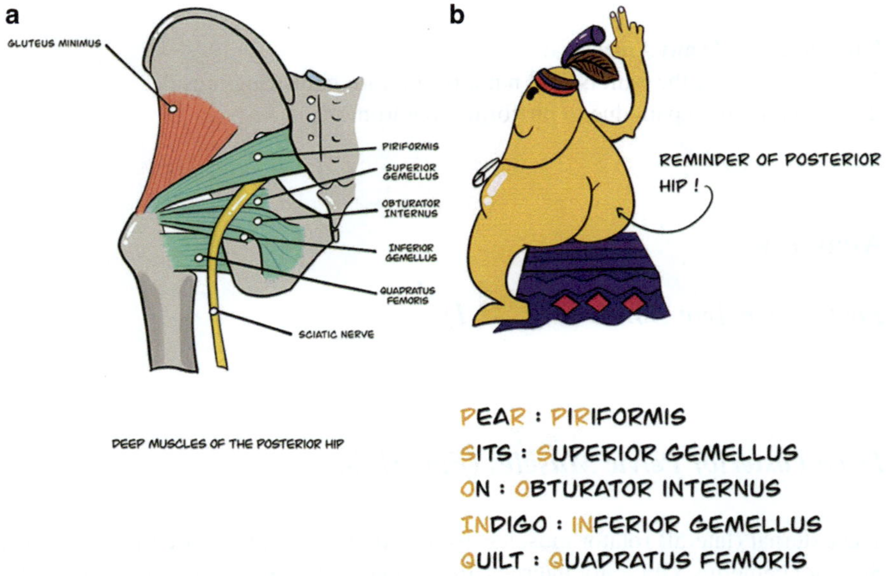

Fig. 21.2 (**a**) Deep posterior pelvic muscles (**b**): Deep posterior pelvic muscles mnemonic

Posterior Thigh Muscles (Fig. 21.3)

A way to remember the location of the hamstring muscles in the thigh is that the long head of the biceps femoris muscle (**L**HBFM) is *lateral.* The semitendinosus muscle (STM) is *medial* to the LHBFM and has a long, thin distal tendon, hence the name "semitendinosus." The LHBFM and the STM share a common (conjoined) tendon on the ischial tuberosity (IT) [2]. This conjoined tendon (CT) is sometimes called the biceps femoris-semitendinosus tendon. The semimembranosus muscle (SMM) also originates from the IT, just *anterior* (deep) to and slightly lateral to the CT. The SMM runs deep to the STM.

Fig. 21.3 Left posterior thigh muscles

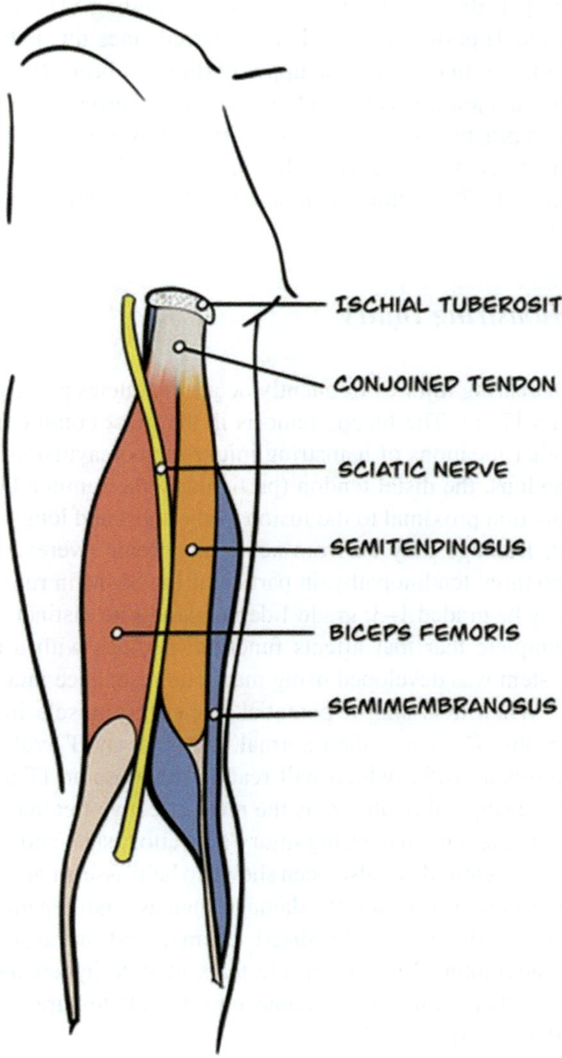

Clinical Comments

Sacroiliac Joint

The sacroiliac joint (SIJ) is sonographically accessible. However, insufficient evidence supports using ultrasound (US) to diagnose sacroiliitis in spondyloarthropathy [3]. By contrast, US is useful for guiding SIJ injections. Injection of the lower (caudal) portion of the SIJ is preferred since this is the synovial component of the joint [4]. In contrast, the upper portion is fibrous (ligamentous) and not a true diarthrodial joint. Successful SIJ injection occurred in 60% of patients in one study, and with practice, it improved to 93.5% [5]. Another study compared US-guided versus fluoroscopy-guided sacroiliac intra-articular injections and demonstrated success rates of 87% with US compared with 98% with fluoroscopy [6].

Hamstring Injury

Hamstring injuries frequently occur in athletes at the proximal myotendinous junction [7, 8]. The biceps femoris is the most commonly injured hamstring muscle. Other locations of hamstring injury include avulsion injury at the IT, the proximal tendons, the distal tendon (particularly the semitendinosus), and at the myofascial junction proximal to the fusion of the short and long heads of the biceps femoris [7, 8]. Damage may be from acute or chronic overexertion, most commonly due to proximal tendinopathy, in particular the SMM in runners [7, 8]. Hamstring injuries may be graded 1–3: grade 1 demonstrates no distinct muscle tear, while grade 3 is a complete tear that affects function, perhaps with a noticeable gap. This grading system was developed using magnetic resonance imaging (MRI) [9].

When assessing a potential hamstring muscle injury, begin with plain radiographs. These are often normal, although an IT avulsion may be evident. The next step is an MRI, which will readily diagnose an IT avulsion, an important finding since surgical treatment is the most effective treatment [10]. Additionally, the MRI will assess the hamstring injury's location, size, and extent.

Ultrasound has also been shown to help assess hamstring injuries [11]. Ultrasound may demonstrate tendon damage such as tendinopathy, peritendinous fluid, fibrillar pattern disruption, localized edema, and intrasubstance anechoic clefts [12]. Sonographic signs of muscle tears include hyperechoic muscle fibers at ruptured ends, hematoma, hypoechoic muscle echotexture, fascial disruption, and muscle edema [13].

Limitations of US in assessing hamstring injuries include imperfect visualization of tendon origin injuries and assessment of re-injuries due to soft tissue scarring. In addition, US may not detect milder muscle strains, particularly in larger patients [11]. MRI is more sensitive than US in detecting tendinopathy, peritendinous fluid, and ischial tuberosity avulsion. An MRI is suggested to be performed within the first 24 hours of an injury; US is most useful after 72 hours and for follow-up [11].

Poor prognostic indicators are proximal tendon injuries, grade 3 injuries, and tears near the IT [8, 14]. The recurrence rate of hamstring muscle injuries is approximately 20% [7, 9, 14]. With posterior thigh pain and a nondiagnostic MRI, consider referred pain from the lumbosacral spine, fascial injury, and gluteal trigger points [8].

Piriformis Syndrome

The sciatic nerve (SN) may be compromised as it emerges between the piriformis and obturator internus muscles. Resultant buttock pain with sciatica mimics lumbar radiculopathy and is termed *piriformis syndrome* [15, 16]. Ultrasound may reveal a hypoechoic, swollen piriformis muscle on the symptomatic side compared with the unaffected side [15, 17]. The SN may be compressed due to piriformis hypertrophy, spasm, inflammation, or an anatomic variant [15].

Since electrophysiologic studies in this deep location may be unhelpful, piriformis syndrome is a diagnosis of exclusion made mainly by ruling out lumbar radiculopathy [18]. Treatment is a corticosteroid injection into the fascial plane between the piriformis and the more superficial gluteus maximus muscles. One study comparing fluoroscopic and sonographic cadaveric injections of piriformis muscles revealed 95% accuracy with ultrasound-guided injection but only 30% accuracy with fluoroscopically-guided injection [19]. This result attests to the ability of US to delineate fascial planes in this location. Be aware that the SN runs deep to the piriformis muscle in 89%, through the muscle in nearly 9%, and superficial to the piriformis muscle in 2–3% [20]. Thus, it is essential to identify the SN before the injection since it may lie at or near the intended injection site.

Ischiogluteal Bursitis

Ischiogluteal bursitis is also known as "weaver's bottom" since prolonged sitting is a causative factor [21]. An enlarged ischiogluteal bursa can be detected on MRI [22]. However, US may likewise demonstrate anechoic fluid in the ischiogluteal bursa adjacent to the IT [17]. An ultrasound-guided ischial bursal injection is technically feasible and may help avoid SN damage [15, 23].

Pitfalls

1. Do not neglect to perform radiographs to look for bony changes such as an IT avulsion.
2. Use an MRI to assess hamstring injuries within the first 24 hours of the injury. Ultrasound is helpful for follow-up.

3. In treating piriformis syndrome, use US guidance to inject between the fascial plane of the piriformis muscle and the gluteus maximus muscle. Take care to avoid injecting the SN, recalling that the nerve is not always located deep to the piriformis muscle.
4. Use US to guide ischial bursal injections; visualizing the SN will avoid injury.

Method

Review radiographs for bony changes, such as an IT avulsion. The patient is prone. Use a linear probe (perhaps with virtual convex at a lower frequency), although many require a curvilinear probe to increase depth acquisition. A pillow beneath the abdomen will assist in comfort and decrease lumbar lordosis. Topographically, the PSIS is seen as a skin dimple. Mark off the PSIS on each side. Confirm all sonographic findings with orthogonal views. Depending on the patient's body habitus, image acquisition may be challenging.

Protocol Image 1: Cephalad Sacroiliac Joint, Transverse (Fig. 21.4)

With the transducer in TAX orientation to the femoral shaft, place the lateral probe end on the hyperechoic hill-like PSIS and look for the hyperechoic sacrum medially near the base of the ilium. Between these two bones is the SIJ.

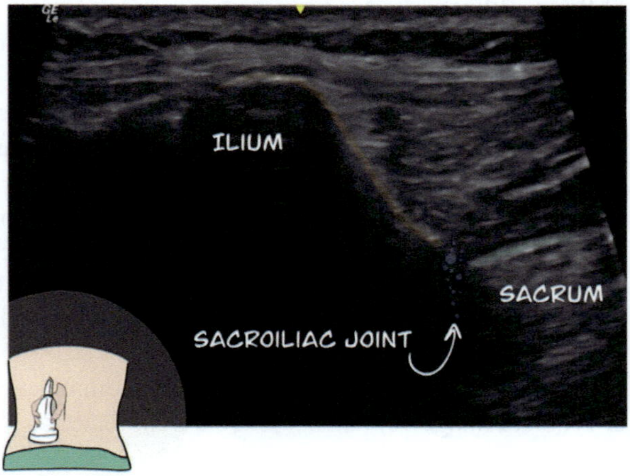

Fig. 21.4 Protocol Image 1: Cephalad sacroiliac joint, transverse

Protocol Image 2: Caudal Sacroiliac Joint, Transverse (Fig. 21.5)

Next, with the probe still in TAX orientation, perform a TAX slide in a caudal direction to image the lower SIJ. As you move the transducer in a caudal direction, note that the height of the ilium decreases and the SIJ narrows.

Protocol Image 3: Piriformis, Longitudinal ± Dynamic (Fig. 21.6)

A curvilinear probe may be required to visualize the piriformis muscle connecting the sacrum to the GT. With the probe still in the same TAX orientation and anchored to the sacrum, rotate the lateral end in a downward (caudal) direction to see the GT's hyperechoic superior/posterior aspect. This requires approximately 40° of angulation plus some fine-tuning by tilting the transducer to visualize the piriformis tendon insertion on the GT. The hypoechoic, longitudinal piriformis muscle is medial. The gluteus maximus muscle is superficial to the piriformis muscle. The two muscles are separated by a longitudinal hyperechoic fascial plane, a common injection target for treating piriformis syndrome.

A dynamic maneuver to assist in identifying the piriformis is to have the patient bend the knee. At the same time, the examiner passively rotates the hip internally and externally by directing the foreleg to the left and then to the right, like a windshield wiper. Since the piriformis muscle connects the GT to the sacrum, external hip rotation will cause the piriformis muscle to stretch out, thus confirming its identity and

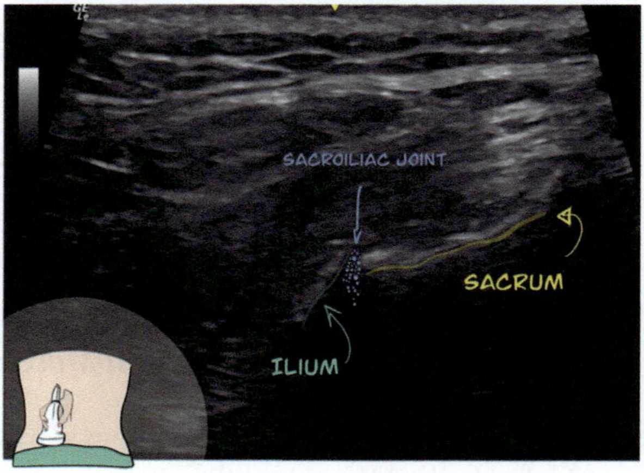

Fig. 21.5 Protocol Image 2: Caudal sacroiliac joint, transverse

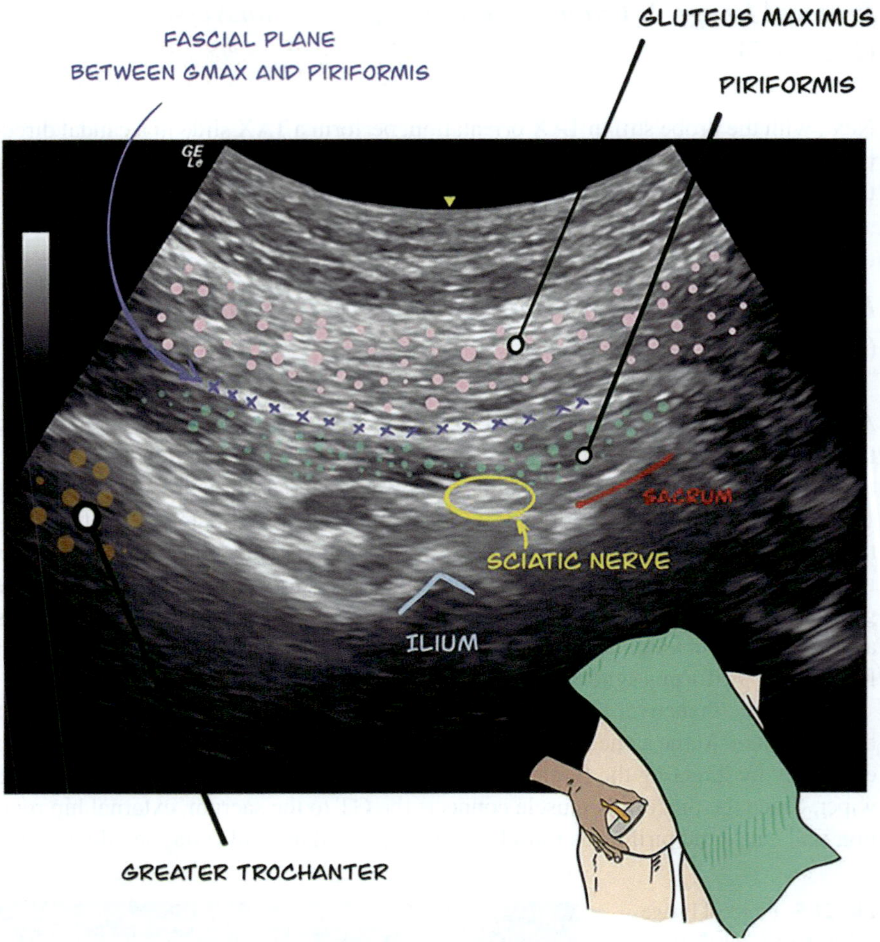

Fig. 21.6 Protocol Image 3: Piriformis, longitudinal ± dynamic

enhancing visualization. The honeycombed SN may be found more medially, just deep to the piriformis muscle, although, in some individuals, it penetrates directly through this muscle. Occasionally, the SN runs superficial to the piriformis.

Protocol Image 4 (Optional): Gluteal Muscles, Transverse

By slowly moving the transducer caudally in TAX, the five posterior pelvic muscles can also be identified if necessary. Recall the external rotator muscles, from superior (cepahlad) to inferior (caudal): piriformis, superior gemellus, obturator internus, inferior gemellus, quadratus femoris. Remember that each muscle lies deep to the gluteus maximus. A detailed sonographic evaluation of these muscles can be found elsewhere [24].

Protocol Image 5: Ischial Tuberosity, Hamstring Muscles, Transverse (Fig. 21.7)

Next, place the transducer in the TAX position, just cephalad to the gluteal fold. The medial end of the probe should visualize the curved, hyperechoic, bony IT. The IT is the origin of the hamstrings and may be palpable in many people. The ischial bursa lies between the hamstring tendons and the gluteus maximus; the bursa may become more prominent when inflamed or distended.

Protocol Image 6: Conjoined Tendon, Longitudinal (Fig. 21.8)

Next, center the probe on the CT and rotate 90° to visualize the CT in LAX orientation. The transducer is vertical. The semimembranosus tendon insertion may be seen deep to the CT.

Protocol Image 7: Semimembranosus Tendon, Longitudinal (Fig. 21.9)

Perform a slight lateral TAX slide with the transducer tilted, aiming towards the midline, to visualize the insertional origin of the SMT on the IT. The SMT is seen at a slightly deeper level than the CT.

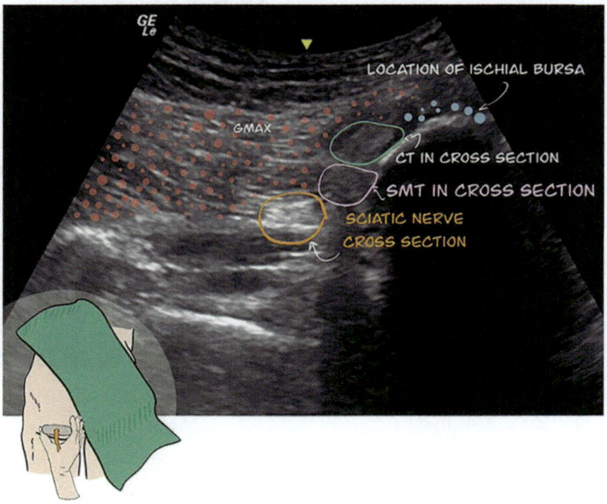

Fig. 21.7 Protocol Image 5: Ischial tuberosity, hamstring muscles, transverse

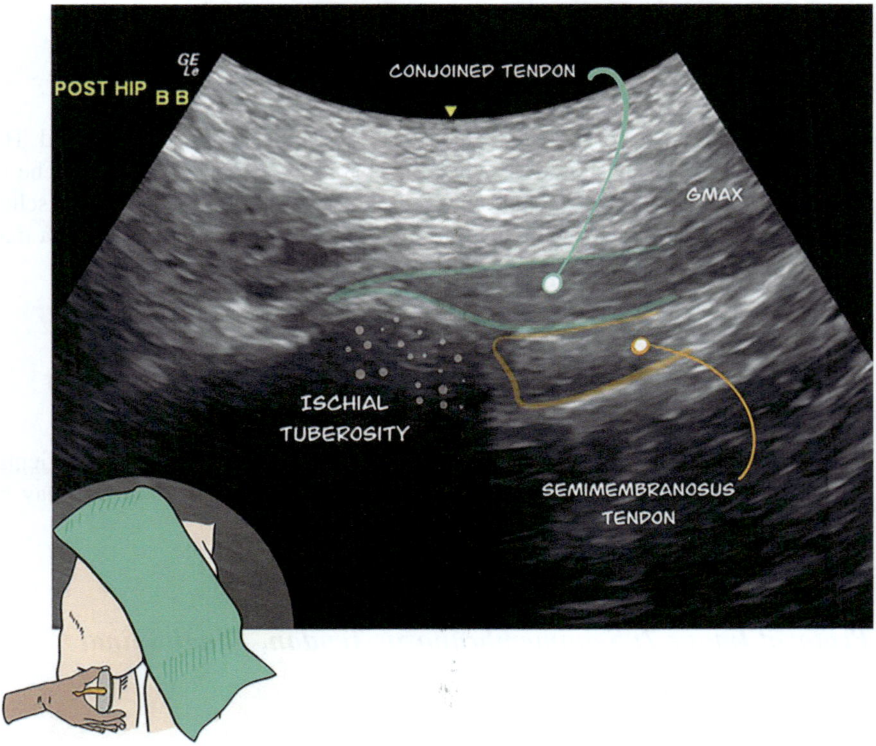

Fig. 21.8 Protocol Image 6: Conjoined tendon, longitudinal

Fig. 21.9 Protocol Image 7: Semimembranosus tendon, longitudinal

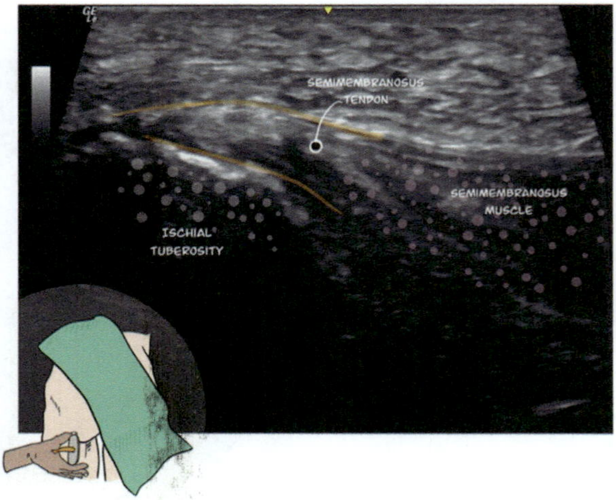

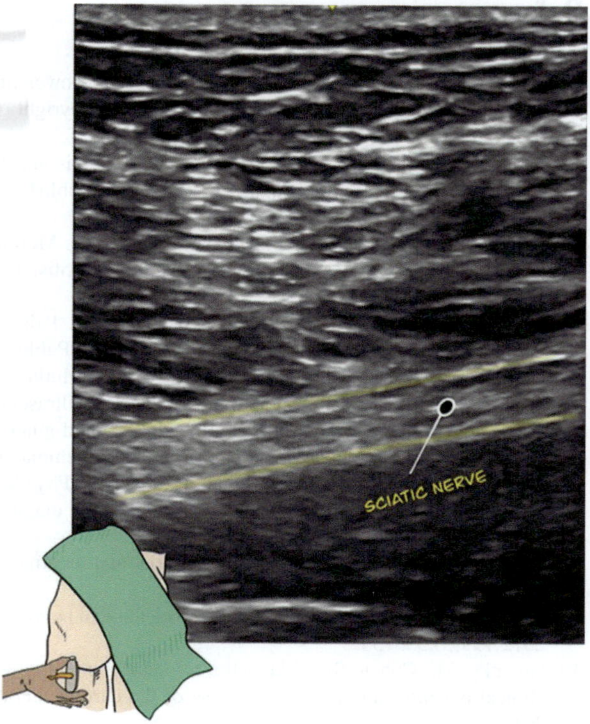

Fig. 21.10 Protocol Image 7: Sciatic nerve, longitudinal

Protocol Image 8: Sciatic Nerve, Longitudinal (Fig. 21.10)

Perform yet another lateral TAX slide with the transducer parallel to the long axis of the thigh. The transducer slides off the bony hyperechoic IT; the SN should be visible in the LAX view. Look for the emergence of the typical tram track appearance of the large SN in the LAX view. If you do not visualize the SN, rotate the transducer 90° to identify the SN in the TAX view.

Complete Posterior Hip Ultrasonic Examination Checklist

☐ Protocol Image 1: Cephalad sacroiliac joint, transverse
☐ Protocol Image 2: Caudal sacroiliac joint, transverse
☐ Protocol Image 3: Piriformis, longitudinal ± dynamic
☐ Protocol Image 4 (optional): Gluteal muscles, transverse
☐ Protocol Image 5: Ischial Tuberosity, hamstring muscles, transverse
☐ Protocol Image 6: Conjoined tendon ± semimembranosus, longitudinal
☐ Protocol Image 7: Semimembranosus tendon, longitudinal
☐ Protocol Image 8: Sciatic nerve, longitudinal

References

1. Lezak B, Massel DH. Anatomy, bony pelvis and lower limb, gemelli muscles. StatPearls. Treasure Island, FL: StatPearls Publishing. Copyright © 2022, StatPearls Publishing LLC.; 2022.
2. Vaughn JE, Cohen-Levy WB. Anatomy, bony pelvis and lower limb, posterior thigh muscles. StatPearls. Treasure Island, FL: StatPearls Publishing. Copyright © 2022, StatPearls Publishing LLC.; 2022.
3. Gutierrez M, Rodriguez S, Soto-Fajardo C, Santos-Moreno P, Sandoval H, Bertolazzi C, et al. Ultrasound of sacroiliac joints in spondyloarthritis: a systematic review. Rheumatol Int. 2018;38(10):1791–805.
4. Raj MA, Ampat G, Varacallo M. Sacroiliac Joint Pain. StatPearls. Treasure Island, FL: StatPearls Publishing. Copyright © 2022, StatPearls Publishing LLC.; 2022.
5. Pekkafahli MZ, Kiralp MZ, Başekim CC, Silit E, Mutlu H, Oztürk E, et al. Sacroiliac joint injections performed with sonographic guidance. J Ultrasound Med. 2003;22(6):553–9.
6. Jee H, Lee JH, Park KD, Ahn J, Park Y. Ultrasound-guided versus fluoroscopy-guided sacroiliac joint intra-articular injections in the noninflammatory sacroiliac joint dysfunction: a prospective, randomized, single-blinded study. Arch Phys Med Rehabil. 2014;95(2):330–7.
7. Chu SK, Rho ME. Hamstring injuries in the athlete: diagnosis, treatment, and return to play. Curr Sports Med Rep. 2016;15(3):184–90.
8. Brukner P. Hamstring injuries: prevention and treatment-an update. Br J Sports Med. 2015;49(19):1241–4.
9. Rubin DA. Imaging diagnosis and prognostication of hamstring injuries. AJR Am J Roentgenol. 2012;199(3):525–33.
10. Greenky M, Cohen SB. Magnetic resonance imaging for assessing hamstring injuries: clinical benefits and pitfalls - a review of the current literature. Open Access J Sports Med. 2017;8:167–70.
11. Pedret C. Hamstring muscle injuries: MRI and ultrasound for diagnosis and prognosis. J Belg Soc Radiol. 2021;105(1):91.
12. Lungu E, Michaud J, Bureau NJ. US assessment of sports-related hip injuries. Radiographics. 2018;38(3):867–89.
13. Minagawa H, Wong KH. Musculoskeletal ultrasound: Echo anatomy & scan technique: Amazon digital services; 2017.
14. Malliaropoulos NG. Non contact hamstring injuries in sports. Muscles Ligaments Tendons J. 2013;2(4):309–11.
15. Blaichman JI, Chan BY, Michelin P, Lee KS. US-guided musculoskeletal interventions in the hip with MRI and US correlation. Radiographics. 2020;40(1):181–99.
16. Hallin RP. Sciatic pain and the piriformis muscle. Postgrad Med. 1983;74(2):69–72.
17. Lin Y-T, Wang T-G. Ultrasonographic examination of the adult hip. J Med Ultrasound. 2012;20:201–9.
18. Ro TH, Edmonds L. Diagnosis and Management of Piriformis Syndrome: a rare anatomic variant analyzed by magnetic resonance imaging. J Clin Imaging Sci. 2018;8:6.
19. Finnoff JT, Hurdle MF, Smith J. Accuracy of ultrasound-guided versus fluoroscopically guided contrast-controlled piriformis injections: a cadaveric study. J Ultrasound Med. 2008;27(8):1157–63.
20. Lewis S, Jurak J, Lee C, Lewis R, Gest T. Anatomical variations of the sciatic nerve, in relation to the piriformis muscle. Transl Res Anatomy. 2016;5:15–9.
21. Anderson CR. Weaver's bottom. JAMA. 1974;228(5):565.
22. Cho KH, Lee SM, Lee YH, Suh KJ, Kim SM, Shin MJ, et al. Non-infectious ischiogluteal bursitis: MRI findings. Korean J Radiol. 2004;5(4):280–6.
23. Wisniewski SJ, Hurdle M, Erickson JM, Finnoff JT, Smith J. Ultrasound-guided ischial bursa injection: technique and positioning considerations. PM R. 2014;6(1):56–60.
24. Battaglia PJ, Mattox R, Haun DW, Welk AB, Kettner NW. Dynamic ultrasonography of the deep external rotator musculature of the hip: a descriptive study. PM R. 2016;8(7):640–50.

Chapter 22
Lateral Hip

Reasons to Do the Study
1. Lateral hip pain
2. Swelling of the lateral hip

Questions We Want Answered
1. Is the cause of the pain greater trochanteric pain syndrome (GTPS), snapping iliotibial band (ITB), tensor fascia latae (TFL) tendinopathy, or another soft tissue structural problem? If so, what is the specific underlying pathology?
2. If the ultrasound (US) exam of the lateral hip soft tissue is unremarkable, could the lateral hip pain be referred from the lumbosacral spine, the femur, the hip joint, the labrum, or the lateral femoral cutaneous nerve?

Basic Anatomy

The greater trochanter (GT) is a bony protuberance of the proximal posterolateral femur (Fig. 22.1). The ITB is a longitudinal fibrous band that originates from the topmost external iliac crest and runs along the lateral aspect of the thigh superficial to the GT, finally inserting on Gerdy's tubercle of the tibia (Fig. 22.2). It receives fibers from two muscles: anteriorly from the TFL and posteriorly from the gluteus maximus (Gmax) [1]. These muscular contributions enable ITB tension, thus stabilizing the knee with ambulation.

The contents do not represent the views of the U.S. Department of Veterans Affairs or the United States Government.

M. H. Greenberg et al., *Manual of Musculoskeletal Ultrasound*,
https://doi.org/10.1007/978-3-031-37416-6_22

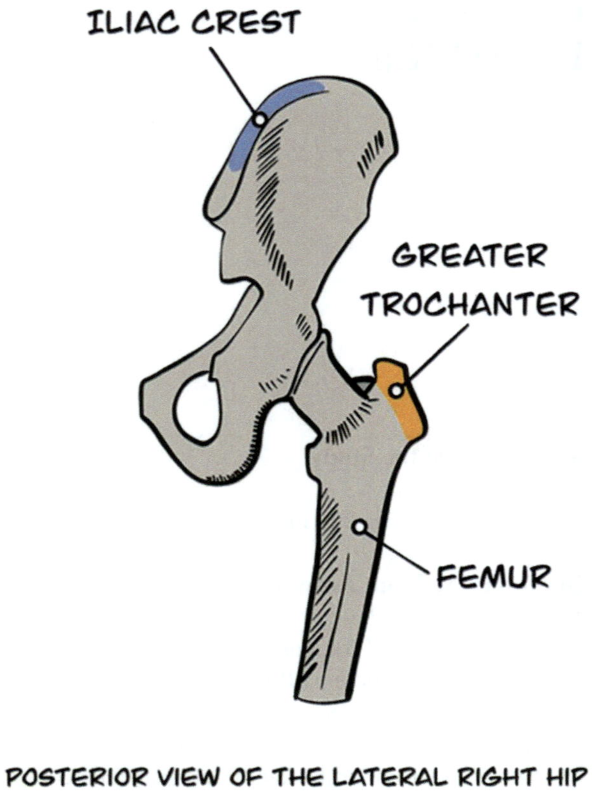

Fig. 22.1 Simplified lateral hip bony anatomy

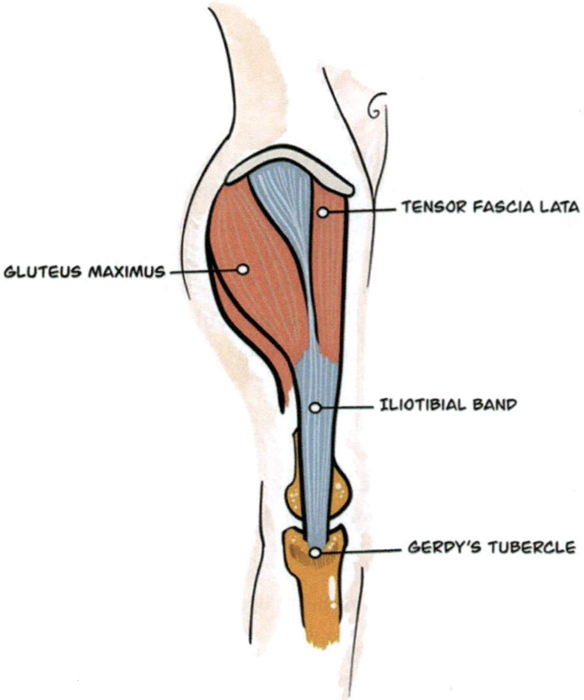

Fig. 22.2 Iliotibial band

The Gmax originates from the posterior external iliac bone and inserts into the posterior ITB (as noted above) and the gluteal (femoral) tubercle, *distal* to the GT (Fig. 22.3). The gluteus minimus (Gmin) and gluteus medius (Gmed) muscles both originate from the external posterior iliac bone, with the Gmin being the deeper of the two. The Gmin flexes, abducts, and medially rotates the thigh depending on the hip position. The Gmed extends and abducts the thigh. The Gmin inserts on the anterior facet (AF) of the GT, while the Gmed inserts on the lateral (LF) and supero-posterior facets of the GT (Fig. 22.4).

A mnemonic for these two insertions is "Aunt Minnie Loves Mead."

Anterior facet: Gluteus **min**imus; **L**ateral facet: Gluteus **me**dius.

A mnemonic for the five soft tissue structures of the lateral hip (gluteus minimus, gluteus medius, gluteus maximus, iliotibial band, and tensor fascia latae) is as follows:

"**Minim**al **medi**tation **maxim**ally **i**mproves **tens**ion."

It is easy to remember the names and locations of the bursae surrounding the GT. Each bursa has the prefix "sub" before the name of the muscle and is located

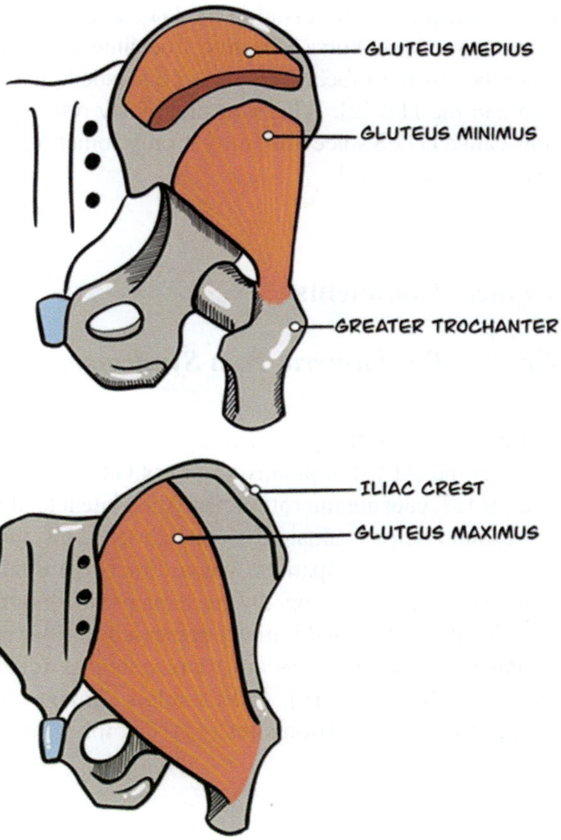

Fig. 22.3 Simplified muscular anatomy of the posterior hip

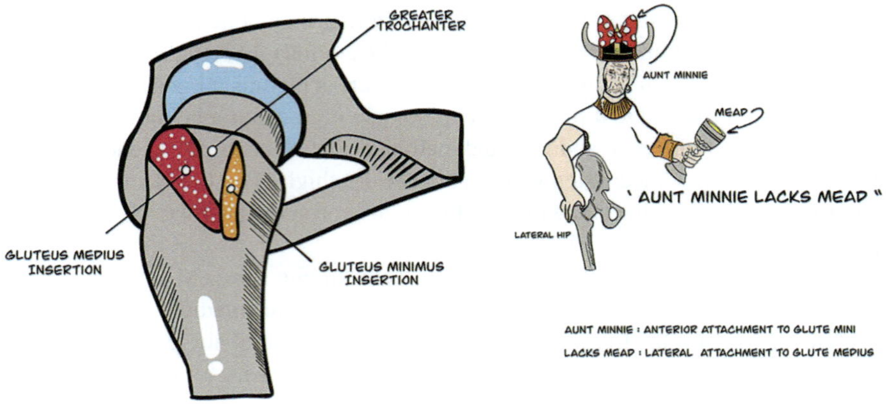

Fig. 22.4 Simplified tendon insertions on the lateral greater trochanter

deep to the respective tendon or muscle near its insertion: subgluteus minimus, subgluteus medius, and subgluteus maximus bursae.

Again, note that the Gmax inserts along the posterior portion of the ITB and the bony gluteal tubercle (distal to the GT) and *not* on the GT. Also note that the subgluteus maximus bursa (the true trochanteric bursa) separates the Gmax *muscle* from the posterior facet (PF) and, if distended, may extend between the Gmed tendon and the ITB [2]. What was previously referred to as "trochanteric bursitis" is now called GTPS since this entity is only sometimes associated with true subgluteus maximus bursitis [3].

Clinical Comments

Greater Trochanteric Pain Syndrome

In this condition, the patient complains of pain or tenderness at the posterolateral hip over the GT [3]. A putative cause of GTPS is repetitive friction between the ITB and the GT, causing microtrauma to the gluteal tendons at their GT insertions [4, 5]. Ultrasound helps evaluate the abductor tendons and bursae at the GT and has the advantage of sonopalpation, which directs the clinician to the painful area [6, 7]. Gluteus medius tendinopathy may manifest sonographically as hypoechogenicity with focal swelling and initially preserved fibrillar echotexture [8–10].

Partial-thickness tears are represented as focal anechoic areas with loss of fibrillar echotexture [8]. With tendon damage, hyperechoic calcification may occur at tendon insertion sites [11]. Complete tears may demonstrate a retracted tendon with anechoic fluid accumulation and a bare GT [12]. Bursal distension of the subgluteal bursae of the Gmed and the Gmin may be seen deep to each respective tendon as hypoechoic fluid collections just proximal to the insertion on the GT. The subgluteal bursa of the Gmax, the true trochanteric bursa, is located deep to the Gmax *muscle* and ITB, superficial to the lateral and posterior facets of the GT [13].

A large retrospective study in patients with GTPS revealed the most common sonographic finding to be Gmed or Gmin tendinosis, followed by ITB thickening [3]. Isolated bursitis was the *least* common finding. The Gmax tendon or underlying subgluteus maximus bursa (the "true trochanteric bursa") is *not* commonly involved in GTPS [3]. In summation, friction from the ITB may cause local microtrauma of the abductor tendons and may be the leading cause of GTPS [9, 10]. However, be aware that gluteal tendinosis or bursitis may be present asymptomatically [14]. Perhaps impingement of the ITB against the Gmed and Gmin is the cause of many cases of GTPS; this would be analogous to shoulder rotator cuff impingement by the acromion [9, 15]. Another school of thought is that the bursal distension noted in GTPS may be *secondary* to a gluteal tendon abnormality, similar to the fluid accumulation in subacromial-subdeltoid bursitis associated with a rotator cuff tear in the shoulder [16].

Predisposing conditions to the GTPS include an external snapping hip, blunt trauma to the hip, iatrogenic injury during hip arthroplasty, increased age, obesity, osteoarthritis of the knee or hip, low back pain, and leg length discrepancy [10]. The differential diagnosis of lateral hip pain includes osteoarthritis, avascular necrosis of the femoral head, labral tears, femoral acetabular impingement, femoral neck stress fractures, loose bodies, radiation from lumbar stenosis, and meralgia paresthetica (lateral femoral cutaneous neuropathy). It is controversial whether cortical bone irregularity at the GT is helpful with the diagnosis of GTPS or not [10]. Radiographic imaging is best used to recognize alternative sources of pain such as osteoarthritis, avascular necrosis, lumbar spondylosis, and femoral acetabular impingement [10].

Magnetic resonance imaging (MRI) is still the gold standard for evaluating GTPS, with a sensitivity/specificity of 93/92% for diagnosing abductor tendon tears [10]. Sonographic sensitivity has been described as 79%. Therefore, US is a less reliable diagnostic modality, and MRI is necessary before considering surgical intervention.

Treatment is conservative, including nonsteroidal anti-inflammatory medications, physical therapy, and attention to remediable risk factors such as leg length discrepancy, obesity, and contributing knee, hip, and foot disorders [4]. Local injection can be performed with US guidance, but debate persists about injecting corticosteroids versus platelet-rich plasma since corticosteroids may potentially weaken damaged tendons. Low-energy shock wave therapy has been used as well. Surgery is often the last resort [4].

Tendinosis and Tendon Tears

The gluteus minimus and medius tendons may also be abnormal at their respective trochanteric insertions, ranging from tendinosis to tendon tears. Tendinosis is evident by thickening (swelling) and hypoechoic loss of fibrillar echotexture. Partial or full-thickness tears may be noted as well. Comparison with the contralateral (presumably unaffected) side may be helpful. Again, pain at the GT is more likely from a gluteal tendon abnormality than true bursitis.

The gold standard for the diagnosis of abductor tendon pathology is MRI. However, one study reports US to be highly sensitive for detecting Gmed tendon pathology (tendinosis, partial-or full-thickness tear), albeit with a tendency to incorrectly *overstate* some normal tendons as being damaged [17]. However, the same study revealed that MRI tended to *understate* the presence of tendon pathology. Both imaging techniques poorly differentiate tendinosis from partial-and full-thickness tears [17]. Despite the improved resolution of modern US machines, deep-lying structures such as the gluteal tendons may not be sharply delineated.

Given the strengths and weaknesses of US and MRI modalities, our practice for uncertain diagnoses has been to follow-up US examinations with an MRI or MR arthrogram to confirm gluteal pathology suggested by US and to evaluate the hip further since other conditions may mimic GTPS. Such conditions include degenerative change, avascular necrosis, labrum tears, and femoral acetabular impingement.

Snapping Iliotibial Band and External Lateral Snapping Hip Syndrome

When the thigh is flexed (with the hip in adduction), the ITB *normally* glides anteriorly over the GT [13]. This is logical since the distal ITB is tethered to the proximal *anterior* tibia at Gerdy's tubercle. External lateral snapping hip (ELSH) is described as intermittent impingement of the ITB (or the anterior Gmax or the TFL) over the GT; this may be painful or painless. The patient may experience a clicking or popping sensation [13]. As the hip moves back into extension, there may be a similar impediment to the movement of the ITB.

Static US may reveal the ITB to be thickened and hypoechoic [11]. Bursal distension may also be involved [13]. Evaluate for ELSH dynamically with the patient standing or lying in a lateral decubitus position. The hip is extended and then slowly flexed. The transducer is placed in TAX over the GT, and the ITB is identified. The ITB (or Gmax) may catch over the GT and abruptly lurch anteriorly during flexion [11].

The snapping sensation may be best demonstrated with the patient standing [13]. Pain is postulated to be from secondary tendinopathy and bursitis due to repetitive snapping. Again, painless snapping may be present and considered normal. Predisposing conditions for ELSH include athletic activities such as ballet dancing, variant body structures affecting biomechanics, femoral osteochondroma, and postoperative changes [13]. Dynamic US is the best technique to demonstrate this pathologic entity [18].

Morel-Lavallee Lesion

Morel-Lavallee (M-L) lesions occur when blunt, soft tissue trauma causes shearing between the subcutaneous tissue and superficial muscle fascial layers, creating a space that fills with serosanguineous fluid and necrotic fat [13, 19, 20]. This lesion often results from an injury sustained in football, baseball, or skiing [21]. Eventually, localized inflammation produces a fibrous pseudocapsule, which may result in persistent or recurrent fluid collection [19].

The most common locations are near the GT and the anterior proximal thigh, and M-L lesions may be associated with pelvic fractures [13]. Acutely, the patient presents with an enlarged, painful mass with decreased cutaneous sensation [22]. Alternatively, the M-L lesion may initially be clinically inapparent, growing slowly without pain, thus mimicking a soft tissue tumor [13]. An acute M-L lesion on US may demonstrate heterogeneous hyperechogenicity due to debris such as necrotic lobules and irregular margins [19].

The US picture, although variable, evolves with time into a compressible, homogeneous lesion with well-defined, smooth margins containing anechoic or hypoechoic fluid and no power Doppler activity [19]. However, rebleeding within a chronic lesion may alter the sonographic appearance. Diagnosing M-L lesions as early as possible is essential to avoid chronic pain and a potential space for infection [19].

Tensor Fascia Latae Tendinopathy

Anterior groin pain in athletes may be associated with tendon injuries, hernias, avulsion injuries, referred pain, or tumors [23]. Injury to the TFL has increasingly become recognized as a cause of anterior groin pain in athletes and is sonographically detectable; point tenderness may correlate with pathology [24]. It is postulated that recurrent microtrauma damages the TFL.

Sonographic signs of tendinopathy include enlargement of the TFL, focal areas of hypoechogenicity, and linear or other anechoic areas. Anechoic regions are assumed to be intrasubstance tears [24]. Tendinopathy may be at or proximal to the insertion [25]. A sonographic comparison with the presumably unaffected contralateral side is suggested [24].

A TFL tear in the muscle belly may mimic GTPS and be sonographically visible [26]. Treatment for TFL injuries includes rest, massage, nonsteroidal anti-inflammatory medications, corticosteroid injections, surgery, and, more recently, tendon fenestration with platelet-rich plasma injections [24, 27].

Bursal Abnormalities

The three gluteal bursae may or may not be distended with GTPS. Bursal distention may be secondary to tendinopathy or primary, perhaps associated with an underlying inflammatory condition such as rheumatoid arthritis. Bursal fluid is typically anechoic but *may also be hypoechoic, particularly if there is associated synovial hypertrophy.*

Proximal Iliotibial Band Syndrome

Not to be confused with snapping ITB syndrome or *distal* "ITB syndrome," proximal ITB syndrome has been described most often in athletes as causing localized pain at the iliac crest due to enthesopathy, presumably from chronic repetitive microtrauma of the ITB at or near its insertion [28]. Proximal ITB syndrome is either uncommon or uncommonly recognized, occurs more frequently in females, and may be due to overuse, trauma, or degenerative change [5]. Pain increases with physical activity and may mimic hip and GT-related conditions [26]. The imaging modality most often used to define this enthesopathy is MRI [26, 28]. Sonographic findings are that of a thickened (swollen) hypoechoic ITB at the IT and crest, particularly in comparison with the normal contralateral side [13].

Pitfalls

1. When describing pain in the GT area, the designation "GTPS" is used rather than the outdated "trochanteric bursitis."
2. Do not unquestionably accept US or MRI findings regarding the Gmed tendon at the GT. Ultrasound tends to overreport, while MR may underreport Gmed pathology [17].
3. Likewise, US and MRI may poorly distinguish tendinosis, partial thickness tears, and full-thickness tears of the Gmed tendon at the GT [17].
4. Don't forget to consider using a combination of US and MR modalities to define the cause of lateral hip pain since MRI will additionally delineate other hip pathologies mimicking GTPS.
5. Be aware that treatment paradigms for GTPS are evolving, including injection protocols and the products being injected.
6. Don't forget to consider TFL and proximal ITB pathology as causes of pain proximal to the hip.

Method

The patient is in a lateral decubitus position, lying on the unaffected side. Palpate the bony prominence, the GT at the posterolateral hip, and mark this area with a ballpoint pen. Use a curvilinear probe since the lower frequency is necessary to achieve adequate depth. The linear transducer (with a virtual convex setting and a lower frequency) will suffice for some thin patients to visualize the GT.

Protocol Image 1: Greater Trochanter, Transverse (Fig. 22.5)

Place the probe in the transverse axis (TAX) over the GT and look for the sharp hyperechoic bony peak that separates the AF and LF. The AF is closer to the anterior thigh. Look at the tendon cross-sectional insertions in TAX of the Gmin on the AF and the Gmed on the LF. One or even both of these tendons may be poorly seen due to anisotropy. In cross-section, the ITB is the thin, curved, hyperechoic structure superficial to the tendons and muscles. The ITB is delineated as the patient slowly extends and flexes the hip to demonstrate the movement of the ITB from posterior to anterior with hip flexion.

Fig. 22.5 Protocol Image 1: Greater trochanter, transverse

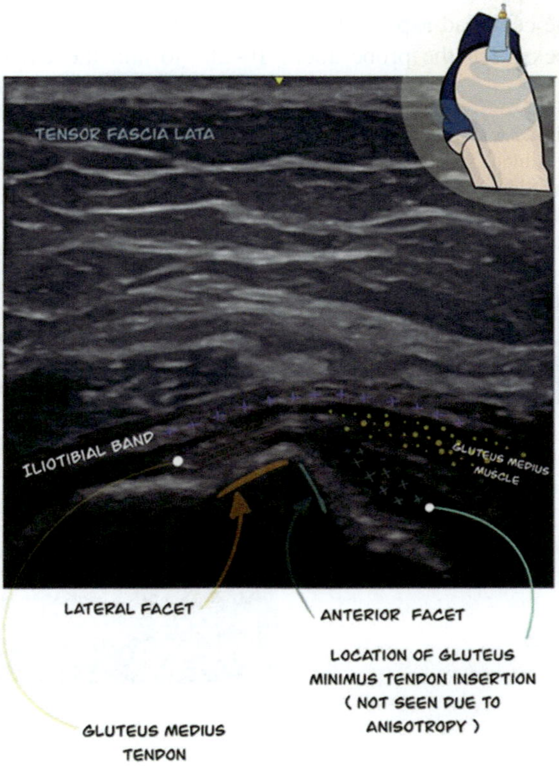

Protocol Image 2: Gluteus Minimus Insertion, Anterior Facet, Longitudinal (Fig. 22.6)

Perform a longitudinal axis (LAX) slide anteriorly to center the probe on the AF in TAX orientation. Visualize the hyperechoic Gmin (in cross-section) inserted into the LF. This may require slight rotation and tilting of the transducer. Then, rotate the probe 90° to look for the insertion of Gmin on the AF in LAX.

To best view the Gmin insertion, rotate the cephalad end of the probe anteriorly about 20–30° (aiming the cephalad end of the transducer in the direction of the umbilicus). Also, tilt the probe about 10–20° to direct the US beam posteriorly to sharpen visualization of the tendon insertion. The insertion on the AF should look like an elongated "bird's beak." Using time-gain control to focus on the area may be very helpful. The ITB is superficial to the Gmed muscle, which itself is superficial to the Gmin. Look for a subgluteus minimus bursa deep to the Gmin, as well as tendinopathy and frank tendon tears.

Protocol Image 3: Gluteus Medius Insertion, Lateral Facet, longitudinal (Fig. 22.7)

Next, perform a TAX slide in the posterolateral direction to the LF and then rotate the cephalad aspect of the probe about 20–30° (from the midline) toward the buttocks. Tilt the probe about 10–20° to aim the US beam anteriorly. Look at the

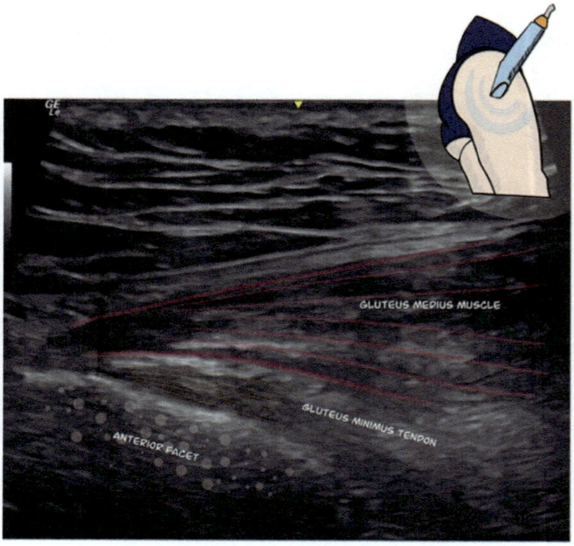

Fig. 22.6 Protocol Image 2: Gluteus minimus insertion, anterior facet, longitudinal

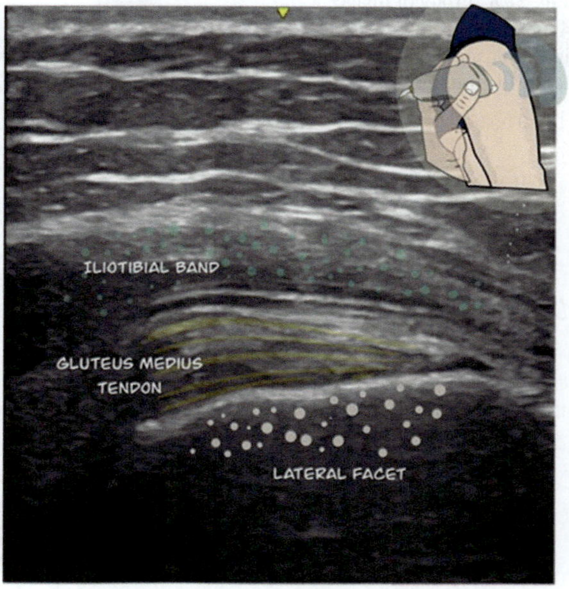

Fig. 22.7 Protocol Image 3: Gluteus medius insertion, lateral facet, longitudinal

anterior fibers of the Gmed as it inserts on the LF in LAX in yet another "bird's beak" configuration. The ITB is superficial to the tendon. Again, ensure the probe is perpendicular to the facet by slightly tilting and rocking the transducer to optimally visualize the tendon insertion. Look for fluid deep to the Gmed tendon, representing a subgluteus medius bursa. The ITB will be seen as a hyperechoic band superficial to the Gmed.

Protocol Image 4: Gluteus Medius Insertion, Superoposterior Facet, Longitudinal (Fig. 22.8)

Next, perform a slight TAX slide in a posterolateral direction to see the insertion of the posterior fibers of the Gmed on the superior portion of the posterior facet (PF). This view will reveal the more superficial ITB and, if present, a distended Gmax bursa superficial to the Gmed. The Gmax bursa is the true trochanteric bursa. The trochanteric bursa, when distended, may spread laterally to become sandwiched between the Gmed tendon and the overlying ITB.

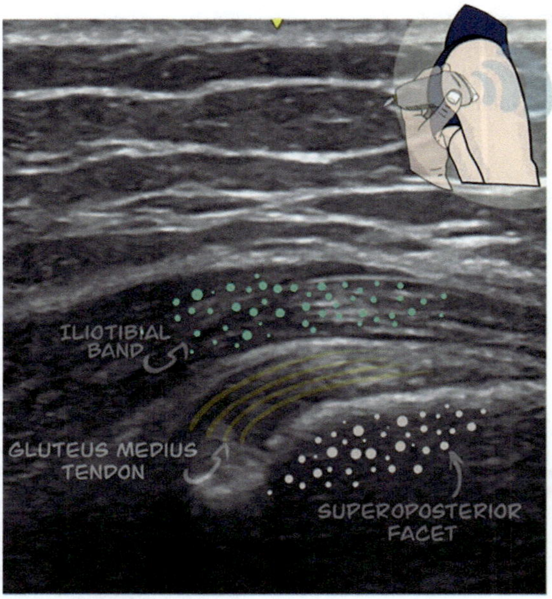

Fig. 22.8 Protocol Image 4: Gluteus medius insertion, superoposterior facet, longitudinal

Protocol Image 5: Proximal Iliotibial Band, longitudinal (Fig. 22.9)

If clinically indicated, evaluate the TFL by locating the anterior superior iliac spine (ASIS) in TAX orientation. The ASIS is a curvilinear hyperechoic bony protuberance most easily located from an anterior approach. The ASIS appears as a broad hill or comet-shaped formation with posterior acoustic shadowing causing an anechoic appearance deep to the hyperechoic curved bony cortex.

Center the probe on the *posterior* slope of the ASIS and then rotate the transducer 90° to visualize in LAX the fibrillar hyperechoic echotexture of the common origin of the TFL and the ITB. The TFL muscle is deeper. The fibers of the TFL merge into the anterior portion of the ITB.

Protocol Image 6: Iliotibial Band, Transverse + Longitudinal (Fig. 22.10)

Place the transducer in TAX orientation across the GT. The ITB is easily identified dynamically. As the patient slowly extends and flexes the hip, the ITB moves from posterior to anterior; the examiner may evaluate for a snapping ITB over the GT. Moving the probe in a cephalad TAX slide will demonstrate the more proximal

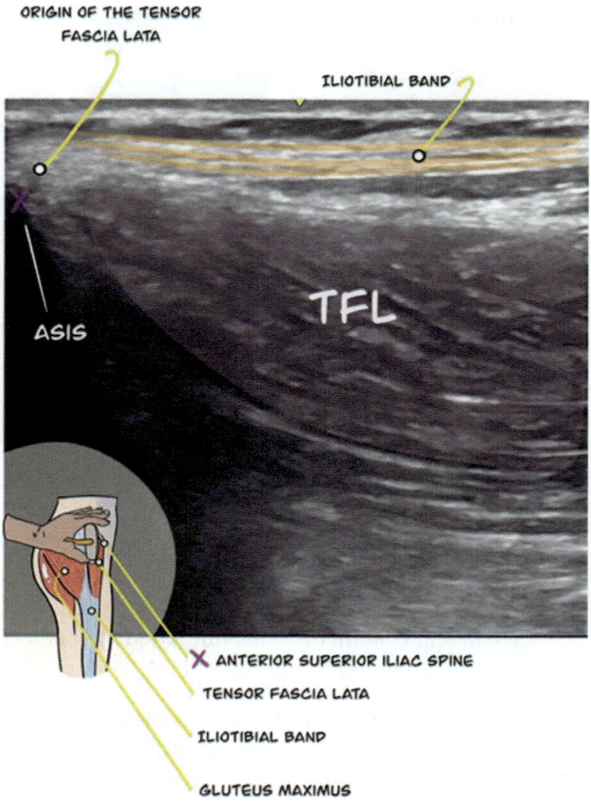

Fig. 22.9 Protocol Image 5: Proximal iliotibial band, longitudinal

ITB. When the iliac crest is reached, the transducer can be rotated 90° to the LAX orientation. Then, TAX slides can be done anteriorly and posteriorly on the iliac crest to evaluate the ITB origin, including the iliac tubercle, which has a slight bony prominence.

Protocol Image 7 (Optional): Iliotibial Band, Transverse + Dynamic

The patient should stand to be evaluated for an external snapping hip at the GT. Return to the GT and turn the probe to TAX (**Protocol Image 6**). Have the patient move the hip from extension to flexion while the hip is in adduction. Look for the ITB (or the TFL or Gmax) suddenly skipping from posterior to anterior over the GT (which may or may not be painful). The ITB may be thickened from repetitive trauma. Use only mild transducer pressure and feel for a snapping sensation through the probe. The ITB moves anteriorly because it is anchored distally at Gerdy's tubercle of the tibia.

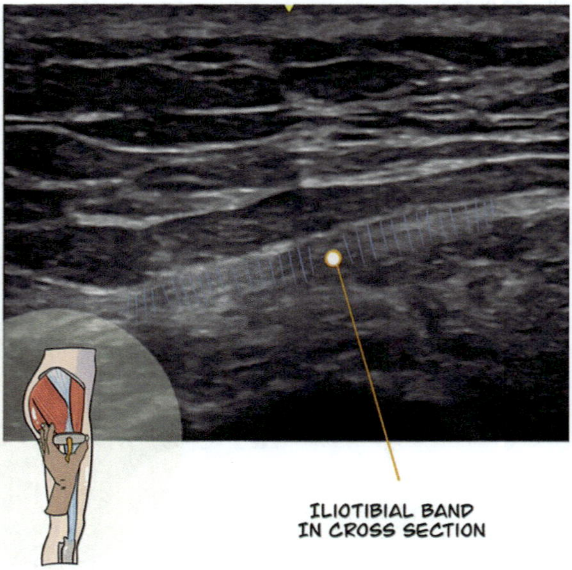

Fig. 22.10 Protocol Image 6: Iliotibial band, transverse + longitudinal

(See Clinical Comments: Snapping iliotibial band (ITB) or external (extraarticular) lateral snapping hip syndrome (ELSH).)

Complete Lateral Hip Ultrasonic Examination Checklist

☐ Protocol Image 1: Greater trochanter, transverse
☐ Protocol Image 2 Gluteus minimus insertion, anterior facet, longitudinal
☐ Protocol Image 3: Gluteus medius insertion, lateral facet, longitudinal
☐ Protocol Image 4: Gluteus medius insertion, superoposterior facet, longitudinal
☐ Protocol Image 5: Proximal iliotibial band, longitudinal
☐ Protocol Image 6: Iliotibial band, transverse + longitudinal
☐ Protocol Image 7 (optional): Iliotibial band, transverse + dynamic

References

1. Hyland S, Graefe SB, Varacallo M. Anatomy, Bony Pelvis and Lower Limb, Iliotibial Band (Tract). StatPearls. Treasure Island, FL: StatPearls Publishing. Copyright © 2022, StatPearls Publishing LLC.; 2022.
2. Pfirrmann CW, Chung CB, Theumann NH, Trudell DJ, Resnick D. Greater trochanter of the hip: attachment of the abductor mechanism and a complex of three bursae--MR imaging and MR bursography in cadavers and MR imaging in asymptomatic volunteers. Radiology. 2001;221(2):469–77.
3. Long SS, Surrey DE, Nazarian LN. Sonography of greater trochanteric pain syndrome and the rarity of primary bursitis. AJR Am J Roentgenol. 2013;201(5):1083–6.
4. Reid D. The management of greater trochanteric pain syndrome: a systematic literature review. J Orthop. 2016;13(1):15–28.

5. Arend C. Sonography of the iliotibial band: Spectrum of findings. Radiol Bras. 2014;47:33–7.
6. Kong A, Van der Vliet A, Zadow S. MRI and US of gluteal tendinopathy in greater trochanteric pain syndrome. Eur Radiol. 2007;17(7):1772–83.
7. Garcia FL, Picado CH, Nogueira-Barbosa MH. Sonographic evaluation of the abductor mechanism after total hip arthroplasty. J Ultrasound Med. 2010;29(3):465–71.
8. Connell DA, Bass C, Sykes CA, Young D, Edwards E. Sonographic evaluation of gluteus medius and minimus tendinopathy. Eur Radiol. 2003;13(6):1339–47.
9. Klauser AS, Martinoli C, Tagliafico A, Bellmann-Weiler R, Feuchtner GM, Wick M, et al. Greater trochanteric pain syndrome. Semin Musculoskelet Radiol. 2013;17(1):43–8.
10. Pianka MA, Serino J, DeFroda SF, Bodendorfer BM. Greater trochanteric pain syndrome: evaluation and management of a wide spectrum of pathology. SAGE Open Med. 2021;9:20503121211022582.
11. Lin Y-T, Wang T-G. Ultrasonographic examination of the adult hip. J Med Ultrasound. 2012;20:201–9.
12. Kagan A 2nd. Rotator cuff tears of the hip. Clin Orthop Relat Res. 1999;368:135–40.
13. Lungu E, Michaud J, Bureau NJ. US assessment of sports-related hip injuries. Radiographics. 2018;38(3):867–89.
14. Blankenbaker DG, Ullrick SR, Davis KW, De Smet AA, Haaland B, Fine JP. Correlation of MRI findings with clinical findings of trochanteric pain syndrome. Skelet Radiol. 2008;37(10):903–9.
15. Bunker TD, Esler CN, Leach WJ. Rotator-cuff tear of the hip. J Bone Joint Surg Br. 1997;79(4):618–20.
16. Bird PA, Oakley SP, Shnier R, Kirkham BW. Prospective evaluation of magnetic resonance imaging and physical examination findings in patients with greater trochanteric pain syndrome. Arthritis Rheum. 2001;44(9):2138–45.
17. Docking SI, Cook J, Chen S, Scarvell J, Cormick W, Smith P, et al. Identification and differentiation of gluteus medius tendon pathology using ultrasound and magnetic resonance imaging. Musculoskelet Sci Pract. 2019;41:1–5.
18. Choi YS, Lee SM, Song BY, Paik SH, Yoon YK. Dynamic sonography of external snapping hip syndrome. J Ultrasound Med. 2002;21(7):753–8.
19. Neal C, Jacobson JA, Brandon C, Kalume-Brigido M, Morag Y, Girish G. Sonography of morel-Lavallee lesions. J Ultrasound Med. 2008;27(7):1077–81.
20. Mellado JM, Bencardino JT. Morel-Lavallée lesion: review with emphasis on MR imaging. Magn Reson Imaging Clin N Am. 2005;13(4):775–82.
21. Hegazi TM, Belair JA, McCarthy EJ, Roedl JB, Morrison WB. Sports injuries about the hip: what the radiologist should know. Radiographics. 2016;36(6):1717–45.
22. Hak DJ, Olson SA, Matta JM. Diagnosis and management of closed internal degloving injuries associated with pelvic and acetabular fractures: the Morel-Lavallée lesion. J Trauma. 1997;42(6):1046–51.
23. Renström PA. Tendon and muscle injuries in the groin area. Clin Sports Med. 1992;11(4):815–31.
24. Bass CJ, Connell DA. Sonographic findings of tensor fascia lata tendinopathy: another cause of anterior groin pain. Skelet Radiol. 2002;31(3):143–8.
25. Rey GA, Señorans CF, Jaén TF. 12 non-insertional tensor fascia Lata tendinopathy: atypical presentation and undescribed us findings. Br J Sports Med. 2014;48(Suppl 2):A8–A.
26. Hung CY, Chang KV, Özçakar L. Reappraisal on the role of ultrasound imaging for lateral hip pain in a case with tensor fascia Lata tear. Am J Phys Med Rehabil. 2015;94(7):e63–4.
27. Mautner K, Colberg RE, Malanga G, Borg-Stein JP, Harmon KG, Dharamsi AS, et al. Outcomes after ultrasound-guided platelet-rich plasma injections for chronic tendinopathy: a multicenter, retrospective review. PM R. 2013;5(3):169–75.
28. Sher I, Umans H, Downie SA, Tobin K, Arora R, Olson TR. Proximal iliotibial band syndrome: what is it and where is it? Skelet Radiol. 2011;40(12):1553–6.

Chapter 23
Medial Hip

Reasons to Do the Study
1. Medial thigh pain
2. Groin pain

Questions We Want Answered
1. What is the cause of groin or thigh pain? Is it athletic pubalgia, adductor tendinopathy, muscle strain, or "thigh splints?"
2. Is an additional imaging modality different from ultrasound necessary to confirm the cause of pain, such as with an inguinal hernia or femoroacetabular impingement?

Basic Anatomy

Bones (Fig. 23.1)

Muscles (Fig. 23.2)

Five adductor muscles originate from the pubic bone; all but the gracilis insert on the femur. The adductor muscles are arranged in three layers. The anterior (superficial) layer is composed of the **p**ectineus, adductor **l**ongus (AL), and **g**racilis. The more posterior middle layer is the adductor **b**revis (AB), and the most posterior deep layer is the adductor **m**agnus (AM). Thus, a mnemonic to remember the adductor muscles from anterior to posterior is "**P**atty's **L**egs **G**ot **B**ig **M**uscles."

The contents do not represent the views of the U.S. Department of Veterans Affairs or the United States Government.

M. H. Greenberg et al., *Manual of Musculoskeletal Ultrasound*,
https://doi.org/10.1007/978-3-031-37416-6_23

The AL originates from the body of the pubis near the pubic tubercle; the AB originates from the body of the pubis and inferior pubic ramus; and the AM originates from the inferior pubic ramus and ischial tuberosity.

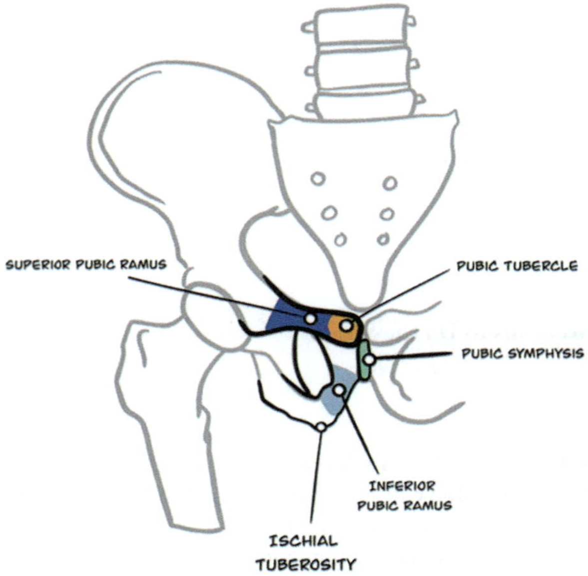

Fig. 23.1 Focused bony anatomy of the medial hip

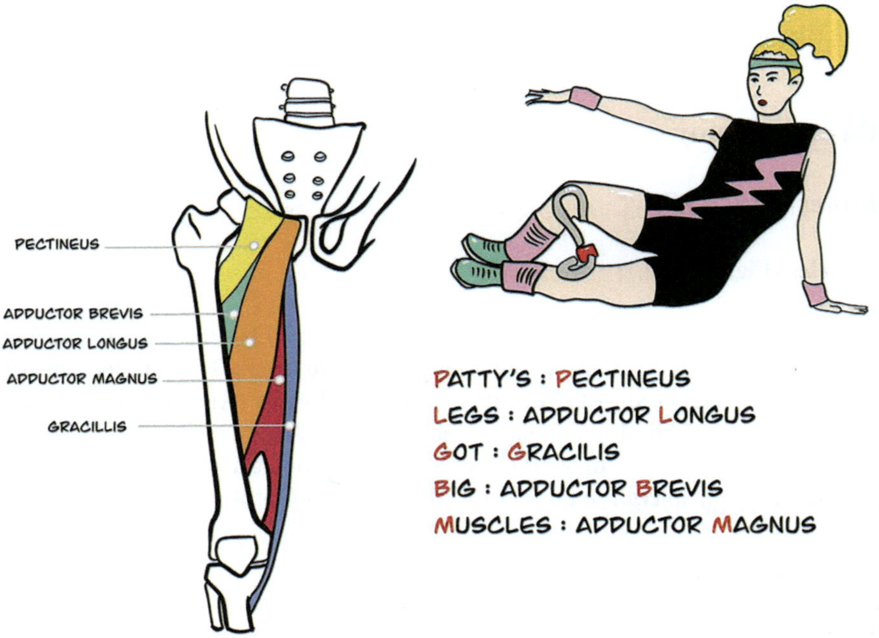

Fig. 23.2 Muscular anatomy of the medial hip

Clinical Comments

Athletic Pubalgia

Sports-related injuries may result in groin pain. Athletic pubalgia (sports or sportsman's hernia) is a general term describing conditions from a variety of causes, including tears of the (abdominal) conjoined tendon (common aponeurosis of the internal abdominal oblique and transverse abdominis muscles) or other soft tissue injuries in this area [1]. The AL and the rectus abdominis share a common aponeurosis anterior (superficial) to the pubic bone, and both attach to the pubic symphysis [2]. The putative precipitating injury is physical activity-related trauma to the common aponeurosis of the rectus abdominis and AL [3]. Participation in tennis, hockey, and football is often a factor.

Symptoms include unilateral groin pain at rest and during exercise. There is often a preceding episode of acute tearing sensation in this location [3]. Pain may be reproduced with resisted hip abduction and sit-ups. Point tenderness may be noted at the pubic attachments of the rectus abdominis and AL. Ultrasound (US) may reveal enthesopathy, tendinosis, or a complete or partial tear of the common aponeurosis [3]. Magnetic resonance imaging (MRI) may reveal reactive bony changes at the symphysis pubis, including bone marrow edema. Some patients may have inflammation of the pubic bone, termed osteitis pubis [3].

Tendinosis and Partial-Thickness Tears

Adductor tendinopathy and tears (partial or complete) may also occur separately from the common aponeurosis of the rectus abdominis and AL. These may be precipitated by an acute event or chronic overuse [3]. Acute tears of the AL muscle may occur in football players and present with pain in the upper medial thigh, worsened by resisted hip adduction. Adductor injuries typically occur with forced hip hyperadduction during contraction of the adductor muscle group [4]. The AL and gracilis are most commonly affected by tendinopathy [5].

Ultrasound may reveal hypoechoic thickening of the tendon compared to the asymptomatic side [6]. Ultrasound findings may include insertional enthesopathy and possibly hyperemia detected by Doppler [7]. Pain due to sonopalpation may also be present. Intratendinous tears may be seen as anechoic clefts [7]. Tendon rupture is suspected if the muscle is retracted. Bony avulsion may occur with tendon rupture, particularly in adolescents [7]. Magnetic resonance imaging is more sensitive than US for detecting low-grade injuries and chronic abductor tendinopathy [5].

Muscle Strain

The AL and gracilis may also be subject to muscle strain. An MRI with contrast is the best modality to visualize a deep adductor muscle tear.

Adductor Insertion Avulsion Syndrome

At the adductor insertion onto the posteromedial femur, chronic repetitive stress trauma has been termed "thigh splints" [8, 9]. This may result in periostitis and stress fractures. It occurs in cheerleading, soccer, hockey, and football participants, as well as military recruits. Symptoms occurring after physical activity include pain in the groin, hip, and thigh, which may be improved with rest [8, 9]. There may be a palpable mass due to periostitis. Radiographs may be normal or show periosteal proliferation or even a stress fracture. Magnetic resonance imaging and nuclear bone scans have proven diagnostically helpful [8, 9]. Ultrasound may show cortical irregularity, localized insertional hypoechogenicity, hyperemia on power Doppler (PD), and pain reproduction with sonopalpation over the affected area of the posterior medial femur. Differential diagnoses include articular hip disease, femoral stress fracture, infection, and neoplasm [8, 9].

Other Sources of Pain

These may include inguinal hernia and femoroacetabular impingement [10]. This Manual does not cover the evaluation of inguinal hernias, but information about *femoroacetabular* impingement is found in Chap. 20. Clinicians treating athletes frequently utilize ultrasound for point-of-care determination of injury type and severity.

Method

The patient's leg should be in the "frog leg" position (abduct and externally rotate the hip, and bend the knee). The curvilinear (low frequency) probe is usually necessary, but a linear transducer with an extended field of view (i.e., virtual convex) may be beneficial for very thin subjects. Once identified in orthogonal views, each medial thigh muscle can be visualized along a short axis from proximal to distal.

Protocol Image 1: 6 cm Distal to the Superior Pubic Ramus, Transverse (Fig. 23.3)

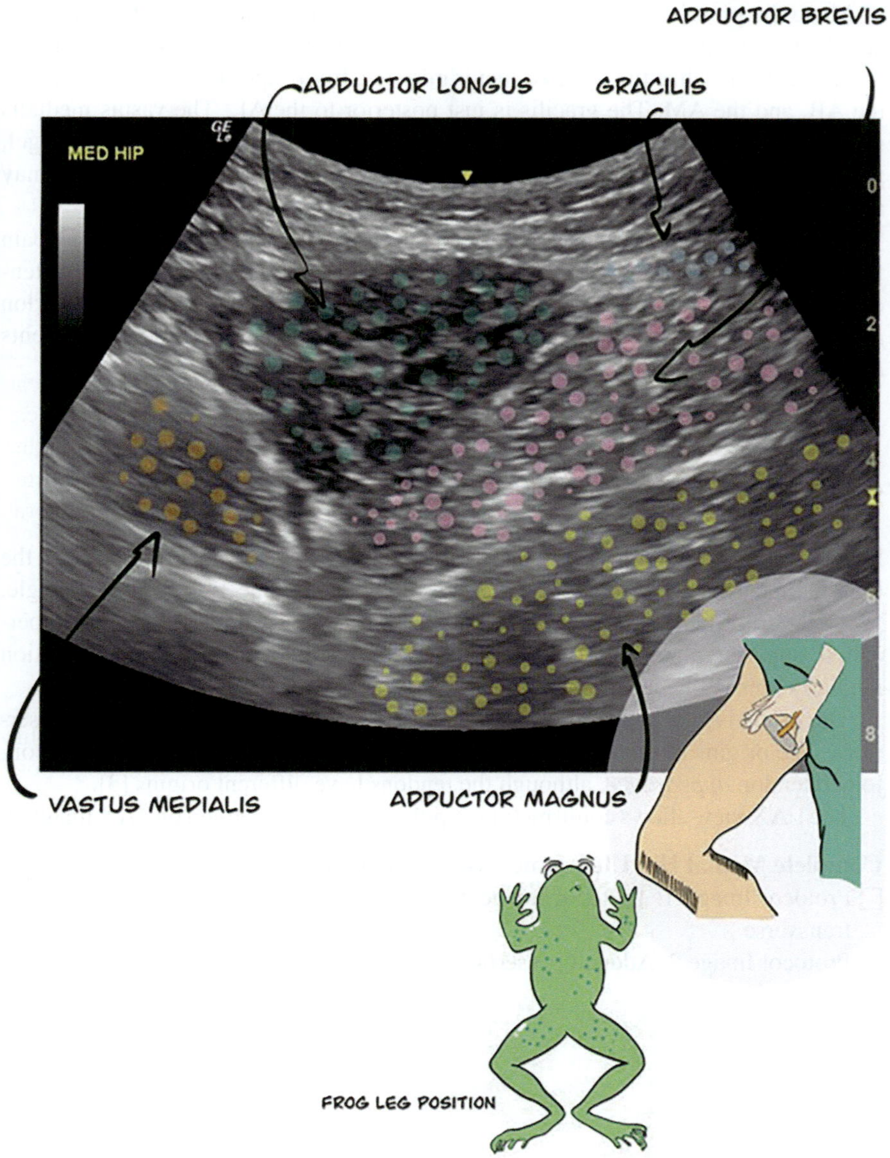

Fig. 23.3 Protocol Image 1: 6 cm distal to the superior pubic ramus, transverse

The origin of the muscles topographically creates a discernible triangle at the medial thigh. Place the transducer on this triangle in the transverse axis (TAX) position. The posteromedial end of the transducer should be angled in a caudal direction of about 35 degrees, which should parallel the superior pubic ramus.

Move the probe cephalad (TAX slide) until the hyperechoic pubic bone is seen, and then, maintaining the same angle, move the transducer in a TAX slide caudally about 6 cm. Visualize the three adductor muscles, from superficial to deep: the AL, the AB, and the AM. The gracilis is just posterior to the AL. The vastus medialis (VM) is anterior to the AL at about the same depth as the AB. Use copious gel, particularly with a curvilinear probe, to preserve contact with the skin. This may also require exerting some downward pressure on the transducer.

Perform TAX slides as necessary to scan the area of complaint and note any pain from sonopalpation. Sonographically-detectable injuries include enthesopathy, tendinosis, complete or partial tears, bony avulsion, muscle tears, muscle or tendon retraction, cortical irregularity, and hyperemia on Doppler. (See Clinical Comments above for specific pathology.)

Protocol Image 2: Adductor Insertions, longitudinal (Fig. 23.4)

Rotate the probe 90 degrees to visualize the insertion of the adductor muscles in the longitudinal axis (LAX). The transducer should bisect the topographic triangle. Move the transducer in TAX slides medially and laterally along the curved hyperechoic superior pubic ramus until you see the hill-like most prominent portion medially; this is the pubic tubercle.

The probe is now in LAX orientation with respect to the adductor tendon insertions. The origins of the AL and the deeper AB near the pubic tubercle give a conjoined tendon *appearance,* although the tendons have different origins [3].

The LAX view allows confirmation of putative pathology noted on TAX imaging.

Complete Medial Hip Ultrasonic Examination Checklist

☐ Protocol Image 1: Protocol Image 1: 6 cm distal to the superior pubic ramus, transverse

☐ Protocol Image 2: Adductor insertions, longitudinal

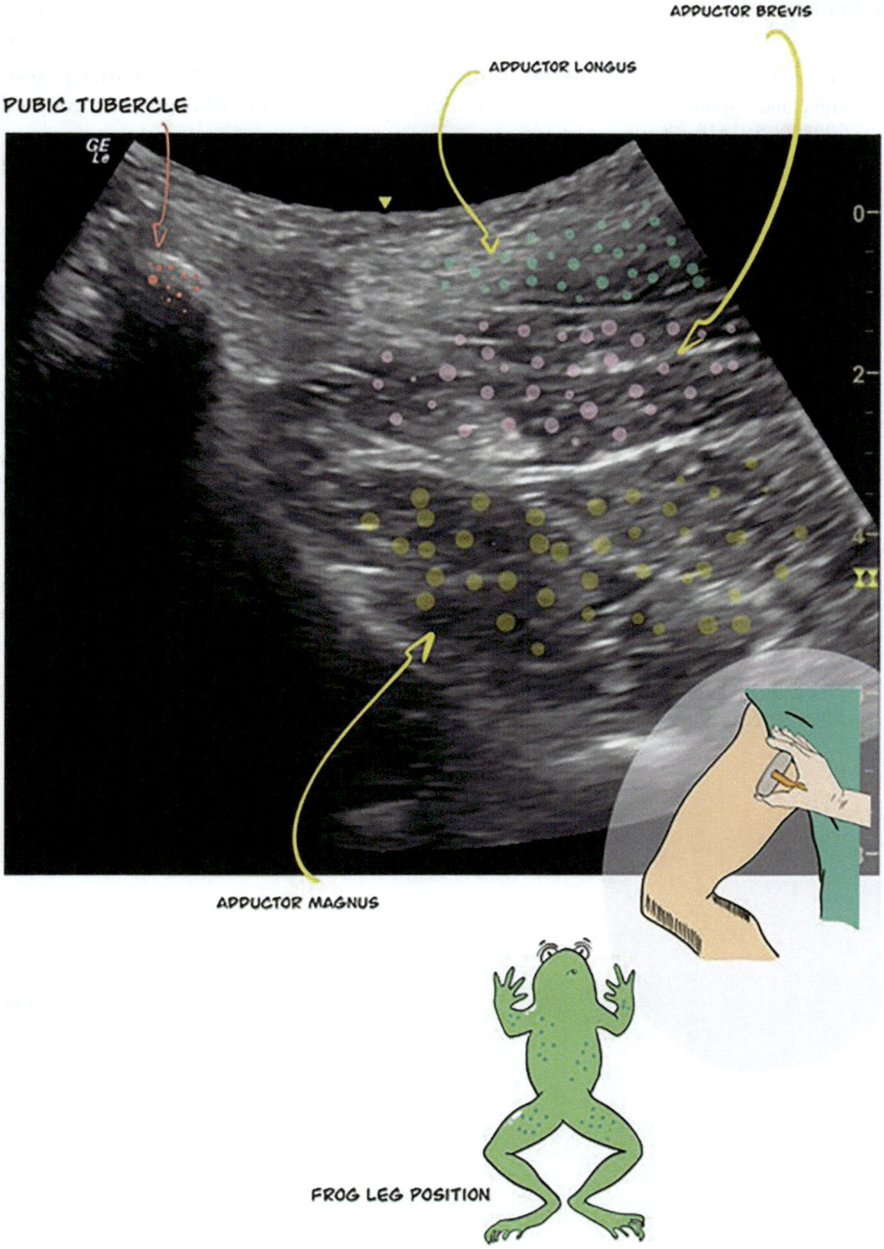

Fig. 23.4 Protocol Image 2: Adductor insertions, longitudinal

References

1. Omar IM, Zoga AC, Kavanagh EC, Koulouris G, Bergin D, Gopez AG, et al. Athletic pubalgia and "sports hernia": optimal MR imaging technique and findings. Radiographics. 2008;28(5):1415–38.
2. Morley N, Grant T, Blount K, Omar I. Sonographic evaluation of athletic pubalgia. Skelet Radiol. 2016;45(5):689–99.
3. Lungu E, Michaud J, Bureau NJ. US assessment of sports-related hip injuries. Radiographics. 2018;38(3):867–89.
4. Rizio L 3rd, Salvo JP, Schürhoff MR, Uribe JW. Adductor longus rupture in professional football players: acute repair with suture anchors: a report of two cases. Am J Sports Med. 2004;32(1):243–5.
5. Robinson P, Barron DA, Parsons W, Grainger AJ, Schilders EM, O'Connor PJ. Adductor-related groin pain in athletes: correlation of MR imaging with clinical findings. Skelet Radiol. 2004;33(8):451–7.
6. Lin Y-T, Wang T-G. Ultrasonographic examination of the adult hip. J Med Ultrasound. 2012;20:201–9.
7. Pesquer L, Reboul G, Silvestre A, Poussange N, Meyer P, Dallaudière B. Imaging of adductor-related groin pain. Diagn Interv Imaging. 2015;96(9):861–9.
8. Weaver JS, Jacobson JA, Jamadar DA, Hayes CW. Sonographic findings of adductor insertion avulsion syndrome with magnetic resonance imaging correlation. J Ultrasound Med. 2003;22(4):403–7.
9. Sofka CM, Marx R, Adler RS. Utility of sonography for the diagnosis of adductor avulsion injury ("thigh splints"). J Ultrasound Med. 2006;25(7):913–6.
10. Naal FD, Dalla Riva F, Wuerz TH, Dubs B, Leunig M. Sonographic prevalence of groin hernias and adductor tendinopathy in patients with femoroacetabular impingement. Am J Sports Med. 2015;43(9):2146–51.

Chapter 24
Crystalline Disease

Our primary focus is on two crystals: monosodium urate (MSU) [which may cause gout] and calcium pyrophosphate dihydrate (CPP) [which may cause CPPD deposition disease, or CPPD disease, for short]. In the past, CPPD disease has been referred to as pseudogout [1]. Refer to OMERACT definitions of ultrasound (US) pathology for MSU and CPPD crystals in other sections: **Getting Started**, **Crystal Disease**, and **Hand Arthropathy**, as well as Reference 1 [2]. Please refer to additional references for a clinical review of gout and CPPD [3].

Reasons to Do the Study

1. Evaluate for the possibility of gout or acute CPPD disease arthritis causing joint inflammation, particularly acute monoarthritis.
2. Evaluate the possibility of chronic gout or CPPD disease causing structural changes in joints as well as bone erosion.
3. Evaluate for the presence of MSU crystals and tophi in patients with hyperuricemia.
4. Evaluate for CPP crystals in patients with polyarthropathy who have associated conditions such as aggressive osteoarthritis, hyperparathyroidism, hemochromatosis, hypomagnesemia, hypophosphatasia, or hypothyroidism [4].
5. To detect a small effusion or bursitis for injection or aspiration.

Questions We Want Answered

1. Is crystal disease the primary cause of acute or chronic arthropathy?
2. Is crystal disease mimicking another inflammatory arthritis, such as rheumatoid or psoriatic arthritis?
3. Is crystal disease a concurrent cause of polyarthropathy along with another condition?

The contents do not represent the views of the U.S. Department of Veterans Affairs or the United States Government.

M. H. Greenberg et al., *Manual of Musculoskeletal Ultrasound*,
https://doi.org/10.1007/978-3-031-37416-6_24

4. Is crystal disease present, active, or inactive and contributing to structural changes or patient discomfort?
5. Are CPP or MSU crystals present but simply "innocent bystanders" in a patient with polyarthritis?

Clinical Comments

High-frequency US excels in the evaluation of gout and CPPD. There are *similarities* between these two conditions [1, 5].

1. Both may mimic other forms of arthritis in any joint, including small joints.
2. Each may have three phases: acute (severe inflammation), intercritical (clinically quiescent), and chronic.
3. Both may deposit crystals in joints and soft tissue but may NOT be causing any clinical problems.
4. Both tend to cause joint inflammation in a monoarticular pattern, less often oligoarticular (2–4 joints), but occasionally polyarticular.
5. Ultrasound may be supportive of one or the other diagnosis by demonstrating distinct patterns consistent with the diagnosis.
6. Both require joint aspiration and rapid synovial fluid analysis for a definitive diagnosis.
7. Gout and CPPD disease each have predilections for certain joints, although overlap may occur.
   (a) Gout: First metatarsophalangeal joint (MTPJ), midfoot, ankle, and knee
   (b) CPPD disease: Knee and wrist

Gout

The most valuable US sign of gout is the **double contour sign (DCS)** (Fig. 24.1). DCS occurs when MSU crystals deposit on the external surface of cartilage, forming a band [2]. On US, the DCS is a hyperechoic layer, continuous or discontinuous, nearly as thick as the underlying bony cortex, and does not change with different insonation angles. A DCS may cover any cartilage surface but favors the first MTPJ, the tibiotalar joint, and the distal femoral cartilage [6]. The other helpful sonographic signs of MSU crystals are **tophi** and **MSU aggregates**.

Tophi are hyperechoic and are either oval or amorphous (Fig. 24.2) [2]. They vary in size and echogenicity and may have an acoustic shadow. There may also be a hypoechoic halo surrounding the tophus. Tophi are often present within joint capsules and other soft tissues such as bursae, tendons, and soft tissue surrounding joints and tendons, but they may occur anywhere. **MSU aggregates** are small hyperechoic foci found in synovial fluid, synovium, and other soft tissues, usually

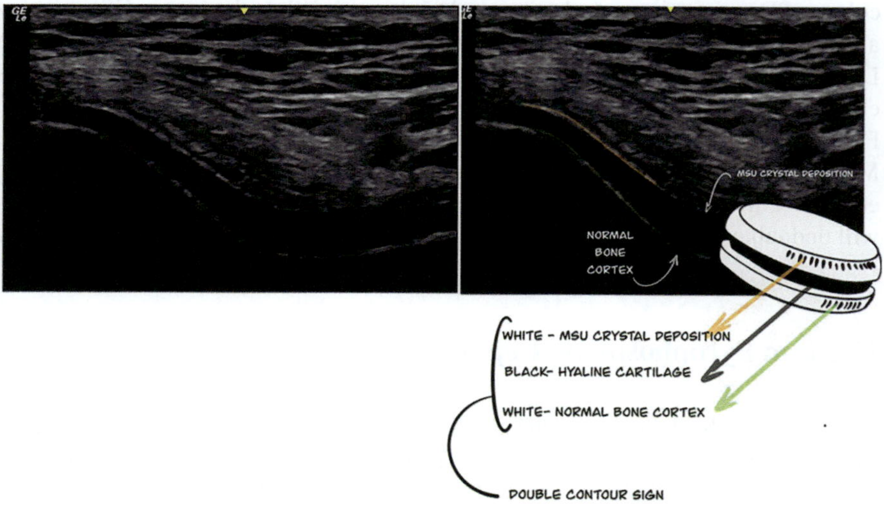

Fig. 24.1 Double contour sign with illustration

Fig. 24.2 Tophus

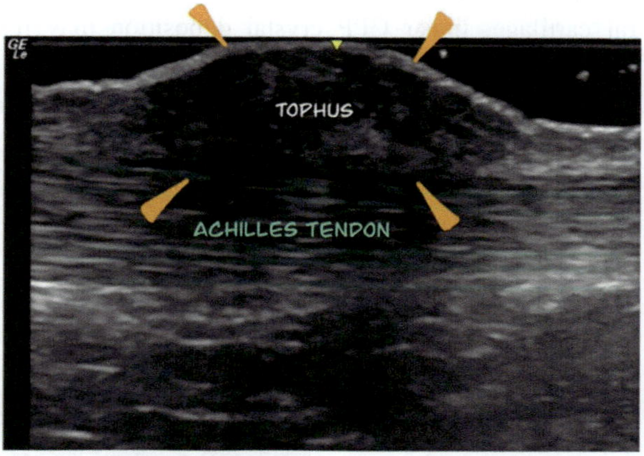

without acoustic shadowing and exhibiting no change with insonation angle [2]. MSU aggregates mimic "sugar clumps" or "snowstorm" patterns. In the proper clinical setting, MSU cloudy areas, or "snowstorms," along with synovial hypertrophy and power Doppler (PD) activity, argue for acute gout.

A positive DCS is listed as a criterion for gout classification [7]. Regarding the use of sonography for treatment decisions, US findings do not directly influence treatment decisions for gout. When considering gout treatment, the presence of *a DCS alone is not a sufficient reason for treatment* with uric acid-lowering therapy [8]. However, since the DCS may help establish the diagnosis of gout, US plays a vital role in determining which patients are under consideration for treatment.

Chronic gout may cause bony erosion, an abnormal discontinuity of the smooth cortex of the bone that should be verified in two planes. Gout may also demonstrate color or PD activity in joints, bursae, and tendons [9, 10]. In the future, US may play a more significant role in gout treatment. Sequential sonographic follow-up of the DCS and tophi may help monitor treatment efficacy and adherence since US changes occur as early as 3 months after urate level normalization [11, 12]. Furthermore, patients with asymptomatic hyperuricemia may demonstrate tophi, MSU aggregates, joint capsule swelling (with or without PD activity), and bone erosion. Current gout guidelines do not address such subclinical activity, but this will undoubtedly be the subject of future investigation [13–16].

Calcium Pyrophosphate Deposition Disease

For CPP crystals, the sonographic sign is calcium deposition within cartilage (hyaline and fibrocartilage) (Figs. 24.3 and 24.4). CPP crystals appear as hyperechoic discontinuous or continuous areas of varying size, which may cause an acoustic shadow if sufficiently large or dense [2]. This calcification may occur within any cartilage but, in particular, is noted in the distal anterior femoral cartilage, medial and lateral menisci, and the triangular fibrocartilage of the wrist. In the distal femoral cartilage, linear CPP crystal deposition may resemble a sandwich cookie.

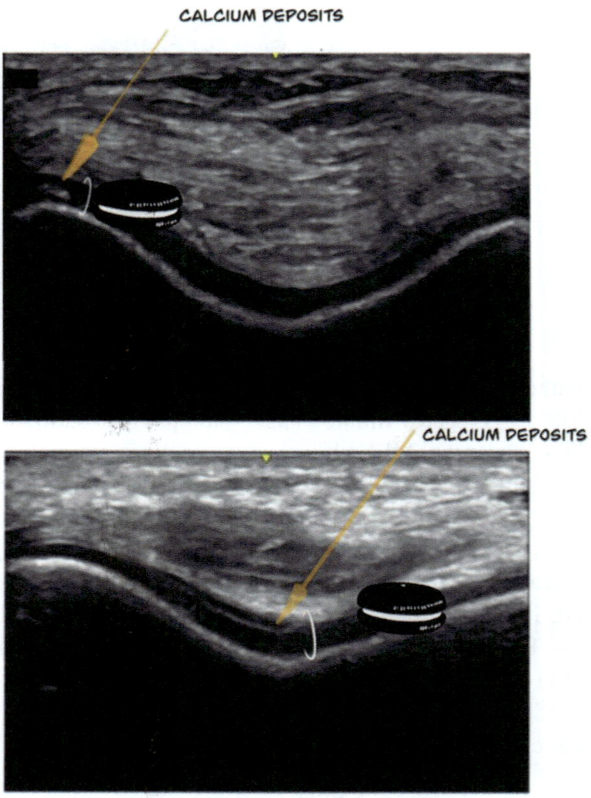

Fig. 24.3 Calcium deposition in hyaline cartilage

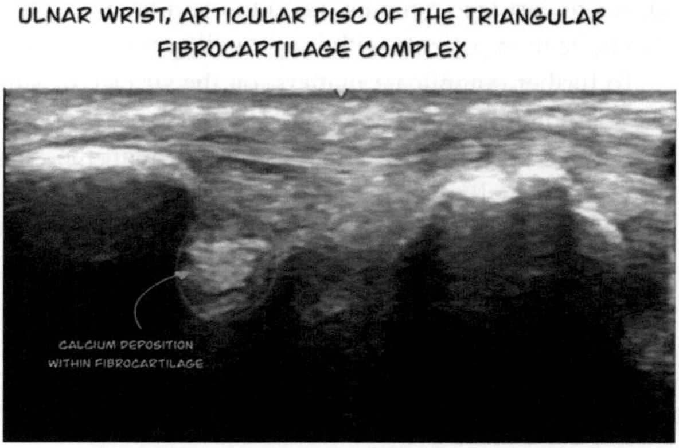

Fig. 24.4 Calcium deposition in fibrocartilage

Calcium deposition may also occur in tendons, ligaments, and entheses. Cartilage calcification may be seen on radiographs, but US is a more sensitive detector of soft tissue calcium [17]. The finding of CPP crystal aggregates on US, particularly in cartilage, is considered to be corroborative but not diagnostic for CPPD disease [1, 4]. Thus, similar to gout, the sonographic findings of CPP crystals do not direct therapeutic decisions.

Be aware that other causes of soft tissue calcium deposition may mimic CPP crystals on US. The domain of CPPD disease is primarily within the cartilage. Still, other causes or conditions associated with cartilage calcification (with or without CPPD disease) include hyperparathyroidism, hemochromatosis, gout, Wilson's disease, ochronosis, hypophosphatasia, hypothyroidism, and hypomagnesemia [18]. In addition, there are causes of soft tissue calcification NOT in cartilage: dystrophic calcification, calcinosis cutis, and basic calcium phosphate. One variety of basic calcium phosphate is hydroxyapatite deposition.

Hydroxyapatite deposition disease (HADD) may cause calcium deposits in the shoulder and other tendons and joints [19]. HADD deposits are not typically in cartilage, occur most often in middle-aged females, and are of unclear origin. There are several HADD stages: pre-calcific, formative, resorptive, and post-calcific [20, 21]. The resorptive, or "slurry," stage is the most inflammatory, but many individuals are asymptomatic. Although typically asymptomatic, HADD may cause structural damage in the shoulder, such as rotator cuff tears, glenohumeral joint narrowing, and bone destruction [22]. The destructive change of HADD has been called Milwaukee shoulder syndrome. More detailed reviews of HADD beyond the scope of this manual are available [23, 24].

Sonographic Overlap

One location of potential sonographic confusion is the dorsal metacarpophalangeal joint (MCPJ). Chronic CPPD disease may cause cartilage destruction, resulting in structural changes mimicking an erosive polyarthropathy such as rheumatoid arthritis [25]. However, CPPD crystals may be seen within the MCPJ cartilage if sufficient cartilage remains. In addition, a DCS on the MCPJ cartilage surface may also be seen in patients with longstanding polyarticular gout.

To further complicate matters, on the surface of smooth, intact MCPJ cartilage, one may find the interface reflex (IR) artifact. An IR is easily differentiated from a DCS since the fragile IR line disappears with the change of the insonation angle [26]. The IR will move along the cartilage surface as you passively flex the finger at the MCPJ and is only present when the probe is 90 degrees to a particular section of the cartilage. Also, the IR is not nearly as thick as the DCS.

Pitfalls

1. According to the 2015 gout classification criteria, the ultrasonic finding of DCS is counted if the DCS is found *in an inflamed joint or a different joint if that joint was previously symptomatic.*
2. There is increased sensitivity for the detection of a DCS by looking at contralateral and noninflamed joints. However, for the DCS to count towards a gout classification, the joint must be currently or previously symptomatic.
3. Always remember that infection or other causes of arthritis may concurrently be present with crystal disease. It is critical to examine synovial fluid whenever possible, not just for the presence of crystal disease but also for infection.
4. Be careful when looking at joints where an IR may mimic a DCS.
5. When looking for calcium deposits, be aware that not all calcium deposits may be from CPPD.
6. When looking at the triangular fibrocartilage and knee menisci, be mindful that the bulk of calcium deposition may appear superficial since deeper portions of the structures can be hypoechoic from acoustic shadowing artifacts.
7. DCS and MSU aggregates may be found in patients without gout or even with normal uric acid levels [27].
8. Consider that some hyperuricemic patients may have subclinical activity, which may be delineated as synovitis by PD or by the presence of bone erosion.
9. Since US is more sensitive at detecting calcium deposition and bone erosion than radiographs, do not rely solely on radiographs to exclude these conditions.
10. Ultrasound demonstration of CPPD crystals at the knee is an accurate tool compared to surgical histopathologic evaluation as the gold standard [28].

Method

The following protocol is a general crystalline US examination. It will need to be customized based on the clinical suspicion of gout or CPPD disease. It is essential to evaluate asymptomatic and contralateral joints since crystal deposition may be present in joints that are not inflamed. Ask the patient which joints have ever been

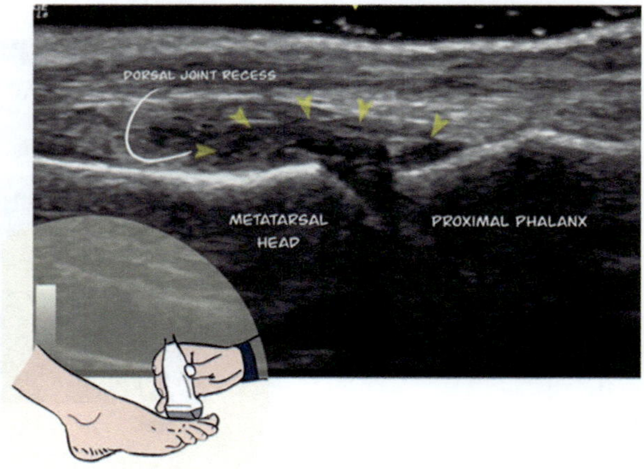

Fig. 24.5 Protocol Image 1a: Dorsal 1st metatarsophalangeal joint, longitudinal ± power Doppler

inflamed, and focus on those joints. Capture an additional PD view of the joint or structure for almost every image listed below.

Protocol Image 1a: Dorsal First Metatarsophalangeal Joint, Longitudinal ± Power Doppler (Fig. 24.5)

The patient is seated and semi-upright. On the affected side, the hip should be partially flexed and the inferior heel in contact with the examination table. The first toe should be in a neutral position, flat on the exam table. Position the probe in a longitudinal axis (LAX) orientation using a high-frequency preset. The joint should be still when evaluated with PD to avoid motion artifacts.

Center the probe over the first MTPJ. Evaluate the distal metatarsal (MT) head and proximal phalanx for bone erosion. Turn on the PD to look for active soft tissue inflammation. Note the anechoic or hypoechoic cartilage overlying the MT head and look for a hyperechoic coating on the surface of the cartilage, which may indicate a DCS. Remember that an IR accentuated by synovial fluid in the dorsal recess may mimic a DCS. Again, the IR is hyperechoic but not nearly as thick as the DCS, and the IR will change dramatically with the angle of insonation in contradistinction to the DCS [10, 26]. Examine the dorsal recess for distention, crystal aggregates, and tophi. Examine the bony surface of the MT neck.

Protocol Image 1b: Medial First Metatarsophalangeal Joint, Longitudinal ± Power Doppler (Fig. 24.6)

Have the patient externally rotate the hip, keeping the knee flexed into a "frog-leg" position. This fully exposes the medial and plantar first MTPJ. Perform a medial transverse axis (TAX) slide to reach the medial MTPJ. Repeat the

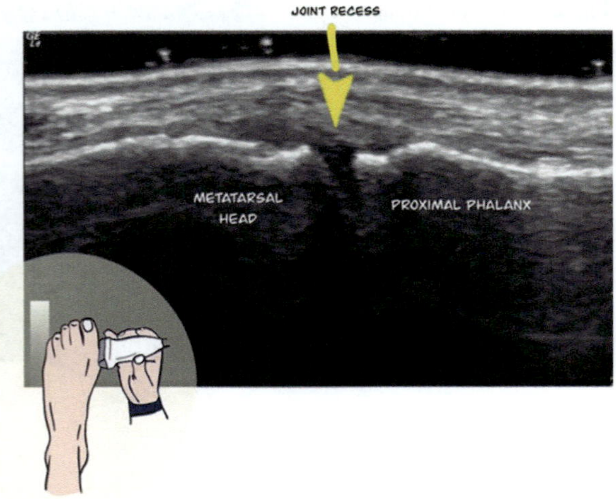

Fig. 24.6 Protocol Image 1b: Medial 1st metatarsophalangeal joint, longitudinal ± power Doppler

examination described in Image 1A, including using PD to look for active soft tissue inflammation. Again, the joint should be supported to avoid motion artifacts with PD.

Protocol Image 1c: Plantar First Metatarsophalangeal Joint, Longitudinal ± Power Doppler (Fig. 24.7)

Slide the probe (still in LAX orientation) to the plantar surface of the first MTPJ to evaluate the structures previously described. The plantar MTPJ is a common site for an IR, so probe movements varying the insonation angle help distinguish between an IR and a DCS. Use PD to look for active soft tissue inflammation.

Protocol Image 2: Tibiotalar Joint, Longitudinal ± Power Doppler (Fig. 24.8)

The patient's foot is flat on the examination table, or the leg is extended with a cushion beneath the calf to expose the anterior ankle. Set the preset for "ankle" and place the probe on the dorsum of the foot in LAX. Locate the hyperechoic tibia and the more distal talus, with the joint space in between. The cartilage surface of the proximal talus is another location to look for a DCS. The tibiotalar joint may exhibit distention with synovitis and possible PD activity. Slide the probe to the medial and lateral aspects of the tibiotalar joint to better encompass the entire joint.

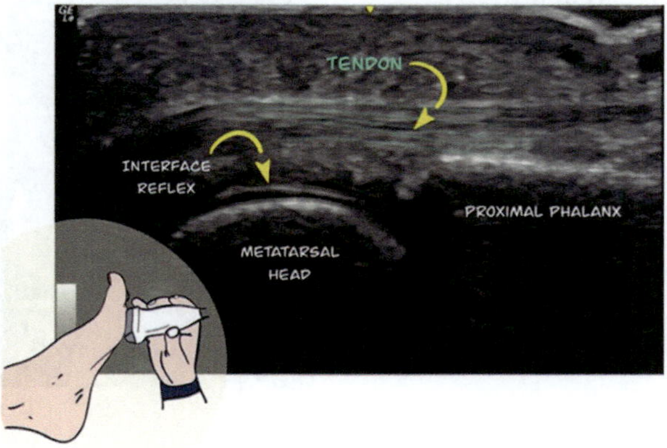

Fig. 24.7 Protocol Image 1c: Plantar 1st metatarsophalangeal joint, longitudinal ± power Doppler

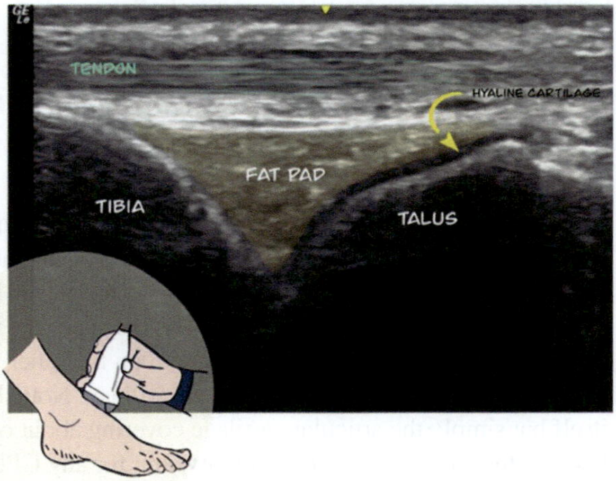

Fig. 24.8 Protocol Image 2: Tibiotalar joint, longitudinal ± power Doppler

Protocol Image 3: Achilles Tendon, Longitudinal ± Power Doppler (Fig. 24.9)

The patient is prone with a bolster underneath the anterior ankle. The foot hangs off the table to enable passive manipulation. With the probe in LAX, visualize the calcaneus, and then focus on the more proximal Achilles tendon. Look for tophaceous deposits or MSU aggregates within and surrounding the tendon (Fig. 24.2). Also, look for calcium deposits within the tendon. Power Doppler may be noted within or surrounding the tendon and in the subcutaneous calcaneal and retrocalcaneal bursae. The PD activity may correlate with visible MSU or CPPD crystals. Look at the enthesis to see if MSU or CPP crystal deposits may be associated with PD activity. Manually flex and extend the ankle, but remember that PD activity diminishes with increased tension on the Achilles tendon [29].

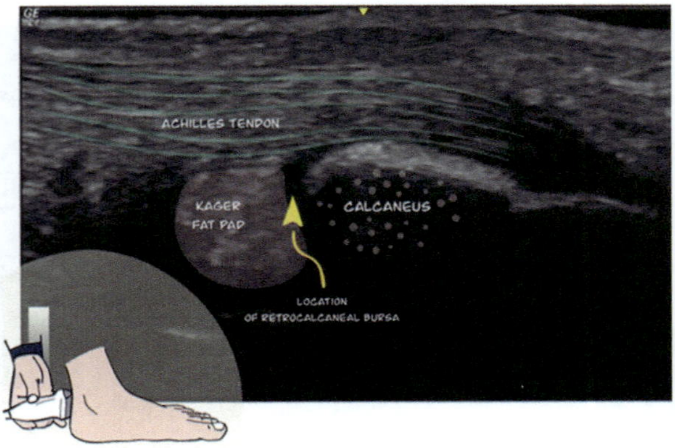

Fig. 24.9 Protocol Image 3: Achilles tendon, longitudinal ± power Doppler

Protocol Image 4: Distal Anterior Femoral Cartilage, Transverse ± Power Doppler (Fig. 24.10)

The patient is semi-upright or supine, and the knee should be maximally flexed. Place a linear probe with a mid-depth setting in TAX orientation across the femoral notch. The transducer should be perpendicular to the table. Move the examination proximally until the femoral notch appears, represented as a hypoechoic "V-shaped" structure. Tilt the probe in one direction and then the other while moving it medially and laterally to delineate the distal femoral cartilage. Note that this is NOT the joint space itself but simply the articular cartilage covering about one-third of the anterior femur. Look at the cartilage thickness, clarity, and for any CPP crystals within the cartilage. Also, examine the superficial surface of the cartilage for a DCS. Again, DCS thickness rivals that of the femoral cortex and persists with different insonation angles.

Protocol Image 5: Medial Meniscus, Longitudinal (Fig. 24.11)

The patient extends the knee, and a bolster is placed behind the knee to bend it to approximately 30 degrees. Rotate the probe to an LAX orientation and place it medially to the distal half of the patella. Slowly move the transducer posteriorly with a TAX slide until the joint space emerges. Tilt the probe to clarify the triangular, slightly hyperechoic medial meniscus. Also, fix the proximal end of the transducer and slightly rotate the distal portion in either an anterior or posterior direction while tilting the probe in either direction to sharpen the image of the medial meniscus. The probe can also be moved posteriorly with a TAX slide to examine the posterior portion of the medial meniscus. Evaluate for calcium deposits within the medial meniscus and the medial collateral ligament. The deeper part of the medial

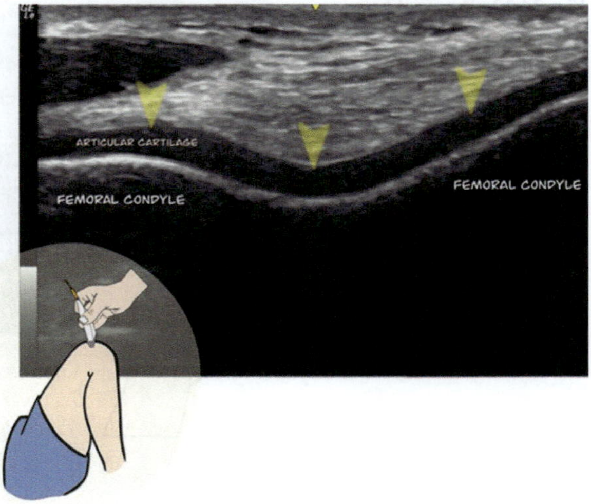

Fig. 24.10 Protocol Image 4: Distal anterior femoral cartilage, transverse ± power Doppler

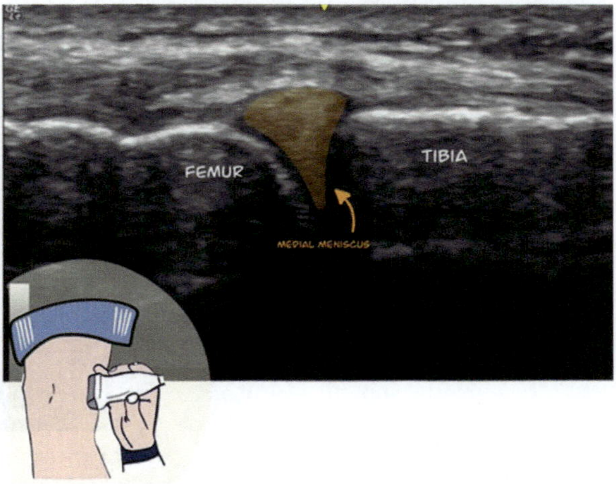

Fig. 24.11 Protocol Image 5: Medial meniscus, longitudinal

meniscus may not exhibit as much calcium due to possible acoustic shadowing from calcium deposition in the superficial portion [30].

Protocol Image 6: Lateral Meniscus, Longitudinal (Fig. 24.12)

With the knee in the same position as in Protocol Image 5, place the probe in LAX orientation just lateral to the distal half of the patella. Slowly perform a TAX slide posteriorly until the joint space is recognized; center the joint space in the field. Tilt the transducer back and forth to identify a sharp image of the lateral meniscus, which is also hyperechoic and triangular. Fixing the proximal portion of the probe and fanning the distal portion anteriorly or posteriorly while slightly tilting in both

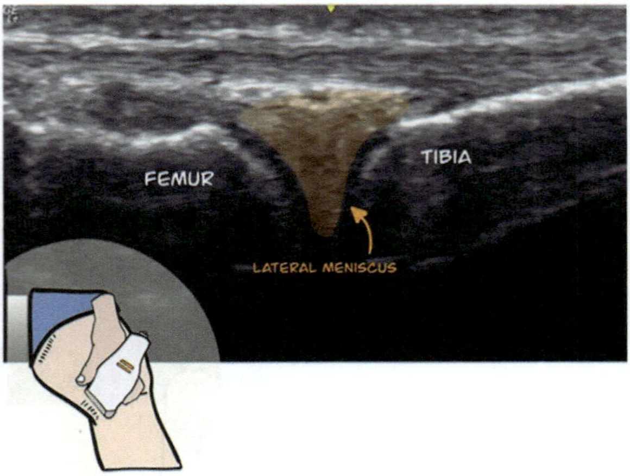

Fig. 24.12 Protocol Image 6: Lateral meniscus, longitudinal

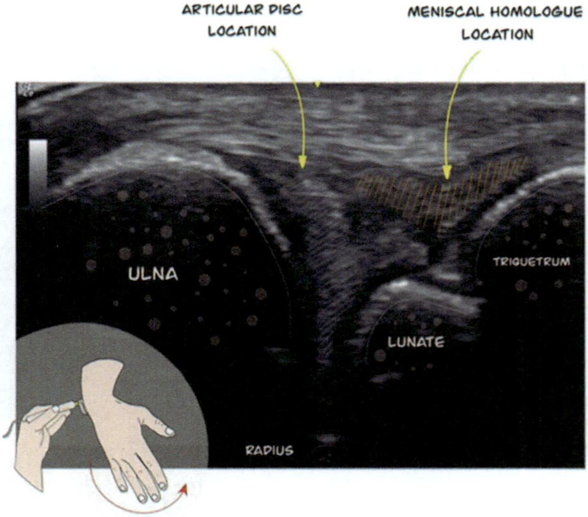

Fig. 24.13 Protocol Image 7: Triangular fibrocartilage complex, longitudinal ± power Doppler

directions will help optimize the image. Again, look for calcifications within the meniscus and the lateral collateral ligament. The probe can be moved posteriorly with a TAX slide to look at the posterior portion of the lateral meniscus.

Protocol Image 7: Triangular Fibrocartilage Complex, Longitudinal ± Power Doppler (Fig. 24.13)

With the patient's hand lying prone and in slight radial flexion, place the high-frequency probe at the ulnar aspect of the wrist in an LAX orientation. One end of the transducer is on the curved, hyperechoic distal ulna, and the other is on the triquetrum, thus encompassing the TFCC. Radial flexion uncovers the triangular fibrocartilage, which is adherent to the distal ulna. Distal to the articular disc is the meniscal homolog, and superficial is the extensor carpi ulnaris tendon. Evaluate the triangular fibrocartilage complex for CPP crystals. If significant superficial calcium deposition is present, you may not be able to visualize deeper calcium deposits due to an acoustic shadowing artifact [30]. The triangular fibrocartilage complex is typically the province of CPPD rather than MSU crystals. Look for PD activity, which may indicate acute CPPD disease arthritis.

Protocol Image 8a: Dorsal Second Metacarpophalangeal Joint, Longitudinal ± Power Doppler (Fig. 24.14)

With the hand prone on a flat surface, place the high-frequency probe in LAX orientation over the second MCPJ. Evaluate the metacarpal head (MCH) for erosions, but do not mistake the normal indentation at the base of the MCH for a true erosion. Look at the hypoechoic or anechoic cartilage covering the MCH. On the cartilage surface, there may be a DCS whose appearance persists despite any change in the insonation angle. The DCS is typically as thick as the underlying cortex. Within the cartilage, there may be chondrocalcinosis associated with CPPD. There may be damage to the cartilage due to chronic CPPD disease or other polyarthropathies. On the cartilage surface, there may be an IR that can mimic a DCS. The IR changes dramatically with any change in insonation angle and is thin and fragile compared to the robust DCS. By gently grasping the patient's fingers and bending the MCPJ, the examiner may demonstrate the extent of the dorsal cartilage covering the MCH. Evaluate the joint capsule for grayscale enlargement and turn on the PD to look for acute synovitis.

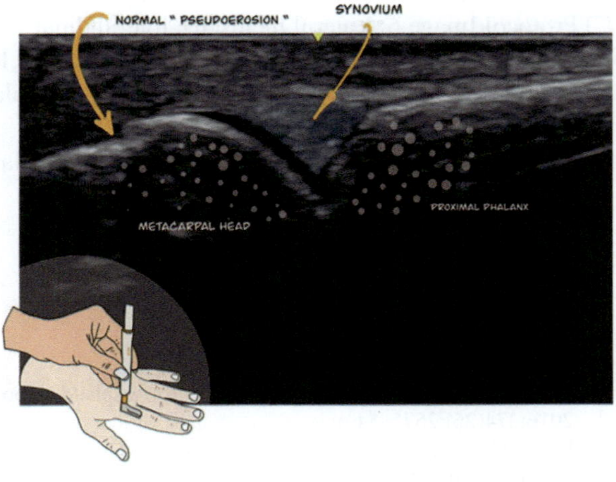

Fig. 24.14 Protocol Image 8a: Dorsal 2nd metacarpophalangeal joint, longitudinal ± power Doppler

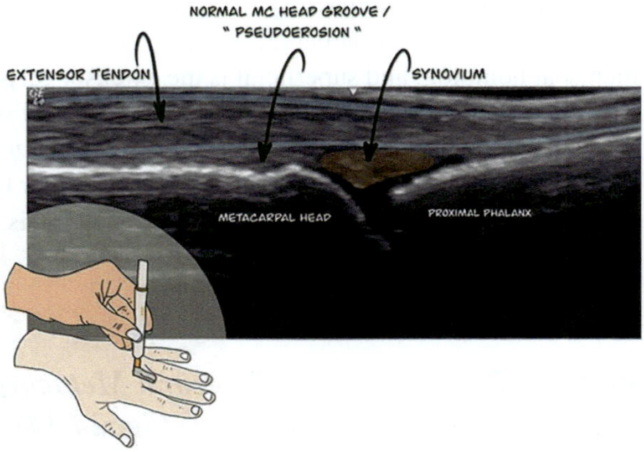

Fig. 24.15 Protocol Image 8b: Dorsal 3rd metacarpophalangeal joint, longitudinal ± power Doppler

Protocol Image 8b: Dorsal Third Metacarpophalangeal Joint, Longitudinal ± Power Doppler (Fig. 24.15)

Repeat the same evaluation as for the second MCPJ.

Crystalline Arthritis Ultrasonic Examination Checklist

☐ Protocol Image 1a: Dorsal first metatarsophalangeal joint, longitudinal ± power Doppler
☐ Protocol Image 1b: Medial first metatarsophalangeal joint, longitudinal ± power Doppler
☐ Protocol Image 1c: Planter first metatarsophalangeal joint, longitudinal ± power Doppler
☐ Protocol Image 2: Tibiotalar joint, longitudinal ± power Doppler
☐ Protocol Image 3: Achilles tendon, longitudinal ± power Doppler
☐ Protocol Image 4: Distal anterior femoral cartilage, transverse ± power Doppler
☐ Protocol Image 5: Medial meniscus, longitudinal
☐ Protocol Image 6: Lateral meniscus, longitudinal
☐ Protocol Image 7: Triangular fibrocartilage complex, longitudinal ± power Doppler
☐ Protocol Image 8a: Dorsal second metacarpophalangeal joint, longitudinal ± power Doppler
☐ Protocol Image 8b: Dorsal third metacarpophalangeal joint, longitudinal ± power Doppler

References

1. Rosenthal AK, Ryan LM. Calcium pyrophosphate deposition disease. N Engl J Med. 2016;374(26):2575–84.
2. Bruyn GA, Iagnocco A, Naredo E, Balint PV, Gutierrez M, Hammer HB, et al. OMERACT definitions for ultrasonographic pathologies and elementary lesions of rheumatic disorders 15 years on. J Rheumatol. 2019;46(10):1388.

3. Greenberg MH. Is the Patient's arthritis from crystal disease. In: Borneman PH, editor. Ultrasound for primary care. Wolters; 2020.
4. Zhang W, Doherty M, Bardin T, Barskova V, Guerne PA, Jansen TL, et al. European league against rheumatism recommendations for calcium pyrophosphate deposition. Part I: terminology and diagnosis. Ann Rheum Dis. 2011;70(4):563–70.
5. Perez-Ruiz F, Castillo E, Chinchilla SP, Herrero-Beites AM. Clinical manifestations and diagnosis of gout. Rheum Dis Clin N Am. 2014;40(2):193–206.
6. Bhadu D, Das SK, Wakhlu A, Dhakad U, Sharma M. Ultrasonographic detection of double contour sign and hyperechoic aggregates for diagnosis of gout: two sites examination is as good as six sites examination. Int J Rheum Dis. 2018;21(2):523–31.
7. Neogi T, Jansen TL, Dalbeth N, Fransen J, Schumacher HR, Berendsen D, et al. 2015 gout classification criteria: an American College of Rheumatology/European league against rheumatism collaborative initiative. Arthritis Rheumatol. 2015;67(10):2557–68.
8. FitzGerald JD, Dalbeth N, Mikuls T, Brignardello-Petersen R, Guyatt G, Abeles AM, et al. 2020 American College of Rheumatology Guideline for the Management of Gout. Arthritis Rheumatol. 2020;72(6):879–95.
9. Stewart S, Dalbeth N, Vandal AC, Allen B, Miranda R, Rome K. Ultrasound features of the first metatarsophalangeal joint in gout and asymptomatic hyperuricemia: comparison with normouricemic individuals. Arthritis Care Res (Hoboken). 2017;69(6):875–83.
10. Girish G, Melville DM, Kaeley GS, Brandon CJ, Goyal JR, Jacobson JA, et al. Imaging appearances in gout. Arthritis. 2013;2013:673401.
11. Ebstein E, Forien M, Norkuviene E, Richette P, Mouterde G, Daien C, et al. Ultrasound evaluation in follow-up of urate-lowering therapy in gout: the USEFUL study. Rheumatology (Oxford). 2019;58(3):410–7.
12. Christiansen SN, Østergaard M, Slot O, Keen H, Bruyn GAW, D'Agostino MA, et al. Assessing the sensitivity to change of the OMERACT ultrasound structural gout lesions during urate-lowering therapy. RMD Open. 2020;6(1):e001144.
13. Neogi T. Asymptomatic hyperuricemia: perhaps not so benign? J Rheumatol. 2008;35(5):734–7.
14. Pineda C, Amezcua-Guerra LM, Solano C, Rodriguez-Henríquez P, Hernández-Díaz C, Vargas A, et al. Joint and tendon subclinical involvement suggestive of gouty arthritis in asymptomatic hyperuricemia: an ultrasound controlled study. Arthritis Res Ther. 2011;13(1):R4.
15. Puig JG, de Miguel E, Castillo MC, Rocha AL, Martínez MA, Torres RJ. Asymptomatic hyperuricemia: impact of ultrasonography. Nucleosides Nucleotides Nucleic Acids. 2008;27(6–7):592–5.
16. Valea AR. SP0090 ultrasound findings in patients with asymptomatic hyperuricemia. Ann Rheum Dis. 2015;74(Suppl 2):23.
17. Gutierrez M, Di Geso L, Filippucci E, Grassi W. Calcium pyrophosphate crystals detected by ultrasound in patients without radiographic evidence of cartilage calcifications. J Rheumatol. 2010;37(12):2602–3.
18. Abhishek A, Doherty M. Update on calcium pyrophosphate deposition. Clin Exp Rheumatol. 2016;34(4 Suppl 98):32–8.
19. Ruban TN, Albert L. Wrist involvement of calcium hydroxyapatite deposition disease. J Rheumatol. 2015;42(9):1724–5.
20. Goller SS, Hesse N, Dürr HR, Ricke J, Schmitt R. Hydroxyapatite deposition disease of the wrist with intraosseous migration to the lunate bone. Skelet Radiol. 2021;50(9):1909–13.
21. Hongsmatip P, Cheng KY, Kim C, Lawrence DA, Rivera R, Smitaman E. Calcium hydroxyapatite deposition disease: imaging features and presentations mimicking other pathologies. Eur J Radiol. 2019;120:108653.
22. Merolla G, Bhat MG, Paladini P, Porcellini G. Complications of calcific tendinitis of the shoulder: a concise review. J Orthop Traumatol. 2015;16(3):175–83.
23. Garcia GM, McCord GC, Kumar R. Hydroxyapatite crystal deposition disease. Semin Musculoskelet Radiol. 2003;7(3):187–93.
24. Beckmann NM. Calcium apatite deposition disease: diagnosis and treatment. Radiol Res Pract. 2016;2016:4801474.

25. Ivory D, Velázquez CR. The forgotten crystal arthritis: calcium pyrophosphate deposition. Mo Med. 2012;109(1):64–8.
26. Lai K-L, Chiu Y-M. Role of ultrasonography in diagnosing gouty arthritis. J Med Ultrasound. 2011;19(1):7–13.
27. Abhishek A, Courtney P, Jenkins W, Sandoval-Plata G, Jones AC, Zhang W, et al. Brief report: monosodium urate monohydrate crystal deposits are common in asymptomatic sons of patients with gout: the sons of gout study. Arthritis Rheumatol. 2018;70(11):1847–52.
28. Filippou G, Scanu A, Adinolfi A, Toscano C, Gambera D, Largo R, et al. Criterion validity of ultrasound in the identification of calcium pyrophosphate crystal deposits at the knee: an OMERACT ultrasound study. Ann Rheum Dis. 2021;80(2):261–7.
29. Zappia M, Cuomo G, Martino MT, Reginelli A, Brunese L. The effect of foot position on power Doppler ultrasound grading of Achilles enthesitis. Rheumatol Int. 2016;36(6):871–4.
30. Filippou G, Adinolfi A, Iagnocco A, Filippucci E, Cimmino MA, Bertoldi I, et al. Ultrasound in the diagnosis of calcium pyrophosphate dihydrate deposition disease. A systematic literature review and a meta-analysis. Osteoarthr Cartil. 2016;24(6):973–81.

Chapter 25
Enthesopathy

Reasons to Do the Study
1. Suspicion of enthesitis
2. Pain at an enthesis

Questions We Want Answered
1. Is there a systemic process causing enthesitis?
2. What precisely is causing pain near an enthesis?
3. Is the pain near or at an enthesis due to a localized problem rather than a systemic one?

Clinical Comments

The enthesis is the insertional point for attachment to bone of tendons, ligaments, fascia, muscles, and joint capsules [1]. OMERACT defines enthesitis as hypoechoic or thickened insertion of the tendon <2 mm from bony surfaces, which may exhibit a Doppler signal if active and may show erosions [2, 3]. Enthesophytes and calcifications may be signs of structural damage.

Enthesitis is predominantly the domain of spondyloarthropathy (SpA), particularly psoriatic arthritis (PsA) [4]. Differentiating seronegative rheumatoid arthritis (RA) from PsA is a challenge compounded by the fact that approximately 15% of patients with PsA exhibit no psoriasis before the arthritis presentation [5]. In patients with seronegative peripheral polyarthritis, sonographic confirmation of enthesitis supports the diagnosis of PsA.

The contents do not represent the views of the U.S. Department of Veterans Affairs or the United States Government.

M. H. Greenberg et al., *Manual of Musculoskeletal Ultrasound*,
https://doi.org/10.1007/978-3-031-37416-6_25

Recall that SpA in adults classically includes ankylosing spondylitis (AS), PsA, reactive arthritis, and enteropathic arthritis (the latter associated with inflammatory bowel disease). These four types may overlap and, more recently, may be classified as axial and/or peripheral SpA [6]. In this newer classification system, enthesitis is a prominent criterion for diagnosing both peripheral and axial SpA [6]. However, this system is based on *clinical* enthesitis detection without the aid of ultrasound (US) or other imaging modalities.

Bear in mind that enthesis involvement is *not* the exclusive realm of SpA. It may also be less commonly associated with RA, gout, other crystal diseases, sarcoidosis, juvenile idiopathic arthritis, diffuse idiopathic skeletal hyperostosis (DISH), and osteoarthritis (OA). Almost any enthesis can be evaluated sonographically, and the US exam should be tailored to the symptoms and clinical suspicion of systemic disease. The more common sonographically accessible areas for tendon enthesis evaluation include the following [7]:

1. Achilles tendon insertion on the calcaneus
2. Plantar fascia insertion on the calcaneus
3. Patellar tendon insertions on the patella and tibial tuberosity
4. Quadriceps tendon insertion on the patella
5. Common extensor tendon insertion on the lateral elbow epicondyle
6. Common flexor tendon insertion on the medial elbow epicondyle
7. Triceps tendon insertion on the olecranon
8. Supraspinatus tendon insertion on the greater tuberosity
9. Flexor and extensor tendons insertions on the phalanges of the hand and foot
10. Tibialis anterior tendon insertion on the medial cuneiform and first metatarsal

The enthesis and the surrounding structures (fibrocartilage, bursae, fat pad, adjacent bone, and fascia) are termed the *enthesis organ,* whose function is to dissipate repetitive physical stress [4, 8]. Mechanical problems can also affect the enthesis, including repetitive injury [1, 9, 10]. Such repetitive mechanical forces may trigger an inflammatory response, leading to localized inflammation [1]. Adjacent to the site of enthesitis, the bone may develop spurs, and eventually, there may be destruction of superficial fibrocartilage, vascular invasion (neovascularization), and inflammatory cell infiltration. In addition to the enthesis being a site of repetitive mechanical stress and localized inflammation, an underlying systemic inflammatory process may reveal itself in this location.

Enthesitis is clinically underdiagnosed by physical examination [1]. Clinical methods to score enthesopathy include the Manders enthesis index (MEI) and the Maastricht ankylosing spondylitis enthesitis score (MASES) [11, 12]. Both scoring systems lack accuracy and depend on the examiner physically exerting direct pressure over an enthesis to elicit pain. One study looked at clinical assessment using MASES compared to US as the gold standard. Depending on the specific MASES enthesis site, the physical examination had a sensitivity and specificity of 71.4–87% and 47.4–75%, respectively [13].

Plain radiographs may show late chronic bony changes (enthesophytes, new bone formation, bony irregularity) and likewise lack sensitivity [1, 14, 15]. Conversely, magnetic resonance imaging (MRI) is quite helpful in detecting enthesitis but remains cost-prohibitive [1, 15]. Compared with physical examination, US is very sensitive for detecting enthesitis, particularly subclinical enthesitis [4, 16, 17].

Ultrasound is diagnostically accurate and helpful for early the diagnosis of SpA, notably PsA [15, 18]. Neovascularization (as demonstrated by the Doppler signal) of the enthesis provides good predictive value for diagnosing SpA [19]. It is also suggested that US be used when monitoring response to treatment since changes with treatment have been sonographically demonstrated [17, 20, 21]. Despite advances, there is *no current gold standard* modality for enthesitis detection [15, 22]. Trials are ongoing to determine the best modality and scoring system to monitor the treatment of enthesitis in PsA [15].

Ultrasound is emerging as the preferred method of enthesitis identification due to its sensitivity (subclinical enthesitis), functional evaluation (Doppler activity), morphologic delineation (including new bone formation), and low cost [4]. Enthesitis reportedly occurs in 35–50% of PsA patients, a more frequent rate than in RA, AS, and OA. However, one study demonstrated that although enthesitis with power Doppler (PD) activity occurs more frequently with PsA and psoriasis, it may also be present in patients with fibromyalgia [23]. On the other hand, the higher the number of US-documented enthesitis sites, the greater the likelihood of PsA being the correct diagnosis rather than fibromyalgia [23, 24].

A scoring system for enthesitis has been developed using a 0–3 grading system [3]:

- Grayscale (GS) (Fig. 25.1):

  Grade 0 = normal
  Grade 1 = hypoechogenicity (loss of fibrillar pattern)
  Grade 2 = hypoechogenicity + thickening + calcifications/enthesophytes
  Grade 3 = Grade 2 features + bone erosions

- Power Doppler (Fig. 25.2):

  Grade 0 = no PD signal
  Grade 1 = <2 punctate signals
  Grade 2 = 2–4 punctate signals or 1 confluent signal
  Grade 3 = >4 punctate signals or >1 confluent signal

Current efforts are being made to study, refine, and validate an enthesitis scoring system [25].

A more recently proposed enthesitis scoring system identified six entheseal sites that effectively differentiated patients with PsA from normal: proximal and distal insertions of the patella tendon, Achilles tendon, plantar fascia, common extensor tendon, and supraspinatus tendon [26]. The elements and grading of this scoring system are as follows:

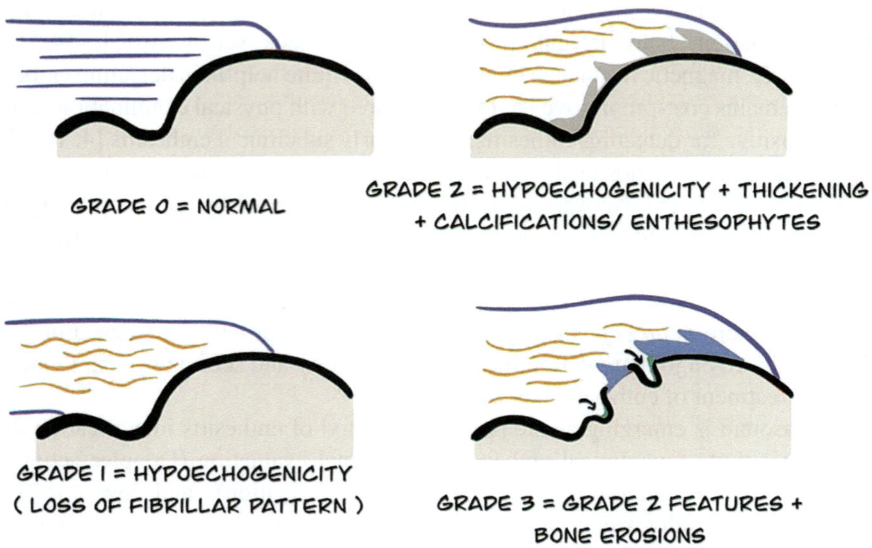

Fig. 25.1 Grayscale enthesitis scoring

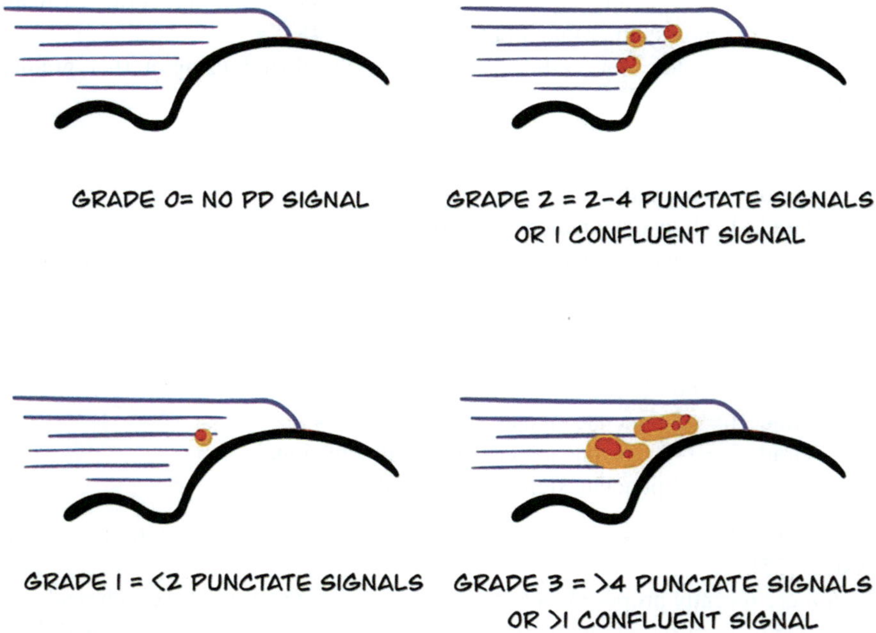

Fig. 25.2 Power Doppler enthesitis scoring

- GS:

  Hypoechogenicity: 0 (absent), 1 (present)
  Tendon/ligament thickening: 0 (absent), 1 (present)
  Bone erosion: 0 (absent), 1 (present)
  "Fluffy" bone irregularities: 0 (absent), 1 (present)
  Enthesophytes: 0 (absent), 1 (small) 2 (medium) 3 (large)
  Calcifications within the tendon/ligament: 0 (absent), 1 (present)
  Bursitis: 0 (absent), 1 (present)

- PD:

  The Doppler signal within 2 mm of the bony cortex is arbitrarily evaluated, although there may be a PD signal within 2–5 mm of cortical bone. Power Doppler grading is 0 (absent), 1 (mild), 2 (moderate), and 3 (severe).

Special Clinical Consideration: Inflammatory Arthritis Differentiation

Differentiating seronegative early rheumatoid arthritis (ERA) from PsA is challenging, particularly in patients without overt psoriasis. Ultrasound may be revealing since enthesitis is a key feature of PsA [14, 27, 28]. Do not neglect sonographic evaluation for enthesitis, particularly in the fingers of such patients since this may tip the diagnostic scale.

Specific US findings supporting a diagnosis present in PsA [29]:

1. At the metacarpophalangeal joint, extensor tendon paratenonitis was 2.5% in ERA vs. 54.1% in early PsA.
2. At the finger proximal interphalangeal joint, enthesitis of the central slip did not occur with ERA *but only in PsA*.

Dactylitis, which is diffuse swelling of a finger or toe ("sausage digit"), most often occurs in SpA and, in particular, PsA; however, it may also be present in sarcoid, gout, syphilis, tuberculosis, and sickle cell disease [30]. Additionally, digits in patients with RA may sometimes demonstrate diffuse swelling. Patients who display dactylitis, distal interphalangeal joint pain, or fingernail changes merit a diligent search for enthesitis [31]. If clinical evaluation reveals nail disease compatible with psoriasis (such as pitting, onycholysis, and subungual hyperkeratosis), perform an US examination of the extensor tendon enthesis at the distal phalanx. Since the nail root fascia is an extension of the extensor tendon, nail disease in PsA is associated with enthesitis [32–34].

Pitfalls

1. *A gentle examiner's touch* is needed to prevent dampening of the Doppler signal. The joint should be in a *relaxed*, neutral position since putting tension on the tendon (or ligament) at the enthesis might diminish PD activity [35].

2. Any US-accessible enthesis can be inspected, even those not included in this suggested protocol. The more superficial the enthesis, the better for sonographic evaluation. Always perform an US exam of any painful area that contains an enthesis.
3. Again, be aware that enthesitis occurs mainly with SpA (and particularly PsA) but also with other conditions such as RA, sarcoidosis, and possibly fibromyalgia. Put this information into proper clinical context if your sonographic examination documents enthesitis with positive PD activity. A sonographic demonstration of enthesitis supports a SpA diagnosis but is not pathognomonic.
4. Evaluation of the hands for enthesitis also bypasses the pitfall of having increased PD activity in weight-bearing entheses, particularly in patients with a high BMI [29].
5. For the fingers and toes, use a high-definition probe set at a minimum of 16 MHz to look at the extensor and flexor entheses.
   (See Chap. 5 for more information.)
6. Do not neglect sonographic evaluation for enthesitis in the hands or feet if symptoms or signs are present, since this may tip the diagnostic scale.
   (See Chap. 5 for more information.)
7. Since nail disease in PsA is associated with enthesitis, clinical findings of nail pitting or thickening should prompt meticulous sonographic assessment of the extensor tendon enthesis at the distal phalanges of the hands and feet.
8. Grade enthesitis according to a scoring system and devote adequate time when pursuing PD activity. However, looking at the contralateral side may detect *subclinical enthesitis* in the presence of systemic disease.
9. In practice, US is best used to confirm a clinical impression of enthesitis and to detect subclinical enthesitis. Ultrasound by itself is not diagnostic.
10. If an inferior calcaneal enthesophyte is noted, examine a plain radiograph; an indistinct margin or "fluffy" periostitis may indicate SpA, as described in one study of PsA [36].
11. Plantar fasciitis may cause PF swelling (>4 mm), loss of fibrillar echotexture, cortical irregularities, and PD activity proximally at the calcaneal insertion [37]. These sonographic features technically meet the enthesitis criteria. However, enthesitis found in other body areas argues for the presence of a systemic condition.

Method

The enthesopathy protocol below is merely a suggestion and should be tailored to the particular patient. Select one of the scoring systems described in Clinical Comments. Often, we are requested to distinguish PsA from normal; the GRAPPA US Working Group Study influenced our site selection based on its success in this clinical situation [26]. We note disagreement in the literature regarding naming the patellar tendon vs. ligament. For consistency, we use the term patellar *tendon* in this Manual.

When scanning the width of a tendon or ligament with PD, (a) use light probe pressure, (b) give sufficient time for a PD signal to register an accurate PD signal, and (c) make certain the tendon or ligament is not too taut. Protocol Images describe the longitudinal axis (LAX) evaluation with respect to tendons or fascia; however, the examiner should confirm GS and PD abnormalities in orthogonal views. Perform slow transverse axis (TAX) slides for both longitudinal and short-axis views for both GS and PD. Evaluate the enthesis in GS with the highest possible frequency and note any hypoechoic or thickened insertion of the tendon <2 mm from bony surfaces, bony enthesophytes, or erosions. Then turn on PD to look for activity. High-frequency transducers enhance the detection of PD activity. Use a minimal Pulse Repetition Frequency (PRF) that does not produce excess noise in surrounding tissue. The protocol image examples are scanned with PD; however, initial GS evaluation will better elucidate echotexture and structural detail.

Set the presets as necessary for the region of interest.

Protocol Image 1: Achilles Tendon at the Calcaneal Insertion, Longitudinal ± Power Doppler (Fig. 25.3)

If able, the patient should be prone with the foot in slight dorsiflexion and hanging off the table. Place the probe in LAX with the distal end over the superior posterior calcaneus to see the insertion of the Achilles tendon (AT). Move the transducer laterally and medially to encompass the entire width of this flat tendon. The approximately 1 cm-long insertion point is typically hypoechoic or anechoic due to the anisotropy of AT fibers. A heel-to-toe maneuver of the probe and slight foot dorsiflexion may diminish anisotropy at the calcaneal insertion. Look for AT tears, swelling, hypoechogenicity, calcifications, MSU aggregates, or tophi adjacent to or within the AT. Look for bony posterior calcaneal erosion; a hyperechoic protrusion might be an enthesophyte.

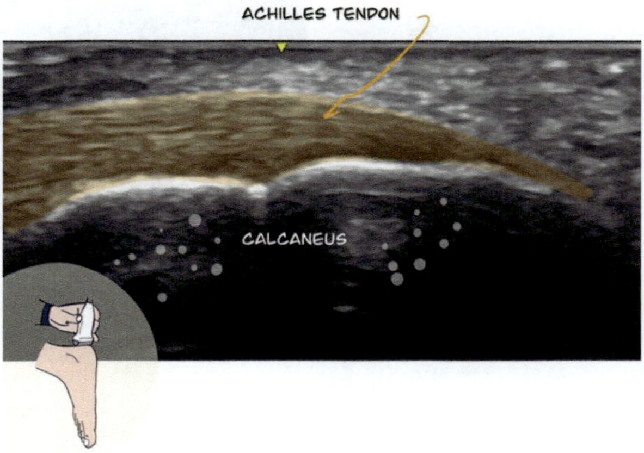

Fig. 25.3 Protocol Image 1: Achilles tendon at calcaneal insertion, longitudinal ± power Doppler

After examining using GS ultrasound, turn on the PD at the insertion. Recall that extreme dorsiflexion stretches the AT, thus enhancing the normal fibrillar echotexture at the cost of a diminished PD signal. Rather than a tenosynovium, the AT has a paratenon, best seen posteriorly.

Protocol Image 2: Plantar Fascia Insertion at the Calcaneus, Longitudinal ± Power Doppler (Fig. 25.4)

Place the probe in LAX at the center of the inferior calcaneus, and slowly move the transducer medially and laterally to examine the full extent of the fibrillar plantar fascia (PF). Manually dorsiflexing the ankle may clarify the PF image. Using a lower frequency and time-gain control enhances this deeper image. Probe tilting and using a heel-to-toe maneuver (with plenty of transmission gel) diminish anisotropy, which might mimic insertional tearing or enthesitis.

Note that the PF will typically lose fibrillar definition near the calcaneal insertion; however, if there is no attendant PF thickening, this is normal [38]. The PF is usually 3-4 mm thick near the calcaneal insertion [39]. Measure the PF thickness to look for thickening >4 mm, which may indicate plantar fasciitis, a mimic of enthesitis. Look for more distal PF thickening to confirm plantar fasciitis. The finding of enthesitis in other body areas favors a systemic disease. Positive PD activity may be from SpA but may occur with the more acute phase of plantar fasciitis [40]. The ankle should *not* be maximally dorsiflexed since this may ablate PD activity.

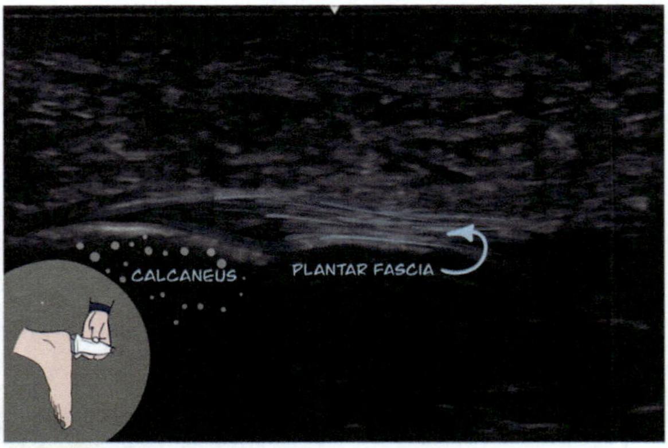

Fig. 25.4 Protocol Image 2: Plantar fascia insertion at calcaneus, longitudinal ± power Doppler

Protocol Image 3: Tibialis Anterior Tendon Insertion, Longitudinal ± Power Doppler (Fig. 25.5)

The patient is recumbent with the foot placed flat on the examination table, exposing the foot dorsum. Place the probe in TAX, centered on the tibialis anterior tendon (TAT) on the medial aspect of the dorsal foot. Turn the probe 90° into LAX. Move the transducer distally in an LAX slide, shifting it slightly laterally and medially as you go to see the complete TAT in LAX. As you look at the talus from proximal to distal, you will sequentially visualize the tibiotalar joint, the talar dome, the talar neck, the talar head, and finally, the talonavicular joint. Continue to move the probe in LAX distally, following the TAT. The joint distal to the navicular is the naviculocuneiform joint. Technically, you are looking at the navicular-*medial* cuneiform joint. Continue to move the probe distally, following the TAT, to see its insertion on the distal medial cuneiform and the base of the first metatarsal.

Protocol Image 4: Patellar Tendon at the Distal Patella, Longitudinal ± Power Doppler (Fig. 25.6)

The patient is recumbent with a bolster placed against the popliteal fossa to achieve 30-degree knee flexion [26]. Place the probe in LAX over the distal patella and the proximal insertion of the patella tendon. Tilt and rock the transducer to enhance the image of the tendon at the insertion. Sweep the probe in slow TAX slides medially and then laterally to scrutinize for swelling, hypoechogenicity, and enthesophytes.

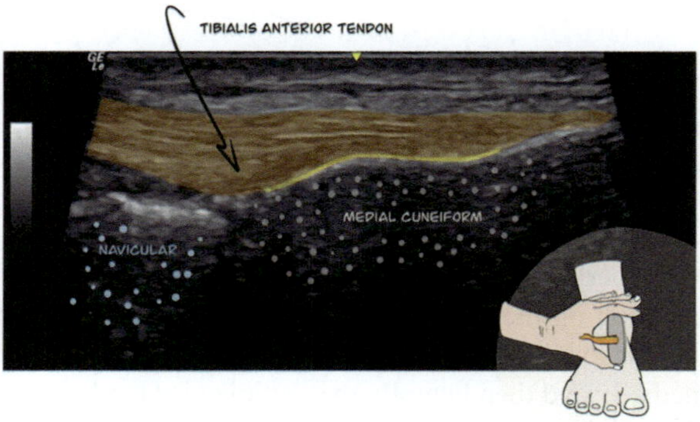

Fig. 25.5 Protocol Image 3: Tibialis anterior tendon insertion, longitudinal ± power Doppler

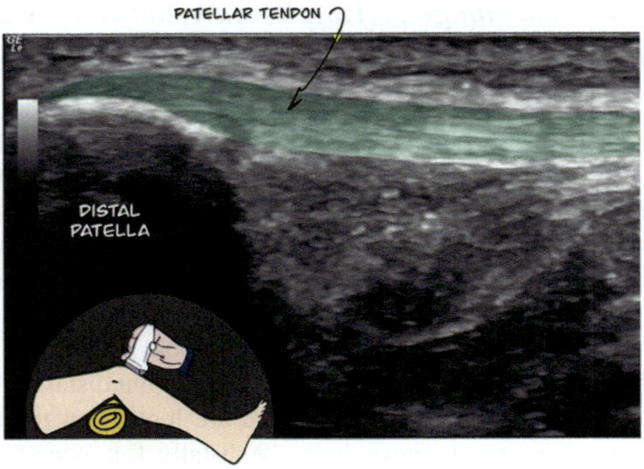

Fig. 25.6 Protocol Image 4: Patellar Tendon at distal patella, longitudinal ± power Doppler

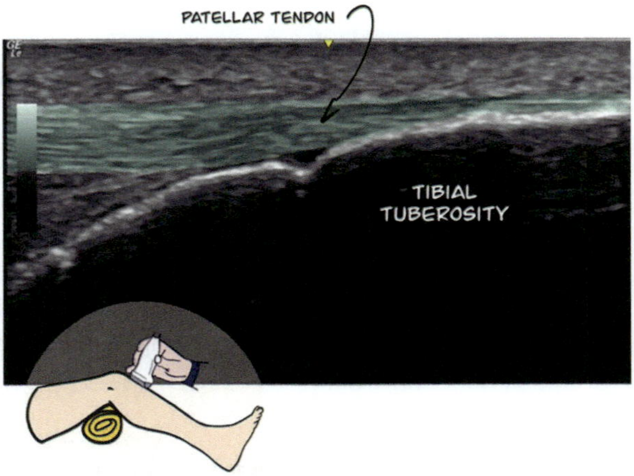

Fig. 25.7 Protocol Image 5: Patellar tendon at proximal tibia, longitudinal ± power Doppler

Protocol Image 5: Patellar Tendon at the Proximal Tibia, Longitudinal ± Power Doppler (Fig. 25.7)

Next, perform an LAX slide distally, following the fibrillar patellar tendon to the insertion on the curved bony insertion point, the tibial tuberosity. Again, enhance the image if needed by slightly tilting or rocking the probe. Sweep the probe slowly medially and then laterally to scrutinize for tendon swelling, hypoechogenicity, and enthesophytes. Be aware that bony irregularity at the tibial tuberosity may be from Osgood-Schlatter disease.

Protocol Image 6: Quadriceps Tendon at the Proximal Patella ± Power Doppler (Fig. 25.8)

Place the probe in LAX at the midline of the knee with the distal end of the probe over the proximal patella. The quadriceps tendon is seen in LAX, with the tendon fibrils creating a paintbrush-like pattern. Be sure to move the probe medially and laterally to evaluate the entire quadriceps tendon insertion on the patella.

Protocol Image 7: Common Extensor Tendon at the Lateral Epicondyle, Longitudinal ± Power Doppler (Fig. 25.9)

The patient is recumbent, resting the arm comfortably with a pronated forearm on the abdomen. The elbow is bent at 90° [26]. Place the proximal end of the probe in LAX on the lateral epicondyle. It may be a bit more posterior than you think. Look for the hyperechoic gentle slope of the lateral epicondyle, the more superficial fibrillar common extensor tendon (CET), and possible cortical irregularities and enthesophytes. Near the enthesis, look for loss of fibrillar echotexture/hypoechogenicity and swelling that may indicate enthesitis.

The lateral collateral ligament (LCL), also called the radial collateral ligament (RCL), is deep to the CET. The LCL is usually challenging to differentiate from the CET. The LCL extends distally to the annular ligament, which is the hyperechoic structure just superficial to the radial head; the CET more distally becomes muscle [41]. Once the CET and LCL are visualized, sweep the probe medially and laterally to encompass the full width of their insertions. Remember, the LCL also inserts on the lateral epicondyle and may be subject to enthesitis.

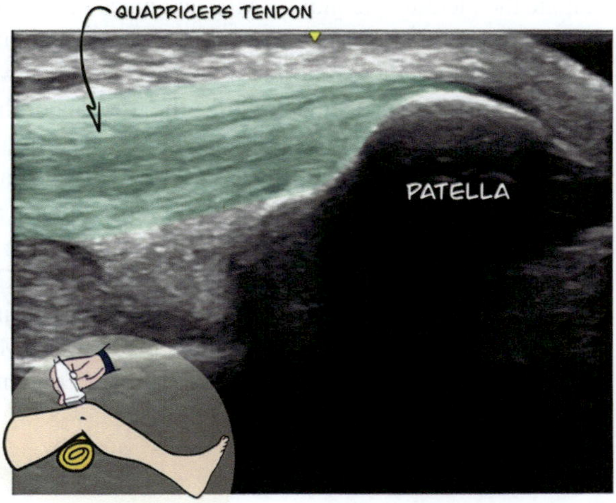

Fig. 25.8 Protocol Image 6: Quadriceps tendon at proximal patella ± power Doppler

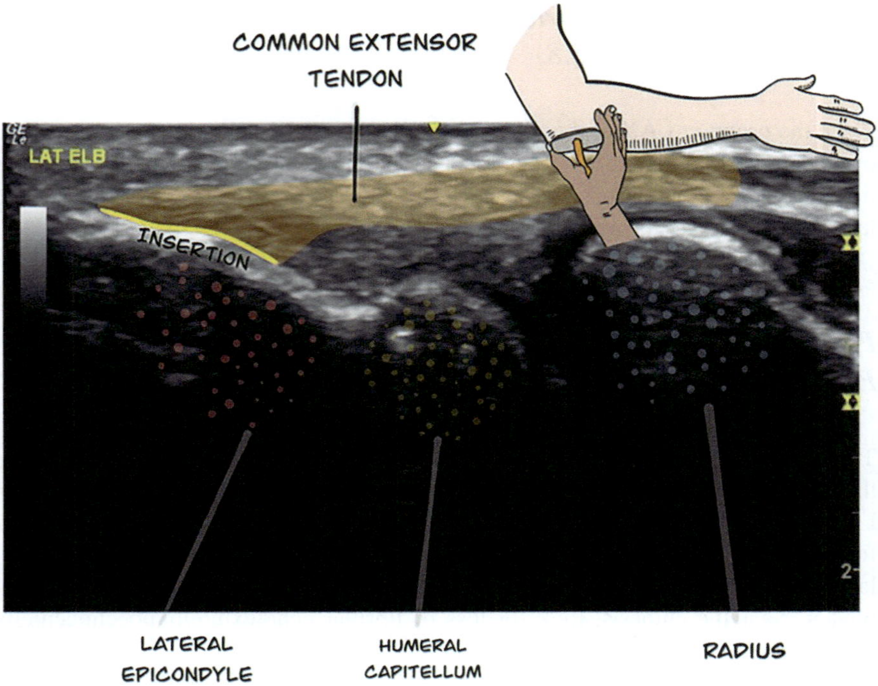

Fig. 25.9 Protocol Image 7: Common extensor tendon at lateral epicondyle, longitudinal ± power Doppler

Protocol Image 8: Common Flexor Tendon at the Medial Epicondyle, Longitudinal ± Power Doppler (Fig. 25.10)

Palpate the medial epicondyle (ME) and place the probe in LAX over the joint along the anterior medial elbow to visualize the humeroulnar joint (HUJ). See the bony hyperechoic curved **"ski-slope"**of the ME leading down the curved valley, the coronoid recess, and then a smaller "hill," the trochlea of the humerus. Look at the common flexor tendon (CFT), its muscle, and the insertion on the ME. Elbow extension enhances CFT visualization. Deep to the CFT and muscles is the ulnar collateral ligament's anterior bundle, which we refer to simply as the UCL. To delineate the separation of the CFT from the UCL, look at the more hypoechoic CFT muscle just superficial to the ligament. In elbow extension, the UCL is hypoechoic. Flexing the elbow to 90° will create tension in the UCL, enhancing echotexture visualization. Look at the enthesis of the CFT and the UCL on the ME for evidence of enthesopathy, muscle tears, and bony avulsions. Compare this image to the (unaffected) contralateral side if needed.

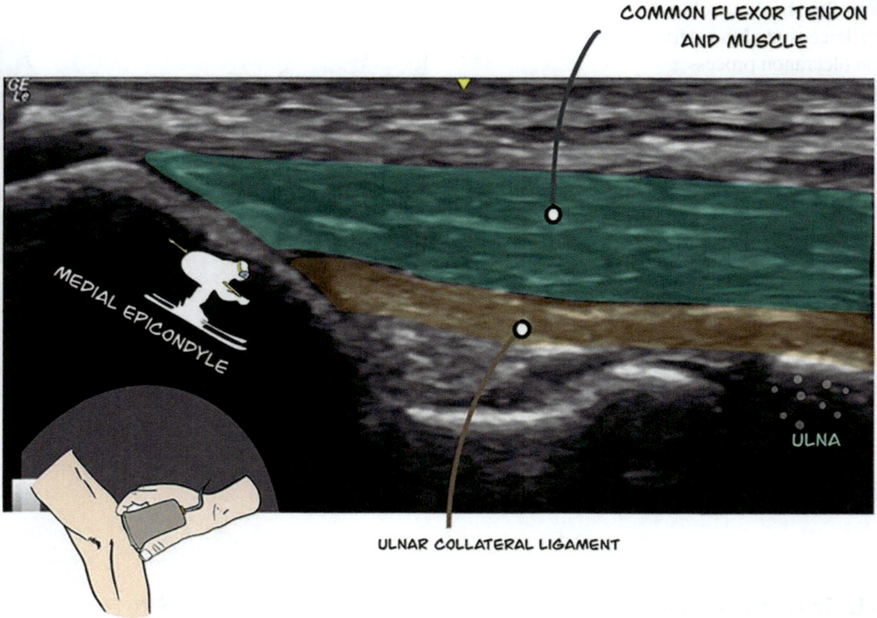

Fig. 25.10 Protocol Image 8: Common flexor tendon at medial epicondyle, longitudinal ± power Doppler

Protocol Image 9: Triceps Tendon Insertion on Olecranon Process ± Power Doppler (Fig. 25.11)

The patient is recumbent with the hand placed over the umbilicus, elbow flexed at 90°, and pointed downward at least 45°. Place the transducer in LAX along the edge of the olecranon process to visualize the bird's beak configuration of the distal triceps tendon (DTT) insertion on the bony olecranon. Use adequate gel, and perform heel-to-toe maneuvers in either direction to sharply visualize the "birds-beak" insertion of the distal triceps tendon onto the olecranon process. Tilt the probe to decrease anisotropy. Small medial and lateral TAX slides help evaluate the full extent of the DTT. Slightly extend the elbow to avoid extreme tension on the tendon, which may ablate PD activity.

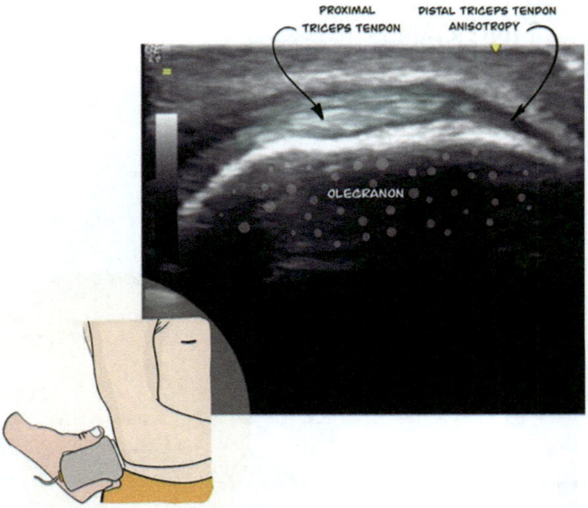

Fig. 25.11 Protocol Image 9: Triceps tendon insertion on olecranon process ± power Doppler

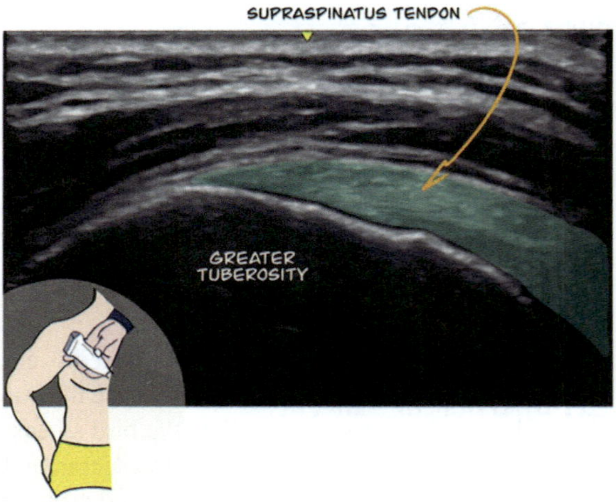

Fig. 25.12 Protocol Image 10: Supraspinatus Tendon at greater tuberosity, longitudinal + power Doppler

Protocol Image 10: Supraspinatus Tendon at Greater Tuberosity, Longitudinal ± Power Doppler (Fig. 25.12)

The patient internally rotates the shoulder by placing the hand into the ipsilateral back pocket. This is the modified Crass position, an excellent position to search for enthesitis [26]. Find the supraspinatus tendon (SST) insertion by placing the probe in TAX over the biceps tendon's round or oval long head (LHBT) in the bicipital groove. Rotate one end of the transducer toward the middle of the clavicle to visualize as much of the LHBT in LAX as possible. Then move the probe laterally to see the large "bird's beak" of the fibrillar SST inserted on the greater tuberosity (GT).

This is the distal anterior portion of the SST at the insertion on the superior facet (SF) of the GT.

The hyperechoic bony SF is concave, with the distal portion of the concavity being somewhat elongated and flattened. Focus on the insertion by alternatively performing heel-to-toe and tilting probe maneuvers. At the insertion of the anterior part of SST on the superior facet, there is a very thin hypoechoic line due to the downward curve of the distal SST fibers, causing anisotropy. This is *not* a tendon tear or enthesopathy.

Next, perform a TAX slide posteriorly to look at the middle facet (MF) and the posterior SST while the probe remains in the LAX orientation. Here, the bird's beak configuration is smaller and sharper, and the MF is flatter. A hypoechoic area of the SST near the insertion on the GT is due to anisotropy from the interdigitation of the insertional fibers of the infraspinatus tendon (IST) coming from a different direction. Do not mistake this for a tear or tendinosis. Again, perform all the probe maneuvers necessary to minimize anisotropy and evaluate the enthesis. Be aware that the SST is about 2.5 cm wide, and you must scan this wide enthesis [42].

Protocol Image 11: Extensor Tendon Insertion on Hand and Feet Phalanges, Longitudinal ± Power Doppler (Fig. 25.13)

Evaluate fingers or toes that are symptomatic with pain, stiffness, and swelling, particularly those digits demonstrating dactylitis, distal interphalangeal joint pain, or fingernail changes (such as pitting, subungual hyperkeratosis, or onycholysis).

The patient's hand or foot rests on a flat surface, exposing the dorsal surface. Place the highest-frequency probe available (at least 16 MHz) on the dorsum of the distal phalanx using sufficient transmission gel. Adjust the PRF downward to enhance the detection of Doppler signals but avoid excess Doppler "noise." Note the normal reverberation effect produced by the fingernail or toenail. Look for PD activity on the distal phalanx's hyperechoic bone, where the extensor tendon's fibers are inserted. Any Doppler signal deep to the bone is an artifact and should be discounted. A Doppler signal not in the distribution of the tendon insertion may be due to normal blood vessels and should not be interpreted as active enthesitis. Recall that while the extensor tendon inserts on the bony distal phalanx, some fibers extend superficially to form the fingernail bed matrix.

Protocol Image 12: Flexor Tendon Insertion on the Hand and Feet Phalanges, Longitudinal ± Power Doppler (Fig. 25.14)

The hand is placed palm up, or the foot is placed in a neutral position, exposing the toes' undersurface. Place a small footprint probe (with a frequency of at least 16 MHz) in LAX over the flexor surface of the distal phalanx to evaluate the insertional fibers of the flexor tendon. Small TAX slides in the ulnar and radial directions

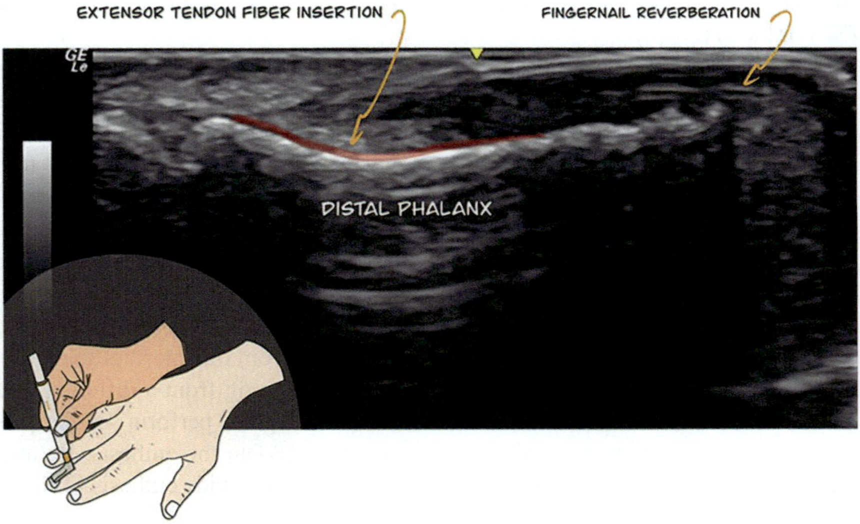

Fig. 25.13 Protocol Image 11: Extensor tendon insertion on hand and foot phalanges, longitudinal ± power Doppler

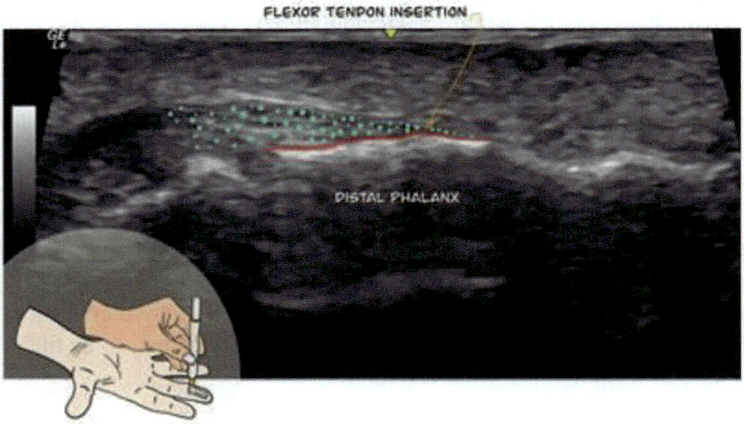

Fig. 25.14 Protocol Image 12: Flexor tendon insertion on hand and feet phalanges, longitudinal ± power Doppler

enhance the search for PD activity indicative of active enthesitis. Remember, a light touch and abundant transmission gel are mandatory.

Enthesis Ultrasonic Examination Checklist

☐ Image 1: Achilles tendon at the calcaneal insertion, longitudinal ± power Doppler
☐ Image 2: Plantar fascia insertion at the calcaneus, longitudinal ± power Doppler
☐ Image 3 Tibialis anterior tendon insertion, longitudinal ± power Doppler
☐ Image 4 Patellar tendon at the distal patella, longitudinal ± power Doppler
☐ Image 5 Patellar tendon at the proximal tibia, longitudinal ± power Doppler
☐ Image 6 Quadriceps tendon at the proximal patella ± power Doppler
☐ Image 7 Common extensor tendon at the lateral epicondyle, longitudinal ± power Doppler
☐ Image 8. Common flexor tendon at the medial epicondyle, longitudinal ± power Doppler
☐ Image 9. Triceps tendon insertion on olecranon process ± power Doppler
☐ Image 10. Supraspinatus Tendon at greater tuberosity, longitudinal ± power Doppler
☐ Image 11. Extensor tendon insertion on hand and feet phalanges, longitudinal ± power Doppler
☐ Image 12. Flexor tendon insertion on the hand and feet phalanges, longitudinal ± power Doppler

References

1. Kehl AS, Corr M, Weisman MH. Review: Enthesitis: new insights into pathogenesis, diagnostic modalities, and treatment. Arthritis Rheumatol. 2016;68(2):312–22.
2. Bruyn GA, Iagnocco A, Naredo E, Balint PV, Gutierrez M, Hammer HB, et al. OMERACT definitions for ultrasonographic pathologies and elementary lesions of rheumatic disorders 15 years on. J Rheumatol. 2019;46(10):1388.
3. Terslev L, Naredo E, Iagnocco A, Balint PV, Wakefield RJ, Aegerter P, et al. Defining enthesitis in spondyloarthritis by ultrasound: results of a Delphi process and of a reliability reading exercise. Arthritis Care Res (Hoboken). 2014;66(5):741–8.
4. Kaeley GS, Eder L, Aydin SZ, Gutierrez M, Bakewell C. Enthesitis: a hallmark of psoriatic arthritis. Semin Arthritis Rheum. 2018;48(1):35–43.
5. Catanoso M, Pipitone N, Salvarani C. Epidemiology of psoriatic arthritis. Reumatismo. 2012;64(2):66–70.
6. Rudwaleit M, van der Heijde D, Landewé R, Akkoc N, Brandt J, Chou CT, et al. The assessment of SpondyloArthritis international society classification criteria for peripheral spondyloarthritis and for spondyloarthritis in general. Ann Rheum Dis. 2011;70(1):25–31.
7. D'Agostino MA, Said-Nahal R, Hacquard-Bouder C, Brasseur JL, Dougados M, Breban M. Assessment of peripheral enthesitis in the spondylarthropathies by ultrasonography combined with power Doppler: a cross-sectional study. Arthritis Rheum. 2003;48(2):523–33.

8. Benjamin M, McGonagle D. The enthesis organ concept and its relevance to the spondyloarthropathies. Adv Exp Med Biol. 2009;649:57–70.
9. Ball J. Enthesopathy of rheumatoid and ankylosing spondylitis. Ann Rheum Dis. 1971;30(3):213–23.
10. Sudoł-Szopińska I, Kwiatkowska B, Prochorec-Sobieszek M, Maśliński W. Enthesopathies and enthesitis. Part 1. Etiopathogenesis. J Ultrason. 2015;15(60):72–84.
11. Mander M, Simpson JM, McLellan A, Walker D, Goodacre JA, Dick WC. Studies with an enthesis index as a method of clinical assessment in ankylosing spondylitis. Ann Rheum Dis. 1987;46(3):197–202.
12. Heuft-Dorenbosch L, Spoorenberg A, van Tubergen A, Landewé R, van ver Tempel H, Mielants H, et al. Assessment of enthesitis in ankylosing spondylitis. Ann Rheum Dis. 2003;62(2):127–32.
13. Klauser AS, Wipfler E, Dejaco C, Moriggl B, Duftner C, Schirmer M. Diagnostic values of history and clinical examination to predict ultrasound signs of chronic and acute enthesitis. Clin Exp Rheumatol. 2008;26(4):548–53.
14. McGonagle D. Imaging the joint and enthesis: insights into pathogenesis of psoriatic arthritis. Ann Rheum Dis. 2005;64(Suppl 2):ii58–60.
15. Bakewell C, Aydin SZ, Ranganath VK, Eder L, Kaeley GS. Imaging techniques: options for the diagnosis and monitoring of treatment of Enthesitis in psoriatic arthritis. J Rheumatol. 2020;47(7):973–82.
16. Bandinelli F, Prignano F, Bonciani D, Bartoli F, Collaku L, Candelieri A, et al. Ultrasound detects occult entheseal involvement in early psoriatic arthritis independently of clinical features and psoriasis severity. Clin Exp Rheumatol. 2013;31(2):219–24.
17. Zhang H, Liang J, Qiu J, Wang F, Sun L. Ultrasonographic evaluation of enthesitis in patients with ankylosing spondylitis. J Biomed Res. 2017;31(2):162–9.
18. de Miguel E, Muñoz-Fernández S, Castillo C, Cobo-Ibáñez T, Martín-Mola E. Diagnostic accuracy of enthesis ultrasound in the diagnosis of early spondyloarthritis. Ann Rheum Dis. 2011;70(3):434–9.
19. D'Agostino MA, Aegerter P, Bechara K, Salliot C, Judet O, Chimenti MS, et al. How to diagnose spondyloarthritis early? Accuracy of peripheral enthesitis detection by power Doppler ultrasonography. Ann Rheum Dis. 2011;70(8):1433–40.
20. Naredo E, Batlle-Gualda E, García-Vivar ML, García-Aparicio AM, Fernández-Sueiro JL, Fernández-Prada M, et al. Power Doppler ultrasonography assessment of entheses in spondyloarthropathies: response to therapy of entheseal abnormalities. J Rheumatol. 2010;37(10):2110–7.
21. D'Agostino MA, Breban M, Said-Nahal R, Dougados M. Refractory inflammatory heel pain in spondylarthropathy: a significant response to infliximab documented by ultrasound. Arthritis Rheum. 2002;46(3):840–1. author reply 1-3
22. Micu MC, Fodor D. Concepts in monitoring enthesitis in patients with spondylarthritis--the role of musculoskeletal ultrasound. Med Ultrason. 2016;18(1):82–9.
23. Macchioni P, Salvarani C, Possemato N, Gutierrez M, Grassi W, Gasparini S, et al. Ultrasonographic and clinical assessment of peripheral enthesitis in patients with psoriatic arthritis, psoriasis, and fibromyalgia syndrome: the ULISSE study. J Rheumatol. 2019;46(8):904–11.
24. Marchesoni A, De Lucia O, Rotunno L, De Marco G, Manara M. Entheseal power Doppler ultrasonography: a comparison of psoriatic arthritis and fibromyalgia. J Rheumatol Suppl. 2012;89:29–31.
25. Eder L, Kaeley GS, Aydin SZ. The GRAPPA sonographic Enthesitis workshop. J Rheumatol Suppl. 2019;95:51–3.
26. Tom S, Zhong Y, Cook R, Aydin SZ, Kaeley G, Eder L. Development of a preliminary Ultrasonographic Enthesitis score in psoriatic arthritis - GRAPPA ultrasound working group. J Rheumatol. 2019;46(4):384–90.

27. Eder L, Kaeley GS, Aydin SZ. Development and validation of a sonographic Enthesitis instrument in psoriatic arthritis: the GRAPPA diagnostic ultrasound Enthesitis tool (DUET) project. J Rheumatol Suppl. 2020;96:50–2.
28. Zabotti A, Idolazzi L, Batticciotto A, De Lucia O, Scirè CA, Tinazzi I, et al. Enthesitis of the hands in psoriatic arthritis: an ultrasonographic perspective. Med Ultrason. 2017;19(4):438–43.
29. Zabotti A, Salvin S, Quartuccio L, De Vita S. Differentiation between early rheumatoid and early psoriatic arthritis by the ultrasonographic study of the synovio-entheseal complex of the small joints of the hands. Clin Exp Rheumatol. 2016;34(3):459–65.
30. Healy PJ, Helliwell PS. Dactylitis: pathogenesis and clinical considerations. Curr Rheumatol Rep. 2006;8(5):338–41.
31. Dubash SR, De Marco G, Wakefield RJ, Tan AL, McGonagle D, Marzo-Ortega H. Ultrasound imaging in psoriatic arthritis: what have we learnt in the last five years? Front Med (Lausanne). 2020;7:487.
32. Aydin SZ, Castillo-Gallego C, Ash ZR, Marzo-Ortega H, Emery P, Wakefield RJ, et al. Ultrasonographic assessment of nail in psoriatic disease shows a link between onychopathy and distal interphalangeal joint extensor tendon enthesopathy. Dermatology. 2012;225(3):231–5.
33. Tan AL, Benjamin M, Toumi H, Grainger AJ, Tanner SF, Emery P, et al. The relationship between the extensor tendon enthesis and the nail in distal interphalangeal joint disease in psoriatic arthritis--a high-resolution MRI and histological study. Rheumatology (Oxford). 2007;46(2):253–6.
34. Tan AL, Grainger AJ, Tanner SF, Emery P, McGonagle D. A high-resolution magnetic resonance imaging study of distal interphalangeal joint arthropathy in psoriatic arthritis and osteoarthritis: are they the same? Arthritis Rheum. 2006;54(4):1328–33.
35. Gutierrez M, Filippucci E, Grassi W, Rosemffet M. Intratendinous power Doppler changes related to patient position in seronegative spondyloarthritis. J Rheumatol. 2010;37(5):1057–9.
36. Gladman DD, Abufayyah M, Salonen D, Thavaneswaran A, Chandran V. Radiological characteristics of the calcaneal spurs in psoriatic arthritis. Clin Exp Rheumatol. 2014;32(3):401–3.
37. Aggarwal P, Jirankali V, Garg SK. Evaluation of plantar fascia using high-resolution ultrasonography in clinically diagnosed cases of plantar fasciitis. Pol J Radiol. 2020;85:e375–e80.
38. Cardinal E, Chhem RK, Beauregard CG, Aubin B, Pelletier M. Plantar fasciitis: sonographic evaluation. Radiology. 1996;201(1):257–9.
39. Fessell DP, Vanderschueren GM, Jacobson JA, Ceulemans RY, Prasad A, Craig JG, et al. US of the ankle: technique, anatomy, and diagnosis of pathologic conditions. Radiographics. 1998;18(2):325–40.
40. Walther M, Radke S, Kirschner S, Ettl V, Gohlke F. Power Doppler findings in plantar fasciitis. Ultrasound Med Biol. 2004;30(4):435–40.
41. Jacobson JA. Elbow. Fundamentals of musculoskeletal ultrasound. 3rd Edition ed. Philadelphia, PA: Elsevier; 2018.
42. Jacobson JA. Shoulder US: anatomy, technique, and scanning pitfalls. Radiology. 2011;260(1):6–16.

Index

M. H. Greenberg et al., *Manual of Musculoskeletal Ultrasound*,
https://doi.org/10.1007/978-3-031-37416-6

G

H

I

J

K

L

M

T

GPSR Compliance

The European Union's (EU) General Product Safety Regulation (GPSR) is a set of rules that requires consumer products to be safe and our obligations to ensure this.

If you have any concerns about our products, you can contact us on ProductSafety@springernature.com

In case Publisher is established outside the EU, the EU authorized representative is:

Springer Nature Customer Service Center GmbH
Europaplatz 3
69115 Heidelberg, Germany

Batch number: 10732076

Printed by Printforce, the Netherlands